Geotechnical Engineering

Principles and Practices

Donald P. Coduto

Professor of Civil Engineering
California State Polytechnic University, Pomona

PRENTICE HALL, Upper Saddle River, NJ 07458

Library of Congress Cataloging-in-Publication Data

Coduto, Donald P.
    Geotechnical engineering : principles and practices / Donald P.
Coduto.
      p. cm.
    Includes bibliographical references and index.
    ISBN 0-13-576380-0 (hb)
    1. Engineering geology.   I. Title.
  TA705.C62   1998
  624.1'51—dc21   98-22191
                                 CIP

Publisher: **ALAN APT**
Editor-in-chief: **MARCIA HORTON**
Managing editor: **BAYANI MENDOZA DE LEON**
Director of production and manufacturing: **DAVID W. RICCARDI**
Production editor: **KATHARITA LAMOZA**
Art director: **MICHELE GIUSTI**
Cover design: **JAYNE CONTE**
Manufacturing buyer: **DONNA SULLIVAN**

The author and publisher of this book have used their best efforts in preparing this book. These efforts include the development, research, and testing of the theories and programs to determine their effectiveness. The author and publisher make no warranty of any kind, expressed or implied, with regard to these programs or the documentation contained in this book. The author and publisher shall not be liable in any event for incidental or consequential damages in connection with, or arising out of, the furnishing, performance, or use of these programs.

Printed in the United States of America

10 9 8 7 6 5

ISBN 0-13-576380-0

Prentice-Hall International (UK) Limited, London
Prentice-Hall of Australia Pty. Limited, Sydney
Prentice-Hall Canada Inc., Toronto
Prentice-Hall Hispanoamericana, S.A., Mexico
Prentice-Hall of India Private Limited, New Delhi
Prentice-Hall of Japan, Inc., Tokyo
Prentice-Hall Asia Pte. Ltd., Singapore
Editora Prentice-Hall do Brasil, Ltda., Rio de Janeiro

Contents

Preface

Geotechnical Engineering: Principles and Practices is primarily intended for use as a textbook for undergraduate civil engineering students enrolled in an introductory course. It also serves well as a reference book for students in follow-on courses and for practicing engineers. As the title infers, this book covers both "principles" (the fundamentals of soil mechanics) and "practices" (the application of these principles to practical engineering problems). This integrated approach gives the reader a broader understanding of geotechnical engineering and provides a foundation for future studies.

This book is the product of thirteen years experience teaching undergraduate geotechnical engineering courses. It is an expanded version of course notes I originally developed for my students, and thus reflects teaching methods that have worked well at Cal Poly. In addition, the manuscript for this book was extensively tested in the classroom before going to press. This classroom testing allowed me to evaluate and refine the text itself, the example problems, the homework problems, and the software.

Key features of this book include:

- An entire chapter on engineering geology (Chapter 2). This chapter is especially helpful for readers who have not taken a geology course and is a good review for those who have.
- Another chapter on geoenvironmental engineering (Chapter 9) that includes discussions of contaminant transport and remediation, and sanitary landfills.
- Clear and concise explanations of the theories and assumptions behind geotechnical analyses.
- Frequent discussions of the sources and magnitudes of uncertainties in geotechnical analyses.
- Use of both English and SI units, because engineers in North America and many other parts of the world need to be conversant in both systems.
- Easy-to-use Windows software developed specifically for this book. This software may be downloaded from the Prentice Hall web site. It has been carefully integrated into the text, and is designed as a tool to enhance learning. In each case, the student must first solve homework problems by hand to master the analysis. Then he or she is introduced to the software, which allows exploration of more difficult problems that would otherwise be too tedious to solve by hand.
- Extensive use of example problems to illustrate the various analyses.
- Carefully developed homework problems distributed throughout the chapters, with comprehensive problems at the end of each chapter.
- Discussions of recent developments in geotechnical engineering, including geosynthetics, soil improvement, and geotechnical earthquake engineering.

An instructor's manual is available to faculty who adopt their text for their course. It may be obtained from your Prentice Hall campus representative.

Another book by the same author, *Foundation Design: Principles and Practices*, 2nd ed., is coordinated with this volume and is intended to be used in a follow-on course. The two books use consistent notation, terminology, and problem-solving methods.

Acknowledgments

Although only the author's name appears on the cover, this book would not have been possible without important contributions from many other individuals. These include the many engineers who have labored to bring our profession to maturity and have published the results of their work. The author also is indebted to the many teachers and engineers who have mentored him and molded his engineering skills and philosophy.

Many friends, colleagues, and counterparts at other universities reviewed all or part of the text, and offered constructive comments and suggestions. These peer reviews produced many important improvements in the text, as reflected in this published version. The reviewers included:

Cemal Basaran
State University of New York at Buffalo

Terry Bening
Cal Poly University, Pomona

Paul Chan
New Jersey Institute of Technology

Samuel P. Clemence
Syracuse University

Sanjay Govil
Leighton and Associates

Frank Janger
Cal Poly University, Pomona

Richard L. Handy
Iowa State University

John Lohman
Kleinfelder

Chris Magdosku
Geotechnical Professionals, Inc.

Kyle M. Rollins
Brigham Young University

Donald Wells
Cal Poly University, Pomona

Jerry A. Yamamuro
Clarkson University

Hundreds of civil engineering students at Cal Poly University, Pomona, also contributed to this book by using various drafts as makeshift textbooks in our geotechnical engineering courses. This classroom testing helped me find and resolve rough spots in the text and the accompanying software. The university itself also contributed by providing a sabbatical leave. Bill Stenquist, Katharita Lamoza, Irwin Zucker, and the others at Prentice Hall provided a great deal of support, encouragement, and advice, all of which contributed to the quality of the final product. Finally, I thank my family for their patient endurance of the writing and editing process.

I welcome constructive comments and suggestions from those who use this book. Please mail them to me at the Civil Engineering Dept., Cal Poly Univ., Pomona, CA 91768.

Donald Coduto
Yucaipa, California

Notation and Units of Measurement

There is no universally accepted notation in geotechnical engineering. However, the notation used in this book, as described in the following table, is generally consistent with popular usage.

Symbol	Description	Typical Units English	Typical Units SI	Defined on Page
A	Cross-sectional area	ft^2	m^2	210
A	Base area of foundation	ft^2	m^2	620
A	Percentage of soil passing #200 sieve	percent	percent	279
A_f	Cross-sectional area at failure	in^2	mm^2	505
A_0	Initial cross-sectional area	in^2	mm^2	505
a	Cross-sectional area of standpipe	ft^2	cm^2	226
a	Length of pseudorectangle in flow net	ft	m	246
a_{max}	Peak horizontal ground acceleration	in/s^2	cm/s^2	688
a_y	Yield acceleration	in/s^2	cm/s^s	551
B	Width of loaded area (such as a footing)	ft	m	327
b	Width of pseudorectangle in flow net	ft	m	246
b	Unit length	ft	m	584
C	Hazen's coefficient	—	$cm/s/mm^2$	226
C	Concentration	lb_m/ft^3	kg/m^3	299
C_A	Aging factor	Unitless	Unitless	104
C_B	SPT borehole diameter correction	Unitless	Unitless	72
C_c	Coefficient of curvature	Unitless	Unitless	122
C_c	Compression index	Unitless	Unitless	385
C_{OCR}	Overconsolidation correction factor	Unitless	Unitless	104
C_P	Grain size correction factor	Unitless	Unitless	104
C_R	SPT rod length correction	Unitless	Unitless	72
C_R	Relative compaction	Percent	Percent	183
C_r	Recompression index	Unitless	Unitless	386
C_S	SPT sampler correction	Unitless	Unitless	72
C_u	Coefficient of uniformity	Unitless	Unitless	122
C_α	Secondary compression index	Unitless	Unitless	411
c_r'	Residual effective cohesion	lb/ft^2	kPa	495
c_T	Total cohesion	lb/ft^2	kPa	479
c'	Effective cohesion	lb/ft^2	kPa	470
c_v	Coefficient of consolidation	ft^2/day	m^2/day	448
D	Depth of foundation	ft	m	619

Symbol	Description	Typical Units English	Typical Units SI	Defined on Page
D	Particle diameter	in	mm	119
D	Depth to failure surface	ft	m	535
D_d	Diffusion coefficient	ft^2/day	m^2/day	299
D_r	Relative density	Percent	Percent	103
D_w	Depth from ground surface to groundwater table	ft	m	211
D_{10}	Grain size at which 10% is finer (comparable definition for D values with other subscripts)	—	mm	122
d	Diameter of capillary rise tube	in	mm	234
d	Diameter of vane	in	mm	510
d	Moment arm	ft	m	536
d	Closest distance to fault trace	—	km	689
d_{85}	Grain size at which 85% of the soil to be filtered is finer	—	mm	279
E	Modulus of elasticity	lb/ft^2	kPa	319
E	Normal side force	lb	kN	531
E_D	DMT modulus	lb/ft^2	kPa	82
E_m	SPT hammer efficiency	Unitless	Unitless	72
e	Void ratio	Unitless	Unitless	102
e	Base of natural logarithms	2.7183	2.7183	
e_{max}	Maximum void ratio	Unitless	Unitless	103
e_{min}	Minimum void ratio	Unitless	Unitless	103
e_p	Void ratio at end of primary consolidation	Unitless	Unitless	411
e_0	Initial void ratio	Unitless	Unitless	381
F	Fines content (% passing #200 sieve)	Percent	Percent	140
F	Factor of Safety	Unitless	Unitless	476
f_{sc}	CPT cone side friction	T/ft^2	MPa or kg/cm^2	76
f_{sc1}	CPT cone side friction corrected for overburden stress	T/ft^2	MPa or kg/cm^2	79
f	Mass flux	$lb_m/ft^2/day$	$kg/m^2/day$	299
G	Shear modulus	lb/ft^2	kPa	319
G_B, G_C	Coefficients in Boore's equation	Unitless	Unitless	689
G_h	Equivalent fluid density	lb/ft^3	kN/m^3	602
G_L	Specific gravity of soil-water mixture	Unitless	Unitless	119
G_s	Specific gravity of solids	Unitless	Unitless	101
g	Acceleration of gravity	ft/s^2	m/s^2	100
H	Thickness of soil strata or soil layer	ft	m	228
H	Height of wall	ft	m	584
H_a	Saturated thickness of aquifer	ft	m	230
H_{dr}	Length of longest drainage path	ft	m	424

Symbol	Description	Typical Units		Defined on Page
		English	SI	
H_{fill}	Thickness of proposed fill	ft	m	370
h	Total head	ft	m	212
h_c	Height of capillary rise	ft	m	234
h_p	Pressure head	ft	m	212
h_v	Velocity head	ft	m	212
h_w	Total head inside well casing during pumping	ft	m	260
h_z	Elevation head	ft	m	212
h_0	Total head in aquifer before pumping	ft	m	260
h_1	Total head in farthest observation well	ft	m	264
h_2	Total head in nearest observation well	ft	m	264
I_D	DMT material index	Unitless	Unitless	82
I_L	Liquidity index	Unitless	Unitless	130
I_P	Plasticity index	Unitless	Unitless	130
i	Hydraulic gradient	Unitless	Unitless	213
j	Seepage force per unit volume of soil	lb/ft^3	kN/m^3	356
K	Coefficient of lateral earth pressure	Unitless	Unitless	581
K_a	Coefficient of active earth pressure	Unitless	Unitless	585
K_D	DMT horizontal stress index	Unitless	Unitless	82
K_d	Depth factor	Unitless	Unitless	638
K_p	Coefficient of passive earth pressure	Unitless	Unitless	588
K_0	Coefficient of lateral earth pressure at rest	Unitless	Unitless	583
k	Hydraulic conductivity	ft/s	cm/s	221
k_n	Hydraulic conductivity normal to fabric	ft/s	cm/s	281
k_x	Horizontal hydraulic conductivity	ft/s	cm/s	228
k_z	Vertical hydraulic conductivity	ft/s	cm/s	228
L	Length perpendicular to cross-section	ft	m	230
L	Length of loaded area (such as a footing)	ft	m	327
LL	Liquid limit (see w_L)	Unitless	Unitless	128
l	Distance the water travels	ft	m	213
l	Length along shear surface	ft	m	532
M	Mass	lb$_m$	kg	97
M_b	Body wave magnitude	Unitless	Unitless	682
M_c	Mass of can	lb$_m$	kg	98
M_L	Local magnitude	Unitless	Unitless	682
M_S	Surface wave magnitude	Unitless	Unitless	682
M_s	Mass of solids	lb$_m$	kg	97
M_w	Moment magnitude	Unitless	Unitless	682
M_w	Mass of water	lb$_m$	g	97
M_1	Mass of moist sample and can	lb$_m$	g	98

Symbol	Description	Typical Units		Defined on Page
		English	SI	
M_2	Mass of dry sample and can	lb_m	g	98
m	Slice width	ft	m	539
N	SPT blow count recorded in field	Blows/ft	Blows/300 mm	70
N	Normal force	lb	kN	531
N_c, N_q, N_γ	Bearing capacity factors	Unitless	Unitless	622
N_D	Number of equipotential drops	Unitless	Unitless	247
N_F	Number of flow tubes	Unitless	Unitless	247
N_F	Factor in Cousin's Charts	Unitless	Unitless	546
N_{60}	SPT blow count corrected for field procedures	Blows/ft	Blows/300 mm	72
$\bar{N}_{60}$	Average SPT N_{60} value	Blows/ft	Blows/300 mm	638
$(N_1)_{60}$	SPT blow count corrected for field procedures and overburden stress	Blows/ft	Blows/300 mm	74
n	Porosity	Percent	Percent	102
n_a	Air porosity	Percent	Percent	103
n_e	Effective porosity	Percent	Percent	232
n_w	Water porosity	Percent	Percent	102
O_{95}	Equivalent opening size of geotextile	—	mm	281
OCR	Overconsolidation ratio	Unitless	Unitless	392
P	Normal load	lb	kN	316
P_f	Normal load at failure	lb	kN	505
P_a	Normal force acting on a wall under active conditions	lb	kN	592
P_p	Normal force acting on a wall under passive conditions	lb	kN	592
P_0	Normal force acting on a wall under at-rest conditions	lb	kN	584
PI	Plasticity index (see I_P)	Unitless	Unitless	130
PL	Plastic limit (see w_P)	Unitless	Unitless	129
Q	Flow rate	ft^3/s	m^3/s	210
Q_c	Compressibility factor	Unitless	Unitless	105
q	Flow rate per unit width	$ft^3/s/ft$	$m^3/s/m$	230
q	Bearing pressure (or gross bearing pressure)	lb/ft^2	kPa	326 & 620
q_A	Allowable bearing pressure	lb/ft^2	kPa	641
q_a	Allowable bearing capacity	lb/ft^2	kPa	625
q_c	CPT cone resistance	T/ft^2	MPa or kg/cm^2	76
q_{c1}	CPT cone resistance corrected for overburden stress	T/ft^2	MPa or kg/cm^2	79
q_u	Unconfined compressive strength	lb/ft^2	kPa	505
q_{ult}	Ultimate bearing capacity	lb/ft^2	kPa	621
R	Distance from load to point	ft	m	316

Symbol	Description	Typical Units		Defined on Page
		English	SI	
R_f	Friction ratio (cone penetration test)	Percent	Percent	76
r	Horizontal component of distance from load to point	ft	m	316
r_d	Stress reduction factor	Unitless	Unitless	696
r_w	Radius of well casing	ft	m	260
r_0	Radius of influence	ft	m	260
r_1	Radius from pumped well to farthest observation well	ft	m	264
r_2	Radius from pumped well to nearest observation well	ft	m	264
S	Number of stories in a building			
S	Degree of saturation	Percent	Percent	98
S	Shear side force	lb	kN	531
S_t	Sensitivity	Unitless	Unitless	494
s	Shear strength	lb/ft^2	kPa	470
s_u	Undrained shear strength	lb/ft^2	kPa	491
T	Transmissivity	ft^2/s	m^2/s	230
T	Tangential force	lb	kN	531
T_f	Torque at failure	in-lb	N-m	510
T_v	Time factor	Unitless	Unitless	429
t	Time	s	s	
t	Thickness of geotextile	in	mm	282
t_{adj}	Adjusted time (settlement computations)	yr	yr	444
t_c	Duration of construction period	yr	yr	444
t_p	Time required to complete primary consolidation	years	years	412
t_{90}	Time to complete 90% of primary consolidation	years	years	448
U	Degree of consolidation	Percent	Percent	437
u	Pore water pressure	lb/ft^2	kPa	217
u_e	Excess pore water pressure	lb/ft^2	kPa	373
u_h	Hydrostatic pore water pressure	lb/ft^2	kPa	217
V	Volume	ft^3	m^3	97
V_a	Volume of air	ft^3	m^3	97
V_a	Shear force acting on a wall under active conditions	lb	kN	592
V_p	Shear force acting on a wall under passive conditions	lb	kN	592
V_{cone}	Volume of sand cone below valve	ft^3	m^3	187
V_f	Volume of fill	yd^3	m^3	195
V_m	Volume of Proctor mold	ft^3	m^3	179
V_s	Volume of solids	ft^3	m^3	97

Symbol	Description	Typical Units		Defined on Page
		English	SI	
V_v	Volume of voids	ft^3	m^3	97
V_w	Volume of water	ft^3	m^3	97
v	Velocity	ft/s	m/s	210
v_s	Seepage velocity	ft/s	m/s	232
W	Weight	lb	kN	97
W_f	Weight of foundation	lb	kN	620
W_m	Weight of Proctor mold	lb	kN	179
W_{ms}	Weight of Proctor mold + soil	lb	kN	179
W_s	Weight of solids	lb	kN	97
W_w	Weight of water	lb	kN	97
W_1	Initial weight of sand cone apparatus	lb	kN	187
W_2	Final weight of sand cone apparatus	lb	kN	187
w	Moisture content	Percent	Percent	97
w_L	Liquid limit	Percent	Percent	128
w_o	Optimum moisture content	Percent	Percent	180
w_P	Plastic limit	Percent	Percent	129
x	Horizontal distance or coordinate	ft	m	211
x_f	Horizontal distance from load	ft	m	316
y	Horizontal distance or coordinate	ft	m	211
y_f	Horizontal distance from load	ft	m	316
z	Depth below ground surface	ft	m	211
z_{dr}	Vertical distance from point to nearest drainage boundary	ft	m	429
z_f	Depth below loaded area	ft	m	211
z_w	Depth below groundwater table	ft	m	211
α	Horizontal angle between strike and the vertical plane on which an apparent dip is to be computed	deg	deg	31
α	Inclination of shear surface	deg	deg	539
α	Inclination of wall from vertical	deg	deg	599
β	Inclination of ground surface from horizontal	deg	deg	545
γ	Shear strain	Unitless	Unitless	318
γ	Unit weight	lb/ft^3	kN/m^3	99
γ'	Effective unit weight	lb/ft^3	kN/m^3	624
γ_b	Buoyant unit weight	lb/ft^3	kN/m^3	100
γ_c	Unit weight of concrete	lb/ft^3	kN/m^3	620
γ_d	Dry unit weight	lb/ft^3	kN/m^3	99
γ_{fill}	Unit weight of fill	lb/ft^3	kN/m^3	370
$(\bar{\gamma}_d)_c$	Average dry unit weight in cut area	lb/ft^3	kN/m^3	195
$(\bar{\gamma}_d)_f$	Average dry unit weight in fill area	lb/ft^3	kN/m^3	195
$(\gamma_d)_{max}$	Maximum dry unit weight	lb/ft^3	kN/m^3	180

Symbol	Description	Typical Units English	Typical Units SI	Defined on Page
γ_{sand}	Unit weight of sand in sand cone test	lb/ft^3	kN/m^3	187
γ_w	Unit weight of water	lb/ft^3	kN/m^3	100
Δe	Change in void ratio	Unitless	Unitless	381
Δh	Head loss	ft	m	213
ΔV	Change in volume during grading	yd^3	m^3	195
$\Delta\sigma_z$	Change in vertical total stress	lb/ft^2	kPa	373
δ	Dip	deg	deg	29
δ	Total settlement	in	mm	368
δ_a	Apparent dip	deg	deg	31
δ_a	Allowable settlement	in	mm	630
δ_c	Consolidation settlement	in	mm	368
$(\delta_c)_{ult}$	Ultimate consolidation settlement	in	mm	396
δ_D	Differential settlement	in	mm	630
δ_{Da}	Allowable differential settlement	in	mm	630
δ_d	Distortion settlement	in	mm	368
δ_s	Secondary compression settlement	in	mm	368
ϵ	Normal strain	Unitless	Unitless	318
ϵ_f	Normal strain at failure	Unitless	Unitless	505
ϵ_z	Vertical normal strain	Unitless	Unitless	379
$\epsilon_\parallel$	Normal strain parallel to load	Unitless	Unitless	319
$\epsilon_\perp$	Normal strain perpendicular to load	Unitless	Unitless	319
η	Dynamic viscosity of soil-water mixture		Poise	119
θ	Angle (stress analysis)	deg	deg	348
θ	Angle (slope stability analysis)	deg	deg	536
θ_z	Angle between σ_1 and σ_z	deg	deg	350
λ	Vane shear correction factor	Unitless	Unitless	510
$\lambda_{c\phi}$	Factor in slope stability computations	Unitless	Unitless	545
μ	Coefficient of friction	Unitless	Unitless	467
ν	Poisson's ratio	Unitless	Unitless	319
ρ	Density	lb_m/ft^3	kg/m^3	100
ρ_d	Dry density	lb_m/ft^3	kg/m^3	100
ρ_w	Density of water	lb_m/ft^3	kg/m^3	100
σ	Normal stress	lb/ft^2	kPa	316
σ	Normal pressure acting on a wall	lb/ft^2	kPa	591
σ'	Effective stress	lb/ft^2	kPa	338
σ_c'	Preconsolidation stress	lb/ft^2	kPa	383
σ_D'	Effective stress at depth D below the ground surface	lb/ft^2	kPa	622
σ_d	Deviator stress	lb/in^2	kPa	506
σ_m'	Overconsolidation margin	lb/ft^2	kPa	391
σ_x	Horizontal total stress	lb/ft^2	kPa	316

Symbol	Description	Typical Units		Defined on Page
		English	SI	
$\sigma_x{'}$	Horizontal effective stress	lb/ft^2	kPa	339
σ_y	Horizontal total stress	lb/ft^2	kPa	316
$\sigma_y{'}$	Horizontal effective stress	lb/ft^2	kPa	339
σ_z	Vertical total stress	lb/ft^2	kPa	316
$(\sigma_z)_{induced}$	Induced vertical total stress	lb/ft^2	kPa	370
$\bar{\sigma}_z$	Average vertical total stress	lb/ft^2	kPa	333
$\sigma_z{'}$	Vertical effective stress	lb/ft^2	kPa	338
$\sigma_{zf}{'}$	Final vertical effective stress	lb/ft^2	kPa	370
$\sigma_{z0}{'}$	Initial vertical effective stress	lb/ft^2	kPa	370
σ_1	Major principal stress	lb/ft^2	kPa	349
σ_2	Intermediate principal stress	lb/ft^2	kPa	349
σ_3	Minor principal stress	lb/ft^2	kPa	349
τ	Shear stress	lb/ft^2	kPa	316
τ	Shear stress acting an a wall	lb/ft^2	kPa	593
τ_{cyc}	Cyclic shear stress	lb/ft^2	kPa	696
τ_{max}	Maximum shear stress	lb/ft^2	kPa	351
ϕ	Potential function	ft^2/s	m^2/s	243
ϕ'	Effective friction angle	deg	deg	467
$\phi_r{'}$	Residual effective friction angle	deg	deg	495
ϕ_T	Total friction angle	deg	deg	479
ϕ_w	Wall-soil interface friction angle	deg	deg	596
ψ	Flow function (or stream function)	ft^2/s	m^2/s	243
ψ	Permittivity of geotextile filter	s^{-1}	s^{-1}	281
ψ	Factor in Bishop's Equation	Unitless	Unitless	542
ψ	Magnitude scaling factor	Unitless	Unitless	697

1

Introduction to Geotechnical Engineering

Virtually every structure is supported by soil or rock. Those that aren't either fly, float, or fall over.

Richard L. Handy (1995)

Geotechnical engineering is the branch of civil engineering that deals with soil, rock, and underground water, and their relation to the design, construction, and operation of engineering projects. This discipline is also called *soils engineering* or *ground engineering*. Nearly all civil engineering projects must be supported by the ground, and thus require at least some geotechnical engineering.

Typical issues addressed by geotechnical engineers include:

- Can the soils and rocks beneath a construction site safely support the proposed project?
- What groundwater conditions currently exist, how might they change in the future, and what impact do they have on the project?
- What will be the impact of any planned excavation, grading, or filling?
- Are the natural or proposed earth slopes stable? If not, what must we do to stabilize them?
- What kinds of foundations are necessary to support planned structures, and how should we design them?
- If the project requires retaining walls, what kind would be best and how should we design them?
- How will the site respond to potential earthquakes?

- Has the ground become contaminated with chemical or biological materials? Do these materials represent a health or safety hazard? If so, what must we do to rectify the problem?

Sometimes these issues are simple and straightforward, and require very little geotechnical engineering. However, in other cases they are very complex and require extensive exploration, testing, and analysis. At difficult sites, geotechnical concerns may even control the project's technical and economic feasibility.

Geotechnical engineering is closely related to *engineering geology*, which is a branch of geology, as shown in Figure 1.1. Individuals from both professions often work together, each making contributions from his or her own expertise to solve practical problems. The combined efforts of these two professions is sometimes called *geotechnics*.

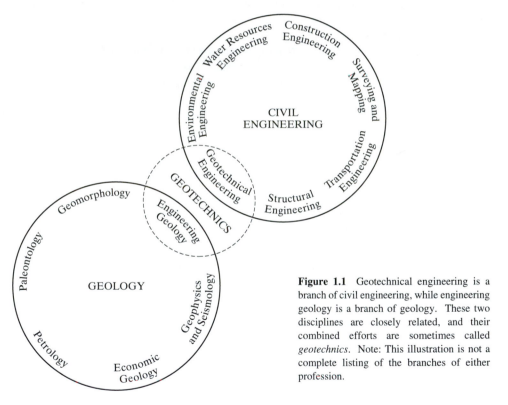

Figure 1.1 Geotechnical engineering is a branch of civil engineering, while engineering geology is a branch of geology. These two disciplines are closely related, and their combined efforts are sometimes called *geotechnics*. Note: This illustration is not a complete listing of the branches of either profession.

Geotechnical engineers usually begin by assessing the underground conditions and the engineering properties of the various strata. We call this process *site exploration and characterization*. It usually involves drilling vertical holes called *exploratory borings* into the ground, obtaining soil and rock samples, and testing these samples in a laboratory. It also may involve conducting tests *in-situ* (in-place).

The next step is to perform engineering analyses based on the information gained from the site exploration and characterization program. The analytical tools we use to perform these analyses are collectively known as *soil mechanics* and *rock mechanics*. Thus, soil mechanics and rock mechanics are to geotechnical engineering what structural mechanics is to structural engineering. In both fields, "mechanics" refers to the analytical tools, while "engineering" is a broader term that also includes the rest of the design and construction process.

We then use the analysis results to develop geotechnical input for design purposes. The design process also includes engineering judgement, experience from previous projects, and a sense of economics. However, regardless of the results from these analyses, geotechnical engineers are reluctant to deviate too far from design criteria that have proven worthy in the past. This is why understanding customary standards of practice is so important.

Geotechnical engineers work as part of a team, which also includes other professionals, such as structural engineers, civil engineers, architects, and others. Many design issues can be resolved only through a group effort, so the final design drawings and specifications reflect the combined expertise of many individuals.

Our work does not stop at the end of the design phase: It is very important to be involved in the construction phase as well. Geotechnical services during construction typically include:

- **Examining the soil and rock conditions actually encountered and comparing them with those anticipated in the design.** This is especially useful when the project includes large excavations, because they expose much more of the subsurface conditions than were seen in the exploratory borings. Sometimes the conditions encountered during construction are different, and this may dictate appropriate changes in the design.
- **Comparing the actual performance with that anticipated in the design.** We may do this by installing special instruments that measure movements, groundwater levels, and other important characteristics. This process, which we call the *observational method*, can also produce changes in the design.
- **Providing quality control testing**, especially in compacted fills and structural foundations.

Occasionally geotechnical services continue beyond the end of construction. For example, sites prone to long-term settlements may require monitoring for months or years after construction. Post-construction activities also can include investigations of facilities that have not performed satisfactorily, and development of remedial measures.

1.1 HISTORICAL DEVELOPMENT

Although this book primarily focuses on current methods of evaluating soil and rock for engineering purposes, it also occasionally discusses the historical development of geotechnical engineering. Part of the reason for these historical vignettes is to gain an appreciation for our heritage as engineers. Another, perhaps more important reason is to help us understand how technical advances have occurred in the past, because this guides us in developing future advancements.

Early Methods (Prior to 1850)

People have been building structures, dams, roadways, aqueducts, and other projects for thousands of years. However, until recently these projects did not include any rational engineering assessment of the underlying soil or rock. Early construction was based on common sense, experience, intuition, and rules-of-thumb, and builders passed this collective wisdom orally from generation to generation, often through trade guilds. Early scientists were concerned with more lofty matters, and generally considered the study of soil and rock beneath their dignity.

Sometimes builders used crude tests to assess the soil conditions. For example, the Italian architect Palladio (1508–1580) wrote that firm ground could be confirmed "if the ground does not resound or tremble if something heavy is dropped. In order to ascertain this, one can observe whether some drum skins placed on the ground vibrate and give off a weak sound or whether the water in a vessel placed on the ground gets into motion" (Flodin and Broms, 1981). The primary objective of such assessments seems to have been the identification and subsequent avoidance of sites with poor soil conditions.

These design methods were usually satisfactory so long as the construction projects were modest in scope, similar to previous projects (and thus tied to experience), and built away from obviously poor sites. Using these methods, the ancient builders sometimes accomplished amazing feats of construction, some of which still exist. For example, some dams in India have been in service for more than two thousand years. Unfortunately, the ancient builders also experienced some dramatic failures.

During the Middle Ages, builders began constructing larger and more sophisticated structures such as the cathedrals and related buildings in Europe. These projects pressed beyond the limits of experience, so the old rules-of-thumb did not always apply and unfortunate failures sometimes occurred. The failures produced new rules-of-thumb that guided subsequent projects. These trial-and-error methods of developing design criteria continued through the Renaissance and into the beginning of the Industrial Revolution, but it became increasingly evident that they were very tedious and expensive ways to learn.

The Leaning Tower of Pisa, shown in Figure 1.2, is the most famous example

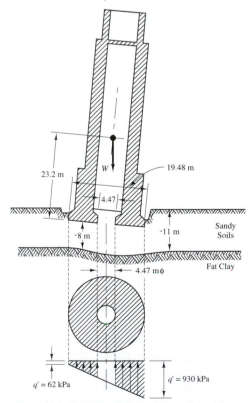

Figure 1.2 The leaning tower of Pisa (adapted from Terzaghi, 1934a).

of soil-related problems from this era. Construction began in AD 1173 and continued off-and-on for nearly 200 years. The tower began to tilt during construction, so the builders attempted to compensate by providing a slight taper to the upper stories. The movement continued after construction, and by 1982 the top of the 58.4 m (192 ft) tall structure was 5.6 m (18.4 ft) off plumb.

Modern investigations of the subsurface conditions have found a weak clay stratum about 11 m below the ground surface. This clay is very compressible, and has settled under the concentrated weight of the tower. The south side has settled more than the north, which has caused the tower to tilt. *Foundation Design: Principles and Practices*, the companion volume to this book, describes the tower in more detail.

Slowly, builders began to apply the scientific method to various aspects of construction. These efforts attempted to determine why the things we build behave the way they do, express this behavior mathematically, and develop analysis and design methods based on these understandings. Initially these efforts focused almost exclusively on structural issues. Leonardo daVinci (1452–1519), the artist/scientist, was one of the few to briefly study the behavior of soils. He observed the angle of repose in sands, proposed test methods to determine the bearing capacity of soils, and speculated on the processes of groundwater hydrology. However, daVinci's ideas on soils do not appear to have extended beyond the pages of his notebooks, and had no impact on design or construction methods.

Engineers and scientists began to address the engineering behavior of soil more seriously during the seventeenth and eighteenth centuries. Most of this early work focused on the analysis and design of retaining walls (Skempton, 1979). It was generally dictated by military needs, and was mostly performed by individuals associated with the army, especially in France. Henri Gautier, B.F. Belidor, Charles Augustin Coulomb, and others developed methods of predicting the forces imparted by soil onto retaining walls, which led to more rational design methods. Coulomb's work, which he published in 1776, is often considered the first example of rational soil mechanics, and still forms the basis for computation of earth pressures acting on walls. Unfortunately, much of this work extended well beyond eighteenth century abilities to measure relevant engineering properties in soil, and thus was difficult to apply to practical problems.

Some scientific investigations of soil behavior continued during the early nineteenth century, including studies of the stability of earth slopes and other topics. However, this work had limited usefulness, was not widely disseminated, and had very little impact on the vast majority of construction projects.

Late Nineteenth Century Developments

The last half of the nineteenth century was a period of rapid industrialization, which produced tremendous growth in both the amount and scope of construction projects. Railroad lines were expanding, urban areas were growing (with the resulting need for infrastructure construction), ports were being enlarged, and larger buildings were being built. Iron and steel had become common civil engineering materials, and reinforced concrete was beginning to appear. These projects drove advances in structural engineering, hydraulic engineering, and other fields. Some of these advancements later became useful to geotechnical engineers. For example, Henri Darcy's work on flow through sand filters for water purification purposes later became the basis for analyses of groundwater flow, and various developments in mechanics of materials, such as that of Otto Mohr, later would be applied to soil. However, progress in geotechnical engineering still lagged behind.

When working at sites with potentially problematic soils, especially soft clays, some engineers drove steel rods into the ground to roughly assess the soil conditions. These tests were called *soundings*. Sometimes engineers used exploratory borings to gain soil samples, although the information gained from them was purely qualitative. Some advances in analysis and design methods also were developed during this period, including empirical formulas for determining the load capacity of pile foundations. However, we had not yet developed important unifying concepts and did not understand how soils behave, so efforts at assessing subsurface conditions were of limited value.

The increasing size of civil engineering projects, especially after 1880, raised more concerns about the consequences of failure, yet overly conservative designs were too expensive. The time was ripe for geotechnical engineering to emerge as a clearly defined discipline within civil engineering, for inventing better techniques of assessing soil and rock conditions, and for developing sound methods of integrating them into civil engineering practice.

Geotechnical Engineering in Sweden, Early Twentieth Century

The first large-scale attempts at geotechnical engineering occurred in Sweden during the early decades of the twentieth century (Bjerrum and Flodin, 1960). The Swedes even introduced the word "geotechnical" (in Swedish, Geotekniska) during this period.

Sweden was a likely place for these early developments, because extremely poor soil conditions underlie much of the country. Soft, weak clays are present beneath the most populated areas, and they are the source of many problems, including excessive settlement and catastrophic landslides. Many of these clays are very *sensitive*, which means they lose strength when disturbed, and thus are prone to dramatic failures.

Old place names that translate to "Earth Fall," "Land Fall," and "Clay Fall," illustrate the long history of landslides in this region. In Norway, which has similar soil problems, landslides killed an average of 17 persons per year between 1871 and 1940 (Flodin and Broms, 1981). These problems became much worse when construction projects created cuts and fills that further destabilized marginal ground.

The city of Göteborg, Sweden suffered from extensive soil-related problems during the early development of its port facilities. These projects required dredging soils, building quays (facilities for docking ships) and other works on the soft clays that existed in the harbor. This was very difficult, and several major landslides occurred during and after construction of these facilities. Port engineers began to assess these landslides and develop methods of safely building port facilities.

Meanwhile, the Swedish State Railways needed to make cuts and fills to provide alignments for new tracks, and these steepened slopes prompted more landslides. One particularly disastrous failure occurred in 1913 when 185 m (600 ft) of track slipped into Lake Aspen. This event prompted the formation of the Geotekniska Kommission

(Geotechnical Commission) of the Swedish State Railways to study the problem and develop solutions. They intended the new term "geotechnical" to reflect the commission's reliance on both geology and civil engineering.

The most prominent engineer in this effort was Wolmar Fellenius (1876–1957). A graduate of the Royal Institute of Technology in Stockholm, Fellenius had become familiar with soil problems when he served as the port engineer in Göteborg. When the commission was formed, he was a professor of hydraulics at the Royal Institute, a position he held until his retirement in 1942. Fellenius became the chair of the commission, and John Olsson, also a civil engineer, did much of the day-to-day work.

Figure 1.3 Wolmar Fellenius was the port engineer in Göteborg, Sweden, a professor at the Royal Institute of Technology in Stockholm, and the chairman of the Geotechnical Commission of the Swedish State Railways. He and his colleagues conducted the first large-scale geotechnical studies. In the process, they developed many of the exploration, sampling, testing, and analysis techniques we use today. (Photo courtesy of Professor Bengt Fellenius)

The commission had to begin by developing new methods of drilling and sampling soils, which was very difficult in the soft clays. They were the first to develop methods of obtaining undisturbed samples of these soils. Then they developed laboratory test equipment, studied the behavior of these soils, and produced new methods of analysis and design. They investigated more than 300 sites and collected 20,000 soil samples. This work was truly a pioneering effort, and is a testimony to the resourcefulness and insights of these men. Their soil mechanics laboratory, established in 1914, appears to have been the first of its kind in the world.

More railroad failures occurred while the commission's work was in progress, most notably the 1918 landslide at Vita Sikudden, which killed forty-one people. These failures further emphasized the importance of their work.

The commission's final report, completed in 1922, was the world's first comprehensive geotechnical report. It presented their methods of investigation and analysis, and contained recommendations on how to avoid future landslides. Afterward, many committee members continued to develop new test equipment and refine their analysis and design methods.

These early developments in Sweden represented the first significant efforts at geotechnical engineering, and they influenced subsequent construction in Scandinavia. However, the rest of the world had little or no knowledge of this work until much later. The task of promoting geotechnical engineering on a widespread international level required more people to spread the message, and one of them soon became recognized as a leader in this effort: Karl Terzaghi.

Karl Terzaghi

Karl Terzaghi (1883–1963) has often been called "the father of soil mechanics." Although he was only one of many people who ushered in the new profession we now call geotechnical engineering, his influence and early leadership were especially noteworthy. Terzaghi, more than any other, set the tone and direction of the profession and promoted it as a legitimate branch of civil engineering.

Terzaghi was born in Prague, which was then part of Austria. His initial academic work was in mechanical engineering, and he earned an undergraduate degree in that subject. However, he found it was not to his liking, so his first engineering job was with a civil engineering firm in Vienna that specialized in reinforced concrete. He worked at construction sites in many European locations, which also gave him opportunities to pursue one of his favorite subjects: geology. He later earned a doctorate based on his work in reinforced concrete design.

Throughout this period, Terzaghi became increasingly interested in the ignorance of civil engineers in matters relating to earthwork and foundation design. Although structural design had already reached a high level of sophistication, the design of earthwork and foundations was based on unreliable empirical rules. He felt this topic needed a more scientific approach, and decided to focus his attentions on developing rational design methods.

In 1916 he accepted a teaching position at the Imperial School of Engineers in Istanbul (then known as Constantinople), and later moved to Robert College, also in Istanbul. There he began research into the behavior of soils, including studies of piping failures in sands beneath dams and settlement in clays. The work on clays eventually led to his theory of consolidation, which we will study in Chapters 11 and 12. This theory, which has since been verified, is considered one of the most significant milestones in civil engineering.

If we wish to define a certain time as the "birth" of geotechnical engineering as a widely recognized discipline, it would be the year 1925, for that was when Terzaghi published the first comprehensive book on the subject. He gave it the title *Erdbaumechanik auf Bodenphysikalischer Grundlage* (German for *The Mechanics of Earth Construction Based on Soil Physics*; Terzaghi, 1925a) and published it in Vienna. *Erdbaumechanik* addressed various aspects of what we would now call geotechnical engineering, and did so from a rational perspective that recognized the importance of field observations.

In 1925 Terzaghi also accepted a visiting lectureship position at the Massachusetts Institute of Technology, where he soon became recognized as the leader of a new branch of civil engineering. The same year, he published a series of English articles in the American journal *Engineering News Record* (Terzaghi, 1925b) and a paper in the *Journal of the Boston Society of Civil Engineers* (Terzaghi, 1925c). He also expanded his research interests to include frost heave, pavement design, and other topics, along with continuing his interests in foundations and dams.

In 1929 he returned to Vienna and began serving as a professor at the Technical University. During the next several years he continued his research activities, along with an active speaking and consulting schedule that brought him to many places in Europe, Asia, Africa, and North America. This work generated extensive interest in soil mechanics. Then, in 1939, he returned to the United States and accepted a professorship at Harvard University. He continued teaching, consulting, and lecturing around the world, but Harvard remained his home for the rest of his life.

Figure 1.4 Karl Terzaghi in 1951 (Photograph courtesy of Margaret Terzaghi-Howe).

Terzaghi had a remarkable ability to develop rational and practical solutions to real engineering problems from a jumble of what had previously been a maze of incoherent facts and observations. He came onto the engineering scene at the right time and with the necessary skills, rising from obscurity to lead the establishment of geotechnical engineering as a rational and legitimate branch of civil engineering.

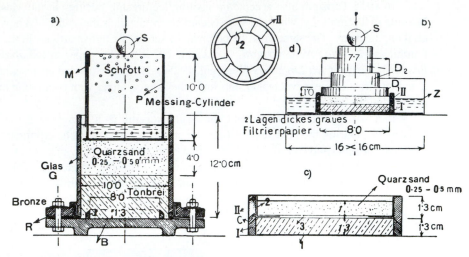

Figure 1.5 Terzaghi's 1925 book *Erdbaumechanik* included this illustration of a consolidometer, which is a laboratory device for measuring the settlement of soils. Terzaghi used devices like this to develop his theory of consolidation, which we will discuss in Chapters 11 and 12.

Additional Twentieth Century Developments

Other prominent engineers also made important contributions to the fledgling profession of geotechnical engineering during the 1920s and 1930s. Some of them were from academic circles, while others were practicing engineers and/or contractors. Those who had the most impact on engineering practice in the United States included:

- *William Housel*, a professor at the University of Michigan, one of the first Americans to study soil mechanics. His work was contemporary with, but independent of, Karl Terzaghi. Housel developed methods of soil sampling, analysis, and design, and gathered large volumes of data from field observations. He also taught the first university soil mechanics course in the United States, which began at the University of Michigan in 1927.
- *Gregory Tschebotarioff*, a German who eventually came to the United States and taught at Princeton University. He made important contributions on earth retaining structures.
- *Arthur Casagrande,* a disciple of Terzaghi and a professor at Harvard University. He made many contributions to the analysis of soft clays, soil composition and classification, seepage, earth dams, and other topics.
- *Fred Converse,* a professor at the California Institute of Technology who began teaching soil mechanics there in the mid 1930s. Converse also was co-founder of one the first geotechnical consulting firms.
- *Daniel Moran*, a foundation engineer and builder who worked on the foundations for many large buildings in New York, major bridges across the country, and other monumental projects. He pioneered new methods of constructing bridge foundations.

- *Lazarus White*, a foundation engineer and builder who developed new methods of design and construction, underpinning, and other advances. He also was co-founder of the firm Spencer, White, and Prentis.
- *R.R. Proctor*, a practicing engineer who made important advances in the assessment of compacted fills during construction.

We reached another important milestone in 1936 when the First International Conference on Soil Mechanics and Foundation Engineering met in Cambridge, Massachusetts. It was the first significant professional conference devoted exclusively to this topic, and its published proceedings represented more technical material on this subject than all of the material published before the conference. The International Society of Soil Mechanics and Foundation Engineering (ISSMFE) was founded during that conference. Both the society and its conferences continue to be an important means of disseminating knowledge.

In spite of these advancements, Terzaghi was presenting lectures with titles like "Soil Mechanics—A New Chapter in Engineering Science" as late as 1939 to professional audiences who apparently had very little familiarity with the subject (Terzaghi, 1939). This increased awareness of the usefulness of geotechnical engineering, along with the massive construction projects of the 1950s and 1960s, finally established geotechnical engineering as a routine part of nearly all significant civil engineering projects.

1.2 MODERN GEOTECHNICAL ENGINEERING

We have made substantial progress during the past century. Our abilities to assess subsurface conditions, predict soil behavior, and accommodate this behavior using appropriate designs are now far better than before. A large number of consulting firms specialize in geotechnical engineering, many government agencies have geotechnical engineering departments, and nearly all civil engineers have occasion to work with geotechnical engineers on their projects. About 15 percent of American Society of Civil Engineers members now identify geotechnical engineering as their primary or secondary area of interest. Geotechnical engineers also have expanded into new areas, most notably *geoenvironmental engineering*, which deals with underground environmental problems.

As geotechnical engineering matured, it also developed a "personality" that is slightly different from other civil engineering disciplines. These personality traits include the following:

- We work with soil and rock, which are natural materials. As such, their engineering properties are more complex and difficult to characterize than those of manufactured materials such as steel. Soil properties also vary significantly from one project site to another, and even at different locations within a single site. Therefore, we devote a significant part of our work and budget to site characterization. Unlike structural engineers, who can simply look up material properties in a book, geotechnical engineers must obtain samples from each project site and test them in a laboratory. To accomplish these tasks, geotechnical engineers and their staffs spend a great deal of time in the field and laboratory.

- Practical economic constraints limit the number of exploratory borings we can drill and the number of laboratory tests we can perform. As a result, we have direct knowledge of only a very small portion of the soil or rock beneath a project site. This introduces many potential sources of error: What are the subsurface conditions between and beyond the borings? Are the samples truly representative of the field conditions? How much sample disturbance has occurred during recovery and transport to the laboratory? What effect does this sample disturbance have on the measured engineering properties?

- Because of the potentially large errors in our site characterization programs, we use a large measure of engineering judgement when bringing the laboratory and field data into our analyses. In addition, our ability to perform quantitative analyses far exceeds the accuracy of the data on which they are based. Therefore, the results of these analyses are usually not very precise. As a result, we typically use larger factors of safety and more conservative designs.

- We rely more heavily on "engineering judgement," which is a combination of experience, subjectivity, reliance on precedent, and other factors.

- We have a more extensive involvement during construction, and frequently revise our design recommendations when conditions encountered during construction are different from those anticipated.

Geotechnical engineers also spend a great deal of time interacting with others, including general civil engineers, structural engineers, architects, building officials, geologists, contractors, attorneys, and owners. As a result, good written and oral communication skills are very important.

Geotechnical engineers continue to face new technical challenges. The high cost of real estate, especially in urban areas, often dictates the need to build on sites with poor soil conditions – sites we would have rejected in the past. These difficult sites pose special problems, and have resulted in the development of new construction materials and techniques, such as ground improvement methods. However, the construction industry, which includes all branches of civil engineering, also has become very competitive, and is largely driven by the marketplace. Clients demand high-quality services and expect to receive them quickly and inexpensively. Thus, it has become very important to work efficiently. It also has enhanced the demand for innovative construction methods, such as mechanically stabilized earth (MSE) walls, that are more cost-effective than previous solutions.

1.3 ACCURACY OF GEOTECHNICAL ENGINEERING ANALYSES

Although the many advances in geotechnical engineering over the past century have greatly improved our ability to predict the behavior of soil and rock, we still need to maintain a healthy sense of skepticism. Most of our analyses are handicapped by the uncertainties introduced by the site exploration and characterization program. In addition, our mathematical models of soil behavior are only approximate, and often do not explicitly consider important factors. Simply because an equation is available to describe a certain process does not mean that we can expect to perform precise computations!

One of the most common mistakes among students studying geotechnical engineering, and even among some practicing engineers, is to overestimate the accuracy of geotechnical analyses. The widespread availability of digital computers and the related software has made this problem even worse, because our ability to perform analyses has far surpassed the technical and economic realities of obtaining the underlying soil and rock data. This often leads to overconfidence, and ultimately may result in construction failures.

Most of the example problems in this book have been solved to a precision of three significant figures. This has been done for clarity and to avoid excessive round-off errors. However, few if any geotechnical analyses are really this accurate. In reality, the actual behavior often varies from the predicted behavior by 50 percent or more. Therefore, it is best to perform most geotechnical analyses to no more than two or three significant figures and recognize the true precision is really much less.

There are occasions when more precise analyses are useful, especially when conducting "what-if" studies or when actual performance data is available from the field. More precise analyses also may be appropriate for very sophisticated projects that have a correspondingly intense site exploration and characterization programs. However, it is very important to avoid placing too much confidence in the results, for it is very easy to perform analyses to much greater levels of precision than are justified by the data.

KEY TO COLOR PHOTOGRAPHS

The color photographs at the end of this chapter show the role of geotechnical engineering in various projects, and geotechnical engineers at work.

First Page

Virtually all civil engineering projects require at least some geotechnical engineering. Here are some examples:

A. Buildings — The Sears Tower in Chicago is one of the tallest buildings in the world. It needs massive foundations to transmit the structural loads into the ground. The design of these foundations depends on the nature of the underlying soils. Geotechnical engineers are responsible for assessing these soil conditions and developing suitable foundation designs.

B. Bridges — The foundation for the south pier of the Golden Gate Bridge in San Francisco had to be built in the open sea. It extends down to bedrock, some 30 m (100 ft) below the water level and 12 m (40 ft) below the channel bottom. This was especially difficult to build because of the tremendous tidal currents at this site.

C. Dams — Oroville Dam in California is one of the largest earth dams in the world. It is made of 61,000,000 m^3 (80,000,000 yd^3) of compacted soil. The design and construction of such dams requires extensive geotechnical engineering.

D. Tunnels — The Ted Williams Tunnel is part of the Central Artery Project in Boston. This prefabricated tunnel section was floated to the job site, then sunk into a prepared trench in the bottom of the bay. Its integrity depends on proper support from the underlying soils.

Second Page

Geotechnical engineers try to avoid failures like these:

E. This house was built near the top of a slope and had a beautiful view of the Pacific Ocean. Unfortunately, a landslide occurred during a wet winter, undermining the house and causing part of its floor to fall away.

F. Teton Dam in Idaho failed in 1976, only a few months after the embankment had been completed and the reservoir began to be filled. This failure killed 11 to 14 people, and caused about $400 million of property damage.

G. The 1964 Niigata Earthquake in Japan caused extensive liquefaction in this port city. These apartment buildings rotated when the underlying soils liquefied.

H. The approach fill to this highway bridge has settled because the underlying soils are soft clays and silts. However, the bridge has not settled because it is supported on piles. Although this "failure" is not as dramatic as the others, it is a source of additional maintenance costs, and can be a safety hazard to motorists and pedestrians.

Third Page

We use a variety of techniques to assess the subsurface conditions. These include:

I. Performing a field reconnaissance. This is the top of a recent landslide, and the man in the photograph is examining the soil and rock exposed in the scarp.

J. Drilling exploratory borings to obtain soil and rock samples. This rig drills holes up to 30 m (100 ft) deep.

K. Testing samples in a soil mechanics laboratory. These tests help us determine the engineering properties of the soil or rock.

L. Monitoring geotechnical instruments. These instruments measure groundwater levels and pressures, soil movements, and other similar attributes.

Fourth Page

Geotechnical engineers also are actively involved in construction. Examples of geotechnical construction include:

M. This rig is drilling a hole in the ground that will be filled with reinforced concrete to form a drilled shaft foundation.

N. This 11 m (35 ft) deep excavation extends 10 m (30 ft) below the groundwater table. In addition, a river is present just beyond the excavation on the left side of the photograph. Therefore, it was necessary to first install an extensive dewatering system to draw down the groundwater table.

O. This rig is installing a series of wick drains, which help accelerate the settlements that will occur as a result of a proposed fill.

P. The fill for this highway near Fort St. John, British Columbia is being reinforced with geogrids, thus allowing the side slopes to be steeper than would be possible with an unreinforced fill.

◀ Plate A
The Sears Tower in Chicago

▼ Plate B
The Golden Gate Bridge in San Francisco

▼ Plate D
The Ted Williams Tunnel
in Boston *(Central Artery/
Tunnel Project)*

▲ Plate C
Oroville Dam in California
(California Department of Water Resources)

◄ Plate E
A house that has been
undermined by a landslide

Plate F ►
The failure of Teton Dam in Idaho

◄ Plate G
The consequences of liquefaction
in Niigata, Japan *(Earthquake
Engineering Research Center Library,
University of California, Berkeley,
Steinbrugge Collection)*

Plate H ►
Settlement of the approach
fill to a bridge

◀ Plate I
Conducting a field reconnaissance

▼ Plate J
Drilling an exploratory boring

▲ Plate K
Conducting laboratory tests

Plate L ▶
Monitoring instruments in the field

◀ Plate M
Drilled shaft foundation
under construction *(Association
of Drilled Shaft Contractors)*

▼ Plate N
Dewatered excavation
for a pipeline project
(Foothill Engineering)

▼ Plate O
Wick drains
being installed
*(American Wick
Drain Corporation)*

▲ Plate P
Geogrid reinforced fill for a highway
(Tensar Earth Technologies, Inc.)

2

Engineering Geology

And so geology, once considered mostly a descriptive and historical science, has in recent years taken on the aspect of an applied science. Instead of being largely speculative as perhaps it used to be, geology has become factual, quantitative, and immensely practical. It became so first in mining as an aid in the search for metals; then in the recovery of fuels and the search for oil; and now in engineering in the search for more perfect adjustment of man's structures to nature's limitations and for greater safety in public works.

Charles P. Berkey, Pioneer Engineering Geologist, 1939

Geology is the science of rocks, minerals, soils, and subsurface water, including the study of their formation, structure, and behavior. As the quotation above indicates, geology was once confined to purely academic studies, but it has since expanded into a practical science as well. *Engineering geology* is the branch that deals with the application of geologic principles to engineering works.

Unlike geotechnical engineers, whose training is in civil engineering, engineering geologists have a background in geology. Their work includes mapping, describing, and characterizing the rock at a construction site; assessing stability issues, such as landslides; and appraising local seismicity and earthquake potentials. These two professions are complementary, and work together as a team. Nevertheless, it is important for the geologist to have some understanding of engineering, and the engineer to have some understanding of geology. Some individuals have even acquired full professional credentials in both fields.

This chapter explores fundamental principles of geology and their application to geotechnical engineering, with extra emphasis on the geological origin of soils. These principles are important to geotechnical engineers because they help us understand the nature of the subsurface conditions and form much of the basis for interpreting data gathered from exploratory borings.

2.1 ROCK AND SOIL

Both geologists and engineers frequently divide earth materials into two broad categories: *rock* and *soil*. Although this may seem to be a simple distinction, in reality it is not and has often been a source of confusion. To a geologist, rock is "any naturally formed aggregate or mass of mineral matter, whether or not coherent, constituting an essential and appreciable part of the earth's crust" (American Geological Institute, 1976). This definition focuses on the modes of origin and structure of the material. Conversely, engineers (and contractors) sometimes consider rock to be a "hard, durable material that cannot be excavated without blasting," a definition based on strength and durability.

Unfortunately, these two definitions sometimes produce conflicting classifications, especially in intermediate materials. For example, some materials that are rock in terms of their geologic origin are soft enough to be easily excavated with the same equipment used for soil. They may even look like soil. Siltstone is good example. Conversely, some cemented soils, such as caliche, are "hard as rock" and very difficult to excavate. This difficulty in classifying some materials has often led to construction lawsuits, because contractors are typically paid more to excavate "rock." It also can be a problem when piles are to be driven to "rock."

Therefore, it is important for both engineering geologists and geotechnical engineers to properly communicate the nature of earth materials (rock vs. soil) to other members of the design and construction teams. Our thought processes tend to use the geologists' definitions because they help us interpret the subsurface conditions, but contractors and other engineers usually interpret our comments in light of the engineers' definitions.

Sometimes this difficulty can be overcome by using the terms *hard rock* and *soft rock*, where the latter is capable of being excavated by conventional earthmoving equipment. However, this definition also can lead to confusion, and is not entirely satisfactory. In Chapter 6 we will discuss more specific classification methods to be used in excavation specifications.

Another aspect of dividing earth materials into rock and soil is that this distinction often determines the kinds of subsurface data we need to acquire, the tests we will perform, and the analyses we will conduct. This is because there are important differences between these two materials, including the following (Goodman, 1990):

- Rocks are generally cemented; soils are rarely cemented
- Rocks usually have much lower porosity than soils
- Rocks can be found in states of decay with greatly altered properties and attributes; effects of weathering on soils are more subtle and generally less variable

- Rock masses are often discontinuous; soil masses usually can be represented as continuous
- Rocks have more complex, and generally unknowable stress histories. In many rock masses, the least principal stress is vertical; in most soils the greatest principal stress is vertical.

Although there are times when soil mechanics techniques can be applied to rock mechanics problems, and vice-versa, any such sharing must be done cautiously.

2.2 ROCK-FORMING MINERALS

Minerals are naturally formed elements or compounds with specific structures and chemical compositions. As the basic constituents of rocks, minerals control much of rock behavior. Some minerals are very strong and resistant to deterioration, and produce rocks with similar properties, while others are much softer and produce weaker rock.

More than 2000 different minerals are present in the earth's crust. They can be identified by their physical and chemical properties, by standardized tests, or by examination under a microscope. Only a few of them occur in large quantities, and they form the material for most rocks. The most common minerals include:

Feldspar—This is the most abundant mineral, and is an important component of many kinds of rock. *Orthoclase* feldspars contain potassium ($KAlSi_3O_8$) and usually range from white to pink. *Plagioclase* feldspars contain sodium ($NaAlSi_3O_8$), calcium ($CaAl_2Si_2O_8$), or both, and range from white to gray to black. Feldspars have a moderate hardness.

Quartz—Also very common, quartz is another major ingredient in many kinds of rock. It is a silicate (SiO_2), and usually has a translucent to milky white color, as shown in Figure 2.1. Quartz is harder than most minerals, and thus is very resistant to weathering. *Chert* is a type of quartz sometimes found in some sedimentary rocks. It can cause problems when used as a concrete aggregate.

Figure 2.1 A large quartz crystal. Quartz crystals in rocks are normally much smaller.

Ferromagnesian minerals—A class of minerals, all of which contain both iron and magnesium. This class includes *pyroxene, amphibole, hornblende,* and *olivine.* These minerals have a dark color and a moderate hardness.

Iron oxides—Another class of minerals, all of which contain iron (Fe_2O_3). Includes *limonite* and *magnetite.* Although less common, these minerals give a distinctive rusty color to some rocks and soils, and can act as cementing agents.

Calcite—A mineral made of calcium carbonate ($CaCO_3$); usually white, pink, or gray. It is soluble in water, and thus can be transported by groundwater into cracks in rock where it precipitates out of solution. It also can precipitate in soil, becoming a cementing agent. Calcite is much softer than quartz or feldspar, and effervesces vigorously when treated with dilute hydrochloric acid.

Dolomite—Similar to calcite, with magnesium added. Less vigorous reaction to dilute hydrochloric acid.

Mica—Translucent thin sheets or flakes. *Muscovite* has silvery flakes, while *biotite* is dark gray or black. These sheets have a very low coefficient of friction, which can produce shear failures in certain rocks, such as schist.

Gypsum—A very soft mineral often occurring as a precipitate in sedimentary rocks. It is colorless to white and has economic value when found in thick deposits. For example, it is used to make drywall. Gypsum is water soluble, and thus can dissolve under the action of groundwater, which can lead to other problems.

When rock breaks down into soil, as discussed later in this chapter, many of these minerals remain in their original form. For example, many sand grains are made of quartz, and thus reflect its engineering properties. Other minerals undergo chemical and physical changes and take on new properties. For example, feldspar often experiences such changes, and forms clay minerals (discussed later in this chapter). Soil also can acquire other materials, including organic matter, man-made materials, and water.

2.3 THE GEOLOGIC CYCLE

The geologic processes acting on the earth's crust are extremely slow by human standards. Even during an entire lifetime, one can expect to directly observe only a minutely small amount of progress in these processes. Therefore, geologists must rely primarily on observations of the earth as it presently exists (i.e., on the *results* of these processes) to develop their theories.

Geologic theories are organized around a framework known as the *geologic cycle.* This cycle, shown in Figure 2.2, includes many processes acting simultaneously. The most important of these begin with molten magma from within the earth forming into rock, then continue with the rocks being broken down into soil, and that soil being converted back into rock.

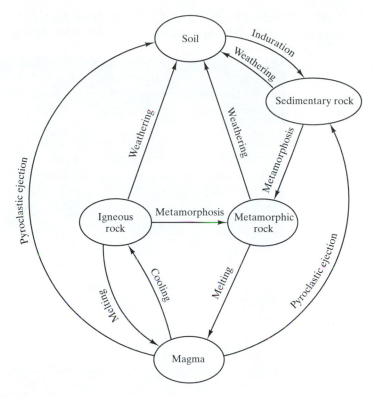

Figure 2.2 Primary processes in the geologic cycle.

Rocks are classified according to their place in the geologic cycle. The three major categories are igneous, sedimentary, and metamorphic, as discussed below.

Igneous Rocks

The geologic cycle begins with *magma*, a molten rock deep inside the earth. This magma cools as it moves upward toward the ground surface, forming *igneous rocks*. There are two primary types of igneous rocks: *Intrusives* (also called *plutonic rocks*) form below the ground surface, where they cool slowly, whereas *extrusives* (also called *volcanic rocks*) arrive at the ground surface in a molten state, such as through a volcano, and then cool very rapidly. Intrusives include both large bodies of rock (known as *plutons*) and smaller sheet-like bodies (known as *sills* and *dikes*) that fill cracks inside other rocks. Extrusives generally have finer grained, smoother surfaces. Some extrusive material, such as volcanic ash, bypasses the rock stage and forms directly into sediment.

Common igneous rocks include:

Granite—An intrusive, granite is one of the most common and familiar igneous rocks. It is found over wide areas, such as the Canadian Shield and the Sierra Nevada (see

Figure 2.3), and in isolated domes, such as Stone Mountain in Georgia. Granite contains primarily orthoclase feldspar and quartz, with some biotite and amphibole.

Figure 2.3 Half dome in Yosemite National Park. The near-vertical face was carved by glaciers. This rock, often classified as granite, is more accurately called grandiorite—a material halfway between granite and diorite.

Basalt—A dark, dense rock; the most abundant extrusive. Very difficult for tunnel construction due to its hardness, yet the rapid cooling associated with all extrusives creates joints in basalt, and slopes made of basalt often fail along these joints.

Diorite—Similar to granite, with plagioclase feldspar instead of orthoclase and little or no quartz.

Andesite—A very hard extrusive.

Rhyolite—The extrusive equivalent of granite.

Gabbro—The intrusive equivalent of basalt. Darker in color than granite or diorite. Unweathered igneous rocks generally have excellent engineering properties and are good materials to build on. Intrusive rocks are especially good. However, the cooling process, along with various tectonic forces within the earth, produce fractures in these rocks, especially in extrusives. The intact rock between these cracks can be very strong, but the fractures form planes of weakness. The rock can slide along these weak planes, potentially causing instability in the rock mass. The engineering properties of weathered igneous rocks are less desirable because the rock is changing into a more soil-like material.

Weathering Processes

Rocks exposed to the atmosphere are immediately subjected to physical, chemical, and biological breakdown through *weathering*. There are many weathering processes, including:

- The erosive action of water, ice, and wind
- Chemical reactions induced by exposure to oxygen, water, and chemicals
- Opening of cracks as a result of unloading due to erosion of overlying soil and rock
- Loosening through the growth of plant roots
- Loosening through the percolation and subsequent freezing (and expansion) of water
- Growth of minerals in cracks, which forces them to open further
- Thermal expansion and contraction from day to day and season to season
- Landslides and rockfalls
- Abrasion from the downhill movement of nearby rock and soil

The rock passes through various stages of weathering, eventually being broken down into small particles, the material we call soil. These soil particles may remain in place, forming a *residual soil*, or they may be transported away from their parent rock through processes discussed later in this chapter, thus forming a *transported soil*. Figure 2.4 shows an accumulation of fallen rock fragments called *talus* at the base of a rock slope, which is the beginning of one process of soil transport.

Figure 2.4 Talus accumulation at the base of a rock slope in eastern Washington state.

Weathering processes continue even after the rock becomes a soil. As soils become older, they change due to continued weathering. The rate of change depends on many factors, including:

- The general climate, especially precipitation and temperature (note that climates in the past were often quite different from those today)
- The physical and chemical makeup of the soil
- The elevation and slope of the ground surface
- The depth to the groundwater table
- The type and extent of flora and fauna
- The presence of microorganisms
- The drainage characteristics of the soil

Sedimentary Rocks

Soil deposits can be transformed back into rock through the hardening process called *induration* or *lithification*, thus forming the second major category of rocks: *Sedimentary rock*. There are two types: Clastic and carbonate.

Clastic Rocks

Clastic rocks form when deep soil deposits become hardened as a result of pressure from overlying strata and cementation through precipitation of water-soluble minerals such as calcium carbonate or iron oxide. Because of their mode of deposition, many clastic rocks are *layered* or *stratified*, which makes them quite different from *massive* formations. The interfaces between these layers are called *bedding planes*. Table 2.1 lists common clastic rocks. Shale and sandstone are the most common.

Often, various types of clastic rocks are *interbedded*. For example, a sequence might contain a 1 m thick bed of sandstone, then 5 m of siltstone, 0.5 m of claystone, and so on.

Most *conglomerate*, *breccia*, *sandstone*, and *arkose* rocks generally have favorable engineering properties. Those cemented with silica or iron oxide are especially durable, but may be difficult to excavate. However, some are only weakly indurated, often cemented only with clay or other water-soluble minerals. These may behave much like a soil, and be much easier to excavate.

Fine and very fine grained clastic rocks are more common, and much more problematic. Sometimes the term *mudstone* is used to collectively describe these rocks, but they are more precisely described as *siltstone* (when the rock is derived from silt), *claystone* (when derived from clay and slightly to mildly indurated) or *shale* (when derived from clay and well indurated). Nearly all of these have distinct bedding planes, as shown in Figure 2.5, and are subject to shearing along these planes. All except shale are usually easy to excavate with conventional earthmoving equipment.

Some fine and very fine grained clastic rocks also are subject to *slaking*, which is a deterioration after excavation and exposure to the atmosphere and wetting-and-drying cycles. Rocks that exhibit strong slaking will rapidly degenerate to soil, and thus can create problems for engineering structures built on them.

TABLE 2.1 COMMON CLASTIC SEDIMENTARY ROCKS (Adapted from Hamblin and Howard, 1975)

Texture and Average Particle Size	Composition	Rock Name
Coarse grained Gravel-size (> 2 mm)	Rounded fragments of any rock type; quartz, quartzite, chert dominant	Conglomerate
	Angular fragments of any rock type; quartz, quartzite, chert dominant	Breccia
Medium grained Sand-size (0.06 - 2 mm)	Quartz with minor accessory minerals	Sandstone
	Quartz with at least 25% feldspar	Arkose
	Quartz, rock fragments, and considerable clay	Graywacke
Fine grained Silt-size (0.002 - 0.06 mm)	Quartz and clay minerals	Siltstone
Very fine grained Clay-size (< 0.002 mm)	Quartz and clay minerals	Claystone and Shale

Figure 2.5 Steeply inclined bedding planes in a sedimentary rock. Shear failures can easily occur along such steep planes, especially when excavations destabilize the adjacent ground. For example, the rock in the foreground has already moved along one of the bedding planes, and has become twisted out of alignment.

Carbonates

A different type of sedimentary rock forms when organic materials accumulate and become indurated. Because of their organic origin, they are called *carbonates*. Common carbonate rocks include:

Limestone—The most common type of carbonate rock, limestone is composed primarily of calcite ($CaCO_3$). Most limestones formed from the accumulation of marine organisms on the bottom of the ocean, and usually extend over large areas. Some of these deposits were later uplifted by tectonic forces in the earth and now exist below land areas. For example, much of Florida is underlain by limestone.

Chalk—Similar to limestone, but much softer and more porous.

Dolomite—Similar to limestone, except based on the mineral dolomite instead of calcite.

Some carbonate rocks also have bedding, but it is usually less distinct than in clastic rocks.

Carbonate rocks, especially limestone, can be dissolved by long exposure to water, especially if it contains a mild solution of carbonic acid. Groundwater often gains small quantities of this acid through exposure to carbon dioxide in the ground. This process often produces *karst topography*, which exposes very ragged rock at the ground surface and many underground caves and passageways. In such topography, streams sometimes "mysteriously" disappear into the ground, only to reappear elsewhere.

Sometimes the rock is covered with soil, so the surface expressions of karst topography may be hidden. Nevertheless, the underground caverns remain, and sometimes the ground above caves into them. This creates a *sinkhole*, such as the one in Figure 2.6. This caving process can be triggered by the lowering of the groundwater table, which often occurs when wells are installed for water supply purposes.

Figure 2.6 This large sinkhole in Winter Park, Florida suddenly appeared on May 8, 1981. Within 24 hours, it was 75 m (250 ft) in diameter (GeoPhoto Publishing Company).

In areas underlain by carbonate rock, especially limestone, geotechnical engineers are concerned about the formation of sinkholes beneath large and important structures. We use exploratory borings, geophysical methods, and other techniques (see Chapter 3) to locate hidden underground caverns, then either avoid building above these features, or fill them with grout.

Metamorphic Rocks

Both igneous and sedimentary rocks can be subjected to intense heat and pressure while deep in the earth's crust. These conditions produce more dramatic changes in the minerals within the rock, thus forming the third type of rock—*metamorphic rock*. The metamorphic processes generally improve the engineering behavior of these rocks by increasing their hardness and strength. Nevertheless, some metamorphic rocks still can be problematic.

Some metamorphic rocks are *foliated*, which means they have oriented grains similar to bedding planes in sedimentary rocks. These foliations are important because the shear strength is less for stresses acting parallel to the foliations. Other metamorphic rocks are *nonfoliated* and have no such orientations.

Common metamorphic rocks include:

Foliated rocks:

> *Slate* — Derived principally from shale; dense; can be readily split into thin sheets parallel to the foliation (such sheets are used to make chalkboards).

> *Schist* — A strongly foliated rock with a large mica content; this type of foliation is called *schistosity*; prone to sliding along foliation planes.

> *Gneiss* — Pronounced "nice"; derived from granite and similar rocks; contains banded foliations.

Nonfoliated rocks:

> *Quartzite* — Composed principally or entirely of quartz; derived from sandstone; very strong and hard.

> *Marble* — Derived from limestone or dolomite; used for decorative purposes and for statues.

Unweathered nonfoliated rocks generally provide excellent support for engineering works, and are similar to intrusive igneous rocks in their quality. However, some foliated rocks are prone to slippage along the foliation planes. Schist is the most notable in this regard because of its strong foliation and the presence of mica. The 1928 failure of St. Francis Dam in California (Rogers, 1995) has been partially attributed to shearing in schist, and the 1959 failure of Malpasset Dam in France (Goodman, 1993) to shearing in a schistose gneiss.

Metamorphic rocks also are subject to weathering, thus forming weathered rock, residual soils, and transported soils and beginning the geologic cycle anew.

2.4 STRUCTURAL GEOLOGY

Structural geology is the study of the configuration and orientation of rock formations. This is an important part of engineering geology because it gives us important insights on how a rock mass will behave. Therefore, engineering geologists routinely develop detailed geologic maps that describe these structures.

Bedding Planes and Schistosity

All sedimentary rocks formed in horizontal or near-horizontal layers, and these layers often reflect alternating cycles of deposition. This process produces parallel *bedding planes* as shown in Figure 2.5. The shear strength along these planes is typically much less than across them, a condition we call *anisotropic strength*. When these rocks were uplifted by tectonic forces in the earth, the bedding planes usually were rotated to a different angle, as shown in Figure 2.7. Because the rock could shear much more easily along these planes, their orientation is important. Many landslides have occurred on slopes with unfavorable bedding orientations. Therefore, engineering geologists and geotechnical engineers are very careful to compare the attitudes of bedding planes with the orientation of proposed slopes.

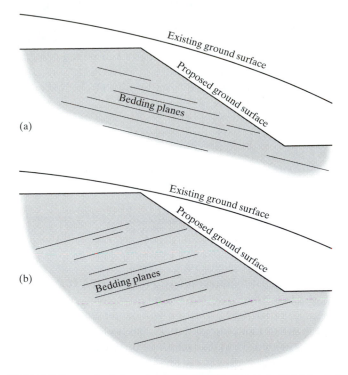

Figure 2.7 Proposed cut slopes in a bedded sedimentary rock. Cut A is much more likely to fail than cut B because it undermines the bedding planes, a condition called *daylighted bedding*.

Some metamorphic rocks have similar planes of weakness. They are called schistosity and are mapped in a similar way.

Folds

Tectonic forces also distort rock masses. When horizontal compressive forces are present, the rock distorts into a wavy pattern called *folds* as shown in Figure 2.8. Sometimes these folds are gradual; other times they are very abrupt. When folds are oriented concave downward they are called *anticlines*; when concave upward they are called *synclines*.

Figure 2.8 Folds in a sedimentary rock (GeoPhoto Publishing Company).

Fractures

Fractures are cracks in a rock mass. Their orientation is very important because the shear strength along these fractures is less than that of the intact rock mass, so they form potential failure surfaces. There are three types of fractures: joints, shear zones, and faults.

Joints are fractures that have not experienced any shear movements. They can be the result of cooling (in the case of igneous rocks), tensile tectonic stresses, or tensile stresses from lateral movement of adjacent rock. Joints usually occur at fairly regular spacings, and a group of such joints is called a *set*.

Shear zones are fractures that have experienced a small shear displacement, perhaps a few centimeters. They are caused by various stresses in the ground, and do not appear in sets as joints do. Shear zones often are conduits for groundwater.

Faults are similar to shear zones, except they have experienced much greater shear displacements. Although there is no standard for distinguishing the two, many geologists would begin using the term "fault" when the shear displacement exceeds about 1 m. Such movements are normally associated with earthquakes, as discussed in Chapter 20.

Faults are classified according to their geometry and direction of movement, as shown in Figure 2.9. *Dip-slip faults* are those whose movement is primarily along the dip. It is a *normal fault* if the overhanging block is moving downward, or a *reverse fault* if it is moving upward. A reverse fault with a very small dip angle is called a *thrust fault*. Conversely, *strike-slip faults* are those whose movement is primarily along the strike. They can be either *right-lateral* or *left-lateral* depending on the relative motion of the two sides. Some faults experience both dip-slip and strike-slip movements. The *fault trace* is the intersection of the fault and the ground surface.

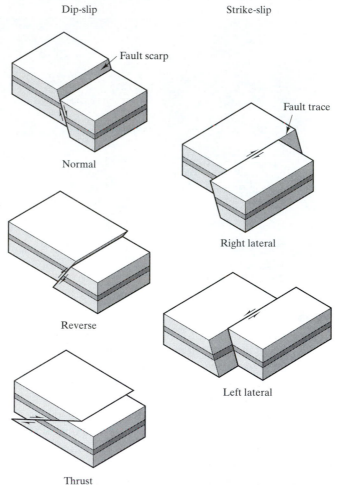

Figure 2.9 Types of faults.

The term *discontinuity* is often used in this context to include bedding planes, schistosity, joints, shear zones, faults, and all other similar defects in rock. Because the orientation of these features is one of the most important engineering aspects of the rock mass, extensive analytical methods have been developed to systematically evaluate discontinuity data gathered in the field (Priest, 1993).

Strike and Dip

When developing geologic maps, we are interested in both the presence of certain geologic structures and their orientation in space. For example, a rock mass may be unstable if it has joints oriented in a certain direction, but much more stable if they are oriented in a different direction. For similar reasons, we also are interested in the orientation of faults, bedding planes, and other geologic structures.

Many of these structures are roughly planar, at least for short distances, and therefore may be described by defining the orientation of this plane in space. We express this orientation using the *strike* and *dip*, as shown in Figure 2.10.

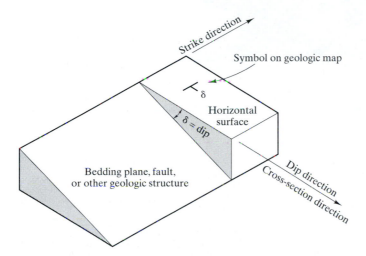

Figure 2.10 Use of strike and dip to define the orientation of a geologic structure (adapted from *Engineering Geology* by Richard E. Goodman, Copyright ©1993. Reprinted by permission of John Wiley and Sons).

The strike is the compass direction of the intersection of the plane and the horizontal, and is expressed as a bearing from true north. For example, if a fault has a strike of N30W, then the intersection of the fault plane with a horizontal plane traces a line oriented 30° west of true north. The dip is the angle between the geologic surface and the horizontal, and is measured in a vertical plane oriented perpendicular to the strike. The dip also needs a direction. For example, a fault with a N30W strike might have a dip of 20° northeasterly. When expressed together, this data is called an *attitude*, and may be written in condensed form as N30W; 20NE. Although the strike direction is "exact," the dip direction is only approximate. In this case, there are only two possibilities for the dip direction, NE or SW, so the purpose of this direction is simply to distinguish between these two possibilities. The "exact" dip direction is 90° from the strike.

Attitudes are usually measured in the field using a *Brunton compass*, as shown in Figure 2.11. This device includes both a compass and a level, and thus can measure both strikes and dips. The measured attitudes are then recorded graphically on geologic maps using the symbol shown in Figure 2.12. This symbol may be modified to indicate the type of structure being identified.

Figure 2.11 A Brunton compass is used to measure bedrock attitudes and other geologic features in the field.

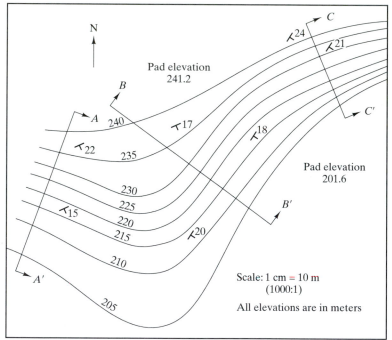

Figure 2.12 Geologic map showing bedrock attitudes. In this case, the attitudes represent the bedding planes in a sedimentary rock.

Sometimes we need to know the dip as it would appear in a vertical plane other than the one perpendicular to the strike. Figure 2.13 shows such a plane. For example, we may have drawn a cross-section that is oriented perpendicular to a slope, but at some angle other than 90° from the strike, and need to know the dip angle as it appears in that cross-section. This dip is called the *apparent dip* and may be computed using:

$$\boxed{\tan\delta_a = \tan\delta\ \sin\alpha}$$ (2.1)

Where:

δ_a = apparent dip

δ = dip

α = horizontal angle between strike and the vertical plane on which the apparent dip is to be computed

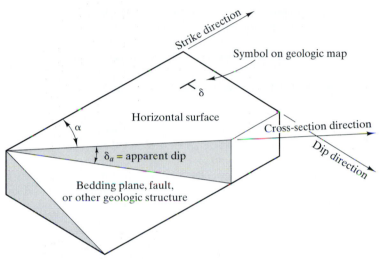

Figure 2.13 The apparent dip is the inclination of a geologic structure as seen on any vertical plane. It is always less than or equal to the true dip, which is determined on the vertical plane perpendicular to the strike (adapted from *Engineering Geology* by Richard E. Goodman, Copyright ©1993. Reprinted by permission of John Wiley and Sons).

Example 2.1

Compute the apparent dip of the bedding planes as they would appear in the central portion of Section B-B′ in Figure 2.12.

Solution

Base analysis on the 17° measured attitude. The angle between its strike and Section B-B′ is 65°. Therefore, using Equation 2.1:

$$\begin{aligned}
\tan\delta_a &= \tan\delta\ \sin\alpha \\
&= \tan 17°\ \sin 65° \\
\delta_a &= \mathbf{15°} \qquad \Leftarrow \textbf{\textit{Answer}}
\end{aligned}$$

Thus, the bedding plane will appear to be flatter than it really is.

QUESTIONS AND PRACTICE PROBLEMS

2.1 Geologists and engineers do not always use the same definitions of "rock" and "soil." Thus, there are some materials that are "rock" in the geologic sense, but not in the engineering sense. For example, some mudstones might be classified as rock by a geologist, yet be weaker than some "soils." Give an example of a situation where this difference could cause problems in the design or construction of a civil engineering project.

2.2 Which would probably provide better support for a large, heavy building, diorite or shale? Why?

2.3 Fossils are imprints in rock of ancient plants and animals. What type of rock might contain fossils? What type would never contain fossils? Explain.

2.4 What type of rock is most prone to contain sinkholes? Why?

2.5 Define "bedding planes" and explain why is it important to assess their orientation as a part of slope stability analyses.

2.6 The bedding planes in a certain sedimentary rock have a strike of N43E and a dip of 38SE, as shown by the attitude in Figure 2.14. A 15 m tall east-west cut slope inclined 34° from the horizontal is to be made in this rock. The ground surface above and below this proposed slope will be nearly level. Compute the apparent dip of the bedding planes as they will appear in cross-section A-A′, then draw this cross-section. Your drawing should show the ground surface and the bedding planes. Do these bedding planes pose a potential slope stability problem? Explain.

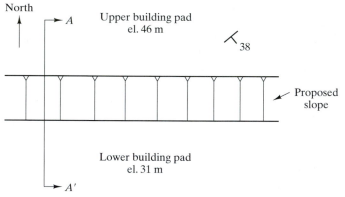

Figure 2.14 Plan view of proposed slope for Problem 2.6.

2.7 Draw cross-section A-A′ in Figure 2.12 and compute the apparent dip of the bedding planes as they would appear in this cross-section. There are two nearby attitudes, so compute the apparent dip for each. Then, sketch in the bedding planes on the cross-section. Do these bedding planes pose a potential stability problem? Why or why not?

2.8 Draw cross-section C-C′ in Figure 2.12 and compute the apparent dip of the bedding planes as they would appear in this cross-section. There are two nearby attitudes, so compute the apparent dip for each. Then, sketch in the bedding planes on the cross-section. Do these bedding planes pose a potential stability problem? Why or why not?

2.5 SOIL FORMATION, TRANSPORT, AND DEPOSITION

Geotechnical engineers work with both rock and soil, and need to be familiar with both. Nevertheless, we focus more of our energies on the engineering behavior of soil because:

- More civil engineering projects are built on soil
- Soil, being generally weaker and more compressible than rock, is more often a source of problems

Therefore, we are especially interested in those portions of the geologic cycle that produce and transport soils. A clear understanding of these processes helps geotechnical engineers interpret data gained from exploratory borings, and thus supports the very important function of engineering judgment. This discussion focuses on the inorganic components within a soil. Organic soils and their origins are discussed in Chapter 4.

Residual Soils

When the rock weathering process is faster than the transport processes induced by water, wind, and gravity, much of the resulting soil remains in place. It is known as a *residual soil*, and typically retains many of the characteristics of the parent rock. The transition with depth from soil to weathered rock to intact rock is typically gradual with no distinct boundaries.

In tropical regions, residual soil layers can be very thick, sometimes extending for hundreds of meters before reaching unweathered bedrock. Cooler and more arid regions normally have much thinner layers, and often no residual soil at all.

The soil type depends on the character of the parent rock. For example, *decomposed granite* (or simply "DG") is a sandy residual soil obtained from granitic rocks. DG is commonly used in construction as a high-quality fill material. Shales, which are sedimentary rocks that consist largely of clay minerals, weather to form clayey residual soils.

Saprolite is a general term for residual soils that are not extensively weathered and still retain much of the structure of the parent rock. Some have used the term "rotten rock" to describe saprolite. They typically include small concretions (harder, less weathered fragments) surrounded by more weathered material. Extensive saprolite deposits exist in the Piedmont area of the eastern United States (the zone between the Appalachian Mountains and the coastal plain) (Smith, 1987).

Laterite is a residual soil found in tropical regions. This type of soil is cemented with iron oxides, which gives it a high dry strength.

The engineering properties of residual soils range from poor to good, and generally improve with depth.

Glacial Soils

Much of the earth's land area was once covered with huge masses of ice called *glaciers*. In North America, glaciers once extended as far south as the Ohio River, as shown in Figure 2.15. In Europe, glaciers once existed as far south as Germany. Many of these areas are now heavily populated, so the geologic remains of glaciation have much practical significance.

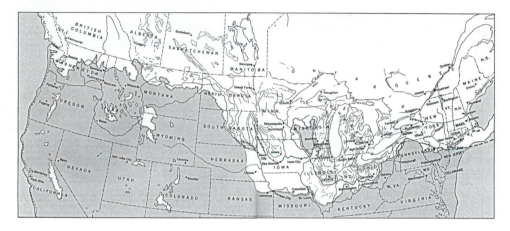

Figure 2.15 Southern extent of glaciation in North America during the various ice ages. The white areas were once covered with glaciers, and the heavy lines in these areas indicate locations of major moraines (adapted from *Physical Geology* by Flint and Skinner, Copyright ©1974. Reprinted by permission of John Wiley and Sons).

Glaciers had a dramatic effect on the landscape and created a category of soils called *glacial soils*. Glacial ice was not stationary; it moved along the ground, often grinding down some areas and filling in others. In some locations, glaciers reamed out valleys, leaving long lakes, such as the Finger Lakes of upstate New York. The Great Lakes also have been attributed to glacial action. Figure 2.16 shows how the moving ice strips away weathered rock, leaving a hard, unweathered surface in its wake.

Glaciers grind down the rock and soil, and transport these materials over long distances, even hundreds of kilometers, so the resulting deposits often contain a mixture of materials from many different sources. These deposits also can have a wide range of hardness and particle size, and are among the most complex and heterogeneous of all soils. The term *drift* encompasses all glacial soils, which then can be divided into three categories: till, glaciofluvial, and glaciolacustrine.

Till is soil deposited directly by the glacier. It typically contains a wide variety of particle sizes, ranging from clay to gravel. Soil that was bulldozed by the glacier, then deposited in ridges or mounds is called *ablation till*, as shown in Figure 2.17. These ridges and mounds are called *moraines* and are loose and easy to excavate. In contrast, soil caught beneath the glacier, called *lodgement till*, has been heavily consolidated under the weight of the ice. Because of these heavy consolidation pressures and the wide range of particle sizes, lodgement till has a very high unit weight and often is nearly as strong as concrete. Lodgement till is sometimes called *hardpan*. It provides excellent support for structural foundations, but is very difficult to excavate.

Figure 2.16 Effects of glaciation on metamorphic rock in Manitoba. The striations, gouging, and polishing of the rock surface are all due to the moving ice (Geological Survey of Canada).

    Geotechnical site assessments need to carefully distinguish between ablation till and lodgement till. Both engineers and contractors need to be aware of the difference and plan accordingly. For example, construction of the St. Lawrence Seaway along the U.S.–Canada border during the 1950s encountered extensive deposits of lodgement till that caused significant problems and delays. This problem was especially acute on the Cornwall Canal section of the seaway, causing one contractor to go bankrupt, another to default, and a third to file a $5.5 million claim on a $6.5 million contract (Legget and Hatheway, 1988).

Figure 2.17 The glacier in the background, which is part of the Athabasca Glacier in Alberta, is retreating and has left these moraines in its wake. The horizontal mounds of soil in the foreground are terminal moraines, and the ridges at the base of the mountain along the sides of the glacier are lateral moraines. Notice the wide range of particle sizes in these moraines.

When the glaciers melted, they generated large quantities of runoff. This water eroded much of the till and deposited it downstream, forming *glaciofluvial soils* (or *outwash*). Because of the sorting action of the water, these deposits are generally more uniform than till, and many of them are excellent sources of sand and gravel for use as concrete aggregates.

The fine-grained portions of the till often remained suspended in the runoff water until reaching a lake or the ocean, where it finally settled to the bottom. These are called *glaciolacustrine soils* and *glaciomarine soils*. Sometimes silts and clays were deposited in alternating layers according to the seasons, thus forming a banded soil called *varved clay*. The individual layers in varved clays are typically only a few millimeters thick, and often are separated by organic strata. These soils are soft and compressible, and thus are especially prone to problems with shear failure and excessive settlement.

Glaciolacustrine soils that formed in seawater are especially problematic because they have a high *sensitivity* (they lose shear strength when disturbed, as discussed in Chapter 13), and thus are prone to disastrous landslides. Such deposits are found in the Ottawa and St. Lawrence river valleys in eastern Canada (known as Champlain, Laurentian or Leda clays) and in southern Scandinavia. Figure 2.18 shows a flowslide in Leda clay adjacent to the South Nation River near Ottawa, Ontario. It resulted in the loss of 50 acres of farmland (Sowers, 1992).

Soils in the Chicago area are good examples of glacial deposits, and are typical of conditions in the Great Lakes region (Chung and Finno, 1992). The bedrock in this area consists of a marine dolomite that was overridden by successive advances and retreats of continental glaciers. At times this area was under ancient Lake Chicago, which varied in elevation from 18 m above to 30 m below the present level of Lake Michigan. These glaciers left both lodgement till and moraines, glaciolacustrine clays (deposited in the ancient lake), and glaciofluvial deposits in the riverbottoms, as shown in Figure 2.19.

Figure 2.18 The 1971 South Nation River flowslide near Ottawa, Ontario. This failure occurred in a soft marine soil called Leda Clay (Geological Survey of Canada).

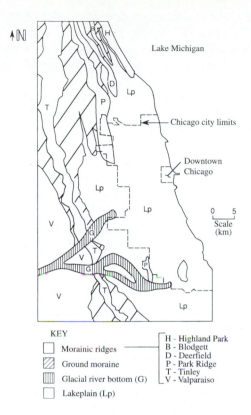

KEY

☐ Morainic ridges

▨ Ground moraine

▥ Glacial river bottom (G)

☐ Lakeplain (Lp)

H - Highland Park
B - Blodgett
D - Deerfield
P - Park Ridge
T - Tinley
V - Valparaiso

Figure 2.19 Soils in the Chicago, Illinois area (Chung and Finno, 1992).

Alluvial Soils

Alluvial soils (also known as *fluvial soils* or *alluvium*) are those transported to their present position by rivers and streams. These soils are very common, and a very large number of engineering structures are built on them. Alluvium often contains extensive groundwater aquifers, so it also is important in the development of water supply wells and in geoenvironmental engineering.

When the river or stream is flowing rapidly, the silts and clays remain in suspension and are carried downstream; only sands, gravels, and boulders are deposited. However, when the water flows more slowly, more of the finer soils also are deposited. Rivers flow rapidly during periods of heavy rainfall or snowmelt, and slowly during periods of drought, so alluvial soils often contain alternating horizontal layers of different soil types.

The water also slows when the stream reaches the foot of a canyon, and tends to deposit much of its soil load there. This process forms *alluvial fans*, as shown in Figure 2.20, which are one of the most obvious alluvial soils. They are especially common in arid areas.

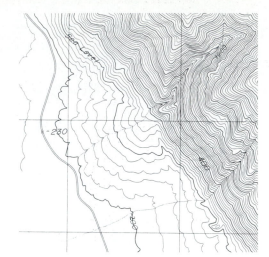

Figure 2.20 Topographic map of an alluvial fan in Death Valley, California. Soils eroded in the mountains are deposited at the foot of the canyon, thus forming a fan-shaped alluvial deposit (USGS Gold Valley quadrangle map).

Large boulders are sometimes carried by water, especially in steep terrain, and deposited in the upper reaches of alluvial deposits as shown in Figure 2.21. Sometimes such boulders are subsequently covered with finer soils and become obscured. However, they can cause extensive problems when engineers attempt to drill exploratory borings or contractors try to make excavations or drive pile foundations.

Rivers in relatively flat terrain move much more slowly and often change course, creating complex alluvial deposits. Some of these are called *braided stream deposits* and *meander belt deposits*, as shown in Figure 2.22. In addition, the deposition characteristics at a given location can change with time, so one type of alluvial soil is often underlain by other types.

Figure 2.21 Most alluvial soils consist of gravel, sand, silt, and clay. However, cobbles and boulders also can be present, especially along the base of mountains. For example, these large boulders were carried here by water and thus are an alluvial soil. They are located near the top of an alluvial fan that spreads out from a steep canyon.

Figure 2.22 The meanders in this river are forming a broad deposit of alluvial soils (GeoPhoto Publishing Company).

In arid areas, evaporation draws most of the water out of soil, leaving any dissolved chemicals behind. The resulting deposits of calcium carbonate, calcium sulfate, and other substances often act as cementing agents, converting the alluvial soil into a very hard material called *caliche*. These deposits are common in the southwestern states, and can be very troublesome to contractors who need to excavate through them.

Most alluvial soils have moderately good engineering properties, and typically provide fair to good support for buildings and other structures.

Lacustrine and Marine Soils

Lacustrine soils are those deposited beneath lakes. These deposits may still be underwater, or may now be exposed due to the lowering of the lake water level, such as the glaciolacustrine soils in Chicago (Figure 2.19). Most lacustrine soils are primarily silt and clay. Their suitability for foundation support ranges from poor to average.

Marine soils also were deposited underwater, except they formed in the ocean. *Deltas* are a special type of marine deposit formed where rivers meet larger bodies of water, and gradually build up to the water surface. Examples include the Mississippi River Delta and the Nile River Delta. This mode of deposition creates a very flat terrain, so the water flows very slowly. The resulting soil deposits are primarily silts and clays, and are very soft. Because of their deposition mode, most lacustrine and marine soils are very uniform and consistent. Thus, although their engineering properties are often poor, they may be more predictable than other more erratic soils.

Some sands also accumulate as marine deposits, especially in areas where rivers discharge into the sea at a steeper gradient. This sand is moved and sorted by the waves and currents, and some of it is deposited back on shore as *beach sands*. These sands typically are very poorly graded (i.e., they have a narrow range of particle sizes), have well-rounded particles, and are very loose. Beach deposits typically move parallel to the shoreline, and this movement can be interrupted by the construction of jetties and other harbor improvements. As a result, sand can accumulate on one side of the jetty, and be almost nonexistent on the other side. Changes in sea level elevations can leave beach deposits oriented along previous shorelines.

Deeper marine deposits are more uniform and often contain organic material from marine organisms. Those that have a large organic content are called *oozes*, one of the most descriptive of all soil names. The construction of offshore oil drilling platforms requires exploration and assessment of these soils.

Some lacustrine and marine soils have been covered with fill. This is especially common in urban areas adjacent to bays, such as Boston and San Francisco. The demand for real estate in these areas often leads to reclaiming such land, as shown in Figure 2.23. However, this reclaimed land is often a difficult place to build upon, because the underlying lacustrine and marine deposits are weak and compressible. Sometimes these soils have special names, such as Boston Blue Clay and San Francisco Bay Mud.

Figure 2.23 When the Puritans first settled in Boston, Massachusetts, the land area was as shown by the black zone in this map. It was connected to the mainland via a narrow isthmus. Since then, the city has been extended by placing fill in the adjacent water, thus forming the shoreline as it now exists.

Aeolian Soils

Aeolian soils (also known as *eolian soils*) are those deposited by wind. This mode of transport generally produces very poorly graded soils (i.e., a narrow range of particle sizes) because of the strong sorting power of wind. These soils also are usually very loose, and thus have only fair engineering properties.

There are three primary modes of wind-induced soil transport (see Figure 2.24):

- *suspension* occurs when wind lifts individual silt particles to high altitudes and transports them for great distances. This process can create large dust storms, such as those that occurred in Oklahoma and surrounding states during the "dust bowl" drought of the 1930s.

- *saltation* (from the Latin *saltatio* — to dance) is the intermediate process where soil particles become temporarily airborne, then fall back to earth. Upon landing, the particle bounces or dislodges another particle, thus initiating another flight. This motion occurs in fine sands, and typical bounce distances are on the order of 4 m. Particles moving by saltation do not gain much altitude; generally no more than 1 m.

- *creep* occurs in particles too large to become airborne, such as medium to coarse sands. This mode consists of rolling and sliding along the ground surface.

There are no distinct boundaries between these processes, so intermediate modes of transport also occur.

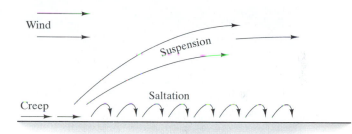

Figure 2.24 Modes of aeolian transport.

Aeolian sands can form horizontal strata, which often are interbedded with alluvial soils, or they can form irregular hills called *sand dunes*. These dunes are among the most striking aeolian deposits, and are found along some beaches and in some desert areas. Sand dunes tend to migrate downwind, and thus can be a threat, as shown in Figure 2.25. Migrations of 3 m/yr are not unusual, but this rate can be slowed or halted by establishing appropriate vegetation on the dune.

Figure 2.25 This sand dune near the beach in Marina, California is slowly migrating to the right and has partially buried the fence.

Aeolian silts often form deep deposits called *loess*. Such deposits are often found downwind of deserts and glacial outwash deposits. Extensive loess deposits are present in the Midwestern states.

Because of its deposition mode, loess typically has a very high porosity. It is fairly strong when dry, but becomes weak when wetted. As a result, it can be stable when cut to a steep slope (where water infiltration is minimal), yet unstable when the slope is flatter and water is able to enter the soil. Figure 2.26 shows a near-vertical cut slope in loess.

Figure 2.26 The slope in the center of this photograph is a cut made in a loess deposit near the Mississippi River in Tennessee. Notice how it is stable in spite of being near-vertical.

Nearly all aeolian soils are very prone to erosion, and often have deep gullies. Good erosion control measures are especially important in these soils.

Colluvial Soils

A *colluvial soil* is one transported downslope by gravity, as shown in Figure 2.27. There are two types of downslope movement, slow and rapid. Both types occur only on or near sloping ground.

Slow movement, which is typically on the order of millimeters per year, is called *creep*. It occurs because of gravity-induced downslope shear stresses, the expansion and contraction of clays, frost action, and other processes. Creep typically extends to depths of 0.3 to 3 m, with the greatest displacements occurring at the ground surface. In spite of the name, this process is entirely different from the "creep" process in aeolian soils.

Such slow movements might first appear to be inconsequential, but in time they can produce significant distortions in structures founded on such soils. Foundations that extend through creeping soils to firm ground below may be subjected to significant downslope forces from these soils, and need to be designed accordingly. In addition, the engineering properties of the soil deteriorate as it moves downhill, thus producing a material that is inferior to the parent soils.

Rapid downslope movements, such as landslides or mudflows, are more dramatic events which we will discuss in Chapter 14. Although these rapid movements can occur in any type of soil, the product is considered to be a colluvial soil.

Although colluvial soils occur naturally, construction activities sometimes accelerate their formation. For example, making an excavation at the toe of a slope may change a slow creep condition into a landslide.

(a) Slow   (b) Rapid

Figure 2.27 Colluvial soils: a) Slowly formed by creep; b) Rapidly formed by landslides or mudflows.

QUESTIONS AND PRACTICE PROBLEMS

2.9   Explain the difference between ablation till and lodgement till. Which would provide better support for heavy civil engineering projects? Why?

2.10   Which would probably provide better support for a proposed structure, an alluvial sand or an aeolian sand? Why?

2.11   A new car dealership has recently been built in an area known for occasional strong winds. Unfortunately, an open field of fine sandy soil exists immediately upwind of the dealership. Soon after construction, a 70 mi/hr wind blew large quantities of this soil onto the new cars, seriously damaging their paint. Could this problem have been anticipated? What mode of aeolian transport brought the sand from the field to the cars? Given the current conditions, how might this problem be avoided in the future?

2.12   Make a copy of Figure 2.19 and indicate the probable lateral limits of Lake Michigan and the probable locations of previous river channels.

SUMMARY

Major Points

1. Engineering geology is a profession closely related to geotechnical engineering. It deals with the application of geologic principles to engineering works, and is especially useful at sites where rock is at or near the ground surface.
2. It is important for geologists to have some understanding of engineering, and for engineers to have some understanding of geology.
3. Earth materials may be divided into two broad categories, rock and soil. Unfortunately, everyone doesn't agree on how to distinguish between the two, especially in intermediate materials.
4. Minerals are naturally formed elements or compounds with specific structures and chemical compositions. They are the basic constituents of rocks and soils.

5. The earth's crust is always changing through a process called the geologic cycle. Although this process is very slow, we must understand it to properly interpret geologic profiles.

6. There are three major categories of rock: igneous, sedimentary, and metamorphic.

7. Properly identifying the configuration and orientation of rock formations is at least as important as identifying the rock types contained in them. This study is called structural geology.

8. Soils are formed through several different geologic processes. Understanding these processes gives us insight into the engineering behavior of these soils.

Vocabulary

ablation till
aeolian soils
alluvial fan
alluvial soil
alluvium
andesite
anisotropic strength
anticline
apparent dip
arkose
attitude
basalt
bedding planes
breccia
Brunton compass
calcite
carbonates
chalk
clastic rocks
claystone
colluvial soil
conglomerate
creep
decomposed granite
diorite
dip
dip-slip fault
discontinuity
dolomite
drift
engineering geology
extrusives
fault
fault trace

feldspar
ferromagnesian minerals
folds
foliations
fracture
gabbro
geologic cycle
geology
glacial soil
glaciofluvial soil
glaciolacustrine soil
glaciomarine soil
gneiss
granite
graywacke
gypsum
hardpan
igneous rocks
intrusives
iron oxides
joint
karst topography
lacustrine soils
laterite
left-lateral fault
limestone
lodgement till
loess
magma
marble
marine soils
metamorphic rocks
mica
minerals

moraine
mudstone
normal fault
quartz
quartzite
residual soil
reverse fault
rhyolite
right-lateral fault
rock
saltation
sand dune
sandstone
saprolite
schist
sedimentary rocks
shale
shear zone
siltstone
sinkhole
slate
soil
strike
strike-slip fault
structural geology
suspension
syncline
thrust fault
till
transported soil
varved clay
weathering

COMPREHENSIVE QUESTIONS AND PRACTICE PROBLEMS

2.13   A proposed construction site is underlain by a sedimentary rock that was formed from rounded gravel-size particles and sand. The gravel-size particles represent about 75 percent of the total mass. What is the name of this rock? Would you expect it to provide good support for the proposed structural foundations? Will it be difficult to excavate?

2.14   As the glaciers in North America melted, the runoff formed a large lake in what is now southern Manitoba, eastern North Dakota, and western Minnesota. Called Lake Agassiz, it was larger than all of the current Great Lakes combined. The present Lake Winnipeg is a remnant of this ancient lake. The City of Winnipeg is located on the ancient lakebed. What kinds of soil would you expect beneath the city, and what is their likely geologic origin?

2.15   Would you expect to find till in Houston, Texas? Why or why not?

2.16   A heavy structure is to be built on a site adjacent to the Hudson River near Albany, NY. This area was once covered with glaciers that left deposits of lodgement till and glaciofluvial soils. Since then, the river has deposited alluvial soils over the glacial deposits. The design engineer wishes to support the structure on pile foundations extending to the lodgement till, and you are planning a series of exploratory borings to determine the depth to these strata. What characteristics would you expect in the lodgment till (i.e., how would you recognize it?).

2.17   New Orleans, Louisiana is located near the mouth of the Mississippi River. What geological process has been the dominant source of the soils beneath this city? What engineering characteristics would you expect from these soils (i.e., overall quality, uniform or erratic, etc.)? Explain.

2.18   A project is to be built on a moderately sloping site immediately below the mouth of a canyon near Phoenix, Arizona. Using the geologic terms described in Section 2.5, what type of soil is most likely to be found? Why?

2.19   A varved clay deposit has been progressively buried by other deposits and eventually has been lithified into a sedimentary rock. What type of rock is it? Would you expect its bedding planes to be distinct or vague? Explain.

3

Site Exploration and Characterization

> *The process of exploring to characterize or define small scale properties of substrata at construction sites is unique to geotechnical engineering. In other engineering disciplines, material properties are specified during design, or before construction or manufacture, and then controlled to meet the specification. Unfortunately, subsurface properties cannot be specified; they must be deduced through exploration.*
>
> Charles H. Dowding (1979)

Most engineers work with manufactured products that have very consistent and predictable engineering properties. For example, when a structural engineer designs a W18×55 beam to be made of A36 structural steel, he or she can be confident the yield strength will be 36 k/in^2, the modulus of elasticity will be 29×10^3 k/in^2, the moment of inertia will be 891 in^4, and so on. There is no need to test A36 steel every time someone wants to design a beam; we simply specify what is to be used and the contractor is obligated to supply it.

Geotechnical engineers do not have this luxury. We work with soil and rock, which are natural materials whose engineering properties vary dramatically from place to place. For example, one site may be underlain by strong, hard deposits, such as lodgement till, and can safely support heavy loads, while another may be underlain by soft, weak deposits, such as varved clay, and thus requires careful design and construction techniques to support even nominal loads. In addition, we need to work with whatever soil or rock is present at our site. Thus, instead of specifying required properties, our task is to determine the existing properties at our site. This process is called *site characterization*.

This distinction is no small matter, because the site characterization efforts typically represent a very large share of the geotechnical engineering budget. We often spend more time and money exploring the subsurface conditions and defining their engineering characteristics than we do performing our analyses and developing our designs.

The objectives of a site exploration and characterization program include:

- Determining the location and thickness of soil and rock strata
- Determining the location of the groundwater table, along with other important groundwater-related issues
- Recovering samples for testing and evaluation
- Conducting tests, either in the field or in the laboratory, to measure relevant engineering properties
- Defining special problems and concerns

Unfortunately, most of what we want to know is hidden underground and thus very difficult to discern. We can explore the subsurface conditions using borings and other techniques, and recover samples for testing and evaluation, but even the most thorough exploration program encounters only a small fraction of the soil and rock below the site. We do not know what soil conditions exist between borings, and must rely on interpolation combined with a knowledge of soil deposition processes. In addition, we never can be completely sure if our samples are truly representative, or if we have missed some important underground feature. These uncertainties represent the single largest source of problems for geotechnical engineers. We overcome them using a combination of techniques, including:

- Recognizing the uncertainties and applying appropriate conservatism and factors of safety to our analyses and designs
- Using a knowledge of the local geology to interpret the available subsurface information
- Observing and monitoring conditions during construction, and being prepared to modify the design based on newly acquired information
- Acknowledging that 100 percent reliability is not attainable, and accepting some risk of failure due to unforeseen conditions

3.1 PROJECT ASSESSMENT

Before planning a site exploration and characterization program, the geotechnical engineer must gather certain information on the proposed development. This information would include such matters as:

- The types, locations, and approximate dimensions of the proposed improvements (i.e., a 9-story building is to be built here, a parking lot there, and an access road to connect the project with the main highway over there)
- The type of construction, structural loads, and allowable settlements
- The existing topography and any proposed grading
- The presence of previous development on the site, if any

All of these factors have an impact on the methods and thoroughness of the program. For example, a proposed nuclear power plant to be built on a difficult site would require a very extensive exploration and characterization, while a one-story wood frame building on a good site may require only minimal effort.

3.2 LITERATURE SEARCH

The first step in gathering information on a site often consists of reviewing published sources. Sometimes these efforts reveal the results of extensive work already performed on the site, and very little additional exploration may be necessary. More often, literature searches provide only a general understanding of the local soils and rocks.

Sources of relevant literature include:

- **Geologic maps**, which are representations of the soil and rock types exposed at the ground surface, and usually show the extent of various geologic formations, alignments of faults, major landslides, and other geologic features. They also may include cross-sections showing subsurface conditions. Published maps usually cover areas much larger than a single project site. Scales of about 1:24,000 (the same as a USGS 7.5 minute quad map) are common.

    Studying the local geology helps alert us to potential problems at the site and helps us interpret the data gathered from our surface and subsurface exploration programs, which are virtually always limited to the site under consideration. If bedrock is exposed at our site, geologic maps help us identify the formation to which it belongs, and thus assist in the identification of potential problems. For example, the Bearpaw Formation in Montana, Saskatchewan, and the surrounding area is a shale with large quantities of montmorillonite. Structures built on this formation often have problems when it becomes wet and swells, so the identification of its presence on a site signals the need for special precautions.

    Sometimes our site characterization efforts involve developing new large-scale geologic maps that describe our site in more detail.

- **Soil survey reports**, which contain maps of the near-surface soil conditions. These maps are developed primarily for agricultural purposes, but can provide useful information for engineers. In the United States, soil survey reports are produced primarily by the Natural Resources Conservation Service (formerly known as the Soil Conservation Service).

    Typical soil surveys encompass areas about the size of a county, with mapping scales of about 1:15,000 to 1:24,000. The soil survey maps are accompanied by reports that include limited test data, along with qualitative evaluations of each soil series. Although these surveys are not sufficiently detailed to develop detailed geotechnical designs, they do identify the general surface soil conditions in the area, and can be helpful in planning more detailed site-specific investigations.

- **Geotechnical investigation reports** from other nearby projects, or even previous projects on our site, are often available, especially in urban areas. These reports can be very valuable because they normally include borings, soil tests, and other relevant data.

- **Historic groundwater data** is sometimes available from maps or reports. This data may be used to predict the worst-case groundwater conditions that might occur during the life of a project.

3.3 REMOTE SENSING

Remote sensing is the process of detecting features on the earth's surface from some remote location, such as an aircraft or spacecraft. This can be done using aerial photographs, radar, and other types of sensors. For geotechnical engineers, aerial photographs are the most useful remote sensing tool.

Conventional Aerial Photographs

Aerial photographs or simply *airphotos* are taken from airplanes using special cameras. Some of these are *oblique*, which means they view the landscape at some angle, while others are *vertical*, or looking straight down. The latter are more common, and generally more useful. Figure 3.1 shows a vertical airphoto. Both black-and-white and color photos are available, usually on 9 in × 9 in (229 mm × 229 mm) negatives. Color photographs are more useful, because they reveal geologic information in more detail.

Sometimes overlapping vertical airphotos are used to form a *stereo pair*. When viewed through a stereoscope, as shown in Figure 3.2, these photos present a three-dimensional image of the ground.

Figure 3.1 Vertical aerial photograph. The ocean is at the bottom of the photo. The land area includes a major highway, residential areas, and agricultural areas. The dark vertical line near the center of the photograph is trees along a creek (Pacific Western, Inc., Santa Barbara, CA).

Figure 3.2 The author using a stereoscope to view aerial photographs in three dimensions.

The scale of airphotos generally is between 1:3,000 and 1:40,000, which allows us to identify important geologic features, such as landslides, faults, and erosion features, and helps us understand site topography and drainage patterns. This technique is especially useful at sites where observations from the ground are blocked by forests. Viewing old airphotos also helps determine the site history, including previous buildings, old cuts and fills, and so on.

Infrared Aerial Photographs

It also is possible to take aerial photographs using a special film that is sensitive to both the visible and infrared spectra. The colors are shifted from that of normal color photographs (yellow objects appear green, etc.) and reflected infrared light is shown as red. This is valuable because vegetation reflects infrared, and thus is easily discernible.

Healthy, vigorous vegetation reflects the most, and is bright red in the photographs. This normally indicates the presence of water, and thus can be used to locate springs and seepage zones. This technique is especially useful in certain slope stability studies, because water from these springs and seeps may cause future landslides, or may be part of the explanation of a past landslide.

3.4 FIELD RECONNAISSANCE AND SURFACE EXPLORATION

The *field reconnaissance* consists of "walking the site" and visually assessing the local conditions. It includes obtaining answers to such questions as:

- Is there any evidence of previous development on the site?
- Is there any evidence of previous grading on the site?
- Is there evidence of landslides or other stability problems?

- Are nearby structures performing satisfactorily?
- What are the surface drainage conditions?
- What types of soil and/or rock are exposed at the ground surface?
- Will access problems limit the types of subsurface exploration techniques that can be used?
- Might the proposed construction affect existing improvements? For example, a fragile old building adjacent to the site might be damaged by vibrations from pile driving.
- Do any offsite conditions affect the proposed development? For example, potential flooding, mudflows, or rockfalls from offsite might affect the property.

This work also includes marking the locations of proposed exploratory borings and trenches. When rock is exposed, the field reconnaissance often will include geologic mapping.

Depending on the site conditions, a field reconnaissance also might include detailed mapping of the surface conditions. For example, if peat bogs (depressions filled with highly organic soils) are present, their lateral extent must be carefully recorded. If rock is exposed at or near the ground surface, geologic mapping by an engineering geologist may be required.

3.5 SUBSURFACE EXPLORATION

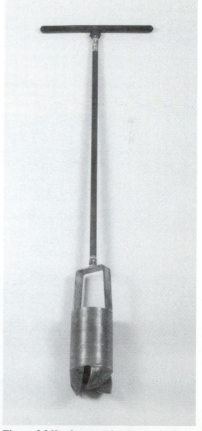

Although information on the soil and rock conditions exposed at the ground surface is very valuable, geotechnical engineers also need to evaluate the subsurface (underground) conditions. The geophysical methods described earlier can provide some insight, but we primarily rely on soil and rock samples obtained by drilling vertical holes known as borings, or by digging exploratory trenches or pits. These subsurface exploration activities usually are the heart of a site characterization program, and typically are the most expensive part because they require the mobilization of both equipment and labor.

Exploratory Borings

The most common method of exploring the subsurface conditions is to drill a series of vertical holes in the ground. These are known as *borings* or *exploratory borings* and are typically 75 to 600 mm (3–24 in) diameter and 2 to 30 m (7–100 ft) deep.

Small, shallow borings can be made with lightweight hand-operated augers, as shown in Figure 3.3. This equipment is inexpensive and

Figure 3.3 Hand-operated auger.

portable, but limited in its capabilities and generally suitable only for very small projects with boring depths less than about 4 m (13 ft). Some additional capacity can be gained by using portable power-operated equipment, but such equipment is still too limited for most projects.

Geotechnical engineers usually use much heavier equipment powered by larger engines. Sometimes it is mounted on skids or small roll-in units, as shown in Figure 3.4, but most often it is truck-mounted as shown in Figure 3.5. These *truck-mounted drill rigs* perform at least 90 percent of geotechnical drilling, and can drill to depths of 30 m (100 ft) with little difficulty. Some truck mounted rigs can drill to 60 m (200 ft) or even more, but such capabilities are rarely needed.

Figure 3.4 Limited access drill rig that can be moved through narrow openings. It is connected to a truck-mounted hydraulic pump via the hoses in the foreground.

Figure 3.5 Truck-mounted drill rig.

Drilling Methods

Different methods are available to advance the boring, depending on the anticipated soil and rock conditions.

Drilling in Firm and Dense Soils

The simplest drilling methods use a *flight auger* or a *bucket auger*, as shown in Figures 3.6 and 3.7, to produce an open hole. With either type, the auger is lowered into the hole and rotated to dig into the soil. Then, it is removed, the soil is discharged onto the ground, and the process is repeated. The hole is free of equipment between these cycles, which allows the driller to insert sampling equipment at desired depths and obtain undisturbed samples.

Figure 3.6 A crew drilling an exploratory boring using a truck-mounted flight auger. These augers typically have an outside diameter of 120–200 mm (3–8 in). (Foremost Mobile Drilling Co.)

These methods are comparatively inexpensive, so long as they are used in suitable conditions, such as firm and dense soils or soft rock. However, they can meet *refusal* (the inability to progress further) when they encounter hard boulders or hard bedrock. This is especially likely when the boring diameter is small, since even large cobbles might block the drilling. Sometimes this problem can be overcome by using a larger diameter auger (i.e., one that is larger than the cobbles and boulders). Alternatively, some rigs can switch to a coring mode and continue as described below. Otherwise, it becomes necessary to use some other type of drilling method.

Drilling in Soils Prone to Caving or Squeezing

Open hole methods encounter problems in soils prone to *caving* (i.e., the sides of the boring fall in) or *squeezing* (the soil moves inwards, reducing the boring diameter). Caving is most likely in loose sands and gravels, especially below the groundwater table, while squeezing is likely in soft saturated silts and clays. In such cases, it becomes necessary to provide some type of lateral support inside the hole during drilling.

One method of supporting the hole is to install *casing* (see Figure 3.8), which is a temporary lining made of steel pipe. This method is especially useful if only the upper soils are prone to caving, because the casing does not need to extend for the entire depth of the boring.

Another, more common method is to use a *hollow stem auger*, as shown in Figure 3.9. Each auger section has a pipe core known as a stem, with a temporary plug on the bottom of the first section. The driller screws these augers into the ground, adding sections as needed. Unlike conventional augers, it is not necessary to remove them to obtain samples. Instead, the driller removes the temporary plug and inserts the sampler through the stem and into the soils below the bottom auger section, as shown in Figure 3.10. Then, the sample is recovered, the plug is replaced, and drilling continues to the next sample depth. When the boring is completed, the augers are removed. Hollow stem drill rigs with 200 mm (8 in) diameter augers are very common, and are often used even when caving is not a problem.

Figure 3.7 Using a bucket auger. The bucket has just come out of the hole and will be tilted back by the drill rig. The driller's helper will then pull the rope, which will open the bottom and release the soil. Most bucket augers have a diameter between 300 and 900 mm (12–36 in).

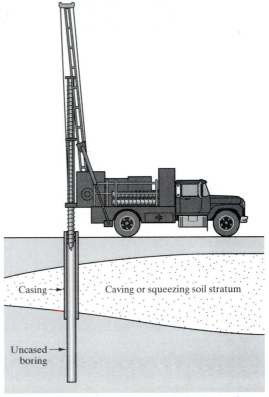

Figure 3.8 Use of casing to prevent caving and squeezing.

Figure 3.9 Use of a hollow-stem auger.

Figure 3.10 Lowering a soil sampler through a hollow-stem auger.

The third method is to fill the boring with *drilling mud* or *slurry*, which is a mixture of bentonite or attapulgite clay and water. This material provides a hydrostatic pressure on the walls of the boring, as shown in Figure 3.11, thus preventing caving or squeezing. These borings are usually advanced using the *rotary wash method*, which flushes the drill cuttings up to the ground surface by circulating the mud with a pump. When samples are needed, the drilling tools are removed from the hole and the sampling tools are lowered through the mud to the bottom. Special drilling tools may be added if the boring reaches hard soils or rock.

Coring

Drilling through rock, especially hard rock, requires different methods and equipment. Engineers usually use *coring*, which simultaneously advances the hole and obtains nearly continuous undisturbed samples. This is fundamentally a different method that consists of grinding away an annular zone with a rotary diamond drill bit, leaving a cylindrical core which is captured by a *core barrel* and removed from the ground. The cuttings are removed by circulating drilling fluid, water, or air. Figure 3.12 shows a core sampler partway through a *core run*, which is the segment sampled during one stroke of the sampler. Coring also can be done in hard soils.

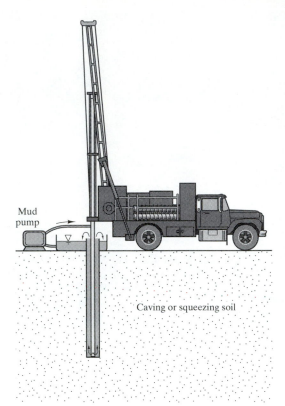

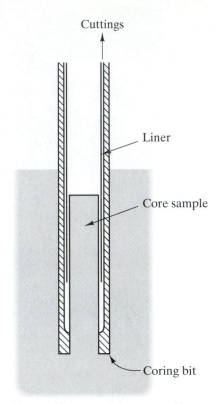

Figure 3.11 Use of drilling mud to prevent caving and squeezing. The mud provides a hydrostatic pressure to the sides of the boring, thus keeping the adjacent soils in place.

Figure 3.12 A sample being recovered by coring. The coring bit cuts an annular-shaped hole as it penetrates into the ground.

After each core run, the sample is brought to the ground surface and placed in a wooden *core box* for examination and storage, as shown in Figure 3.13. This permits detailed logging of the hole and provides high-quality samples for laboratory testing. Most cores are 48 or 54 mm in diameter.

Coring logs often record the *rock quality designation* or RQD, which is the percentage of core in pieces 100 mm or longer. It is a useful measure of rock fracturing, and thus an indicator of stability. RQD values greater than 90 percent typically indicate excellent rock, while values less than 50 percent indicate poor or very poor rock.

The *core recovery*, which is the total sample length recovered from each core run divided by the run length, also should be recorded. Often some of the sample is "lost," especially in weak or friable rocks.

Unfortunately, the weakest and most fractured zones, which are the most important zones to identify, are those most likely to be lost during coring. Also, most coring does not retain the in-situ orientation of the sample, so information on the directions of joints, bedding planes, etc. is lost. Both of these problems can be at least partially overcome by using downhole cameras that take photographs or video recordings of the hole after the core has been removed.

Figure 3.13 Core samples in a core box. These samples are very long, with an RQD of nearly 100 percent.

Boring Logs

The conditions encountered in an exploratory boring are recorded on a *boring log*, such as the one shown in Figure 3.14. The vertical position on these logs represents depth, and the various columns describe certain characteristics of the soil and rock. These logs also indicate the sample locations and might include some of the laboratory and in-situ test results. Usually, a field log is prepared while the boring is being drilled, then "cleaned up" in the office when the lab results become available.

Downhole Logging

Sometimes it is useful to drill large-diameter (500–900 mm) borings so the subsurface conditions can be observed by *downhole logging*. A geologist descends into such holes on a specially fabricated cradle and inspects the exposed walls. This allows thorough mapping of soil and rock types, attitudes of various contacts and bedding planes, etc., and thus is much more reliable and informative than relying solely on samples. Of course, this method is suitable only above the groundwater table in holes not prone to caving or squeezing.

Number, Spacing, and Depth

There are no absolute rules to determine the required number, spacing, and depth of exploratory borings. Such decisions are based on the findings from the field reconnaissance, along with engineering judgement and a knowledge of customary standards of practice.

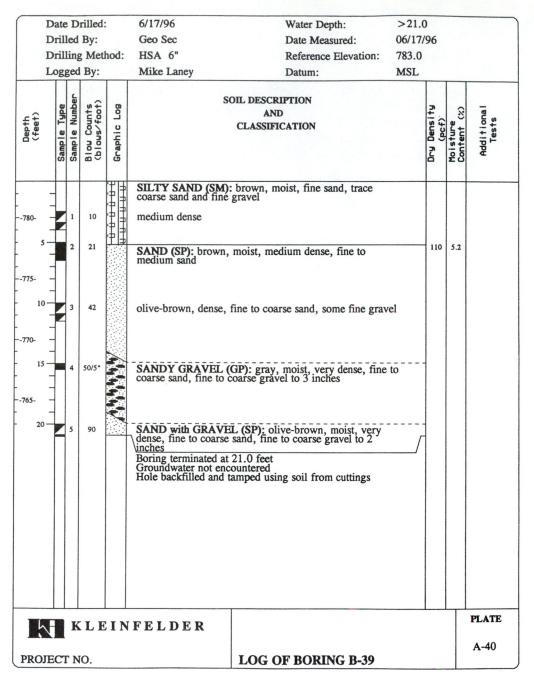

Figure 3.14 A boring log. Samples 2 and 4 were obtained using a heavy-wall sampler, and the corresponding blow counts are the number of hammer blows required to drive the sampler. Samples 1, 3, and 5 are standard penetration tests, and the corresponding blow counts are the N_{60} values, as discussed later in this chapter. (Kleinfelder, Inc.)

This is a subjective process that involves many factors, including:

- How large is the site?
- What kinds of soil and rock conditions are expected?
- Is the soil profile erratic, or is it consistent across the site?
- What is to be built on the site (small building, large building, highway, etc.)?
- How critical is the proposed project (i.e., what would be the consequences of a failure)?
- How large and heavy are the proposed structures?
- Are all areas of the site accessible to drill rigs?

Although we will not know the final answers to some of these questions until the site characterization program is completed, we should have at least a preliminary idea based on the literature search and field reconnaissance.

Table 3.1 presents rough guidelines for determining the normal spacing of exploratory borings. However, it is important to recognize that there is no single "correct" solution for the required number and depth of borings, and these guidelines must be tempered with appropriate engineering judgement.

TABLE 3.1 ROUGH GUIDELINES FOR SPACING EXPLORATORY BORINGS FOR PROPOSED MEDIUM TO HEAVY WEIGHT BUILDINGS, TANKS, AND OTHER SIMILAR STRUCTURES.

Subsurface Conditions	Structure Footprint Area for Each Exploratory Boring	
	(m^2)	(ft^2)
Poor quality and/or erratic	100–300	1,000–3,000
Average	200–400	2,000–4,000
High quality and uniform	300–1,000	3,000–10,000

Borings for buildings and other structures on shallow foundations generally should extend at least to the depths described in Table 3.2. If fill is present, the borings must extend through it and into the natural ground below, and if soft soils are present, the borings should extend through them and into firmer soils below. For heavy structures, at least some of the borings should be carried down to bedrock, if possible, but certainly well below the depth of any proposed deep foundations.

On large projects, the drilling program might be divided into two phases: a preliminary phase to determine the general soil profile, and a final phase based on the results of the preliminary borings.

TABLE 3.2 ROUGH GUIDELINES FOR DEPTHS OF EXPLORATORY BORINGS FOR BUILDINGS ON SHALLOW FOUNDATIONS (Adapted from Sowers, 1979)

Subsurface Conditions	Minimum Depth of Borings (S = number of stories; D = anticipated depth of foundation)	
	(m)	(ft)
Poor	$6\,S^{0.7} + D$	$20\,S^{0.7} + D$
Average	$5\,S^{0.7} + D$	$15\,S^{0.7} + D$
Good	$3\,S^{0.7} + D$	$10\,S^{0.7} + D$

Example 3.1

A three-story steel frame office building is to be built on a site where the soils are expected to be of average quality and average uniformity. The building will have a 30 m × 40 m footprint and is expected to be supported on spread footing foundations located about 1 m below the ground surface. The site appears to be in its natural condition, with no evidence of previous grading. Bedrock is several hundred feet below the ground surface. Determine the required number and depth of the borings.

Solution

Per Table 3.1, one boring will be needed for every 200–400 m^2 of footprint area. Since the total footprint area is 30 × 40 = 1200 m^2, use four borings.

Per Table 3.2, the minimum depth is $5\,S^{0.7} + D = 5\,(3)^{0.7} + 1 = 12$ m. However, it would be good to drill at least one of the borings to a greater depth.

Exploration plan:
    3 borings to 12 m   ⇐ *Answer*
    1 boring to 16 m

Exploratory Trenches and Pits

Sometimes it is only necessary to explore the upper 3 m (10 ft) of soil. This might be the case for lightweight projects on sites where the soil conditions are known to be good, or on sites with old shallow fills of questionable quality. Additional shallow investigations also might be necessary to supplement a program of exploratory borings.

In such cases, geotechnical engineers often dig *exploratory trenches* (also known as *test pits*) using a backhoe, as shown in Figure 3.15. These techniques provide more information than a boring of comparable depth (because more of the soil is exposed), and often are less expensive. The log from a typical exploratory trench is shown in Figure 3.16.

Two special precautions are in order when using exploratory trenches: First, these trenches must be adequately shored or laid back to a sufficiently flat slope before anyone enters them. Many individuals (including one of the author's former colleagues) have been killed by neglecting to enforce this basic safety measure. Second, these trenches must be properly backfilled to avoid creating an artificial soft zone that might affect future construction.

Figure 3.15 This exploratory trench was dug by the backhoe in the background, and has been stabilized using aluminum-hydraulic shoring. An engineering geologist is logging the soil conditions in one wall of the trench. In this case, the purpose of the trench is to locate a fault.

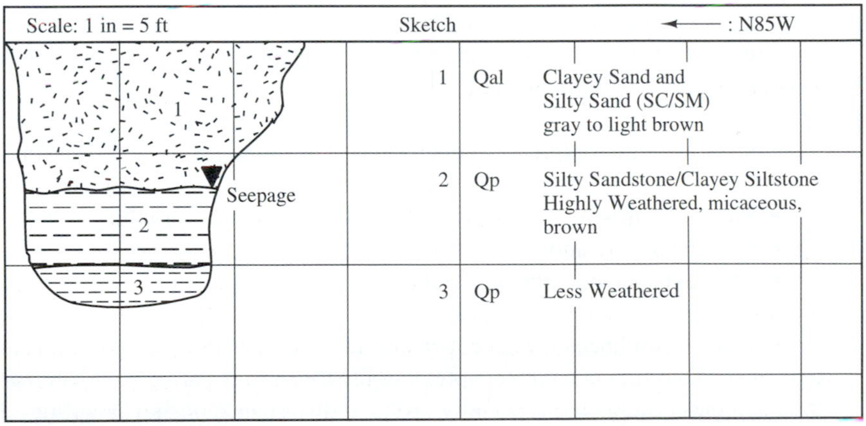

Scale: 1 in = 5 ft		Sketch		◄──── : N85W		
	1	Qal	Clayey Sand and Silty Sand (SC/SM) gray to light brown			
Seepage	2	Qp	Silty Sandstone/Clayey Siltstone Highly Weathered, micaceous, brown			
	3	Qp	Less Weathered			

Figure 3.16 Log from an exploratory trench (Courtesy of Converse Consultants).

3.6 SOIL AND ROCK SAMPLING

The primary purpose of drilling exploratory borings and digging exploratory trenches is to obtain representative soil and rock samples. We use these samples to determine the subsurface profile and to perform laboratory tests. There are two categories of samples: disturbed and undisturbed, each discussed below.

Disturbed Samples

A *disturbed sample* (also called a *bulk sample*) is one obtained with no attempt to retain the in-place structure of the soil or rock. The driller might obtain such a sample by removing cuttings from the bottom of a flight auger and placing them in a bag. Disturbed samples, such as the one in Figure 3.17, are suitable for many purposes, such as classification and compaction tests.

Undisturbed Samples

The greater challenge in soil sampling is to obtain *undisturbed samples*, which are necessary for many soil tests. Except for coring, which recovers undisturbed samples as the hole is advanced, drilling operations must stop periodically to permit insertion of special sampling tools into the hole as shown in Figure 3.10.

Figure 3.17 A typical disturbed sample stored in a plastic bag. The label attached to the bag identifies the sample.

In a truly undisturbed soil sample, the soil is recovered completely intact and its in-place structure and stresses are not modified in any way. Unfortunately, the following problems make it impossible to obtain such samples:

- Shearing and compression that occurs during the process of inserting the sampling tool
- Release of in-situ stresses as the sample is removed from the ground
- Possible drying and desiccation
- Vibrations during recovery and transport

Additional disturbances can occur in the laboratory as the sample is removed from its container. Thus, many engineers prefer to use the term "relatively undisturbed" to describe their samples. Sands are especially prone to disturbance during sampling. Nevertheless, geotechnical engineers have developed various methods of obtaining high-quality samples of most soils.

Shelby Tube Samplers

In the mid 1930s, Mr. H. A. Mohr developed the *Shelby tube sampler*, shown in Figure 3.18a, which soon became the most common soil sampling tool (Hvorslev, 1949). It also is known as a *thin-wall sampler* ("Shelby tubing" is a trade name for the seamless steel tubing from which the sampler is manufactured). Figure 3.18b shows a Shelby tube sampler attached to a standard head assembly. Most Shelby tube samplers have a 3.00 in (76.2 mm) outside diameter and 1/16 in (1.6 mm) wall thickness.

The head assembly is attached to a series of drilling rods, lowered to the bottom of the boring, then smoothly pressed into the natural ground below. This smooth pressing is accomplished by attaching a hydraulic cylinder to the top of the rods and using the drill rig as a reaction. Sometimes it is necessary to pound the sampler in by striking the rods with a 63.5 kg (140 lb) hammer, as shown in Figure 3.25, but this method can produce significantly more sample disturbance. The sampler is then pulled out of the ground with the soil retained inside, capped, and brought to the laboratory.

The standard head assembly has vents to allow water and air trapped above the sample to escape as it is inserted into the ground. However, some backpressure remains, and it can compress the sample. The *piston sampler* in Figure 3.18c avoids this problem by placing a piston inside the Shelby tube sampler. The piston is initially at the bottom of the tube, and remains at a constant elevation as the tube is advanced, thus shielding the soil from the backpressure.

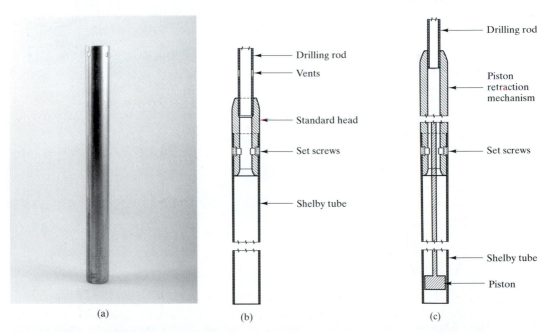

Figure 3.18 (a) A 3 × 36 inch Shelby tube; (b) A Shelby tube attached to a standard head with four screws; (c) A Shelby tube attached to a piston sampler.

Heavy-Wall Samplers

Although Shelby tube samplers generally provide very good results in soft soils, they are difficult to use in hard soils. The tube may bend or collapse due to the heavy loads required to press or drive it into such soils, or it may become jammed into the ground and impossible to retrieve. The usual solution is to use a sampler with heavier walls as shown in Figure 3.19. Although these heavy walls induce more disturbance, they also provide sufficient strength and durability to survive hard soil conditions. These *heavy-wall samplers* are almost always pounded into the bottom of the boring.

Heavy-wall samplers usually contain brass or stainless steel liners as shown in Figures 3.19 and 3.20, and these liners contain the soil sample. After being extracted from the boring, the sampler is opened and the soil and liners are removed and placed in a protective cylinder for transport to the laboratory and storage.

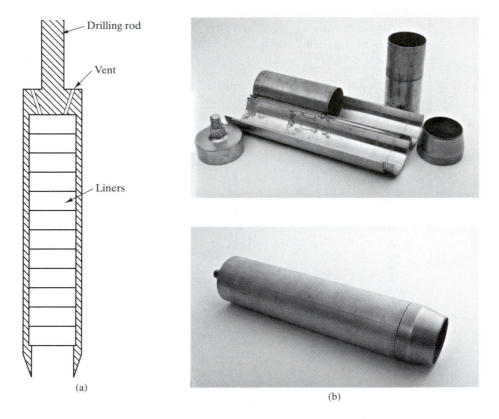

(a)

(b)

Figure 3.19 A heavy-wall sampler. (a) Cross-section showing liners. The dimensions shown are typical, but many different sizes are used. (b) The sampler can be opened to retrieve the liners, and thus is often called a *split barrel sampler*. The liners are available in different lengths.

Figure 3.20 Soil samples from heavy-wall samplers are contained in liners and stored in plastic tubes. In this case, the liners have an outside diameter of 2.5 in, a height of 1.0 in, and are made of brass.

3.7 GROUNDWATER EXPLORATION AND MONITORING

The presence of water in soil pores or rock fissures has a very significant impact on the engineering behavior of the soil or rock, so site characterization programs also need to assess groundwater conditions. When drilling a boring or excavating an trench, we may observe small seeps, with moisture trickling into the hole. These may be due to small non-uniformities in the soil conditions that have trapped water at a certain level. Larger zones of trapped water are known as *perched groundwater*. If we continue drilling to a great enough depth, the *groundwater table* is eventually encountered, which is the level to which water fills an open boring. Soils below the groundwater table are said to be *saturated*, which means all of their voids are filled with water.

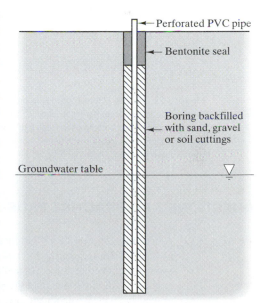

Figure 3.21 An observation well.

Sometimes the water quickly flows into the hole, reaching equilibrium in an hour or less. In these cases, the groundwater table can be located in the open hole before it is backfilled. However, in silty and clayey soils many hours or even days may be required to reach equilibrium, and leaving the hole open that long may pose safety problems. In addition, the groundwater table often changes with time, and we may wish to monitor these changes. The solution to both problems is to install an observation well in the boring as shown in Figure 3.21. It consists of a slotted plastic pipe backfilled with pervious soils (or even the drill cuttings) and sealed with an impervious cap. Groundwater is able to flow freely into or out of this pipe, so the water level inside is the groundwater table. The depth to this water can be measured using the electronic probe shown in Figure 3.22.

We will discuss groundwater in much more detail in Chapters 7 and 8.

Figure 3.22 Using an electronic probe to measure the water level inside an observation well. When the electrodes on the bottom of the tape touch the water, an electrical circuit is closed and a buzzer sounds inside the reel. Also see Figure 7.7.

QUESTIONS AND PRACTICE PROBLEMS

3.1 A one-story, 50 m wide × 90 m long manufacturing building is to be built on a site underlain by medium dense to dense silty sand with occasional gravel. This soil probably has better-than-average engineering properties and average uniformity. There are no indications of previous grading or fill at this site, and the groundwater table is believed to be about 30 m below the ground surface. We anticipate supporting this building on spread footing foundations located about 0.5 m below the ground surface. There are no accessibility problems at this site.

 a. How many exploratory borings will be required, and to what depths should they be drilled?
 b. What type of drilling equipment would you recommend for this project?

3.2 A one-story, 20 m wide × 50 m long concrete tilt-up office building is to be built at a site near a wetlands. Previous exploratory borings at nearby sites encountered about 1 m of moderately stiff clayey fill underlain by about 4 m of very soft organic silts and clays, then 15 m of progressively stiffer sandy clays and clayey sands. Limestone bedrock is located about 20 m below the ground surface. The groundwater table is thought to be at a depth of about 0.5 m. Because of the soft soils, we will probably need to support this building on deep foundations that extend at least into the stiffer soils, and possibly to bedrock. There are no accessibility problems at this site.

   a. How many exploratory borings will be required, and to what depth should they be drilled?
   b. What type of drilling and sampling equipment would you recommend for this project, and what kind of problems should the field crew be prepared to solve?

3.3   A ten-story steel-frame office building with a 200 ft × 200 ft footprint is to be built on a site underlain by alluvial sands and silts. These soils are fairly uniform and probably have good engineering properties. The building will have one 12 ft deep basement and will probably be supported on either a mat foundation[1] located 5 ft below the bottom of the basement, or a deep foundation extending about 60 ft below the bottom of the basement. The groundwater table is about 30 ft below the ground surface and bedrock is several hundred feet below the ground surface. There are no accessibility problems at this site.

   a. How many exploratory borings will be required, and to what depth should they be drilled?
   b. What type of drilling equipment would you recommend for this project?

3.4   A small commercial development consisting of a one-story supermarket and a one-story retail store building is to be built on the site shown in Figure 3.23. The proposed spread footing foundations will be located at a depth of 2 ft below the ground surface. The site has never been developed before, but a study of old aerial photographs indicates a fill was placed in the northeast section. This fill appears to be up to 5 ft thick, probably was not compacted, and most likely will need to be removed during construction. However, we may be able to reuse this material as fill, so long as it does not contain trash or other deleterious substances. The remainder of

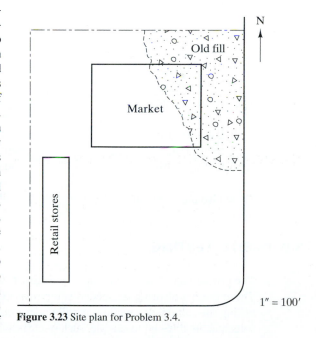

Figure 3.23 Site plan for Problem 3.4.

---

[1] A mat foundation is a type of shallow foundation that encompasses the entire footprint of the building. See Chapter 17 for more details.

the soils are probably stiff clayey silts and sandy silts. The groundwater table is believed to be about 20 ft below the ground surface. There are no accessibility problems at this site.

Develop a subsurface exploration program and present it as a 250–350 word memo to your field crew instructing them what to do. Be sure to include a copy of the site plan marked with the proposed location of each activity.

3.8 EX-SITU TESTING

The most common method of measuring soil and rock properties is to conduct laboratory tests. Some of these tests may be performed on either disturbed or undisturbed samples, while others require undisturbed samples. We call these *ex-situ* testing methods, which is Latin for "out of its original place," and refers to the removal of soil samples from the ground and testing them elsewhere.

We will discuss various laboratory tests throughout this book. For clarity, these discussions are in the chapters related to the engineering properties being measured:

Moisture content test	Chapter 4
Unit weight test	Chapter 4
Specific gravity test	Chapter 4
Relative density test	Chapter 4
Sieve analysis	Chapter 4
Hydrometer analysis	Chapter 4
Atterberg limits tests	Chapter 4
Proctor compaction test	Chapter 6
Hydraulic conductivity test	Chapter 7
Consolidation test	Chapter 11
Direct shear test	Chapter 13
Triaxial compression test	Chapter 13
Unconfined compression test	Chapter 13
Ring shear test	Chapter 13
Swell test	Chapter 18
Collapse test	Chapter 18

There also are many other laboratory tests we will not cover in this book (see Bardet, 1997).

3.9 IN-SITU TESTING

The primary alternative to laboratory testing is to conduct *in-situ* (Latin for in-place) tests. These consist of bringing special equipment to the field, inserting it into the ground, and testing the soil or rock while it is still underground. Such methods are especially useful in soils that are difficult to sample, such as clean sands. In-situ tests are usually less expensive than obtaining samples and performing ex-situ tests, so we can afford to do more of them. This additional data gives us more insight into the soil variability beneath a proposed construction site.

Geotechnical engineers often use the raw data obtained from in-situ tests as general indicators of soil properties. For example, some in-situ tests involve pounding or pressing something into the ground. If this is difficult to do, the soil must be stiff; if it is easy to do, the soil must be soft. We also have developed empirical correlations between in-situ test results and specific engineering properties. We will discuss some of these correlations in the following chapters:

Relative density	Chapter 4
Consistency	Chapter 5
Shear strength	Chapter 13
Settlement of foundations	Chapter 17

The discussions in this chapter are limited to describing the test procedures, adjusting the results, performing basic interpretations, and discussing the advantages and disadvantages of each test method.

Standard Penetration Test

One of the oldest and most common in-situ tests is the *standard penetration test* or *SPT*. It was developed in the late 1920s and has been used extensively in North and South America, the United Kingdom, Japan, and elsewhere. Because of this long record of experience, the SPT is well-established in engineering practice. It is performed inside an exploratory boring using inexpensive and readily available equipment, and thus adds little cost to a site characterization program.

Although the SPT also is plagued by many problems that affect its accuracy and reproducibility, it probably will continue to be used for the foreseeable future, primarily because of its low cost. However, it is partially being replaced by other test methods, especially on larger and more critical projects.

Test Procedure

The test procedure was not standardized until 1958 when ASTM standard D1586 first appeared. It is essentially as follows[2]:

1. Drill a 60–200 mm (2.5–8 in) diameter exploratory boring to the depth of the first test.
2. Insert the SPT sampler (also known as a *split-spoon sampler*) into the boring. The shape and dimensions of this sampler are shown in Figure 3.24. It is connected via steel rods to a 63.5 kg (140 lb) hammer, as shown in Figure 3.25.
3. Using either a rope and cathead arrangement or an automatic tripping mechanism, raise the hammer a distance of 760 mm (30 in) and allow it to fall. This energy drives the sampler into the bottom of the boring. Repeat this process until the sampler has

---

[2] See the ASTM D1586 standard for the complete procedure.

penetrated a distance of 460 mm (18 in), recording the number of hammer blows required for each 150 mm (6 in) interval. Stop the test if more than 50 blows are required for any of the intervals, or if more than 100 total blows are required. Either of these events is known as *refusal* and is so noted on the boring log.

4. Compute the *N*-value by summing the blow counts for the last 300 mm (12 in) of penetration. The blow count for the first 150 mm (6 in) is retained for reference purposes, but not used to compute *N* because the bottom of the boring is likely to be disturbed by the drilling process and may be covered with loose soil that fell from the sides of the boring. Note that the *N*-value is the same regardless of whether the engineer is using English or SI units.

5. Extract the SPT sampler, then remove and save the soil sample.

6. Drill the boring to the depth of the next test and repeat steps 2 through 6 as required.

Thus, *N*-values may be obtained at intervals no closer than 500 mm (20 in). Typically these tests are performed at 1.5–5 m (5–15 ft) intervals.

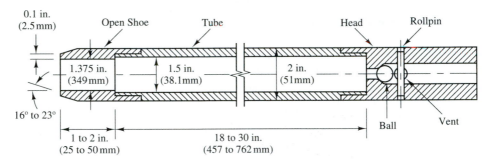

Figure 3.24 The SPT sampler (Adapted from ASTM D1586; Copyright ASTM, used with permission)

Soft or very loose soils typically have *N*-values less than 5; soils of average stiffness generally have $20 < N < 40$; and very dense or hard soils have *N* of 50 or more. For further classification based on the *N*-value, see Tables 5.4 and 5.5. Very high *N*-values (>75) typically indicate very hard soil or rock, but may simply occur because the sampler has hit a cobble or boulder.

Before the test was standardized, the actual procedures and equipment used in the field often varied substantially, which affected the measured *N*-values. As a result, two drillers testing the same strata could obtain *N*-values that differed by as much as 100 percent. Even after standardization, these variations still are significant, which means the test has a poor *repeatability*. The principal variants are:

- Method of drilling
- Cleanliness at the bottom of the hole (lack of loose dirt) before the test
- Presence or lack of drilling mud
- Diameter of the drill hole
- Location of the hammer (surface type or down-hole type)
- Type of hammer, especially whether it has a manual or automatic tripping mechanism

- Number of turns of the rope around the cathead
- Actual hammer drop height (manual types are often as much as 25 percent in error)
- Mass of the anvil that the hammer strikes
- Friction in rope guides and pulleys
- Wear in the sampler drive shoe
- Straightness of the drill rods
- Presence or absence of liners inside the sampler (this seemingly small detail can alter the test results by 10–30 percent)
- Rate at which the blows are applied

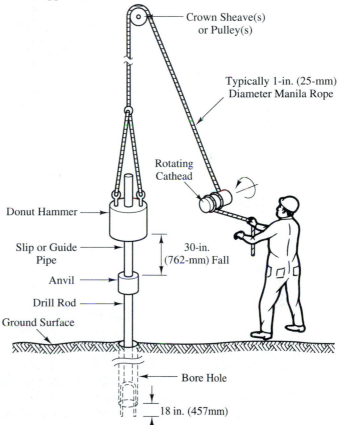

Figure 3.25 The SPT sampler in place in the boring with hammer, rope, and cathead (Adapted from Kovacs, et al., 1981).

Poor workmanship from the drilling crew also may be an important factor. Some crews are more interested in advancing the boring than in performing tests, and thus may tend to rush through the test.

These variations, as well as other aspects of the test, were the subject of increased scrutiny during the 1970s and 1980s, along with efforts to further standardize the "standard" penetration test (DeMello, 1971; Nixon, 1982). Based on these studies, Seed et al. (1985) recommended the following additional criteria be met when conducting standard penetration tests:

- Use the rotary wash method to create a boring that has a diameter between 200 and 250 mm. The drill bit should provide an upward deflection of the drilling mud (tricone or baffled drag bit).
- If the sampler is made to accommodate liners, then these liners should be used so the inside diameter is 35 mm.
- Use A or AW size drill rods for depths less than 15 m, and N or NW size for greater depths.
- Use a hammer that has an efficiency of 60 percent.
- Apply the hammer blows at a rate of 30 to 40 per minute.

Fortunately, automatic hammers are becoming more popular. They are much more consistent than hand-operated hammers, and thus improve the reliability of the test.

In spite of these disadvantages, the SPT does have at least three important advantages over other in-situ test methods: First, it obtains a sample of the soil being tested. This permits direct soil classification. Most of the other methods do not include sample recovery, so soil classification must be based on conventional sampling from nearby borings and on correlations between the test results and soil type. Second, it is very fast and inexpensive because it is performed in borings that would have been drilled anyway. Finally, nearly all drill rigs used for soil exploration are equipped to perform this test, whereas other in-situ tests require specialized equipment that may not be readily available.

Corrections to Test Results

We can improve the raw SPT data by applying certain correction factors, thus significantly improving its repeatability. The variations in testing procedures may be at least partially compensated by converting the N recorded in the field to N_{60} as follows (Skempton, 1986):

$$N_{60} = \frac{E_m \, C_B \, C_S \, C_R \, N}{0.60} \tag{3.1}$$

where:
N_{60} = SPT N-value corrected for field procedures
E_m = hammer efficiency (from Table 3.3)
C_B = borehole diameter correction (from Table 3.4)
C_S = sampler correction (from Table 3.4)
C_R = rod length correction (from Table 3.4)
N = SPT N-value recorded in the field

Many different hammer designs are in common use, none of which is 100 percent efficient. Some common hammer designs are shown in Figure 3.26, and typical hammer efficiencies are listed in Table 3.3. Many of the SPT-based design correlations were developed using hammers that had an efficiency of about 60 percent, so Equation 3.1 corrects the results from other hammers to that which would have been obtained if a 60 percent efficient hammer was used.

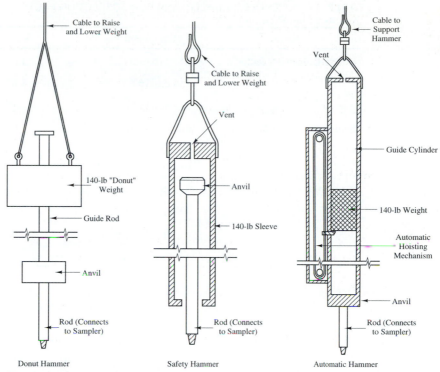

Figure 3.26 Types of SPT hammers.

TABLE 3.3 SPT HAMMER EFFICIENCIES (Adapted from Clayton, 1990).

Country	Hammer Type	Hammer Release Mechanism	Hammer Efficiency E_m
Argentina	Donut	Cathead	0.45
Brazil	Pin Weight	Hand Dropped	0.72
China	Automatic	Trip	0.60
	Donut	Hand dropped	0.55
	Donut	Cathead	0.50
Colombia	Donut	Cathead	0.50
Japan	Donut	Tombi trigger	0.78 - 0.85
	Donut	Cathead 2 turns + special release	0.65 - 0.67
UK	Automatic	Trip	0.73
USA	Safety	2 turns on cathead	0.55 - 0.60
	Donut	2 turns on cathead	0.45
Venezuela	Donut	Cathead	0.43

TABLE 3.4 BOREHOLE, SAMPLER, AND ROD CORRECTION FACTORS
(Adapted from Skempton, 1986).

Factor	Equipment Variables	Value
Borehole diameter factor, C_B	65 - 115 mm (2.5 - 4.5 in)	1.00
	150 mm (6 in)	1.05
	200 mm (8 in)	1.15
Sampling method factor, C_S	Standard sampler	1.00
	Sampler without liner (not recommended)	1.20
Rod length factor, C_R	3 - 4 m (10 - 13 ft)	0.75
	4 - 6 m (13 - 20 ft)	0.85
	6 - 10 m (20 - 30 ft)	0.95
	> 10 m (> 30 ft)	1.00

The SPT data also may be adjusted using an *overburden correction* that compensates for depth effects. Tests performed near the bottom of uniform soil deposits have higher *N*-values than those performed near the top, so the overburden correction adjusts the measured *N*-values to what they would have been if the vertical effective stress, σ_z', was 100 kPa (2000 lb/ft^2). Chapter 10 will discuss σ_z' and how to compute it, but for now think of it as a compressive stress produced by the weight of the overlying soil. Until then, the value of σ_z' will be given in any problem statements.

The corrected value, $(N_1)_{60}$, is (Liao and Whitman, 1986):

$$(N_1)_{60} = N_{60} \sqrt{\frac{2000 \text{ lb/ft}^2}{\sigma_z'}}$$
(3.2 - English)

$$(N_1)_{60} = N_{60} \sqrt{\frac{100 \text{ kPa}}{\sigma_z'}}$$
(3.2 - SI)

where:
$(N_1)_{60}$ = SPT *N*-value corrected for field procedures and overburden stress
σ_z' = vertical effective stress at the test location (kPa or lb/ft^2), as defined in Chapter 10
N_{60} = SPT *N*-value corrected for field procedures

The use of SPT correction factors is often a confusing issue. Corrections for field procedures (Equation 3.1) are always appropriate, but the overburden correction may or may not be appropriate depending on the procedures used by those who developed the analysis method under consideration. We will identify the proper value by using the appropriate subscripts.

Example 3.2

A standard penetration test has been conducted in a coarse sand at a depth of 16 ft below the ground surface. The blow counts obtained in the field were as follows: 0–6 in: 4 blows; 6–12 in: 6 blows; 12–18 in: 6 blows. The tests were conducted using a USA-style donut hammer in a 6 in diameter boring using a standard sampler with the liner installed. The vertical effective stress at the test depth was 1500 lb/ft^2. Determine $(N_1)_{60}$

Solution

$$N = 6 + 6 = 12$$

$E_m = 0.45$ per Table 3.3
$C_B = 1.05$ per Table 13.4
$C_S = 1.00$ per Table 13.4
$C_R = 0.85$ per Table 13.4

$$N_{60} = \frac{E_m C_B C_S C_R N}{0.60} = \frac{(0.45)(1.05)(1.00)(0.85)(12)}{0.60} = 8$$

$$(N_1)_{60} = N_{60} \sqrt{\frac{2000 \text{ lb/ft}^2}{\sigma_z'}}$$

$$= (8) \sqrt{\frac{2000 \text{ lb/ft}^2}{1500 \text{ lb/ft}^2}}$$

$$= 9 \qquad \leftarrow Answer$$

Cone Penetration Test

The *cone penetration test* or CPT [ASTM D3441] is another common in-situ test (Schmertmann, 1978; De Ruiter, 1981; Meigh, 1987; Robertson and Campanella, 1989; Briaud and Miran, 1991). Most of the early development of this test occurred in western Europe in the 1930s and again in the 1950s. Further development has occurred in recent decades in both Europe and North America. Although many different styles and configurations have been used, the current standard grew out of work performed in the

Netherlands, so it is sometimes called the *Dutch cone*. The CPT has been used extensively in Europe for many years and is becoming increasingly popular in North America and elsewhere.

Two types of cones are commonly used: the *mechanical cone* and the *electric cone*, as shown in Figure 3.27. Both have two parts, a 35.7 mm diameter cone-shaped tip with a 60° apex angle and a 35.7 mm diameter × 133.7 mm long cylindrical sleeve. A hydraulic ram pushes this assembly into the ground and instruments measure the resistance to penetration. The *cone resistance*, q_c, is the total force acting on the cone divided by its projected area (10 cm^2); the *cone side friction*, f_{sc}, is the total frictional force acting on the friction sleeve divided by its surface area (150 cm^2). It is common to express the side friction in terms of the *friction ratio*, R_f:

$$R_f = \frac{f_{sc}}{q_c} \times 100\%$$

(3.3)

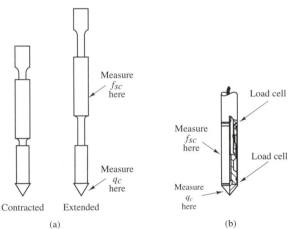

Figure 3.27 Types of cones: (a) mechanical cone (also known as a Begemann cone; (b) electric cone (also known as a Fugro cone).

The operation of the two types of cones differs in that the mechanical cone is advanced in stages and measures q_c and f_{sc} at intervals of about 20 cm, whereas the electric cone includes built-in strain gages and is able to measure q_c and f_{sc} continuously with depth. In either case, the CPT defines the soil profile with much greater resolution than does the SPT.

CPT rigs are often mounted in large three-axle trucks such as the one in Figure 3.28. These are typically capable of producing maximum thrusts of 100–200 kN (10–20 tons). Smaller, trailer-mounted or truck-mounted rigs also are available.

Figure 3.28 A truck-mounted CPT rig. A hydraulic ram located inside the truck pushes the cone into the ground, using the weight of the truck as a reaction.

The CPT has been the object of extensive research and development (Robertson and Campanella, 1983) and thus is becoming increasingly useful to the practicing engineer. Some of this research effort is now being conducted using cones equipped with pore pressure transducers in order to measure the excess pore water pressures that develop while conducting the test. These are known as *piezocones*, and the enhanced procedure is known as a CPTU test. These devices promise to be especially useful in saturated clays.

A typical plot of CPT results is shown in Figure 3.29.

The CPT is an especially useful way to evaluate soil profiles. Since it retrieves data continuously with depth (with electric cones) or at very close intervals (with mechanical cones), the CPT is able to detect fine changes in the stratigraphy. Therefore, engineers often use the CPT in the first phase of subsurface investigation, saving boring and sampling for the second phase.

It also is much less prone to error due to differences in equipment and technique, and thus is more repeatable and reliable than the SPT.

Although the CPT has many advantages over the SPT, there are at least three important disadvantages:

- No soil sample is recovered, so there is no opportunity to inspect the soils
- The test is unreliable or unusable in soils with significant gravel content
- Although the cost per foot of penetration is less than that for borings, it is necessary to mobilize a special rig to perform the CPT

Overburden Correction

Most analysis methods use the CPT results directly from the field, but some require the use of an overburden correction factor. This factor is identical to the one applied to SPT results:

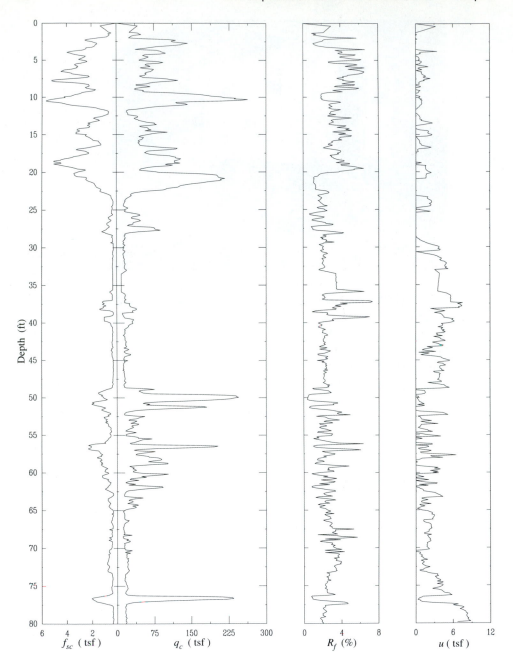

Figure 3.29 Sample CPT test results. These results were obtained from a piezocone, and thus also include a plot of pore water pressure, u, vs. depth. All stresses and pressures are expressed in tons per square foot (tsf). For practical purposes, 1 tsf = 1 kg/cm^2. (Alta Geo Cone Penetrometer Testing Services, Sandy, Utah).

$$q_{c1} = q_c \sqrt{\frac{2000 \text{ lb/ft}^2}{\sigma_z'}}$$

(3.4 - English)

$$q_{c1} = q_c \sqrt{\frac{100 \text{ kPa}}{\sigma_z'}}$$

(3.4 - SI)

$$f_{sc1} = f_{sc} \sqrt{\frac{2000 \text{ lb/ft}^2}{\sigma_z'}}$$

(3.5 - English)

$$f_{sc1} = f_{sc} \sqrt{\frac{100 \text{ kPa}}{\sigma_z'}}$$

(3.5 - SI)

where:

q_c = cone resistance obtained in the field

q_{c1} = cone resistance corrected for overburden stress

f_{sc} = cone side friction obtained in the field

f_{sc1} = cone side friction corrected for overburden stress

σ_z' = vertical effective stress (as defined in Chapter 10)

Soil Classification

The primary difficulty associated with the lack of a soil sample is that we do not know the type of soil being tested. Although we may be able to estimate the soil classification based on nearby borings, this problem is still a handicap.

Engineers have developed empirical correlations between soil type and CPT data, including the one in Figure 3.30. These correlations can be programmed into a computer, and often are printed along with the test results. However, they need to be used with caution, and are not nearly as precise as a visual classification of real soil samples.

Correlation with SPT

Geotechnical engineers also have developed empirical correlations between the CPT and SPT. The one shown in Figure 3.31 presents the q_c/N_{60} ratio as a function of the mean particle size, D_{50} (as defined in Chapter 4).

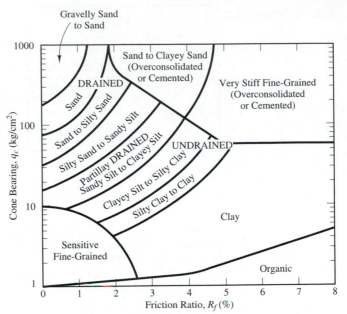

Figure 3.30 Soil classification based on CPT results (Adapted from Robertson and Campanella, 1983).

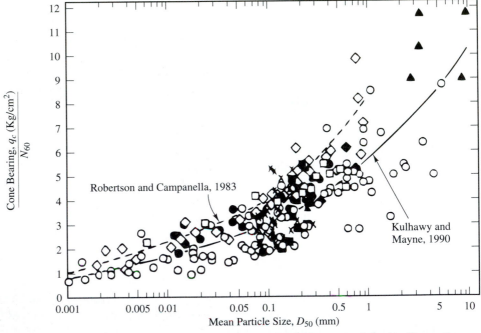

Figure 3.31 Correlation between the q_c/N_{60} ratio and the mean particle size, D_{50} (as defined in Chapter 4). (Adapted from Kulhawy and Mayne, 1990, copyright Electric Power Research Institute, used with permission).

Pressuremeter Test

In 1954, a young French engineering student named Louis Ménard began to develop a new type of in-situ test: the pressuremeter test. Although Kögler had done some limited work on a similar test some 20 years earlier, it was Ménard who made it a practical reality.

The pressuremeter is a cylindrical balloon that is inserted into the ground and inflated, as shown in Figure 3.32 and 3.33. Measurements of volume and pressure can be used to evaluate the in-situ stress, compressibility, and strength of the adjacent soil and thus the behavior of a foundation (Baguelin et al., 1978; Briaud, 1992).

The PMT may be performed in a carefully drilled boring or the test equipment can be combined with a small auger to create a self-boring pressuremeter. The latter design provides less soil disturbance and more intimate contact between the pressuremeter and the soil.

The PMT produces much more direct measurements of soil compressibility and lateral stresses than do the SPT and CPT. Thus, in theory, it should form a better basis for settlement analyses, and possibly for pile capacity analyses. In addition, the applied load from the pressuremeter cell is spread out over a larger area of soil than the SPT or CPT, and thus is less likely to be adversely affected by gravel in the soil. However, the PMT is a difficult test to perform and is limited by the availability of the equipment and personnel trained to use it.

Although the PMT is widely used in France and Germany, it is used only occasionally in other parts of the world. However, it may become more popular in the future.

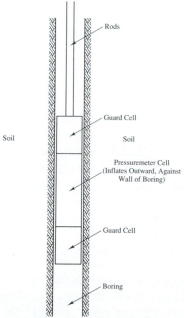

Figure 3.32 Schematic of the pressuremeter test.

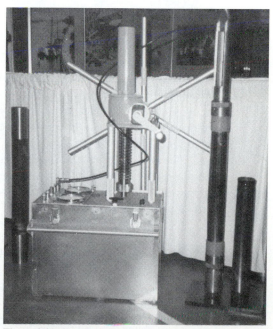

Figure 3.33 A complete pressuremeter set, including three cell assemblies and the control unit.

Dilatometer Test

The dilatometer (Marchetti, 1980; Schmertmann, 1986b, 1988a, and 1988b), which is one of the newest in-situ test devices, was developed during the late 1970s in Italy by Silvano Marchetti. It is also known as a *flat dilatometer* or a *Marchetti dilatometer* and consists of a 95 mm wide, 15 mm thick metal blade with a thin, flat, circular, steel membrane on one side, as shown in Figure 3.34.

The dilatometer test (DMT) is conducted as follows (Schmertmann, 1986a):

1. Press the dilatometer into the soil to the desired depth using a CPT rig or some other suitable device.
2. Apply nitrogen gas pressure to the membrane to press it outward. Record the pressure required to move the center of the membrane 0.05 mm into the soil (the *A* pressure) and that required to move its center 1.10 mm into the soil (the *B* pressure).
3. Depressurize the membrane and record the pressure acting on the membrane when it returns to its original position. This is the *C* pressure and is a measure of the pore water pressure in the soil.
4. Advance the dilatometer 150 to 300 mm deeper into the ground and repeat the test. Continue until reaching the desired depth.

Figure 3.34 The Marchetti dilatometer with its control unit and nitrogen gas bottle. (Courtesy GPE, Inc., Gainesville, FL).

Each of these test sequences typically requires one to two minutes to complete, so a typical *sounding* (a complete series of DMT tests between the ground surface and the desired depth) may require about two hours. In contrast, a comparable CPT sounding might be completed in about thirty minutes.

The primary benefit of the DMT is that it measures the lateral stress condition and compressibility of the soil. These are determined from the *A*, *B*, and *C* pressures and certain equipment calibration factors and expressed as the *DMT indices*, as follows:

$$I_D = \text{material index (a normalized modulus)}$$
$$K_D = \text{horizontal stress index (a normalized lateral stress)}$$
$$E_D = \text{dilatometer modulus (theoretical elastic modulus)}$$

Researchers have developed correlations between these indices and soil classification as well as certain engineering properties (Schmertmann, 1988b; Kulhawy and Mayne, 1990).

The CPT and DMT are complementary tests (Schmertmann, 1988b). The cone is a good way to evaluate soil strength, whereas the dilatometer assesses compressibility and in-situ stresses.   These three kinds of information form the basis for most foundation engineering analyses.   In addition, the dilatometer blade is most easily pressed into the ground using a conventional CPT rig, so it is a simple matter to conduct both CPT and DMT tests while mobilizing only a minimum of equipment.

The dilatometer test is a relative newcomer, and thus has not yet become a common engineering tool.   Engineers have had only limited experience with it and the analysis and design methods based on DMT results are not yet well developed.   However, its relatively low cost, versatility, and compatibility with the CPT suggest that it may enjoy widespread use in the future.   It has very good repeatability, and can be used in soft to moderately stiff soils (i.e., those with $N \le 40$), and provides more direct measurements of stress–strain properties.

Becker Penetration Test

Soils that contain a large percentage of gravel and those that contain cobbles or boulders create problems for most in-situ test methods.   Often, the in-situ test device is not able to penetrate through such soils (it meets refusal) or the results are not representative because the particles are about the same size as the test device.   Frequently, even conventional drilling equipment cannot penetrate through these soils.

One method of penetrating through these very large-grained soils is to use a *Becker hammer drill*.   This device, developed in Canada, uses a small diesel pile-driving hammer and percussion action to drive a 135 to 230 mm (5.5–9.0 in) diameter double-wall steel casing into the ground.   The cuttings are sent to the top by blowing air through the casing. This technique has been used successfully on very dense and coarse soils.

The Becker hammer drill also can be used to assess the penetration resistance of these soils using the *Becker penetration test*, which is monitoring the hammer blow-count.   The number of blows required to advance the casing 1 ft (300 mm) is the Becker blow-count, N_B.   Several correlations are available to convert it to an equivalent SPT N-value (Harder and Seed, 1986).   One of these correlation methods also considers the bounce chamber pressure in the diesel hammer.

Other In-Situ Tests

Many other in-situ tests are available, some of which are discussed in other parts of this book.   These include the field density test (Chapter 6), the hydraulic conductivity test (Chapter 8), and the vane shear test (Chapter 13).

Comparison of In-Situ Test Methods

Each of the in-situ test methods has its strengths and weaknesses.   Table 3.5 compares some of the important attributes of the tests described in this chapter.

TABLE 3.5 ASSESSMENT OF IN-SITU TEST METHODS (Adapted from Mitchell, 1978; used with permission of ASCE)

	Standard Penetration Test	Cone Penetration Test	Pressuremeter Test	Dilatometer Test	Becker Penetration Test
Simplicity and Durability of Apparatus	Simple; rugged	Complex; rugged	Complex; delicate	Complex; moderately rugged	Simple, rugged
Ease of Testing	Easy	Easy	Complex	Easy	Easy
Continuous Profile or Point Values	Point	Continuous	Point	Point	Continuous
Basis for Interpretation	Empirical	Empirical; theory	Empirical; theory	Empirical; theory	Empirical
Suitable Soils	All except gravels	All except gravels	All	All except gravels	Sands through boulders
Equipment Availability and Use in Practice	Universally available; used routinely	Generally available; used routinely	Difficult to locate; used on special projects	Difficult to locate; used on special projects	Difficult to locate; used on special projects
Potential for Future Development	Limited	Great	Great	Great	Uncertain

QUESTIONS AND PRACTICE PROBLEMS

3.5   The vertical effective stress at Sample 3 in Figure 3.14 is 1270 lb/ft^2. Compute $(N_1)_{60}$

3.6   A standard penetration test has been performed at a depth of 6.5 m in a medium sand using a standard sampler and a USA-style donut hammer. The N-value recorded in the field was 16. The boring diameter was about 100 mm, and the vertical effective stress at the test location was 85 kPa. Compute $(N_1)_{60}$.

3.7   Using Figure 3.30, classify the soils between depths of 66 and 80 ft in the CPT results presented in Figure 3.29. Why are there spikes in the q_c, f_{sc}, and R_f curves between depths of 76 and 78 ft?

3.8   A cone penetration test on a sandy soil with mean particle size of 0.5 mm produced a q_c of 80 kg/cm^2. Compute the equivalent SPT N_{60}-value.

3.10 GEOPHYSICAL EXPLORATION

Geophysics is the use of various principles of physics to discern geologic profiles and to measure certain properties of the ground. Many such techniques are available, such as:

- Inducing seismic waves and measuring their propagation
- Assessing natural gravitational and magnetic fields and their variation across the surface of the earth
- Passing electrical currents through the ground and measuring their propagation
- Sending radar waves or other kinds of radiation into the ground and recording its transmission, absorption, or reflection

Geophysical exploration methods were originally developed by mining and petroleum geologists searching for valuable minerals and oil deep below the earth's surface. They were able to distinguish between different geologic strata by observing their physical properties, and thus locate the most promising sites for mining and oil drilling (Dobrin, 1988).

Later, geotechnical engineers and engineering geologists began to use some of these methods to assist in site characterization studies. Although geophysics is not nearly as precise as drilling borings and obtaining samples, it has the benefit of covering large areas at a small cost, and sometimes can locate features that might be missed by conventional borings. Geophysical methods also can be used as a first step in the exploration process, thus guiding the placement and depth of exploratory borings.

The applicability of geophysical methods for geotechnical site characterization is fairly limited, and only a small fraction of practical projects can benefit from them. However, in certain circumstances they can be very valuable as a supplement to, but not instead of, exploratory borings.

Seismic Refraction

The most common geophysical method among geotechnical engineers and engineering geologists is *seismic refraction*, which consists of sending seismic waves into the ground and measuring their arrival at various points. The waves can be generated by striking the ground with a heavy object, such as a sledge hammer, or by detonating a small explosive. The resulting waves are measured by *geophones* aligned in an array and connected to a *seismograph*, as shown in Figure 3.35.

The wave velocity depends on the physical properties of the soil or rock, most notably its density, modulus of elasticity, porosity, and frequency of discontinuities. Dense, massive rocks, such as basalt, have high wave velocities; softer rocks, such as limestone are intermediate; and soils have very low velocities. Seismic refraction uses these differences to discern the subsurface profile.

In Figure 3.35, various strata of soil are underlain by bedrock, and we wish to know the depth to this rock. To gather this information, we discharge a small explosive and record the wave arrivals at each of the geophones. Some waves travel through the soil and arrive at the various geophones. Simultaneously, other waves travel down to the rock, horizontally through it, and back up to the ground surface. Although this path is longer, the wave velocity through the rock is much faster, so at some distance from the wave source, the latter arrives before the former. By comparing the wave arrival times at each geophone, we can determine the depth to bedrock.

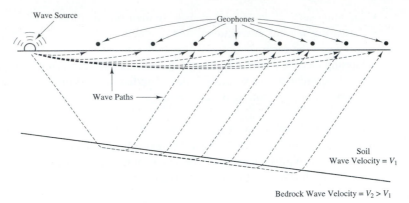

Figure 3.35 Use of seismic refraction to measure the depth to a hard layer, such as bedrock.

Wave velocity data also can be used to evaluate the rippability (ease of excavation) of various soil and rock strata, as discussed in Chapter 6. Finally, geotechnical engineers use this and other dynamic data to analyze the propagation of earthquake motions through the ground, which is part of the process of predicting ground motions due to future earthquakes.

3.11 SYNTHESIS AND INTERPRETATION

Cross-Sections

Site characterization programs often generate large amounts of information that can be difficult to sort through and synthesize. In addition, this data is spread throughout three dimensions, so visualization can be difficult.

One useful method of compiling subsurface data is to draw vertical *cross-sections* across the site, as shown in Figure 3.36. These sections are most easily developed when they intersect the borings, but additional borings slightly off the section also can be used. Some interpretation is always required when developing cross-sections, since we do not know what conditions exist between the borings. Two perpendicular sections can help geotechnical engineers visualize the site in three dimensions.

Sometimes cross-sections show only the subsurface conditions actually encountered in the exploratory borings, as in Figure 3.37. This method leaves the interpretation to the reader.

One-Dimensional Design Profiles

Although cross-sections are important tools for understanding subsurface variations across a site, many geotechnical analyses are based on one-dimensional profiles. For example, the settlement analyses we will conduct in Chapters 11 and 12 are one-dimensional and compute the settlement at a point on the ground surface due to compression of the soils immediately below that point. If we need to know the settlement at other points, the analysis needs to be repeated as necessary.

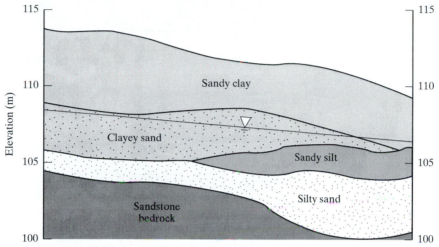

Figure 3.36 Typical cross-section through a site.

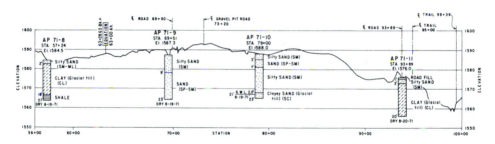

Figure 3.37 Cross-section along a pipeline route (U.S. Bureau of Reclamation).

A one-dimensional design profile is similar to a boring log in that it describes subsurface conditions as a function of depth, as shown in Figure 3.38. However, the profile used for design probably will be a compilation of several borings and not exactly like any one of them. If the subsurface conditions are fairly uniform across the site (at least by geotechnical engineering standards!), then we often use a single representative profile for design.

The development of these design profiles requires a great deal of engineering judgement along with interpolation and extrapolation of the data. It is important to have a feel for the approximate magnitude of the many uncertainties in this process and reflect them in an appropriate degree of conservatism. This judgement comes primarily with experience combined with a thorough understanding of the field and laboratory methodologies.

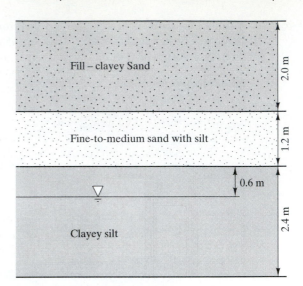

Figure 3.38 A typical one-dimensional design soil profile.

Geotechnical Investigation Reports

The final results of a site characterization program are usually presented in a *geotechnical investigation report* that includes copies of all boring logs, laboratory test results, cross-sections, etc., along with the engineer's interpretations. These reports are virtually always prepared in the context of a specific project, and thus include geotechnical recommendations for design of foundations, slopes, retaining walls, and other features. For example, a report for a proposed building might have an outline similar to the following:

> Scope and Purpose
> Proposed Development
> Field Exploration
> Groundwater Monitoring
> Laboratory Testing
> Analysis of Subsurface Conditions
> Design Recommendations
> > Grading
> > Foundations
> > Retaining walls
> > Pavements
> Closure
> Appendix A - Boring Logs
> Appendix B - Laboratory Test Results
> Appendix C - Recommended Construction Specifications

3.12 ECONOMICS

The site investigation and soil testing phase of foundation engineering is the single largest source of uncertainties. No matter how extensive it is, there is always some doubt whether the borings accurately portray the subsurface conditions, whether the samples are representative, and whether the tests are correctly measuring the soil properties. Engineers attempt to compensate for these uncertainties by applying factors of safety in our analyses. Unfortunately, this solution also increases construction costs.

In an effort to reduce the necessary level of conservatism in the foundation design, the engineer may choose a more extensive investigation and testing program to better define the soils. The additional costs of such efforts will, to a point, result in decreased construction costs, as shown in Figure 3.39. However, at some point, this becomes a matter of diminishing returns, and eventually the incremental cost of additional investigation and testing does not produce an equal or larger reduction in construction costs. The minimum on this curve represents the optimal level of effort.

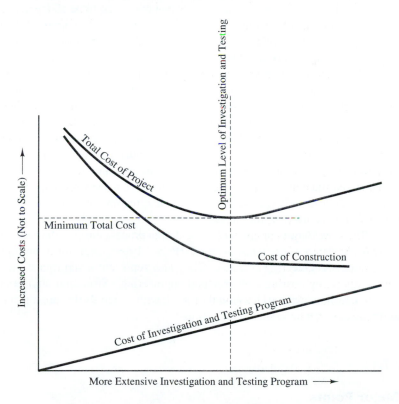

Figure 3.39 Cost-effectiveness of more extensive site characterization programs.

We also must decide whether to conduct a large number of moderately precise tests (such as the SPT) or a smaller number of more precise but expensive tests (such as the

PMT). Handy (1980) suggested the most cost-effective test is the one with a variability consistent with the variability of the soil profile. Thus, a few precise tests might be appropriate in a uniform soil deposit, but more data points, even if they are less precise, are more valuable in an erratic deposit.

3.13 GEOTECHNICAL MONITORING DURING CONSTRUCTION

One of the most important ways geotechnical engineers have to deal with the many uncertainties of site characterization is continued monitoring of subsurface conditions during construction. Often, new information becomes evident during construction, especially if the construction involves making excavations. For example, if a highway cut is to be made into a hillside, geotechnical engineers and engineering geologists base its design on exploratory borings as discussed in this chapter. Then, when the cut is actually made, we examine the newly exposed ground and compare it to the anticipated conditions. If new conditions were found, then the design may need to be changed accordingly.

Another way of dealing with these uncertainties is to conduct full-scale tests in the field. For example, we have methods of predicting the load-bearing capacity of pile foundations, but often conduct full-scale load tests on piles to verify the computed capacity. Such tests involve installing a real pile at the project site and loading it. These tests might be performed before construction, or more often at the beginning of construction. If the load test indicates capacities significantly different than those anticipated, then we modify the design, perhaps by adding new piles.

A third way is to install *geotechnical instrumentation* into the ground. These are devices specifically designed to measure certain attributes in soil or rock. For example, an *inclinometer* is a geotechnical instrument that measures horizontal movements in the ground. We could install one or more inclinometers in a slow-moving landslide and use the resulting data to help assess the depth and direction of movement, and to judge the effectiveness of stabilization measures.

These techniques of continuing the design process through the construction period are known as the *observational method* (Peck, 1969), and form an important part of geotechnical engineering practice. They also represent a significant difference between geotechnical engineering and structural engineering. Structural engineers rarely, if ever, need to use these methods because they work with materials that are much more predictable and thus does not need such verification.

SUMMARY

Major Points

1. An important difference between geotechnical practice and that of most other branches of engineering is that we must work with natural materials, not manufactured products. These materials, soil and rock, vary significantly from place to place, so each building site requires a site characterization program to define the subsurface

profile and relevant engineering properties. We then design our project based on the results of this program.

2. The initial stages of a site characterization program typically consist of gathering published data, reviewing airphotos (if applicable), and conducting a field reconnaissance. These tasks are in preparation for the subsurface exploration.

3. Exploration of the subsurface conditions is generally the most important, and most expensive, component. This is usually accomplished by drilling exploratory borings and obtaining disturbed and undisturbed samples. Exploratory trenches and pits also can be useful at some sites. A wide variety of methods and tools is available, and the proper choice depends on the site conditions, cost, and other factors.

4. Geotechnical engineers divide soil and rock tests into two broad categories: ex-situ and in-situ. Ex-situ tests are conducted in the laboratory, and thus are subject to sample disturbance problems. In-situ tests are conducted in the field, but suffer from less control and less precision. Often we use both methods to take advantage of each one's strengths.

5. In some cases, geophysical methods, such as seismic refraction, can be used to assess subsurface conditions. However, these methods supplement conventional testing, but do not replace it.

6. Once the site characterization data has been collected, geotechnical engineers synthesize it and present the findings and recommendations in a geotechnical investigation report.

7. In well-managed projects, site characterization continues through construction, since further data often becomes available and may dictate changes in the design.

Vocabulary

aerial photographs	exploratory borings	pressuremeter test (PMT)
Becker penetration test	exploratory trenches	remote sensing
boring log	ex-situ testing	rock quality designation
bucket auger	fence diagram	(RQD)
caving	field reconnaissance	seismic refraction
cone penetration test (CPT)	flight auger	Shelby tube sampler
coring	geophysical exploration	site characterization
dilatometer test (DMT)	geotechnical investigation	site exploration
disturbed sample	report	squeezing
downhole logging	heavy-wall sampler	standard penetration test
drill rig	hollow-stem auger	(SPT)
drilling mud	in-situ testing	subsurface exploration
		undisturbed sample

COMPREHENSIVE QUESTIONS AND PRACTICE PROBLEMS

3.9   An engineer is planning to use a 24-inch diameter bucket auger similar to the one in Figure 3.7 to drill several exploratory borings at a site adjacent to a lake. The underlying soils are probably soft clays and silts with N-values of less than 5. Is this a wise choice? Why or why not?

3.10 What type of soil sampling equipment would be most appropriate for the soils described in Problem 3.9? Why?

3.11 A large compacted fill is to be placed on a site underlain by a 15 m thick strata of saturated clay. The weight of this fill will cause the clay strata to consolidate, which will result in large settlements at the ground surface. Since these settlements would have an adverse effect on buildings and other improvements planned for this site, a settlement rate analysis, similar to those we will discuss in Chapter 12, is to be performed to estimate the time required for a certain percentage of the settlement to be completed.

A series of exploratory borings have already been drilled at this site, samples have been recovered, and laboratory tests have been performed to evaluate the consolidation properties of the clay. However, to complete the settlement rate analysis, we need to know if thin horizontal sand seams are present in the clay, and the approximate spacing between these seams. If they exist at all, these seams are probably less than 100 mm thick. Although some of the undisturbed samples contained sand seams, more information is needed.

What kind of additional exploration would you do to determine whether or not more sand seams are present? Be sure to consider both technical feasibility and cost, and explain the reason for your choice.

3.12 A level building pad is to be built at the site shown in Figure 3.40 by cutting and filling as shown. The final pad elevation is to be 215 ft. Then, a three-story steel-frame office building is to be built.

Five exploratory borings have been drilled to determine the subsurface conditions. The logs from these borings were as follows:

BORING 1
Groundwater table depth = 44 ft

Depth (ft)	Soil or Rock Conditions
0 - 18	Sandy clay
18 - 35	Clayey sand
35 - 52	Silty sand
52 - 55	Sandstone bedrock

BORING 2
Groundwater table depth = 31 ft

Depth (ft)	Soil or Rock Conditions
0 - 28	Clayey sand
28 - 36	Silty sand
36 - 39	Sandstone bedrock

BORING 3
Groundwater table depth = 41 ft

Depth (ft)	Soil or Rock Conditions
0 - 34	Clayey sand
34 - 48	Silty sand
48 - 52	Sandstone bedrock

BORING 4
Groundwater table depth = 40 ft

Depth (ft)	Soil or Rock Conditions
0 - 33	Clayey sand
33 - 45	Silty sand
45 - 47	Sandstone bedrock

BORING 5
Groundwater table depth = 49 ft

Depth (ft)	Soil or Rock Conditions
0 - 17	Clayey sand
17 - 25	Silt
25 - 42	Clayey sand
42 - 57	Silty sand
57 - 60	Sandstone bedrock

Develop cross-sections along axes A-A′ and B-B′ and indicate the soil profiles beneath the proposed building. The profiles should be similar to the one in Figure 3.36 and should include the existing and proposed grades, the proposed building, strata boundaries, and the groundwater table. Do not use an exaggerated vertical scale.

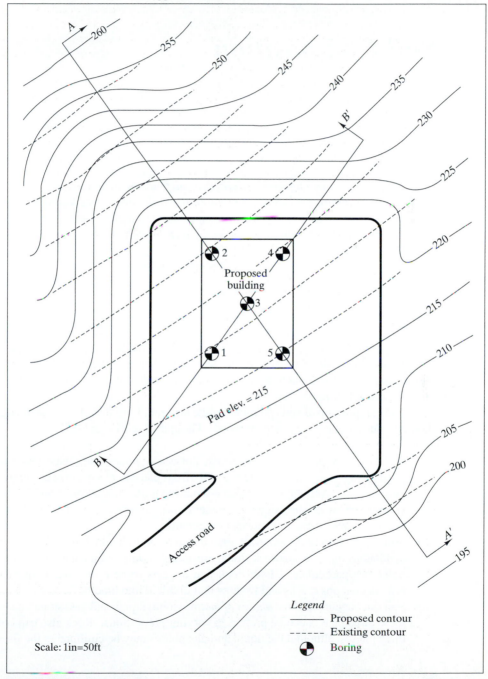

Figure 3.40 Site plan for Problem 3.12.

4

Soil Composition

. . . in engineering practice, difficulties with soils are almost exclusively due not to the soils themselves but to the water contained in their voids. On a planet without any water there would be no need for soil mechanics.

Karl Terzaghi, 1939

Once the soil and rock samples have been brought to the laboratory, we need to conduct appropriate tests to develop data for our analyses. Some of these tests measure familiar engineering properties, such as shear strength, while others focus on the sample's composition and structure.

The composition of soil and rock is quite different from that of other civil engineering materials, such as steel, concrete, or wood. These differences include:

- Soil and rock are *natural materials*, not manufactured products. As such, their engineering properties vary significantly from place to place and even across a single building site. Although wood also is a natural product, it is sorted and graded before being used in construction. In general, we cannot sort soil or rock, and must accommodate whatever is present on our site.
- Soil is a *particulate material* that consists of individual particles. It is not a continuous mass. Some rock strata, especially certain sedimentary rocks, also can be treated as particulates. Even rock that appears to be a continuous mass virtually always contains cracks and fissures that divide it into blocks.
- Soil can contain *all three phases of matter* (solid, liquid, and gas) simultaneously, and these three phases can be present in varying proportions. Rock also can contain all three phases, although the liquid and gas phases may be confined to the fissures.

94

This chapter discusses the methods we use to assess the composition of soils and the parameters we use to describe this composition.

4.1 SOIL AS A PARTICULATE MATERIAL

Most civil engineering materials consist of a continuous mass held together with molecular bonds, and the mechanical properties of such materials depend on their chemical makeup and on the nature of these bonds. For example, the shear strength of steel depends on the strength of the molecular bonds, and shear failure requires breaking them. In contrast, soil is a particulate material that consists of individual particles assembled together as shown in Figure 4.1. Its engineering properties depend largely on the interaction between these particles, and only secondarily on their internal properties. This is especially true in gravels, sands, and silts. For example, when soils fail in shear, they do so because the particles begin to roll and slide past each other, not because the particles break internally. Breakage of individual particles is typically minimal. Thus, the shear strength depends on factors such as the coefficient of friction between the particles, the tightness of packing, and so on, rather than the chemical bonds inside the particles.

Figure 4.1 Microphotograph of a medium sand at 5× magnification.

Clays also have a particulate structure, but the nature of the particles is quite different, as discussed later in this chapter. In clays, there is much more interaction between the particles and the pore water, so their behavior is more complex than that of other soils.

4.2 THE THREE PHASES

Soil also is different from most civil engineering materials in that it can simultaneously contain solid, liquid, and gas phases. The liquid and gas phases are contained in the *voids* or *pores* between the solid particles. In addition, the three phases often interact, and these interactions have important effects on the soil's behavior.

The solid phase is always present in soil, and usually consists of particles derived from rocks, as discussed in Chapter 2. It also can include organic material.

The liquid phase is usually present, and most often consists of water. However, it also can include other materials, such as:

- Gasoline or other chemicals that have leaked out of underground tanks or pipelines, or infiltrated from the ground surface. The cleanup of such contaminants can be a significant problem, and has been the object of many "superfund" sites identified by the Environmental Protection Agency.
- Leachate that has escaped from a sanitary landfill. This leachate can contaminate groundwater, and make it unfit for drinking.
- Sea water moving inland through the soil, which often occurs when wells are installed near the ocean to pump out fresh water for municipal purposes.
- Natural petroleum seeps.

Although these constituents usually represent a small fraction of the liquids in a soil, they can be very important, especially if the groundwater is to be pumped and used for domestic purposes, but becomes contaminated.

If the liquid phase does not completely fill the voids, then the remaining space is occupied by the gas phase. It is usually air, but can include other gasses, such as:

- Methane (CH_4) and carbon dioxide (CO_2) from the decomposition of organic materials. These gasses are often found near sanitary landfills and near highly organic natural soils.
- Petroleum vapors from contaminated soils.
- Hydrogen sulfide (H_2S) from the decomposition of sulfur-bearing materials. It is sometimes found near sanitary landfills, sewage facilities, and coal deposits.

Gasses other than air can be important because they sometimes pose safety hazards to workers in utility vaults, small excavations, mines, and other underground areas. Some, such as hydrogen sulfide, are poisonous, while others, such as methane, are flammable. For example, methane gas permeating up from the ground became trapped in a small room of a department store in Los Angeles. In 1985, this gas accidently ignited, creating an explosion that blew out the windows and collapsed part of the roof. Twenty-three people required hospital treatment as a result of the blast (Hamilton and Meehan, 1992).

Even gasses that are neither toxic nor flammable can be dangerous because they displace oxygen, thus suffocating workers. This problem has caused injury and death, but can be avoided by using special detection equipment and providing adequate ventilation.

Clearly, components other than water and air are very important. However, they generally represent only a small portion of the soil weight and volume. Therefore, for purposes of this chapter, we will simply refer to the liquid and gas phases as "pore water" and "pore air".

4.3 WEIGHT–VOLUME RELATIONSHIPS

It is helpful to identify the relative proportions of solids, water, and air in a soil, because these proportions have a significant effect on its behavior. Therefore, geotechnical engineers have developed quantitative methods of assessing these components.

Phase Diagrams

Phase diagrams, such as those shown in Figure 4.2, indicate the relative proportions of solids, water, and air in a soil. The dimensions on the left side of the diagram indicate the weight or mass of each component, while those on the right side indicate their volumes.

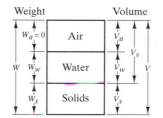 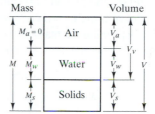

Figure 4.2 Phase diagrams describe the relative proportions of solids, water, and air in a soil. One side presents weights or masses, while the other presents volumes.

Definitions of Weight–Volume Parameters

Geotechnical engineers have defined several *weight–volume parameters* based on the dimensions shown in the phase diagrams. These parameters give important information on the composition of a particular soil, and form part of the basic language of soil mechanics.

Moisture Content and Degree of Saturation

One of the most common soil parameters is the *moisture content, w* (also known as the *water content*), which can be expressed either in terms of weight or mass:

$$w = \frac{W_w}{W_s} \times 100\%$$

(4.1)

$$w = \frac{M_w}{M_s} \times 100\%$$

(4.2)

A small w indicates a dry soil, while a large w indicates a wet one. Values in the field are usually between 3 and 70%, but values greater than 100% are sometimes found in soft soils below the groundwater table, which simply means such soils have more water than solids.

The definitions in Equations 4.1 and 4.2 are slightly different than one might expect, because the denominators are W_s and M_s, not W and M. These are good formulas to memorize, but don't make the common mistake of using the wrong denominator.

The moisture content can be easily measured in the laboratory by conducting a *moisture content test* (ASTM D2216) as follows:

1. Obtain a small can to hold the soil sample and find its mass, M_c.
2. Place a representative sample of the soil into the can and find the total mass, M_1.
3. Place the soil and can into an oven with a constant temperature of $110 \pm 5°C$ and leave it there until completely dry. This usually requires 12 to 16 hours.
4. Determine the mass of the dry sample and can, M_2.
5. Compute the moisture content using:

$$w = \frac{M_1 - M_2}{M_2 - M_c} \times 100\%$$

(4.3)

A similar parameter is the *degree of saturation, S*, which is the percentage of the voids filled with water:

$$S = \frac{V_w}{V_v} \times 100\%$$

(4.4)

This is similar to moisture content in that both are equal to zero when there is no water. However, S has a maximum value of 100%, which occurs when all of the voids are filled with water. We use the term *saturated* to describe this condition. Soils below the groundwater table are generally saturated. Values of S above the groundwater table are usually between 5 and 100%, although values approaching zero can be found in very arid areas.

Capillary effects, discussed in Chapter 7, can draw water upward from the groundwater table, often producing soils with $S = 100\%$ well above the groundwater table. Therefore, do not use saturation computations to determine the location of the groundwater table. Instead, use observation wells as discussed in Section 3.7.

Unit Weight and Density

Geotechnical engineers often need to know the *unit weight*, γ:

$$\gamma = \frac{W}{V} \tag{4.5}$$

The unit weight of undisturbed soil samples can easily be determined in the laboratory by measuring their physical dimensions and weighing them. This method produces reliable assessments of γ for many soils. However, it is affected by sample disturbance, especially in sandy and gravelly soils. Sometimes unit weight measurements are made on supposedly "undisturbed" samples that in reality have significant disturbance. Such measurements are very misleading, so it is best to not even attempt unit weight measurements on poor quality samples. Table 4.1 presents typical ranges of γ for various soils.

TABLE 4.1 TYPICAL UNIT WEIGHTS.

Soil Type and Unified Soil Classification (See Section 5.3)	Typical Unit Weight, γ			
	Above Groundwater Table		Below Groundwater Table	
	(lb/ft^3)	(kN/m^3)	(lb/ft^3)	(kN/m^3)
GP — Poorly graded gravel	110 - 130	17.5 - 20.5	125 - 140	19.5 - 22.0
GW — Well graded gravel	110 - 140	17.5 - 22.0	125 - 150	19.5 - 23.5
GM — Silty gravel	100 - 130	16.0 - 20.5	125 - 140	19.5 - 22.0
GC — Clayey gravel	100 - 130	16.0 - 20.5	125 - 140	19.5 - 22.0
SP — Poorly graded sand	95 - 125	15.0 - 19.5	120 - 135	19.0 - 21.0
SW — Well graded sand	95 - 135	15.0 - 21.0	120 - 145	19.0 - 23.0
SM — Silty sand	80 - 135	12.5 - 21.0	110 - 140	17.5 - 22.0
SC — Clayey sand	85 - 130	13.5 - 20.5	110 - 135	17.5 - 21.0
ML — Low plasticity silt	75 - 110	11.5 - 17.5	80 - 130	12.5 - 20.5
MH — High plasticity silt	75 - 110	11.5 - 17.5	75 - 130	11.5 - 20.5
CL — Low plasticity clay	80 - 110	12.5 - 17.5	75 - 130	11.5 - 20.5
CH — High plasticity clay	80 - 110	12.5 - 17.5	70 - 125	11.0 - 19.5

Two variations of unit weight also are commonly used, the *dry unit weight*, γ_d and the *unit weight of water*, γ_w:

$$\gamma_d = \frac{W_s}{V} \tag{4.6}$$

$$\gamma_w = \frac{W_w}{V_w} \qquad (4.7)$$

Normally we use $\gamma_w = 9.81$ kN/m^3 = 62.4 lb/ft^3 for fresh water and $\gamma_w = 10.1$ kN/m^3 = 64.0 lb/ft^3 for sea water.

For soils below the groundwater table, some computations use the *buoyant unit weight*, γ_b:

$$\gamma_b = \gamma - \gamma_w \qquad (4.8)$$

We also can define similar parameters based on mass instead of weight. These become the *density*, ρ, *dry density*, ρ_d, and *density of water*, ρ_w:

$$\rho = \frac{M}{V} \qquad (4.9)$$

$$\rho_d = \frac{M_s}{V} \qquad (4.10)$$

$$\rho_w = \frac{M_w}{V_w} \qquad (4.11)$$

Design values of $\rho_w = 1000$ kg/m^3 and 62.4 lb$_m$/ft^3 for fresh water.

Based on $F = Ma$, the unit weight and density are related by:

$$\gamma = \rho g \qquad (4.12)$$

where:
g = acceleration due to gravity = 9.81 m/s^2 = 32.2 ft/s^2

For most geotechnical computations, unit weight is more useful than density because we use it to compute stresses due to the weight of the soil. A notable exception is dynamic analyses that need to consider inertial effects that are best presented in terms of mass.

Unfortunately, geotechnical engineers are often careless in our use of these terms.

Many of us use the term *density* when we really mean *unit weight*. This usage is technically incorrect, even though it is common in conversation, reports, and even in technical literature, and can be confusing. As a general rule, one can assume the speaker or writer really means unit weight unless the discussion involves a soil dynamics problem. In this book, we will avoid such confusion by always using the proper terms.

Specific Gravity of Solids

The specific gravity of any material is the ratio of its density to that of water. In the case of soils, we compute it for the solid phase only, and express the results as the *specific gravity of solids*, G_s:

$$G_s = \frac{M_s}{V_s \rho_w} \tag{4.13}$$

$$G_s = \frac{W_s}{V_s \gamma_w} \tag{4.14}$$

This is quite different from the specific gravity of the entire soil mass, which would include solid, water, and air. Therefore, do not make the common mistake of computing G_s as γ/γ_w! Tables 4.2 and 4.3 list G_s values for common soil minerals.

TABLE 4.2 SPECIFIC GRAVITY OF SELECTED NON-CLAY MINERALS

Mineral	G_s
Quartz	2.65
Feldspar	2.54 - 2.76
Hornblende	3.00 - 3.50
Mica	2.76 - 3.20
Calcite	2.71
Hematite	5.20
Limonite	3.6 - 4.0
Gypsum	2.32
Talc	2.70 - 2.80
Olivene	3.27 - 4.50

TABLE 4.3 SPECIFIC GRAVITY OF SELECTED CLAY MINERALS

Mineral	G_s
Kaolinite	2.62 - 2.66
Montmorillonite	2.75 - 2.78
Illite	2.60 - 2.86
Chlorite	2.60 - 2.96

Although a standard laboratory test is available to measure G_s(ASTM D854), it is usually not needed because nearly all real soils have $2.60 < G_s < 2.80$, which is a very narrow range. The additional precision obtained by performing a test is generally not worth the expense.

For most practical problems, it is sufficient to estimate G_s from the following list:

- Clean, light colored sand of quartz and feldspar   2.65
- Dark colored sand   2.72
- Sand-silt-clay mixtures   2.72
- Clay   2.65

Nevertheless, some unusual soils have G_s values well outside these limits. For example, the olivene sands in Hawaii have G_s values as high as 4.50, while organic soils have low G_s values, sometimes less than 2.0.

Void Ratio and Porosity

The relative volumes of voids and solids may be expressed using the *void ratio, e*:

$$e = \frac{V_v}{V_s}$$

(4.15)

Thus, densely packed soils have a low void ratio. Typical values in the field range from 0.1 to 2.5.

The *porosity, n*, is a similar parameter:

$$n = \frac{V_v}{V} \times 100\%$$

(4.16)

It typically is between 9 and 70% and is related to e as follows:

$$n = \frac{e}{1 + e} \times 100\%$$

(4.17)

Sometimes geotechnical engineers divide the porosity into two parts: the *water porosity, n_w*, and the *air porosity, n_a*:

$$n_w = \frac{V_w}{V} \times 100\%$$

(4.18)

$$n_a = \frac{V_a}{V} \times 100\%$$

(4.19)

Relative Density

The *relative density* is a special weight–volume parameter used in sandy and gravelly soils. It is defined as:

$$D_r = \frac{e_{max} - e}{e_{max} - e_{min}} \times 100\%$$

(4.20)

where:

D_r = Relative density

e = Void ratio

e_{min} = Minimum void ratio

e_{max} = Maximum void ratio

The values of e_{min} and e_{max} represent the soil in very dense and very loose conditions, respectively, and are determined by a standard laboratory test (ASTM D4253 and D4254). Thus, loose soils have low values of D_r, while dense soils have high values. In theory, the lowest possible value of D_r is 0% and the highest possible value is 100%. Thus, D_r is often more useful than e because we can easily compare the field value to the lowest and highest possible values. Table 4.4 presents a classification of soil consistency based on its relative density.

TABLE 4.4 CONSISTENCY OF COARSE-GRAINED SOILS VARIOUS RELATIVE DENSITIES (Lambe and Whitman, 1969; Adapted by permission of John Wiley and Sons, Inc.)[a]

Relative Density, D_r (%)	Classification
0 - 15	Very loose
15 - 35	Loose
35 - 65	Medium dense [b]
65 - 85	Dense
85 - 100	Very dense

[a] Other classification systems have been proposed by others that use these terms, but with different values for the corresponding relative densities.

[b] Lambe and Whitman used the term "medium," but "medium dense" is probably better because "medium" usually refers to the grain size distribution.

Another way to determine D_r is by using empirical correlations with in-situ tests. These methods are often preferred, because it is so difficult to obtain sufficiently undisturbed samples to obtain a reliable value of e in sandy soils. In addition, in-situ tests, especially the CPT, give a better representation of the variability of the soil strata, because coarse-grained soils inevitably have some zones that are looser and denser than average.

Equations 4.21 and 4.25 present correlations between D_r and the standard penetration test (SPT) and cone penetration test (CPT) results (Kulhawy and Mayne, 1990). The CPT correlation is probably more reliable than the SPT.

Equations 4.21 and 4.25 include a couple of parameters that have not yet been defined. The *vertical effective stress*, σ_z', is a measure of the compressive stress in the ground at the depth where the test data is being evaluated; the *overconsolidation ratio* (OCR) is a measure of the stress history in the soil; and D_{50} is the grain size at which 50 percent of the soil is finer. All of these will be defined and discussed in more detail later in this book. In the meantime, the required values will simply be given in any problem statements.

$$D_r = \sqrt{\frac{(N_1)_{60}}{C_P C_A C_{OCR}}} \times 100\% \qquad (4.21)$$

$$C_P = 60 + 25 \log D_{50} \qquad (4.22)$$

$$C_A = 1.2 + 0.05 \log\left(\frac{t}{100}\right) \qquad (4.23)$$

$$C_{OCR} = OCR^{0.18} \qquad (4.24)$$

$$D_r = \sqrt{\left(\frac{q_c}{315\, Q_c\, OCR^{0.18}}\right)\sqrt{\frac{2000\ \text{lb/ft}^2}{\sigma_z'}}} \times 100\% \qquad (4.25\ \text{English})$$

$$D_r = \sqrt{\left(\frac{q_c}{315\, Q_c\, OCR^{0.18}}\right)\sqrt{\frac{100\ \text{kPa}}{\sigma_z'}}} \times 100\% \qquad (4.25\ \text{SI})$$

where:

$(N_1)_{60}$ = corrected SPT N-value, as defined in Chapter 3

C_P = grain size correction factor

C_A = aging correction factor

C_{OCR} = overconsolidation correction factor

D_{50} = grain size at which 50 percent of the soil is finer (mm) as defined in Section 4.4

t = age of soil (time since deposition in years). If no age information data is available, use $t = 100$ yr.

OCR = overconsolidation ratio, as defined in Chapter 11. If no information is available to assess the OCR, use a value of 2.

q_c = cone resistance (kg/cm^2 or ton/ft^2), as defined in Chapter 3

Q_c = compressibility factor

= 0.91 for highly compressible sands

= 1.00 for moderately compressible sands

= 1.09 for slightly compressible sands

For purposes of solving this formula, a sand with a high fines content or a high mica content is "highly compressible," whereas a pure quartz sand is "slightly compressible."

σ_z' = vertical effective stress (lb/ft^2; kPa), as defined in Chapter 10

Many people confuse relative density with relative compaction. The latter is defined in Chapter 6. Although the names are similar, and they measure similar properties, these two parameters are numerically different. In addition, some people in other professions use the term "relative density" to describe what we call specific gravity! Geotechnical engineers should never use the term in this way.

Table 4.5 presents typical values of e_{min} and e_{max} for various sandy soils. These are not intended to be used in lieu of laboratory or in-situ tests, but could be used to check test results or for preliminary analyses.

TABLE 4.5 TYPICAL VALUES OF e_{min} AND e_{max} (Hough, 1969; Adapted by permission of John Wiley and Sons, Inc.)

Soil Description	e_{min} (dense)	e_{max} (loose)
Equal spheres (theoretical values)	0.35	0.92
Clean, poorly graded medium sand (Ottawa, Illinois)	0.50	0.80
Clean, fine-to-medium sand	0.40	1.0
Uniform inorganic silt	0.40	1.1
Silty sand	0.30	0.90
Clean fine-to-coarse sand	0.20	0.95
Micaceous sand	0.40	1.2
Silty sand and gravel	0.14	0.85

Derived Equations

By combining the definitions just described and the phase diagrams in Figure 4.2, we can derive a series of new weight–volume equations. For example:

Let $V_s = 1$, as shown in Figure 4.3.

$$e = \frac{V_v}{V_s} \quad \rightarrow \quad \text{gives} \quad V_v = e$$

$$S = \frac{V_w}{V_v} \times 100\% \quad \rightarrow \quad \text{gives} \quad V_w = Se$$

$$\gamma_w = \frac{W_w}{V_w} \quad \rightarrow \quad \text{gives} \quad W_w = Se\gamma_w$$

$$G_s = \frac{W_s}{V_s \gamma_w} \quad \rightarrow \quad \text{gives} \quad W_s = G_s \gamma_w (1)$$

$$w = \frac{W_w}{W_s} \times 100\% = \frac{Se\gamma_w}{G_s\gamma_w} \times 100\% = \frac{Se}{G_s} \times 100\%$$

Rewriting this equation gives:

$$\boxed{S = \frac{w\,G_s}{e} \times 100\%} \qquad (4.26)$$

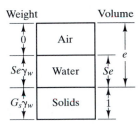

Weight		Volume
0	Air	
		e
$Se\gamma_w$	Water	Se
$G_s\gamma_w$	Solids	1

Figure 4.3 Phase diagram for derivation of $S=wG_s/e$.

Derivations based on any other assumed value of V_s would produce the same equation. Using similar derivations, we also can develop the following equations:

$$\gamma_d = \frac{\gamma}{1 + w} \tag{4.27}$$

$$\gamma_d = \frac{G_s \gamma_w}{1 + w G_s / S} \tag{4.28}$$

$$W_s = \frac{W}{1 + w} \tag{4.29}$$

$$M_s = \frac{M}{1 + w} \tag{4.30}$$

$$e = \frac{G_s \gamma_w}{\gamma_d} - 1 \tag{4.31}$$

$$w = S \left[\frac{\gamma_w}{\gamma_d} - \frac{1}{G_s} \right] \times 100\% \tag{4.32}$$

$$S = \frac{w}{\dfrac{\gamma_w}{\gamma_d} - \dfrac{1}{G_s}} \times 100\% \tag{4.33}$$

For all of these equations, parameters normally expressed as a percentage must be inserted in decimal form. For example, a degree of saturation of 45% in Equation 4.28 would be expressed as 0.45 not 45.

Solving Weight–Volume Problems

We often encounter problems where one or more of the weight-volume parameters is known and others need to be determined. For example, some parameters, such as moisture content, can be measured in the laboratory, while others, such as void ratio, cannot. Therefore, we need to have a means of computing them.

Often we can perform these computations using the derived equations presented on pages 106 and 107. However, if this method does not work, we go back to fundamentals and solve the problem using a phase diagram as follows:

1. Draw a phase diagram and annotate all of the dimensions presented in the problem statement. Remember to set $W_a = 0$. If the soil is saturated, we also can set $V_a = 0$.
2. Sometimes no weights or volumes are given in the problem statement, or the only dimensions given are equal to zero (i.e., no air or no water). If this is the case, then the given data is applicable to the entire soil strata, regardless of the sample size. However, the phase diagram analysis method requires that a certain quantity of soil be specified, and will not work unless we do so. Therefore, we must assume a quantity. Any one dimension may be assumed (but only one!). Usually we assume $V = 1$ m^3 or $V = 1$ ft^3.
3. Using Equations 4.1 through 4.33 and obvious addition and subtraction from the phase diagram (i.e., $V_a = V - V_s - V_w$), determine all of the remaining dimensions.
4. Compute the required parameters from these dimensions using Equations 4.1 through 4.20.

Example 4.1

A 27.50 lb soil sample has a volume of 0.220 ft^3, a moisture content of 10.2%, and a specific gravity of solids of 2.65. Compute the unit weight, dry unit weight, degree of saturation, void ratio, and porosity.

Solution using fundamental and derived equations

$$\gamma = \frac{W}{V} = \frac{27.50 \text{ lb}}{0.220 \text{ ft}^3} = 125.0 \text{ lb/ft}^3 \quad \rightarrow \quad \textbf{125 lb/ft}^3 \quad \Leftarrow Answer$$

$$\gamma_d = \frac{\gamma}{1+w} = \frac{125.0 \text{ lb/ft}^3}{1+0.102} = 113.4 \text{ lb/ft}^3 \quad \rightarrow \quad \textbf{113 lb/ft}^3 \quad \Leftarrow Answer$$

$$S = \frac{w}{\dfrac{\gamma_w}{\gamma_d} - \dfrac{1}{G_s}} \times 100\% = \frac{0.102}{\dfrac{62.4 \text{ lb/ft}^3}{113.4 \text{ lb/ft}^3} - \dfrac{1}{2.65}} \times 100\% = 59\% \quad \Leftarrow Answer$$

$$e = \frac{G_s \gamma_w}{\gamma_d} - 1 = \frac{(2.65)(62.4 \text{ lb/ft}^3)}{113.4 \text{ lb/ft}^3} - 1 = 0.4582 \quad \rightarrow \quad \mathbf{0.458} \quad \Leftarrow \textit{Answer}$$

$$n = \frac{e}{1+e} \times 100\% = \frac{0.4582}{1 + 0.4582} \times 100\% = \mathbf{31.4\%} \quad \Leftarrow \textit{Answer}$$

Solution using a phase diagram

Although the fundamental and derived equations were sufficient to solve this problem, and would be the easiest method, we also will illustrate a solution using a phase diagram.

Step 1: Draw and annotate a phase diagram (see Figure 4.4).

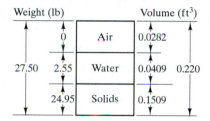

Figure 4.4 Phase diagram for Example 4.1.

Step 2: Assume dimension.

Not applicable to this problem, because weights and volumes have been given.

Step 3: Determine dimensions in phase diagram.

$$W_s = \frac{W}{1+w} = \frac{27.50 \text{ lb}}{1 + 0.102} = 24.95 \text{ lb}$$

$$W_w = W - W_s = 27.50 \text{ lb} - 24.95 \text{ lb} = 2.55 \text{ lb}$$

$$G_s = \frac{W_s}{V_s \gamma_w}$$

$$2.65 = \frac{24.95 \text{ lb}}{V_s (62.4 \text{ lb/ft}^3)} \quad \rightarrow \quad V_s = 0.1509 \text{ ft}^3$$

$$\gamma_w = \frac{W_w}{V_w}$$

$$62.4 \text{ lb/ft}^3 = \frac{2.55 \text{ lb}}{V_w} \quad \rightarrow \quad V_w = 0.0409 \text{ ft}^3$$

$$V_a = V - V_w - V_s = 0.220 \text{ ft}^3 - 0.0409 \text{ ft}^3 - 0.1509 \text{ ft}^3 = 0.0282 \text{ ft}^3$$

Step 4: Compute parameters.

$$\gamma = \frac{W}{V} = \frac{27.50 \text{ lb}}{0.220 \text{ ft}^3} = \textbf{125 lb/ft}^3 \quad \leftarrow \textit{Answer}$$

$$n = \frac{V_v}{V} \times 100\% = \frac{0.0282 \text{ ft}^3 + 0.0409 \text{ ft}^3}{0.220 \text{ ft}^3} \times 100\% = \textbf{31.4\%} \quad \leftarrow \textit{Answer}$$

$$e = \frac{V_v}{V_s} = \frac{0.0409 \text{ ft}^3 + 0.0282 \text{ ft}^3}{0.1509 \text{ ft}^3} = \textbf{0.458} \quad \leftarrow \textit{Answer}$$

$$\gamma_d = \frac{W_s}{V} = \frac{24.95 \text{ lb}}{0.220 \text{ ft}^3} = \textbf{113 lb/ft}^3 \quad \leftarrow \textit{Answer}$$

$$S = \frac{V_w}{V_v} \times 100\% = \frac{0.0409 \text{ ft}^3}{0.0409 \text{ ft}^3 + 0.0282 \text{ ft}^3} \times 100\% = \textbf{59\%} \quad \leftarrow \textit{Answer}$$

Example 4.2

A certain soil has the following properties:

$$G_s = 2.71$$
$$n = 41.9\%$$
$$w = 21.3\%$$

Find the degree of saturation, S, and the unit weight, γ.

Solution

Although this problem could be solved using Equations 4.17 and 4.26-4.28, we will use the phase diagram method to illustrate Step 2 of the procedure described above.

Step 1: Draw and annotate phase diagram (see Figure 4.5).

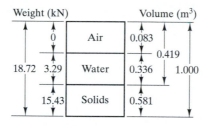

Figure 4.5 Phase diagram for Example 4.2.

Step 2: Assume dimension.

No weights or volumes were stated, so all of the given data applies to the entire soil strata. Therefore, we will develop a phase diagram for an assumed volume of 1 m^3.

Step 3: Determine dimensions in phase diagram.

$$n = \frac{V_v}{V} \times 100\% \quad \rightarrow \quad 0.419 = \frac{V_v}{1} \quad \rightarrow \quad V_v = 0.419 \text{ m}^3$$

$$V_s = V - V_v = 1.000 \text{ m}^3 - 0.419 \text{ m}^3 = 0.581 \text{ m}^3$$

$$G_s = \frac{W_s}{V_s \gamma_w} \quad \rightarrow \quad 2.71 = \frac{W_s}{(0.581 \text{ m}^3)(9.81 \text{ kN/m}^3)} \quad \rightarrow \quad W_s = 15.44 \text{ kN}$$

$$w = \frac{W_w}{W_s} \times 100\% \quad \rightarrow \quad 0.213 = \frac{W_w}{15.44 \text{ kN}} \quad \rightarrow \quad W_w = 3.29 \text{ kN}$$

$$W = W_s + W_w = 15.43 \text{ kN} + 3.29 \text{ kN} = 18.73 \text{ kN}$$

$$\gamma_w = \frac{W_w}{V_w} \quad \rightarrow \quad 9.81 \text{ kN/m}^3 = \frac{3.29 \text{ kN}}{V_w} \quad \rightarrow \quad V_w = 0.335 \text{ m}^3$$

$$V_a = V_v - V_w = 0.419 \text{ m}^3 - 0.335 \text{ m}^3 = 0.084 \text{ m}^3$$

Step 4: Compute parameters.

$$S = \frac{V_w}{V_v} \times 100\% = \frac{0.335 \text{ m}^3}{0.335 \text{ m}^3 + 0.084 \text{ m}^3} \times 100\% = \textbf{80.0 \%} \quad \Leftarrow Answer$$

$$\gamma = \frac{W}{V} = \frac{18.73 \text{ kN}}{1 \text{ m}^3} = \textbf{18.7 kN/m}^3 \quad \Leftarrow Answer$$

Example 4.3

The standard method of measuring the specific gravity of solids (ASTM D854) uses a calibrated glass flask known as a *pycnometer*, as shown in Figure 4.6. The pycnometer is first filled with water and set on a balance to find its mass. Then, it is refilled with a known mass of dry soil plus water so the total volume is the same as before. Again, its mass is determined. From this data, we can compute G_s.

Using this technique on a certain soil sample, we have obtained the following data:

Mass of soil = 81.8 g
Moisture content of soil = 11.2%
Mass of pycnometer+water = 327.12 g
Mass of pycnometer+soil+water = 373.18 g
Volume of pycnometer = 250.00 ml

Compute G_s for this soil.

Figure 4.6 Use of a pycnometer to measure G_s in the laboratory.

Solution

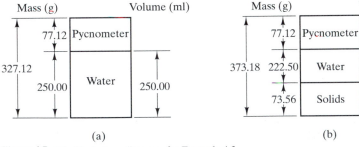

Figure 4.7 Modified phase diagrams for Example 4.3.

$$(M_w)_A = V\rho_w = (250.00 \text{ ml})(1 \text{ g/ml}) = 250.00 \text{ g}$$

$$M_P = M_A - (M_w)_A = 327.12 \text{ g} - 250.00 \text{ g} = 77.12 \text{ g}$$

$$(M_s)_B = \frac{M}{1+w} = \frac{81.8 \text{ g}}{1+0.112} = 73.56 \text{ g}$$

$$(M_w)_B = 373.18 \text{ g} - 77.12 \text{ g} - 73.56 \text{ g} = 222.50 \text{ g}$$

$$(V_w)_B = \frac{(M_w)_B}{\rho_w} = \frac{222.50 \text{ g}}{1 \text{ g/ml}} = 222.50 \text{ ml}$$

$$(V_s)_B = 250.00 \text{ ml} - 222.50 \text{ ml} = 27.50 \text{ ml}$$

$$G_s = \frac{M_s}{V_s \rho_w} = \frac{73.56 \text{ g}}{(27.50 \text{ ml})(1 \text{ g/ml})} = \textbf{2.67} \quad \Leftarrow \textit{Answer}$$

Note: A real laboratory specific gravity test also considers various correction factors to account for the density of water as a function of temperature and other variables. For simplicity, we have not considered these factors here.

QUESTIONS AND PRACTICE PROBLEMS

4.1   A cube of moist soil weighs 330 lb and has a volume of 3.00 ft^3. Its moisture content is 27.0% and the specific gravity of solids is 2.72. Compute the void ratio, porosity, degree of saturation, and unit weight of this soil.

4.2   A sample of soil is compacted into a 9.44×10^{-4} m^3 laboratory mold. The mass of the compacted soil is 1.91 kg and its moisture content is 14.5%. Using a specific gravity of solids of 2.66, compute the degree of saturation, density (kg/m^3) and unit weight (kN/m^3) of this compacted soil.

4.3   A saturated soil has a moisture content of 38.0% and a specific gravity of solids of 2.73. Compute the void ratio, porosity and unit weight (lb/ft^3 or kN/m^3) of this soil.

4.4   A sample of clay was obtained from a point below the groundwater table. A moisture content test on this sample produced the following data:

> Mass of can = 10.88 g
> Mass of can + moist soil = 116.02 g
> Mass of can + dry soil = 85.34 g

a. Compute the moisture content.
b. Assume a reasonable value for G_s, then compute the void ratio and unit weight.

4.5   An undisturbed cylindrical soil sample is 60 mm in diameter and 152 mm long. It has a mass of 816 g. After finding the mass of the entire sample, a small portion was removed and a moisture content test was performed on it. The results of this test on the sub-sample were:

> Mass of can = 22.01 g
> Mass of can + moist soil = 124.97 g
> Mass of can + dry soil = 112.72 g

Using G_s = 2.70, compute w, γ_d, e, and S.

4.6   A strata of clean, light-colored quartz sand located below the groundwater table has a moisture content of 25.6%. The minimum and maximum void ratios of this soil are 0.380 and 1.109. Select an appropriate value of G_s for this soil, compute its relative density, and classify its consistency using Table 4.4.

4.7   A contractor needs 214 yd^3 of aggregate base material for a highway construction project. It will be compacted to a dry unit weight of 130 lb/ft^3. This material is available in a stockpile at a local material supply yard, but is sold by the ton, not by the cubic yard. The moisture content of the stockpile is 7.0%.

a. How many tons of aggregate base material should the contractor purchase to have exactly the correct volume of compacted material?
b. The contractor purchased the material per the computation in part a, and it exactly met the needs at the project site. An intense rainstorm occurred the following week, which delayed further construction and raised the moisture content of the stockpile to 19.0%. Now, the contractor needs to prepare another identical section of aggregate base and is ordering the same number of tons as before. How many cubic yards of compacted aggregate base will be produced from this second shipment? How will it compare with the first shipment? Explain.

4.8   A cone penetration test has been conducted, and has measured a cone resistance of 85 kg/cm^2 at a depth of 10 m. The vertical effective stress at this depth is 150 kPa, and the overconsolidation ratio is 2. The soils at this depth are quartz sands. Compute the relative density, and classify the soil using Table 4.4.

4.9   A standard penetration test has been conducted at a depth of 15 ft in an 8-inch diameter exploratory boring using a USA-style safety hammer and a standard sampler. This test produced an uncorrected N-value of 12. The soil inside the sampler was a fine-to-medium sand with D_{50} = 0.6 mm. The vertical effective stress at this depth is 1100 lb/ft^2. Adjust the N-value as described in Chapter 3, then compute the relative density and classify the soil using Table 4.4.

4.4 PARTICLE SIZE AND SHAPE

The individual solid particles in a soil can have different sizes and shapes, and these characteristics also have a significant effect on its engineering behavior. Therefore, geotechnical engineers often assess particle size and shape.

Particle Size Classification

Several systems have been developed to classify soil particles based on their size. We will examine only one: the ASTM system, as described in Table 4.6. According to ASTM, particles larger than 3 in (75 mm) in diameter are known as *rock fragments*. Smaller particles are defined as *soil*, and are classified according their ability to pass through certain size sieves. A *sieve* is a carefully manufactured mesh of wires with a specified opening size, as shown in Figure 4.8.

Figure 4.8 An 8-in (200 mm) diameter sieve used for soil testing. This one is a 1-inch sieve. Notice how the smaller pieces of gravel have passed through, while the larger pieces have not.

TABLE 4.6 ASTM PARTICLE SIZE CLASSIFICATION (Per ASTM D2487)

Sieve Size		Particle Diameter		Soil Classification	
Passes	Retained on	(in)	(mm)		
	12 in	> 12	> 350	Boulder	Rock Fragments
12 in	3 in	3 - 12	75.0 - 350	Cobble	
3 in	3/4 in	0.75 - 3	19.0 - 75.0	Coarse gravel	
3/4 in	#4	0.19 - 0.75	4.75 - 19.0	Fine gravel	
#4	#10	0.079 - 0.19	2.00 - 4.75	Coarse sand	Soil
#10	#40	0.016 - 0.079	0.425 - 2.00	Medium sand	
#40	#200	0.0029 - 0.016	0.075 - 0.425	Fine sand	
#200		< 0.0029	< 0.075	Fines (silt + clay)	

Figures 4.9 and 4.10 are photographs of samples from each category. Natural deposits often include a mixture of both rock fragments and soil.

Laboratory Tests

Although the distribution of particle sizes can often be estimated by eye, two laboratory tests are commonly used to provide more precise assessments: the *sieve analysis* and the *hydrometer analysis*. In this context, the word "analysis" means a laboratory test, not a series of computations.

Boulder

Cobble

Figure 4.9 Boulders are particles larger than 12 inches in diameter; Cobbles are particles between 3 and 12 inches in diameter.

Sieve Analysis

A *sieve analysis* is a laboratory test that measures the grain-size distribution of a soil by passing it through a series of sieves. The larger sieves are identified by their opening size. For example, a 3/4-inch sieve will barely pass a 3/4-inch diameter sphere. Smaller sieves are numbered, with the number indicating the openings per inch. For example, a #8 sieve has 8 openings per inch or 64 per square inch. However, the size of these openings is less than 1/8 inch because of the width of the wire. Table 4.7 presents opening sizes for standard sieves used in North America.

Coarse gravel

Coarse Sand

Medium Sand

Fine gravel

Fine Sand

Figure 4.10 Full-scale photos of coarse and fine gravel; coarse, medium, and fine sand.

TABLE 4.7 ASTM STANDARD SIEVES (per ASTM D422 and E100)

Sieve Identification	Opening Size (in)	Opening Size (mm)	Sieve Identification	Opening Size (in)	Opening Size (mm)
3 inch	3.00	76.2	#16	0.0465	1.18
2 inch	2.00	50.8	#20	0.0335	0.850
1½ inch	1.50	38.1	#30	0.0236	0.600
1 inch	1.00	25.4	#40	0.0167	0.425
3/4 inch	0.75	19.0	#50	0.0118	0.300
3/8 inch	0.375	9.52	#60	0.00984	0.250
#4	0.187	4.75	#100	0.00591	0.150
#8	0.0929	2.36	#140	0.00417	0.106
#10	0.0787	2.00	#200	0.00295	0.075

Most people can just barely see 0.1 mm diameter objects without using a magnifying glass, and this nearly corresponds to the #200 sieve. This diameter also represents the border between gritty and smooth textures. Thus, referring back to Table 4.6, sands and silts could be visually distinguished by looking at the particles (if you can see them, it's a sand) and by feeling for grittiness (if it feels gritty, it's a sand). Both are best done with wetted soil samples.

The test (ASTM D422) consists of preparing a soil sample with known weight of solids, W_s, and passing it through the sieves as shown in Figure 4.11. The sieves are arranged in order with the coarsest one on top. A pan is located below the finest sieve. The weight retained on each sieve is then expressed as a percentage of the total weight.

The percentage passing the #200 sieve is especially noteworthy. Soils that have less than about 5 percent passing the #200 sieve are called "clean," while those that have more are called "dirty." For example, a *clean sand* is primarily sand, with less than 5 percent silt or clay.

Hydrometer Analysis

Although sieve analyses work very well for particles larger than the #200 sieve (sands and gravels) and they determine the total amount of fines, they do not give any insight on the distribution of finer particles (silts and clays). The smallest clay particles are only about 1×10^{-4} mm in diameter, which is about the same size as a smoke particle. It is impossible to

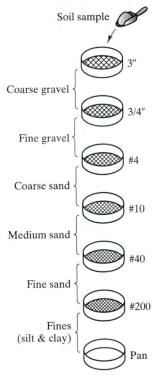

Figure 4.11 A series of sieves for conducting a sieve analysis. The sieve sizes used in a real test are different than those shown here.

manufacture sieves this small, so we need to use another technique: the hydrometer analysis. This procedure (ASTM D422) consists of placing a soil sample with a known W_s into a 1000 ml graduated cylinder and filling it with water. The laboratory technician vigorously shakes the cylinder to place the soil in suspension, then places it upright on a table as shown in Figure 4.12.

Once the cylinder has been set upright, the soil particles begin to settle to the bottom. We describe this downward motion using Stoke's Law:

$$v = \frac{D^2 \gamma_w (G_s - G_L)}{18\,\eta} \tag{4.34}$$

where:
v = velocity of settling soil particle
D = particle diameter
γ_w = unit weight of water
G_s = specific gravity of solid particles
G_L = specific gravity of soil-water mixture
η = dynamic viscosity of soil-water mixture

The velocity is proportional to the square of the particle diameter, so large particles settle much more quickly than small ones. In addition, we can determine the mass of solids still in suspension by measuring the specific gravity of the soil–water mixture, which is done using the hydrometer shown in Figure 4.12. Therefore, by making a series of specific gravity measurements, usually over a period of 24 hours, and employing Stoke's law, we can determine the distribution of particles sizes in the soil sample.

The hydrometer analysis is unsuitable for particles larger than about a #100 sieve because they settle more quickly than we can measure the specific gravity. However, by performing a sieve analysis, hydrometer analysis, or both, we can determine the distribution of particle sizes for virtually any soil.

Figure 4.12 Equipment used in a hydrometer analysis. Part of the hydrometer is protruding from the top of the fluid.

Grain Size Distribution Curves

Real soils rarely fall completely within only one of the categories listed in Table 4.6. They almost always contain a variety of particles sizes mixed together. Therefore, we need to have an effective means of presenting the distribution of particle sizes in a soil. That method is the *grain-size distribution curve*, such as those shown in Figure 4.13. These are plots of the grain diameter (on a logarithmic scale) vs. the percentage of the solids by weight smaller than that diameter. We call the latter "percent finer" or "percent passing," which generates mental images of that portion of the soil passing through a sieve.

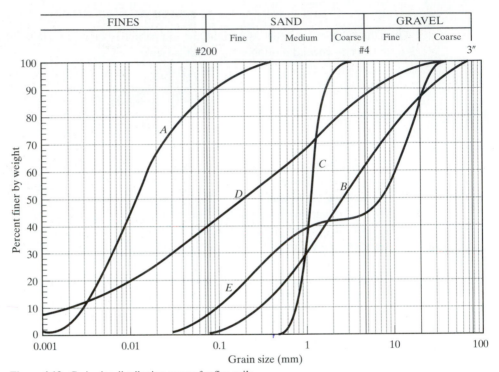

Figure 4.13 Grain size distribution curves for five soils.

Curves on the left side of the diagram, such as soil A, indicate primarily fine-grained soils (silts and clays), while those on the right side, such as soil B, indicate coarse-grained soils (sands and gravels). Steep grain-size distribution curves, such as soil C, reflect soils with a narrow range of particle sizes. These are known as *poorly-graded soils* (or *uniformly-graded soils*). Conversely, soils with flat curves, such as soil D, contain a wide range of particle sizes and are known as *well-graded soils*. Figure 4.14 shows photographs of both types. Literature in the geological sciences often use the terms *well-sorted* and *poorly-sorted*, which have the opposite definitions: Well-sorted = poorly-graded, while poorly-sorted = well-graded.

Some soils have a nearly flat zone in their grain-size distribution curve, such as soil E. These are called *gap-graded* because they are missing particles in a certain size range. Gap-graded soils are sometimes considered a type of poorly graded soil. Aggregates used to make concrete are typically gap graded.

We can determine the percentage of each type of soil by weight by comparing the percents passing the appropriate sieve sizes as listed in Table 4.6. For example, to determine the amount of sand in a soil, subtract the percentage passing the #200 sieve from the percentage passing the #4 sieve.

Figure 4.14 A well graded soil, such as that shown above, has a wide range of particle sizes, in this case ranging from fine sand to coarse gravel. Conversely, a poorly graded soil, such as the one on the left, has a narrow range of particle sizes. This particular soil is a poorly graded gravel that is commercially produced in rock crushing plants. It is called *pea gravel*, even though the particles are slightly larger than most peas. Both photographs are full-scale.

A soils technician who once worked for the author apparently misunderstood the origin of the name pea gravel, because his notes always referred to it as pee gravel!

The particle diameters that correspond to certain percent-passing values for a given soil are known as the *D-sizes*. For example, D_{10} is the grain size that corresponds to 10 percent passing. In other words, 10 percent of the soil is finer then D_{10}. Two additional parameters, the *coefficient of uniformity*, C_u, and the *coefficient of curvature*, C_c, are based on the D-sizes:

$$C_u = \frac{D_{60}}{D_{10}}$$

(4.35)

$$C_c = \frac{(D_{30})^2}{D_{10}\,D_{60}}$$

(4.36)

Steep curves, which reflect poorly graded soils, have low values of C_u, while flat curves (well-graded soils) have high values. Soils with smooth curves have C_c values between about 1 and 3, while irregular curves have higher or lower values. For example, most gap-graded soils have a C_c outside this range.

Example 4.4

Determine the following for soil D in Figure 4.13:
  • Percent gravel, sand, and fines
  • C_u and C_c

Solution

Percent finer data from grain size curve:
    #200 — 40%
    #40   — 88%
    3 in — 100%

Percentage of each type of soil:
    Gravel = 3 in - #4 = 100% - 88% = **12%** ⇐ *Answer*
    Sand = #4 - #200 = 88% - 40% = **48%** ⇐ *Answer*
    Fines = #200 = **40%** ⇐ *Answer*

D-sizes from grain-size curve
    $D_{10} = 0.0019$ mm
    $D_{30} = 0.030$ mm
    $D_{60} = 0.49$ mm

C_u and C_c

$$C_u = \frac{D_{60}}{D_{10}} = \frac{0.49 \text{ mm}}{0.0019 \text{ mm}} = \textbf{260} \quad \leftarrow Answer$$

$$C_c = \frac{D_{30}^2}{D_{10}D_{60}} = \frac{(0.030 \text{ mm})^2}{(0.0019 \text{ mm})(0.49 \text{ mm})} = \textbf{0.97} \quad \leftarrow Answer$$

Note: This is an exceptionally large value of C_u which reflects the very flat grain-size curve. Most C_u values are less than 20.

Particle Shape

The shape of silt, sand, and gravel particles varies from very angular to well rounded, as shown in Figure 4.15 (Youd, 1973). Angular particles are most often found near the rock from which they were formed, while rounded particles are most often found farther away where the soil has experienced more abrasion.

Angular particles have a greater shear strength than smooth ones because it is more difficult to make them slide past one another. This is why aggregate base material used beneath highway pavements is often made of rocks that have been passed through a rock crusher to create a very angular gravel. Clay particles have an entirely different shape, and are discussed later in the next section.

Some non-clay particles are much flatter than any of the samples shown here. One example is mica, which is plate-shaped. Although mica never represents a large portion of the total weight, even a small amount can affect a soil's behavior. Sands that include mica are known as *micaceous sands*.

Very Angular   Angular   Subangular   Subrounded   Rounded   Well Rounded

Figure 4.15 Classification of particle shape for silts, sands, and gravels.

QUESTIONS AND PRACTICE PROBLEMS

4.10 Determine the percent gravel, percent sand, and percent fines for soils A, B, and C in Figure 4.13.

4.11   Determine C_u and C_c for soils A, B, and C in Figure 4.13.

4.12   Plot the grain size distribution curves for each of the soils described below. All three curves should be on the same semilogarithmic diagram.

Sieve Number	% Passing by Weight		
	Lagoon Clay Beufort, SC	Beach Sand Daytona Beach, FL	Weathered Tuff Central America
3/4 in	100	100	100
½ in	100	100	98
#4	100	100	95
#10	100	100	93
#20	100	100	88
#40	100	98	82
#60	100	90	75
#100	95	10	72
#200	80	2	68
Particle Diameter from Hydrometer Analysis (mm)			
0.045	61		66
0.010	42		33
0.005	37		21
0.001	27		10

Data from Sowers (1979).

4.13   Which of the soils in Problem 4.12 is most well graded? Why?

4.14   The American Association of State Highway and Transportation Officials (AASHTO) has defined grading requirements for soils to be used as base courses under pavements (AASHTO designation M 147). The requirements for grading C are as follows:

Sieve Designation	Percent Passing by Weight
1 inch	100
3/8 inch	50 - 85
# 4	35 - 65
# 10	25 - 50
# 40	15 - 30
# 200	5 - 15

An import soil is available from a nearby borrow site, and it has a grain-size distribution as described by curve E in Figure 4.13. Does this soil satisfy the AASHTO grain-size requirements for class C base material?

4.5 CLAY SOILS

Soils that consist of silt, sand, or gravel are primarily the result of physical and mild chemical weathering processes and retain much of the chemical structure of their parent rocks. However, this is not the case with clay soils because they have experienced extensive chemical weathering and have been changed into a new material quite different from the parent rocks. As a result, the engineering properties and behavior of clays also are quite different from other soils.

Formation and Structure of Clay Minerals

Several different chemical weathering processes form *clay minerals* which are the materials from which clays are made. We will examine one of these processes as an example. This process changes orthoclase feldspar, a mineral found in granite and many other rocks, into a clay mineral called kaolinite. The process begins when water acquires carbon dioxide as it falls through the atmosphere and seeps through soil, thus forming a weak carbonic acid (H_2CO_3) solution. This acid reacts with orthoclase feldspar according to the following chemical formula (Goodman, 1993):

$$4KAlSi_3O_8 + 2H_2CO_3 + 2H_2O \rightarrow 2K_2CO_3 + Al_4(OH)_8Si_4O_{10} + 8SiO_2$$
orthoclase + carbonic acid + water → potassium carbonate + kaolinite + silica

Finally, the potassium carbonate and silica are carried off in solution by groundwater, ultimately to be deposited elsewhere, leaving kaolinite clay where orthoclase feldspar once existed.

These various chemical weathering processes form sheet-like chemical structures. There are two types of sheets: *tetrahedral* or *silica* sheets consist of silicon and oxygen atoms; *octahedral* or *alumina* sheets have aluminum atoms and hydroxyls (OH). Sometimes octahedral sheets have magnesium atoms instead of aluminum, thus forming *magnesia* sheets. These sheets then combine in various ways to form dozens of different clay minerals, each with its own chemistry and structure. The three most common ones are:

Kaolinite — Consists of alternating silica and alumina sheets, as shown in Figure 4.16a. These sheets are held together with strong chemical bonds, so kaolinite is a very stable clay. Unlike most other clay minerals, kaolinite does not expand appreciably when wetted, so it is used to make pottery. It also is an important ingredient in paper, paint, and other products, including pharmaceuticals (i.e., kaopectate).

Montmorillonite (also called *smectite*) — Has layers made of two silica sheets and one alumina sheet, as shown in Figure 4.16b. The bonding between these layers is very weak, so large quantities of water can easily enter and separate them, thus causing the clay to swell. This property can be very troublesome or very useful, depending on the situation. Problems with soil expansion include extensive distortions in structures, highways, and other civil engineering projects (see discussion of expansive soils in Chapter 18). However, this expansive behavior and the low permeability of montmorillonite can be useful for sealing borings or providing groundwater barriers. *Bentonite*, a type of montmorillonite, is commercially mined and sold for such purposes.

Illite — Has layers similar to those in montmorillonite, but contains potassium ions between each layer, as shown in Figure 4.16c. The chemical bonds in this structure are stronger than those in montmorillonite but weaker than those in kaolinite, so illite expands slightly when wetted. Glacial clays in the Great Lakes region are primarily illite.

Other clay minerals include *vermiculite, attapulgite,* and *chlorite*.

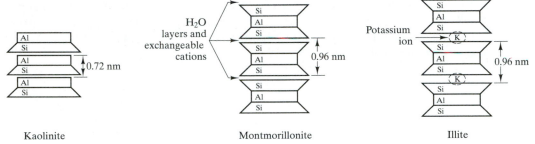

Kaolinite   Montmorillonite   Illite

Figure 4.16 Structures of common clay minerals: a) kaolinite consists of alternating silica and alumina sheets; b) montmorillonite consists of alumina sheets sandwiched between silica sheets. Water and exchangeable cations may be present between the silica sheets; c) illite is similar to montmorillonite, but contains potassium ions between the silica sheets (Adapted from Holtz and Kovacs, 1981).

Individual clay particles are extremely small (less than 2 μm (2×10^{-6} m) in diameter). They cannot be seen with optical microscopes, and require an electron microscope for scientific study (see Figure 4.17). These images show how the shape of clay particles is substantially different from other soils.

Properties of Clays

Because of the small particle diameter and plate-like shape of clays, the surface area to mass ratio is much greater than in other soils   This ratio is known as the *specific surface*. For example, montmorillonite has a specific surface of about 800 m^2/g, which means 3.5 g of this clay has a surface area equal to that of a football field!

     The large specific surface of clays provides more contact area between particles, and thus more opportunity for various interparticle forces to develop. It also provides more places for water molecules to attach, thus giving clays a much greater affinity for absorbing water. Some clays can easily absorb several times their dry weight in water.

Montmorillonite clays have the greatest specific surface, so it is no surprise that they have the greatest affinity for water. The interactions between this water and the clay minerals are quite complex, but the net effect is that the engineering properties vary as the moisture content varies. For example, the shear strength of a given clay at a moisture content of 50% will be less than at a moisture content of 10%.

This behavior is quite different from that in sands, because their specific surface is much smaller and the particles are more inert. Other than changes in pressure within the pore water, which affect all soils and are discussed later in this book, variations in moisture content have very little effect on the behavior of sands.

| Kaolinite | Illite | Chlorite |

Figure 4.17 Electron microscope images of kaolinite, illite, and chlorite clays. Note the sheet-like shapes, which are quite different from the spherically-shaped particles of coarser soils (Images courtesy of Dr. Ralph L. Kugler).

Formation of Clay Soils

On a slightly larger but still microscopic scale, clay minerals are assembled in various ways to form clay soils. These microscopic configurations are called the soil *fabric*, and depend largely on the history of formation and deposition. For example, a residual clay, which has weathered in-place and is still at its original location, will have a fabric much different from a marine clay, which has been transported and deposited by sedimentation.[1] These differences are part of the reason such soils behave differently.

Although we sometimes encounter soil strata that consist of nearly pure clay, most clays are mixed with silts and/or sands. Nevertheless, even a small percentage of clay significantly impacts the behavior of a soil. When the clay content exceeds about 50 percent, the sand and silt particles are essentially floating in the clay, and have very little effect on the engineering properties of the soil.

---

[1] See discussions of residual and marine soils in Chapter 2.

4.6 PLASTICITY AND THE ATTERBERG LIMITS

From the previous discussion, it is clear that silts and clays are two very different kinds of soils. Yet, the classification system described in Table 4.6 used the term "fines" to describe everything that passes through a #200 sieve. It made no attempt to distinguish between silts and clays. Some classification systems draw the line between them based on particle size as determined from a hydrometer test, typically at 0.001 to 0.005 mm. Although such systems can be useful, they also can be misleading because the biggest difference between silt and clay is not their particle sizes, but their physical and chemical structures, as discussed earlier.

It would be impractical to use electron microscopes or other sophisticated equipment to distinguish between clays and silts on a routine basis. Instead, we do so by assessing a property called *plasticity*, which can be determined much more easily and inexpensively. In this context, the term plasticity describes the response of a soil to changes in moisture content. When adding water to a soil changes its consistency from hard and rigid to soft and pliable, the soil is said to be exhibiting plasticity. Clays can be very plastic and silts only slightly plastic, whereas clean sands and gravels do not exhibit any plasticity at all. This assessment can be made using visual–manual procedures, and with experience one can distinguish between clays and silt simply with the hands and a water bottle. More formal assessments of plasticity are performed in the laboratory using the Atterberg limits tests.

The Atterberg Limits

In 1911, the Swedish soil scientist Albert Atterberg (1846-1916) developed a series of tests to evaluate the relationship between moisture content and soil consistency (Atterberg, 1911; Blackall, 1952). Then, in the 1930s, Karl Terzaghi and Arthur Casagrande adapted these tests for civil engineering purposes, and they soon became a routine part of geotechnical engineering. This series includes three separate tests: the *liquid limit test*, the *plastic limit test*, and the *shrinkage limit test*. Together they are known as the *Atterberg limits tests* (ASTM D 427 and D4318).

The liquid limit and plastic limit tests are routinely performed in many soil mechanics laboratories. However, the shrinkage limit test is less useful, and is rarely performed by civil engineers.

The Liquid Limit Test

The liquid limit test uses the *liquid limit device*, a standard laboratory apparatus shown in Figure 4.18. A soil sample is placed in the device, and a groove is cut using a standard tool. The cup is then repeatedly dropped, and the number of drops required for the groove to close for a distance of one-half inch is recorded. The soil is then removed and its moisture content is determined. This test is then repeated at various moisture contents, producing a plot of number of drops vs. moisture content.

By definition, the soil is said to be at its liquid limit when exactly 25 drops are required to close the groove for a distance of one-half inch. The corresponding moisture content is determined from the test data. The liquid limit, LL or w_L, is this moisture content expressed without a percentage sign. For example, if the moisture content is 45%, the liquid limit is 45.

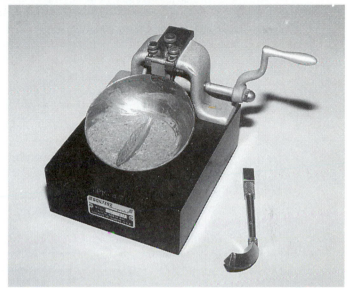

Figure 4.18 This device is used to perform the liquid limit test. The soil pat is in place and has been grooved with the grooving tool shown on the right. The next step is to begin turning the crank, which will repeatedly drop the cup.

The Plastic Limit Test

The plastic limit test procedure involves carefully rolling the soil sample into threads, as shown in Figure 4.19. As this rolling process continues, the thread becomes thinner and eventually breaks. If the soil is dry, it breaks at a large diameter. If it is wet, it breaks at a much smaller diameter. By definition, the soil is at the plastic limit when it breaks at a diameter of one-eighth of an inch (3 mm). The plastic limit, PL or w_p, is this moisture content with the percent sign dropped.

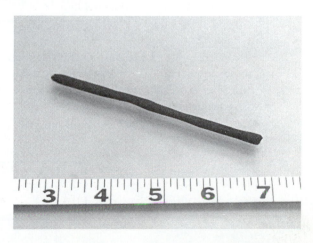

Figure 4.19 A plastic limit test is being performed on this soil sample. It has been rolled until it broke, which occurred at a diameter of 1/8 inch (see scale). Thus, by definition, this soil is at the plastic limit. The value of w_L will be determined by performing a moisture content test. The scale is in inches.

Consistency and Plasticity Assessments Based on Atterberg Limits

The Atterberg limits test results help engineers assess the plasticity of a soil and its consistency at various moisture contents. Figure 4.20 shows the changes in these characteristics with changes in moisture content.

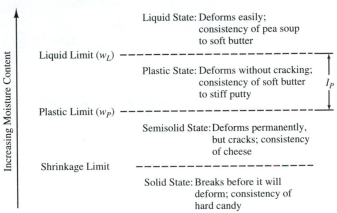

Figure 4.20 Consistency of fine-grained soils at different moisture contents (Sowers, 1979).

When the soil is at a moisture content between the liquid limit and the plastic limit, it is said to be in a *plastic state.* It can be easily molded without cracking or breaking. The children's toy *play-dough* has a similar consistency. This property relates to the amount and type of clay in the soil. The *plasticity index,* PI or I_P, is a measure of the range of moisture contents that encompass the plastic state:

$$I_P = w_L - w_P \qquad (4.37)$$

Soils with a large clay content retain this plastic state over a wide range of moisture contents, and thus have a large plasticity index. The opposite is true of silty soils. Table 4.8 describes soil characteristics at various ranges of plasticity index. Clean sands and gravels are considered to be nonplastic (NP).

The *liquidity index,* I_L, compares the current moisture content of a soil, w, to its Atterberg limits:

$$I_L = \frac{w - w_P}{I_P} \qquad (4.38)$$

Thus, a liquidity index of 0 means the soil is currently at the plastic limit, and 1 means it is at the liquid limit.

TABLE 4.8 CHARACTERISTICS OF SOILS WITH DIFFERENT PLASTICITY INDICES (Sowers, 1979)

Plasticity Index, I_p	Classification	Dry Strength	Visual–Manual Identification of Dry Sample	
0 - 3	Nonplastic	Very low	Falls apart easily	Increasing clay content ↑
3 - 15	Slightly plastic	Slight	Easily crushed with fingers	
15 - 30	Medium plastic	Medium	Difficult to crush with fingers	
> 30	Highly plastic	High	Impossible to crush with fingers	

4.7 STRUCTURED VS. UNSTRUCTURED SOILS

Many soils contain additional physical features beyond a "simple" particulate assemblage. These are known as *structured soils* and include the following:

- *Cemented soils* contain cementing agents that bind the particles together. The most common cementing agents are calcium carbonate ($CaCO_3$) and iron oxides (Fe_2O_3). Both are usually transmitted into the soil in solution within the groundwater.
- *Fissured soils* contain discontinuities similar to fissures in rock. Stiff clays are especially likely to contain fissures.
- *Sensitive clays* are those with a flocculated structure of clay particles that resemble a house of cards. These soils are very sensitive to disturbance, which destroys this delicate structure. The landslide in Figure 2.17 occurred in a sensitive clay.

Unstructured soils are those that do not contain such special features. Most geotechnical analyses are based on unstructured soils, and thus often need to be modified when working with structured soils.

4.8 ORGANIC SOILS

Technically, any material that contains carbon is "organic." However, engineers and geologists use a more narrow definition when we apply the term to soils. An *organic soil* is one that contains a significant amount of organic material recently derived from plants or animals. It needs to be fresh enough to still be in the process of decomposition, and thus retains a distinctive texture, color, and odor.

Some soils contain carbon, but are not recently derived from plants or animals and thus are not considered organic in this context. For example, some sands contain calcium carbonate (calcite), which arrived as a chemical precipitate.

The identification of organic soils is very important, because they are much weaker and more compressible than inorganic soils, and thus do not provide suitable support for most engineering projects. If such soils are present, we usually avoid them, excavate them, or drive piles through them to reach more suitable deposits.

The term *peat* refers to a highly organic soil derived primarily from plant materials. It has a dark brown to black color, a spongy consistency, and an organic odor. Usually plant fibers are visible, but in the advanced stages of decomposition they may not be evident. Peat often occurs in *bogs*, which are pits filled with organic material. Bogs are typically covered with a live growth of moss. Thoroughly decomposed peat is sometimes called *muck*, although this term also is used to describe waste soil, such as the cuttings from tunnels. Peat can be subjected to the same processes that form sedimentary and metamorphic rocks, thus forming *bituminous coal* and *anthracite coal*, respectively.

Swamps are larger than bogs and may contain a wider variety of materials. They are typically fed by slow streams or lakes. The Everglades in Florida is a noteworthy example.

Although many organic soil deposits create obvious topographic features such as swamps and bogs, others are buried underground, having been covered with inorganic alluvial soils. These often are difficult to detect, and can be the source of large differential settlements. For example, such buried deposits are present near the coast in Orange County, California, and have been the source of settlement problems.

Organic deposits also may be mixed with inorganic soils, especially silts and clays, producing soils that are not as bad as peat, but worse than inorganic deposits.

QUESTIONS AND PRACTICE PROBLEMS

4.15 Borings for observation wells, such as the one shown in Figure 3.21, are normally sealed with an impervious cap near the ground surface. This cap prevents significant quantities of surface water from seeping into the well. For convenience, manufacturers supply a pelletized clay that has been dried and formed into 10 mm diameter balls. The driller then pours these balls into the boring and adds water. As the clay absorbs the water, it expands and seals the boring. What type of clay would be most appropriate for this purpose? Why? Would other clays produce less satisfactory results? Why?

4.16 A soil has a liquid limit of 61 and a plastic limit of 30. A moisture content test performed on an undisturbed sample of this soil yielded the following results:

Mass of soil + can before placing in oven	96.2 g
Mass of soil + can after removal from oven	71.9 g
Mass of can	20.8 g

Compute the following:
   a. The plasticity index
   b. The moisture content
   c. The liquidity index

Present a qualitative description of this soil at its in-situ moisture content.

4.17 A soil has $w_P = 30$ and $w_L = 80$. Compute its plasticity index, then describe the probable clay content (i.e., small, moderate, high).

4.18 A soil has $w_P = 22$ and $w_L = 49$. What moisture content corresponds to a liquidity index of 0.5?

4.19 Compute the specific surface (expressed in m^2/g) for a typical fine sand. State any assumptions, and compare your computed value with that quoted for montmorillonite clay in Section 4.5. Discuss the significance of these two numbers.

SUMMARY

Major Points

1. Soil is a particulate material, so its engineering properties depend primarily on the interaction between these particles. This is especially true of gravels, sands, and silts. Clays also are particulates, but their behavior is much more complex because of the interaction between the particles and the pore water.
2. Soil can include all three phases of matter simultaneously, and their relative proportions are important. Geotechnical engineers have developed a series of weight-volume parameters to describe these proportions.
3. The distribution of particle sizes in a soil also is important, and this distribution can be determined by performing a sieve analysis and/or a hydrometer analysis. The results are presented as a grain-size distribution curve.
4. The solid particles also have various shapes, which impact their behavior.
5. Clays are formed by chemical weathering processes. The individual particles are much smaller than sands or silts, and their engineering behavior is much more dependant on the moisture content.
6. Clays and silts are often distinguished from each other by assessing their plasticity, which reflects their affinity for water. The Atterberg limits tests, especially the plastic limit and liquid limit, help us do this.
7. Structured soils are those with special features, such a cementation, fissures, or flocculated structures. They behave differently from unstructured soils, which do not contain these features.
8. Organic soils are those with a significant quantity of organic matter. Their engineering properties are much worse than those of inorganic soils.

Vocabulary

Atterberg limits	degree of saturation	grain-size distribution curve
bog	density	gravel
boulder	dry density	hydrometer analysis
buoyant unit weight	dry unit weight	illite
cemented soil	fines	kaolinite
clay	fissured soil	liquid limit
cobble	gap-graded soil	liquidity index

moisture content	plasticity index	silt
montmorillonite	poorly graded soil	specific gravity of solids
muck	porosity	structured soil
organic soil	pycnometer	unit weight
particulate material	relative density	unstructured soil
peat	sand	void ratio
phase diagram	sensitive clay	weight–volume parameter
plastic limit	shrinkage limit	well-graded soil
plasticity	sieve analysis	

COMPREHENSIVE QUESTIONS AND PRACTICE PROBLEMS

4.20 A sand with $G_s = 2.66$ and $e = 0.60$ is completely dry. It then becomes wetted by a rising groundwater table. Compute the unit weight (lb/ft^3 or kN/m^3) under the following conditions:
   a. When the sand is completely dry
   b. When the sand is 40 percent saturated
   c. When the sand is completely saturated

4.21 A soil initially has a degree of saturation of 95% and a unit weight of 129 lb/ft^3. It is then placed in an oven and dried. After removal from the oven, its unit weight is 109 lb/ft^3. Compute the void ratio, porosity, initial moisture content, and specific gravity of the soil assuming its volume did not change during the drying process.

4.22 A 1.20 m thick strata of sand has a void ratio of 1.81. A contractor passes a vibratory roller over this strata, which densifies it and reduces its void ratio to 1.23. Compute its new thickness.

4.23 A 412 g sample of silty sand with a moisture content of 11.2% has been placed on a #200 sieve. The sample was then "washed" on the sieve, forcing the minus #200 particles to pass through. The soil that remained on the sieve was then oven dried and found to have a mass of 195 g. By visual inspection, it is obvious that all of this soil is smaller than the #4 sieve. Compute the percent sand in the original sample.

4.24 A standard penetration test has been performed on a soil, producing $(N_1)_{60} = 19$. A sieve analysis was then performed on the sample obtained from the SPT sampler, producing curve C in Figure 4.13. Assuming this soil is about 150 years old and has OCR = 1.8, compute its relative density and classify its consistency.

4.25 Develop a formula for relative density as a function of γ_d, $\gamma_{d\text{-}hi}$, and $\gamma_{d\text{-}lo}$, where γ_d is the dry unit weight in the field, $\gamma_{d\text{-}hi}$ is the dry unit weight that corresponds to e_{min} and $\gamma_{d\text{-}lo}$ is the dry unit weight that corresponds to e_{max}.

4.26 All masses for the specific gravity test described in Example 4.3 were determined using a balance with a precision of ±0.01 g. The volume of the pycnometer is accurate to within ±0.5%, and the moisture content measurement is accurate to within ±2.0% (i.e., the real w could be as low as 11.0% or as high as 11.4%). Assuming all measurement errors are random, determine the precision of the computed G_s value.

Hint: All errors are random, so the worst-case would be if none of the errors were compensating (i.e., each measurement had the maximum possible error, and each contributed to making the computed G_s farther from its true value). Therefore, compute the highest possible value of G_s that is consistent with the stated uncertainties, then compare it with the G_s obtained in Example 4.3.

4.27 A 10,000 ft^3 mass of saturated clay had a void ratio of 0.962 and a specific gravity of solids of 2.71. A fill was then placed over this clay, causing it to compress. This compression is called consolidation, a topic we will discuss in Chapters 11 and 12. During this process, some of the water was squeezed out of the voids. However, the volume of the solids remained unchanged. After the consolidation was complete, the void ratio had become 0.758.

    a. Compute the initial and final moisture content of the clay.

    b. Compute the new volume of the clay.

    c. Compute the volume of water squeezed out of the clay.

4.28 What are the three most common clay minerals? Which one usually causes the most problems for geotechnical engineers? Why?

5

Soil Classification

Chinese legends record a classification of soils according to color and structure which was made by the engineer Yu during the reign of Emperor Yao, about 4000 years ago.

This is the earliest known soil classification system
(Thorp, 1936)

Thus far we have studied several ways of categorizing soil, such as by its geologic origin, mineralogy, grain size, plasticity index, and so on. Each of these methods is useful in the proper context, but a more comprehensive system is needed to better classify soils for engineering purposes. These systems need to focus on the characteristics that affect their engineering behavior, and must be standardized so everyone "speaks the same language." This way, engineers can communicate using terms that clearly describe the soil, while still being concise.

Many such soil classification systems have been developed, usually based on the grain-size distribution and the Atterberg limits. They often are supplemented by non-standardized classifications of other properties, such as consistency and cementation.

5.1 USDA SOIL CLASSIFICATION SYSTEM

The United States Department of Agriculture (USDA) soil classification system (Soil Survey Staff, 1975) is used in soil survey reports and other agricultural documents.

136

Although this system differs from those used by geotechnical engineers, we must understand it to interpret their reports, as discussed in Section 3.2.

The first step in using this system is to determine the percentage by dry weight of each constituent, as follows:

- Coarse fragments   (> 2.0 mm)
- Sand   (0.05 - 2.0 mm)
- Silt   (0.002 - 0.05 mm)
- Clay   (< 0.002 mm)

Note that these divisions do not correspond to those defined by ASTM (Table 4.6), and that clays and silts are distinguished by particle size, not Atterberg limits. Since this system is based entirely on particle size, it is a type of *textural classification system*.

Next, find the total weight of sand+silt+clay (do not include coarse fragments in the total!) and convert each weight to a percentage of this total. Then, using the triangle in Figure 5.1, find the soil classification.

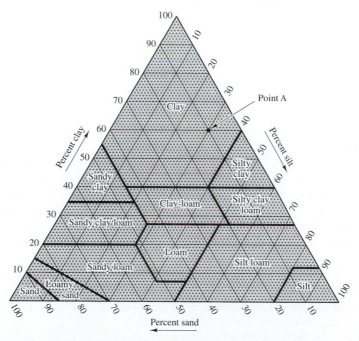

Figure 5.1 USDA soil classification triangle. There are three sets of lines in this diagram: The percent clay is represented by a series of horizontal lines, silt by lines inclined from upper right to lower left, and sand by lines inclined from lower right to upper left. For example, Point A represents 60 percent clay, 30 percent silt, and 10 percent sand. To classify a soil using this diagram, determine the percentages of clay, silt, and sand, and trace the appropriate lines until they meet.

If coarse fragments constitute more than 15 to 20 percent of the entire sample, then add one of the following terms to the classification obtained from the triangle:

Channery: Fragments of thin, flat sandstone, limestone, or schist up to 6 inches along the longer axis.

Cherty: Angular fragments that are less than 3 inches in diameter, at least 75 percent of which are chert (a sedimentary rock that occurs as nodules in limestone and shale).

Cobbly: Rounded or partially rounded fragments of rock ranging from 3 to 10 inches in diameter.

Flaggy: Relatively thin fragments 6 to 15 inches long of sandstone, limestone, slate, shale, or schist.

Gravelly: Rounded or angular fragments, not prominently flattened, up to 3 inches in diameter, with less than 75 percent chert.

Shaly: Flattened fragments of shale less than 6 inches along the longer axis.

Slaty: Fragments of slate less than 6 inches along the longer axis.

Stony: Rock fragments larger than 10 inches in diameter if rounded, or longer than 15 inches along the longer axis if flat.

Example 5.1

Sieve and hydrometer analyses have been performed on a soil sample, and the results of these tests are shown as curve A in Figure 5.2. Determine its USDA classification.

Solution

From the grain-size distribution curve:

Particle Size (mm)	Percent Finer
2.0	92%
0.05	50%
0.002	20%

Coarse fragments = 100 - 92 = 8%
Sand = 92% - 50% = 42%
Silt = 50% - 20% = 30%
Clay = 20%

Adjust percentages as a total of sand+silt+clay
Sand = 42 (100/92) = 46%
Silt = 30 (100/92) = 33%
Clay = 20 (100/92) = 22%

Using Figure 5.1, this soil plots as a **LOAM**   ⟵ *Answer*

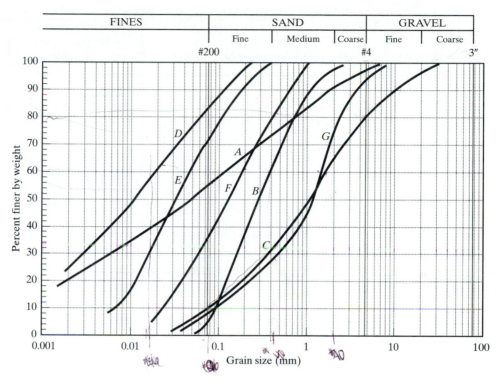

Figure 5.2 Grain-size distribution curves.

5.2 AASHTO SOIL CLASSIFICATION SYSTEM

Terzaghi and Hogentogler developed one of the first engineering classification systems in 1928. It was intended specifically for use in highway construction, and still survives as the American Association of State Highway and Transportation Officials (AASHTO) system (AASHTO, 1993). It rates soils according to their suitability for support of roadway pavements, and continues to be widely used on such projects.

The AASHTO system uses both grain-size distribution and Atterberg limits data to assign a *group classification* and a *group index* to the soil. The group classification ranges from A-1 (best soils) to A-8 (worst soils). Group index values near 0 indicate good soils, while values of 20 or more indicate very poor soils. However, it is important to remember that a soil that is "good" for use as a highway subgrade might be "very poor" for some other purpose.

This system considers only that portion of the soil that passes through a 3-inch sieve. If any plus 3-inch material is present, its percentage by weight should be recorded and noted with the classification.

Use Table 5.1 to determine the group classification. Begin on the left side with A-1-a soils and check each of the criteria. If all have been met, then this is the group classification. If any criterion is not met, step to the right and repeat the process, continuing until all the criteria have been satisfied. Do not begin at the middle of the chart.

TABLE 5.1 AASHTO SOIL CLASSIFICATION SYSTEM (AASHTO, 1993)

General Classification	Granular Materials (35 percent or less passing No. 200 sieve)[c]										Silt-Clay Materials (more than 35 percent passing No. 200 sieve)				Highly Organic
Group Classification	A-1		A-3	A-2				A-4	A-5	A-6	A-7		A-8
	A-1-a	A-1-b		A-2-4	A-2-5	A-2-6	A-2-7				A-7-5 / A-7-6	
Sieve analysis percent passing:												
#10	≤ 50											
#40	≤ 30	≤ 50	≥ 51									
#200	≤ 15	≤25	≤ 10	≤ 35	≤ 35	≤ 35	≤ 35	≥ 36	≥ 36	≥ 36	≥ 36	
Characteristics of fraction passing #40:											[b]	
Liquid limit				≤ 40	≥ 41	≤ 40	≥ 41	≤ 40	≥ 41	≤ 40	≥ 41	
Plasticity index	≤ 6		NP[a]	≤ 10	≤ 10	≥ 11	≥ 11	≤ 10	≤ 10	≥ 11	≥ 11	
Usual types of significant constituent materials	Stone fragments; gravel and sand	Fine sand		Silty or clayey gravel and sand				Silty soils		Clayey soils		Peat or muck
General rating as subgrade	Excellent to good								Fair to poor			Unsuitable

[a] NP indicates the soil is non-plastic (i.e., it has no clay)

[b] The plasticity index of A-7-5 soils is ≤ liquid limit - 30. For A-7-6 soils, it is > than the liquid limit - 30

[c] The placement of A-3 before A-2 is necessary for the "left-to-right elimination process" and does not indicate superiority of A-3 over A-2

Compute the group index using Equation 5.1:

$$\text{Group Index} = (F - 35)[0.2 + 0.005(w_L - 40)] + 0.01(F - 15)(I_P - 10) \tag{5.1}$$

where:

F = fines content (portion passing #200 sieve), expressed as a percentage

w_L = liquid limit

I_P = plasticity index

When evaluating the group index for A-2-6 or A-2-7 soils, use only the second term in Equation 5.1. For all soils, express the group index as a whole number. Computed group index values of less than zero should be reported as zero.

Finally, express the AASHTO soil classification as the group classification followed by the group index in parentheses. For example, a soil with a group classification of A-4 and a group index of 20 would be reported as A-4(20).

Example 5.2

The natural soils along a proposed highway alignment have a grain-size distribution as described by curve A in Figure 5.2, a liquid limit of 44, and a plastic limit of 21. Determine the AASHTO soil classification and rate its suitability for pavement support.

Solution

Referring to the sieve sizes in Table 4.7:
>   Passing #10 sieve (2.00 mm) = 92%
>   Passing #40 sieve (0.425 mm) = 74%
>   Passing #200 sieve (0.075 mm) = 54%

$$I_p = w_L - w_p = 44 - 21 = 23$$

Consider group classifications from left to right:

A-1-a through A-2-7:	No	(>35% passing #200)
A-4:	No	(liquid limit too high)
A-5:	No	(plasticity index too high)
A-6:	No	(liquid limit too high)
A-7:	Yes	(all criteria met)

$I_p > w_L - 30$, so group classification is A-7-6 (per footnote b on Table 5.1)

Compute group index:

$$
\begin{aligned}
\text{Group Index} &= (F - 35)[0.2 + 0.005(w_L - 40)] + 0.01(F - 15)(I_p - 10) \\
&= (54 - 35)[0.2 + 0.005(44 - 40)] + 0.01(54 - 15)(23 - 10) \\
&= 9
\end{aligned}
$$

Final result: **A-7-6 (9)**, which would make a poor subgrade   ⇐ *Answer*

5.3 UNIFIED SOIL CLASSIFICATION SYSTEM (USCS)

Arthur Casagrande developed a new engineering soil classification system for the United States Army during World War II (Casagrande, 1948). Since then, it has been updated and is now standardized in ASTM D2487 as the *Unified Soil Classification System* (USCS). Unlike the AASHTO system, the USCS is not limited to any particular kind of project; it is an all-purpose system and has become the most common soil classification system among geotechnical engineers.

In its original form, the classification consisted only of a two- or four-letter *group symbol*. Later, the system was enhanced by the addition of several *group names* for each group symbol. For example, a typical USCS classification would be:

SM — Silty sand with gravel

where "SM" is the group symbol and "Silty sand with gravel" is the group name.

The position of a soil type in the group name indicates it relative importance, as follows:

Noun = Primary component
Adjective = Secondary component (or further explanation of primary component)
"with ..." = Tertiary component

For example, a *clayey sand with gravel* has sand as the most important component, clay as the second most important, and gravel as the third most important. If very little of a soil type is present, then it is not included in the group name at all. For example, a *clayey sand* is similar to the soil just described, except it has less than 15 percent gravel.

Initial Classification

To use the Unified Soil Classification System, begin with an initial classification as follows:

1. Determine if the soil is *highly organic*. Such soils have the following characteristics:

   - Composed primarily of organic material
   - Dark brown, dark gray, or black color
   - Organic odor, especially when wet, and
   - Soft consistency

   In addition, fibrous material (remnants of stems, leaves, roots, etc) is often evident.
   If the soil does not have these characteristics (and the vast majority do not), then go to Step 2. However, if it does, then classify it as follows:

   Group symbol   Pt
   Group name   Peat

   This completes the unified classification for highly organic soils. These soils are very problematic because of their high compressibility and low strength, so the group symbol Pt on a boring log is a red flag to geotechnical engineers.
2. Conduct a sieve analysis to determine the grain-size distribution curve. For an informal classification, a grain-size distribution curve based on a visual inspection may suffice.
3. Based on the grain-size distribution curve, determine the percent by weight passing the 3-inch, #4, and #200 sieves, then compute the percentage by weight of gravel, sand, and fines using the definitions in Table 4.6.
4. If 100 percent of the sample passes the 3-inch sieve, go to Step 5. If not, base the classification on the part that passes this sieve (usually called the "minus 3-inch

fraction"). To do so, adjust the percentages of gravel, sand, and fines using a procedure similar to that in Example 5.1. Then, perform the classification based on these modified percentages, and note the percentage of cobbles and/or boulders and the maximum particle size with the final classification. For example, if 20 percent of the soil is cobbles, some as large as 8 inches, the USCS classification (after going through the rest of the procedure) might be:

SW — Well-graded sand with gravel and 20% cobbles, max 8 inches

5. If 5 percent or more of the soil passes the #200 sieve, then conduct Atterberg limits tests to determine the liquid and plastic limits.
6. If the soil is fine-grained (i.e., ≥ 50 percent passes the #200 sieve), follow the directions for fine-grained soils. If the soil is coarse-grained (i.e., < 50 percent passes the #200 sieve), follow the directions for coarse-grained soils.

Classification of Fine-Grained Soils

Fine-grained soils are those that have at least 50 percent passing the #200 sieve. Thus, these soils are primarily silt and/or clay. Casagrande developed the *plasticity chart* in Figure 5.3 to assist in the classification of these soils. This chart uses the Atterberg limits to distinguish between clays and silts. Although most fine-grained soils contain both clay and silt, and possibly sand and gravel as well, those that plot above the *A-line* are classified as clays, while those below this line are silts.

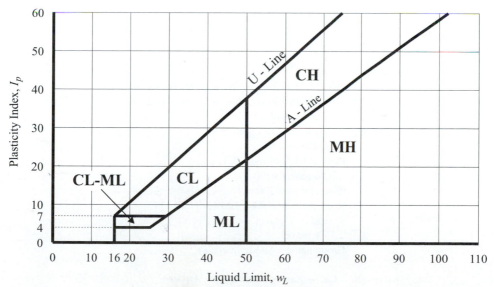

Figure 5.3 Plasticity chart (ASTM D2487). The "A-line" separates silts from clays, while the "U-line" represents the upper limit of recorded test results. Data that plot above the U-line are probably in error. Note how the vertical axis is the plasticity index, not the plastic limit. Soils identified as "non-plastic" (NP) are classified as ML.

We use the plasticity chart to determine the group symbol for fine-grained soils. It usually consists of two letters, which are interpreted as follows:

First Letter		**Second Letter**	
M	Predominantly silt[1]	L	Low plasticity
C	Predominantly clay	H	High plasticity
O	Organic		

CL soils are known as *lean clays*, while CH soils are *fat clays*. The corresponding terms for ML and MH soils are *silt* and *elastic silt*, respectively, even though the stress–strain behavior of MH soils is no more elastic than any other soil.

In this context, an *organic* soil is one that has a noteworthy percentage of organic matter, yet consists primarily of inorganic material. This differs from a *highly organic* soil, as described earlier (group symbol Pt), which contains much more organic material. With experience, one can usually determine whether a fine-grained soil is inorganic (M or C) or organic (O) by visual inspection. Alternatively, we could perform two liquid limit tests, one on an unmodified sample from the field, and another on a sample that is first oven-dried. The drying process alters any organics that might be present, and thus changes the liquid limit. If the liquid limit after oven drying is less than 75 percent of the original value, then the soil is considered to be organic. If not, then it is inorganic.

To classify inorganic fine-grained soils, use the flow chart in Figure 5.4 and the plasticity chart in Figure 5.3. For organic fine-grained soils, use Figures 5.5 and 5.3.

Example 5.3

Classify the inorganic soil from Example 5.2 using the Unified Soil Classification System.

Solution

Initial classification
    100% passes 3-inch sieve, so no adjustments are necessary
    $\geq$ 50% passes #200 sieve, so the soil is fine-grained
Classification of fine-grained soil
    Soil is inorganic, so use Figure 5.4
        Liquid limit < 50
        Plots as CL on Figure 5.3
        < 70% passes #200
        % sand = #4–#200 = 97% - 54% = 43%
        % gravel = 3-in–#4 = 100% - 97% = 3%
        % sand > % gravel
        < 15% gravel
        Group name = Sandy lean clay

Final result: **CL — Sandy lean clay** ⟸ *Answer*

---

[1] M is the first letter in *mo* and *mjala*, the Swedish words for silt and flour.

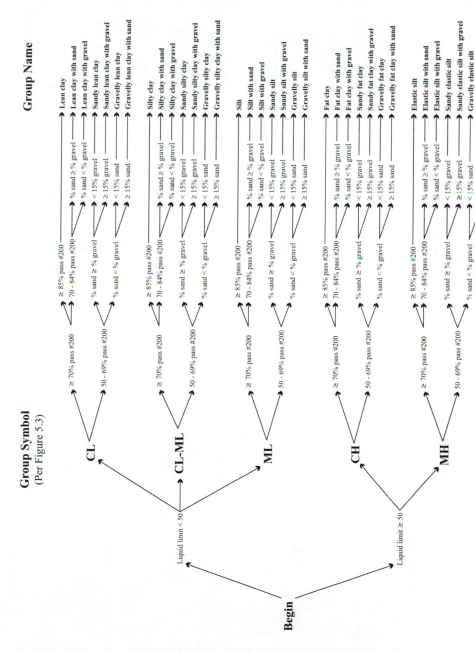

Figure 5.4 Flow chart for classification of inorganic fine-grained soils (≥50% passing #200 sieve) (Adapted from ASTM D2487).

Group Symbol

Group Name

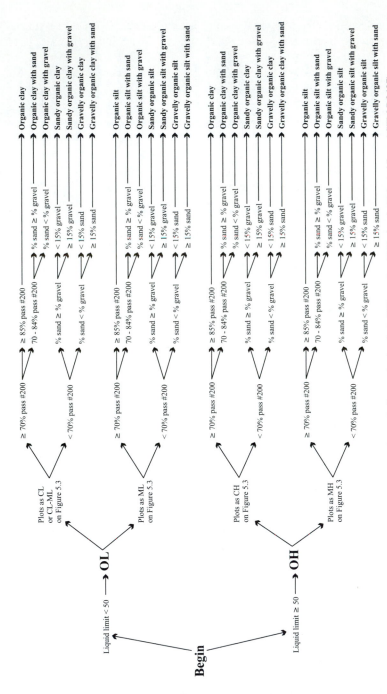

Figure 5.5 Flow chart for classification of organic fine-grained soils (≥50% passing #200 sieve) (Adapted from ASTM D2487).

Classification of Coarse-Grained Soils

Coarse-grained soils are those that have less than 50 percent passing the #200 sieve. Thus, these soils are primarily sand and/or gravel. The group symbols for coarse-grained soils are:

First Letter		**Second Letter**	
S	Predominantly sand	P	Poorly graded
G	Predominantly gravel	W	Well graded
		M	Silty
		C	Clayey

As discussed in Chapter 4, poorly graded soils are those with a narrow range of particle sizes (i.e., a steep grain-size distribution curve), while well-graded soils have a wide range of particle sizes (i.e., a flatter grain-size distribution curve). In this context, silty (M) or clayey (C) indicate a large percentage of silt or clay in a coarse-grained soil.

To classify coarse-grained soils, use the flow chart in Figure 5.6. By inspection of this chart, we see that both sands and gravels are divided into three categories depending on the percentage of fines (fines = percent passing the #200 sieve):

If < 5 percent fines	Use two-letter group symbol to describe gradation (well or poorly graded)
If 5 - 12 percent fines	Use four-letter group symbol to describe both gradation and type of fines
If > 12 percent fines	Use two-letter group symbol to describe type of fines (silt or clay)

Example 5.4

Determine the unified soil classification for the inorganic soil B in Figure 5.2.

Solution

Initial classification
    100% passes 3-in sieve, so no adjustments are necessary
    < 50% passes #200 sieve, so soil is coarse-grained

Classification of coarse-grained soils
    % gravel = 3-in–#4 = 100% - 100% = 0%
    % sand = #4–#200 = 100% - 4% = 96%
    % fines = #200 = 4%
    $D_{10} = 0.10$ mm
    $D_{30} = 0.17$ mm
    $D_{60} = 0.40$ mm
    $C_u = D_{60}/D_{10} = 0.40/0.10 = 4.0$
    $C_c = D_{30}^2/(D_{10}D_{60}) = 0.17^2/[(0.10)(0.40)] = 0.72$

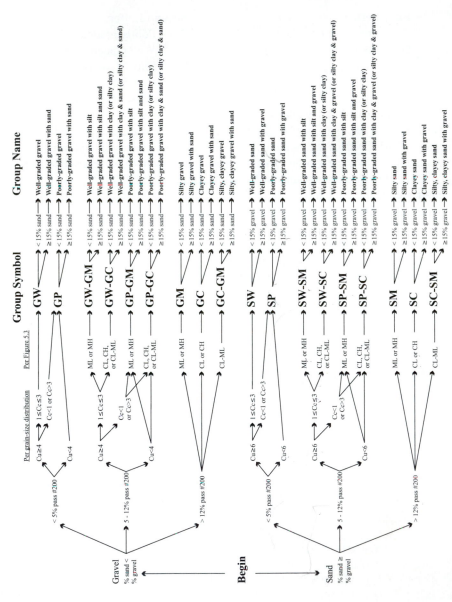

Figure 5.6 Flow chart for classification of coarse-grained soils (< 50% passing #200 sieve) (Adapted from ASTM D2487). C_u and C_c are the coefficients of uniformity and curvature as defined in Chapter 4. The alternative endings for some group names (shown in parenthesis) are for soils that plot as CL-ML on Figure 5.3.

Using Figure 5.6
　　% sand > % gravel
　　< 5% fines
　　$C_u < 6$ and $C_c < 1$, so group symbol is SP
　　< 15% gravel

Final results:　　　**SP — Poorly-graded sand**　　⟸ *Answer*

Example 5.5

The inorganic soil C in Figure 5.2 has a liquid limit of 30 and an plastic limit of 25. Determine its unified soil classification.

Solution

Initial classification
　　100% passes 3-in sieve, so no adjustments are necessary
　　< 50% passes #200 sieve, so soil is coarse-grained

Classification of coarse-grained soils
　　% gravel = 3-in–#4 = 100% - 80% = 20%
　　% sand = #4–#200 = 80% - 10% = 70%
　　% fines = #200 = 10%
　　$D_{10} = 0.075$ mm
　　$D_{30} = 0.39$ mm
　　$D_{60} = 1.7$ mm
　　$C_u = D_{60}/D_{10} = 1.7/0.075 = 23$
　　$C_c = D_{30}^2/(D_{10}\,D_{60}) = 0.39^2/[(0.075)(1.7)] = 1.2$

Using Figure 5.6
　　% sand > % gravel
　　5 - 12% fines
　　$C_u > 6$ and $1 \le C_c \le 3$ (∴ soil is well graded)
　　$I_p = w_L - w_P = 30 - 25 = 5$
　　Fines plot as ML on Figure 5.3, so group symbol is SW-SM
　　> 15% gravel

Final result:　　　**SW-SM — Well-graded sand with silt and gravel**　　⟸ *Answer*

Classification of Borderline Soils

Sometimes a soil classification is very close to the dividing line between two different group symbols. In such cases, it is acceptable to use both symbols in the classification, with the "correct" symbol first, followed by the "almost correct" symbol. For example, a sand–clay combination with slightly less than 50 percent fines could be identified as SC/CL.

Soil Assessment Based on USCS Classification

Geotechnical engineers have many ways to assess the suitability of a soil for particular purposes. For example, if a soil is being considered for use as an "impervious" cap over a sanitary landfill, we would perform hydraulic conductivity tests as described in Chapter 7 to determine how easily water flows through it. Soils that restrict the flow of water are best for landfill caps. However, before we perform these specialized tests, geotechnical engineers assess a soil based on its classification. For example, an SW soil would pass water very easily, and thus would be rejected for the landfill cap even without a hydraulic conductivity test. Table 5.2 presents general soil properties based on the unified group symbol, and may be used to assist in such assessments.

5.4 VISUAL–MANUAL SOIL CLASSIFICATION

Although geotechnical engineers routinely perform sieve, hydrometer, and Atterberg limits tests, it is not cost-effective to do so on every sample obtained from the field, so the remaining samples must be classified without the benefit of laboratory test data. The following methods can assist in this process:

- The #200 sieve approximately corresponds to the smallest particles one can see with the unaided eye. Thus, individual fine sand particles can be distinguished, but individual silt particles cannot. In addition, particles larger than the #200 sieve have a gritty texture, while those smaller are pastey.
- Clay and silt particles often clump together, and may look like sand. These clumps will dissolve when wetted. Therefore, when in doubt, be sure to wet the soil before classifying it.
- Clays (CL and CH) have a higher dry strength, but lose this strength when wetted. In addition, moist clays can be rolled between the fingers into 5-mm diameter threads.
- Silts (ML and MH) have a lower dry strength, and are much more difficult to roll into threads.
- Cementing agents, such as calcium carbonate, are sometimes present in sandy or silty soils. These agents can give the soil a high dry strength, even if no clay is present. Again, the key is to wet the sample before classifying it. Cemented soils will retain their dry strength, while clayey soils will soften when wetted.

5.5 SUPPLEMENTAL SOIL CLASSIFICATIONS

Regardless of the textural classification system being used (AASHTO, USCS, etc.), geotechnical engineers often add certain supplementary classifications. Some of these have been standardized in ASTM D2488, but informal classifications are used at least as often as the standard terms. Either way, they provide important information on the in-situ soil conditions.

Table 5.2 ASSESSMENT OF SOIL PROPERTIES BASED ON GROUP SYMBOL (Adapted from Sowers, 1979)

Group Symbol	Compaction Characteristics	Compressibility and Expansion	Drainage and Hydraulic Conductivity	Value as a Fill Material	Value as a Pavement Subgrade When Not Subject to Frost	Value as a Base Course for Pavement
GW	Good	Almost none	Good drainage; pervious	Very stable	Excellent	Good
GP	Good	Almost none	Good drainage; pervious	Reasonably stable	Excellent to good	Poor to fair
GM	Good	Slight	Poor drainage; semipervious	Reasonably stable	Excellent to good	Fair to poor
GC	Good to fair	Slight	Poor drainage; semipervious	Reasonably stable	Good	Good to fair, not suitable if subject to frost
SW	Good	Almost none	Good drainage; pervious	Very stable	Good	Fair to poor
SP	Good	Almost none	Good drainage; pervious	Reasonably stable when dense	Good to fair	Poor
SM	Good	Slight	Poor drainage; impervious	Reasonably stable when dense	Good to fair	Poor
SC	Good to fair	Slight to medium	Poor drainage; impervious	Reasonably stable	Good to fair	Fair to poor, not suitable if subject to frost
ML	Good to poor	Slight to medium	Poor drainage; impervious	Fair stability, good compaction required	Fair to poor	Not suitable
CL	Good to fair	Medium	No drainage; impervious	Good stability	Fair to poor	Not suitable
OL	Fair to poor	Medium to high	Poor drainage; impervious	Unstable, should not be used	Poor, not suitable	Not suitable
MH	Fair to poor	High	Poor drainage; impervious	Fair to poor stability, good compaction required	Poor	Not suitable
CH	Fair to poor	Very high	No drainage; impervious	Fair stability, expands, weakens, shrinks, cracks	Poor to very poor	Not suitable
OH	Fair to poor	High	No drainage; impervious	Unstable, should not be used	Very poor	Not suitable
Pt	Not suitable	Very high	Fair to poor drainage	Should not be used	Not suitable	Not suitable

Moisture

Knowledge of the moisture condition of a soil can be very useful. Therefore, *moisture content tests*, as discussed in Chapter 4, are among the most common soil tests. In addition, engineers often give a qualitative assessment of soil moisture using descriptors such as those in Table 5.3.

TABLE 5.3 MOISTURE CLASSIFICATION

Classification	Description
Dry	Dusty; dry to the touch
Slightly moist	Some moisture, but still has a dry appearance
Moist	Damp, but no visible water
Very moist	Enough moisture to wet the hands
Wet	Saturated; visible free water

Color

The soil color can vary as its moisture content changes, so it is a less reliable classification. Nevertheless, it is useful as a common supplementary soil classification. Although standardized color description systems, such as the Munsell Color Charts, are available, they express colors using an awkward notation (i.e., 10 YR 5/3). Therefore, most engineers just use common color names, such as brown, tan, gray-brown, etc. Sometimes individual firms or agencies standardize these names, but there is no widely accepted standard.

Consistency

The *consistency* of a soil describes its stiffness, and is a very useful supplementary classification. For example, a hard CH soil is quite different from a very soft CH. Consistency depends on the soil type, moisture content, unit weight, and other factors, and may change in the field with time, especially if the soil becomes wet. Tables 5.4 and 5.5 present classifications of soil consistency, along with qualitative and quantitative descriptions.

When classifying the consistency of coarse grained soils based on standard penetration test data, it is especially important to apply the overburden correction described in Section 3.9, thus obtaining $(N_1)_{60}$.

TABLE 5.4 CONSISTENCY CLASSIFICATION FOR FINE-GRAINED SOILS (Terzaghi, Peck, and Mesri, 1996 and U.S. Navy, 1982; Adapted by permission of John Wiley and Sons, Inc.)

Classification	Description	SPT N_{60} value	Undrained Shear Strength, s_u	
			(kPa)	(lb/ft^2)
Very soft	Thumb penetrates easily; extrudes between fingers when squeezed	< 2	<12	< 250
Soft	Thumb will penetrate soil about 25 mm; molds with light finger pressure	2 - 4	12 - 25	250 - 500
Medium	Thumb will penetrate about 6 mm with moderate effort; molds with strong finger pressure	4 - 8	25 - 50	500 - 1000
Stiff	Thumb indents easily, and will penetrate 12 mm with great effort	8 - 15	50 -100	1000 - 2000
Hard	Thumb will not indent soil, but thumbnail readily indents it	15 - 30	100 - 200	2000 - 4000
Very hard	Thumbnail will not indent soil or will indent it only with difficulty	> 30	> 200	> 4000

[a] The N-value is defined in Chapter 4.

[b] The undrained shear strength is defined in Chapter 13, and is half of the unconfined compressive strength.

TABLE 5.5 CONSISTENCY CLASSIFICATION FOR COARSE-GRAINED SOILS (U.S. Navy, 1982, and Lambe and Whitman, 1969; Adapted by permission of John Wiley and Sons, Inc.)

Classification	Description	SPT $(N_1)_{60}$ value[a]	D_r Relative Density [b]
Very loose	Easy to penetrated with a 12-mm diameter rod pushed by hand	< 4	0 - 15
Loose	Difficult to penetrate with a 12-mm diameter rod pushed by hand	4 - 10	15 - 35
Medium dense	Easy to penetrate 300 mm with a 12-mm diameter rod driven with a 2.3-kg (5-lb) hammer	10 - 17	35 - 65
Dense	Difficult to penetrate 300 mm with a 12-mm diameter rod driven with a 2.3-kg (5-lb) hammer	17 - 32	65 - 85
Very dense	Penetrated only about 150 mm with a 12-mm diameter rod driven with a 2.3-kg (5-lb) hammer	> 32	85 - 100

[a] These values are for sandy soils. If some fine gravel is present, use two-thirds of the field value. If significant quantities of coarse gravel are present, do not use this table.

[b] If CPT data is available, use Equation 4.25 to compute D_r

Cementation

Some soils are cemented with certain chemicals, such as calcium carbonate ($CaCO_3$) or iron oxide (Fe_2O_3). The presence of calcium carbonate can be determined by applying a small amount of hydrochloric acid (HCl) and noting the reaction. A bubbling action indicates the presence of $CaCO_3$. A lack of bubbling indicates some other cementing agent. Iron oxide gives the soil a red-orange tint, similar to rusty steel.

Table 5.6 presents a classification of soil cementation.

TABLE 5.6 CLASSIFICATION OF SOIL CEMENTATION (After ASTM D2488)

Classification	Description
Weak	Crumbles or breaks with handling or little finger pressure
Moderate	Crumbles or breaks with considerable finger pressure
Strong	Will not crumble or break with finger pressure

Structure

Soil particles can be assembled into many different structures. Often these need to be noted in the classification of undisturbed samples as follows (ASTM D2488):

- *Stratified* — Alternating layers of varying material or color with layers at least 6 mm thick
- *Laminated* — Alternating layers of varying material or color with the layers less than 6 mm thick
- *Fissured* — Breaks along definite planes of fracture with little resistance to fracturing
- *Slickensided* — Fracture planes appear polished or glossy, sometimes striated (liner markings showing evidence of past movement)
- *Blocky* — Cohesive soil that can be broken down into small angular lumps which resist further breakdown
- *Lensed* — Inclusions of small pockets of different soils, such as small lenses of sand scattered throughout a mass of clay
- *Homogeneous* — Same color and appearance throughout

5.6 APPLICABILITY AND LIMITATIONS

Standardized soil classification systems are very valuable tools that help geotechnical engineers identify soils and make preliminary assessments of their engineering behavior. However, they also have limitations. Casagrande (1948) said:

> "It is not possible to classify all soils into a relatively small number of groups such that the relation of each soil to the many divergent problems of applied soil mechanics will be adequately presented."

Do not expect too much from a soil classification system. Identifying the proper classification is a good start, but we still need to use other test results, an understanding of soil behavior, engineering judgment, and experience.

SUMMARY

Major Points

1. Standardized soil classification systems are an important part of the language of geotechnical engineering. They help us identify soils and communicate important characteristics to other engineers. In addition, a classification helps us develop preliminary assessments of a soil's behavior.
2. The USDA soil classification system classifies soils based on their grain size distribution, and is used in soil survey reports.
3. The AASHTO soil classification system uses both grain size and Atterberg limits data, and is used to assess soils for use as highway subgrades.
4. The Unified Soil Classification System is an all-purpose system based on grain size and Atterberg limits data.
5. Supplemental soil classifications assist in further describing important characteristics not addressed by the USDA, AASHTO, or unified systems.

Vocabulary

AASHTO soil classification system	group classification	organic soil
blocky	group index	plasticity chart
cementation	group name	slickensided
coarse-grained soil	group symbol	stratified
consistency	highly organic soil	structure
fine-grained soil	homogeneous	Unified soil classification system
fissured	laminated	USDA soil classification system
	lensed	

COMPREHENSIVE QUESTIONS AND PRACTICE PROBLEMS

5.1   What kinds of soil are contained in *loam* as used in the USDA classification system?

5.2   A clean well-graded sand would rate very high in the AASHTO soil classification system, and thus be considered a good soil for use as a highway subgrade. However, it would be a very poor choice for certain other applications. Give one example of a situation where this soil would not be a good choice.

5.3   Is it ever necessary to conduct a hydrometer analysis when classifying soils according to the USCS, or is a sieve analysis always sufficient?

5.4   Explain the difference between a silty sand and a sandy silt in the Unified Soil Classification System.

5.5   How would you determine if a soil was a silty sand or a sandy silt if:
a. Laboratory test equipment was available?
b. Laboratory test equipment was not available?

5.6   A sanitary landfill project needs an import soil for use as an impervious cap over the refuse. The following soils are available for this purpose:

Soil 1: GC
Soil 2: SC
Soil 3: ML
Soil 4: CL

The design engineer desires a soil with a high clay content because it will help keep water out of the landfill. Which of the available soils appears to be most promising? Why?

The following information, and the grain-size distribution data in Figure 5.2, are to be used in Problems 5.7 through 5.11. All of these soils are inorganic.

Soil Sample Identification	Liquid Limit	Plastic Limit
D	60	33
E	40	30
F	30	26
G	Non-plastic	

5.7   Determine the USDA classification for soils D through G

5.8   Determine the AASHTO group classification and group index for soils D through G.

5.9   According to the AASHTO classifications, which of the soils D through G would make the best highway subgrade?

5.10 Determine the USCS group symbol and group name for soils D through G.

5.11 Which of the following USCS soils would probably receive the best rating for use as a highway subgrade per the AASHTO classification: CL, SM, ML, or SW? Why?

5.12 Soil A in Figure 4.13 is organic, but not *highly* organic, and has a plastic limit of 21 and a liquid limit of 40. Determine its unified group symbol and group name.

5.13 Determine the unified group symbol and group name for the inorganic soil E in Figure 4.13. The fines are non-plastic.

6

Excavation, Grading, and Compacted Fill

The wise architect should wash the excavations with the five products of the cow.

The excavation should be made at night and the bricks should be laid in the daytime.

The chief architect should distinguish the two varieties of bricks, namely stony brick and pure brick, and their three genders, and should fix the male bricks in the temples of male deities.

Manasara, a sixth-century architect (Acharya, 1980)

Most civil engineering projects include some *earthwork*, which is the process of changing the configuration of the ground surface. When soil or rock is removed, we have made a *cut* or *excavation*; when it is added, we have made a *fill* or *embankment*. These changes make the site more suitable for the proposed development.

For example, virtually all highway and railroad projects require earthwork to create smooth grades and alignments and to provide proper surface drainage. The value of cuts and fills in highway projects is especially evident in mountainous areas where older two-lane roads were built using as little earthwork as possible, resulting in steep grades and sharp turns. When a modern multi-lane interstate highway is built through the same mountains, design engineers use more generous cuts and fills, thus producing smoother alignments and gentler grades. Figure 6.1 shows a deep cut made during construction of such a highway.

Figure 6.1 This four-lane highway was created by making a cut on the left side of the photo and in the background, and a fill in the foreground.

Earthwork also is important in building construction, especially in hilly terrain. Cuts and fills are used to create level pads for the buildings, parking lots, and related areas, as shown in Figure 6.2, thus using the land area to better advantage. Even building projects in areas with naturally level terrain often require minor earthwork to provide proper surface drainage.

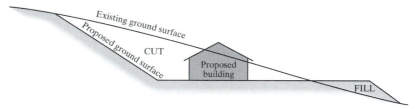

Figure 6.2 Typical cross-section through a proposed building site showing locations of proposed cuts and fills.

Large residential developments in hilly areas often include extensive earthwork, sometimes millions of cubic meters, and can create hundreds of residential lots. Such projects provide level building pads and smooth street alignments, and allow more houses to be built in a given area. They also provide for a safer development, because a properly graded tract is less prone to landslides, excessive settlement, flooding, and other problems.

Finally, the most impressive earthwork projects are earth dams, such as the one in Figure 6.3. They require very large volumes of carefully placed fill, yet are much less expensive than concrete dams because such fills can be placed very economically. When properly designed and constructed, earth dams also are very safe.

Figure 6.3 Oroville Dam on the Feather River in California. This is a 225 m (740 ft) tall earth dam with a volume of 61,000,000 m^3(80,000,000 yd^3) (California Department of Water Resources).

6.1 EARTHWORK CONSTRUCTION OBJECTIVES

The most fundamental objective of earthwork construction is to change the ground surface from some initial configuration, typically described by a topographic map, to some final configuration, as described on a new topographic map known as a *grading plan*. These changes are often necessary to properly accommodate the proposed construction, and to maintain proper surface drainage, which is important to the long-term performance of cuts and fills.

Another important requirement is that earthwork must not create slope stability problems (i.e., landslides). This is especially important in hilly and mountainous areas, since the level building pads or road alignments created by the earthwork are possible only because nearby areas are made steeper and thus less stable.

Compacted fills have additional requirements, including:

1. Fills must have sufficient shear strength to support both their own weight and external loads, such as foundations or vehicles. Lack of sufficient strength can produce landslides, bearing capacity failures, and rutting of pavements.
2. Fills must be sufficiently stiff to avoid excessive settlement. Soft fills permit

foundations to settle excessively, thus damaging buildings, and often produce undesirable changes in surface drainage patterns.

3. Fills must continue to satisfy requirements 1 and 2, even if they become wet.

4. Some fills, such as the core of earth dams or liners for sanitary landfills, must have a sufficiently low hydraulic conductivity[1] to restrict the flow of water. Others, such as aggregate base material below pavements, must have a high hydraulic conductivity to drain water away from critical areas.

5. In areas prone to frost heave (a heaving that occurs when the ground freezes), it is sometimes desirable to have fills made of soils that are not frost-susceptible. Chapter 18 discusses this problem.

Fills for residential, commercial, and industrial projects built before the 1960s often did not fully satisfy these requirements. As a result, these buildings often experienced excessive settlements, and slope failures such as landslides were common. The term "fill" became associated with shoddy construction. However, there is nothing implicitly bad about fills; only poorly constructed fills.

During the 1960s, building codes began to include stricter requirements for fills, and required builders to hire geotechnical engineers. These changes produced significantly better fills, and the problems were significantly diminished. Today, fills built in accordance with accepted standards of practice are at least as good as cuts.

The new grading techniques included many changes, most notably stricter requirements for soil compaction. This means all fills had to be packed tightly using heavy equipment.

6.2 CONSTRUCTION METHODS AND EQUIPMENT

Design engineers need to have a good understanding of construction methods and equipment, because the way something is built often has a significant impact on how it must be designed. This is especially true for geotechnical engineers, because we are usually involved in both the design and construction phases of a project.

Historic Methods

Mankind has been making cuts and fills for thousands of years, often with primitive tools and large numbers of slaves and animals. A number of impressive projects were built long before the advent of modern construction equipment by hauling soil in baskets and carts. Among the most impressive were the earth dams built for irrigation purposes in India and Sri Lanka. For example, Kalaweva reservoir in Sri Lanka was created by a 19 km long, 21 m tall earth dam built in AD 459. Another even larger dam would have created a huge

---

[1] Hydraulic conductivity is a measure of the soil's ability to transmit water. A high value indicates water passes through the soil very easily, while a low value indicates the opposite. Chapter 7 discusses this topic in much more detail.

reservoir, However, this dam was never used because the canal that was to feed the reservoir was built running uphill, suggesting their expertise in earthwork had far surpassed their skills in surveying (Schuyler, 1905).

The first significant advancements came in the nineteenth century with the introduction of steam power. This lead to mechanized earthmoving equipment and began the era of efficient large-scale earthmoving.

Steam-powered equipment was used to build the Panama Canal, which was the largest earthmoving project of its day. The Culebra Cut (now known as the Gaillard Cut), shown in Figure 6.4, was the most difficult part of the construction due to its height and the very difficult geologic conditions. The cut passes through a formation known as the Cucaracha (Cockroach) Shale, a weak sedimentary rock. Unfortunately, the first efforts to build the canal were orchestrated by the arrogant Ferdinand de Lesseps, who despised engineers and began construction without the benefit of a geologic study (Kerisel, 1987). This lead to a misguided attempt to build a sea-level canal, which would have required a 109 m cut at Culebra. Massive landslides (created by the construction activities), yellow fever, and financial insolvency eventually halted the first attempt to build the canal.

Figure 6.4 Construction of the Culebra Cut for the Panama Canal, 1907. The steam-powered excavators placed loads of soil and rock into railroad cars that hauled them away (Library of Congress).

When construction resumed with the benefit of geologic studies, locks had been added to raise the canal 25 m above sea level. Even so, huge landslides continued to be a problem, resulting in enormous excavation quantities. The Culebra Cut alone required 75,000,000 m^3 (98,000,000 yd^3), a massive quantity even by today's standards. The canal was completed in 1914, and stands as both a great civil engineering achievement and a dramatic case study demonstrating the importance of geology.

Hydraulic Fills

During the early decades of the twentieth century, many large earthwork projects were built using a technique called *hydraulic filling*. This method consisted of mixing the soil with large quantities of water, conveying the mixture to the construction site through pipes and flumes, then depositing it at the desired locations. The soil settled in place and the excess water was directed away. No compaction equipment was used. Figure 6.5 shows a hydraulic fill dam under construction.

This technique was popular between 1900 and 1940, especially for earthfill dams, because earthmoving equipment was too small and underpowered for such large projects. Unfortunately, the quality of such fills was poor and they often experienced large settlements and landslides.

The last significant hydraulic fill in the United States was at Ft. Peck Dam in Montana. A very large, 3,800,000 m^3 (5,000,000 yd^3) landslide occurred during construction in 1938. It was blamed on the poor quality of the hydraulic fill. The advent of modern earthmoving equipment was already underway, making hydraulic fills obsolete.

Figure 6.5 Construction of San Pablo Dam in California using hydraulic fill techniques, circa 1918. Two fills are being placed simultaneously, one on each side of the photograph, to form the shells. The finer soils flow to the center and form the impervious core. (US Bureau of Reclamation)

Another serious problem with hydraulic fills became evident in 1971 when the Lower San Fernando Dam near Los Angeles failed during a magnitude 6.4 earthquake. This failure was due to liquefaction of the hydraulic fill soils (see further discussion and photograph in Chapters 15 and 20). As a result, several hydraulic fill dams have been rebuilt or replaced to avoid similar failures.

Modern Earthmoving

The twentieth century brought further advances in earthmoving equipment, especially during the period 1920–1965, so today we can move large quantities of earth at a very low cost. To illustrate these improvements, consider the 1914 excavation costs for the Panama Canal, which were about $0.79/yd^3 (Church, 1981). If we adjust this figure for inflation, it translates to about $12/yd^3 in 1998 money. However, the 1998 cost of similar work using modern equipment is less than half of that adjusted cost.

Equipment

A thorough discussion of modern earthmoving equipment is well beyond the scope of this book. However, we will cover some of the principal machines and their applications.

The key development in modern earthmoving equipment was the introduction of the *tractor* or *crawler* shown in Figure 6.6. Its job is to convert engine power into traction, and thus move both itself and other equipment. The first tractors were developed for agricultural and military purposes during the early twentieth century. They were mounted on tracks to allow mobility over very rough terrain. Modern track-mounted equipment is very powerful and mobile, but operates at slow speeds, generally no more than 11 km/hr (7 mi/hr). Wheel-mounted tractors also are available, and have the advantage of greater operating speeds, often in excess of 50 km/hr (30 mi/hr). However, wheel-mounted equipment has less traction and is not as well suited for rough terrain. Thus, both types have a role in modern earthmoving.

a   b

Figure 6.6 Tractors: a) track-mounted or crawler type; b) wheel-mounted type. (Caterpillar Inc.)

Various kinds of equipment can be attached to tractors to produce productive work. One of the most common accessories is the *bulldozer*, which is a movable steel blade attached to the front of a tractor. For example, the tractor in Figure 6.6a is equipped with a bulldozer, and could be used for cutting, moving, spreading, mixing, and other operations.

Another common attachment is a *loader*, as shown in Figure 6.7. It consists of a bucket attached to the front of a tractor and can be used to pick up, transport, and deposit soil. Loaders also may be used for light excavation.

Figure 6.7 Wheel-mounted loader removing soil from a stockpile. (Caterpillar Inc.)

Figure 6.8 A backhoe is a tractor with a loader on the front and a hoe on the back. They are commonly used to dig trenches, as shown here. (Caterpillar Inc.)

Figure 6.9 An excavator is a large hoe mounted on a special rotating chassis. This excavator is digging a hole on the right side of the picture and dumping the spoils in a pile on the left side. Excavators also can dump directly into trucks. (Caterpillar Inc.)

A *hoe* attachment is a bucket used for digging pits or trenches. Special tractors with a loader in the front and a hoe in the back, as shown in Figure 6.8, are called *backhoes*, and are very common on construction sites. An *excavator*, shown in Figure 6.9, is larger and mounted on a special chassis instead of on a tractor.

Often, tractors are described by the names of their attachments. For example, the term *bulldozer* (or simply *'dozer'*) often refers to a tractor with a bulldozer attachment.

Conventional Earthwork

We will use the term *conventional earthwork* to describe the excavation, transport, placement, and compaction of soil or soft rock in areas where equipment can move freely. This process may be divided into several distinct steps, each requiring appropriate equipment and techniques.

Clearing and Grubbing

The first step in most earthwork projects is to remove vegetation, trash, debris, and other undesirable materials from the areas to be cut or filled. Stumps, roots, buried objects, and contaminated soils also need to be removed. Most of these materials would have a detrimental effect on the fill, and must be hauled off the site. The above-ground portion of this work, as shown in Figure 6.10, is called *clearing* and the underground portion is called *grubbing*.

Figure 6.10 Clearing and grubbing in a former orange grove. The trees are being removed and hauled to the dumpster seen in the background.

The time and money required for clearing and grubbing varies from site-to-site. For example, very little work may be required in arid areas because little vegetation is present, while substantial effort is often required in tropical areas because of their dense forests with thick underbrush. In forests of marketable timber, the value of the trees removed during clearing may generate a net profit.

Vegetation may remain in areas that will not be cut or filled, and grading plans often are designed to save desirable vegetation, especially large trees. These areas often require special protection so they are not damaged by the construction activities.

Sometimes clearing and grubbing is accompanied by *stripping*, which consists of removing and storing the topsoil. Such soils are valuable because they contain nutrients for plants. Once the grading is completed, these soils are returned to the top of the graded surface in areas to be landscaped.

Limited quantities of inorganic debris, such as chunks of concrete, bricks, or asphalt pavement, do not need to be hauled away and may be incorporated into the fill so long as they are no larger than about 250 mm (10 in). Those larger than about 100 mm (4 in) in diameter are called *oversize*, and must be spread out in the fill, because stacking them in one place would leave undesirable voids. In addition, oversized objects must not be placed in fills that are to be penetrated with pile foundations, nor in the upper 3 m (10 ft), as they would cause problems with utility excavations.

Excavation

Excavation, which is the removing of soil or rock, can occur at many different locations. Usually, most of the excavation occurs in areas where the proposed ground surface is lower than the existing ground surface. Normally, these excavated materials are then used to make fills at other portions of the project site. Sometimes additional soil is needed, so it becomes necessary to obtain it by excavating at offsite *borrow pits*, which are places where soil is removed to be used as import (the term "borrow" is a misnomer, since we have no intention of bringing the soil back!). Finally, areas to be filled are often prepared by first excavating loose upper soils, thus exposing firm ground on which to place the fill.

Contractors have various kinds of earthmoving equipment that may be used to excavate soils. Loaders are useful when the excavated soil is to be immediately loaded into a truck or conveyor belt, but are generally not used to transport the soil for long distances. *Scrapers* (sometimes called *pans*), shown in Figure 6.11, are much more efficient for most moderate- to large-size projects because they can load, transport, and unload. Figure 6.12 shows the internal operation of a scraper.

Figure 6.11 A scraper transporting a load of soil across a construction site. This scraper is equipped with an elevating mechanism at the front of the bowl to assist in loading soil.

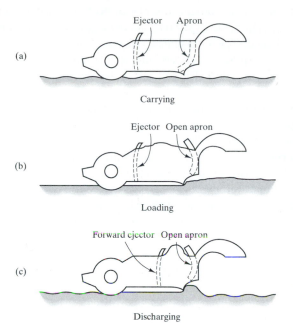

Figure 6.12 Operation of a scraper: a) To load the bowl, the operator opens the apron, moves the ejector plate to the rear, and lowers the front so it digs 100–150 mm (4–6 in) into the ground. As the scraper moves forward, it fills with soil. In harder ground, it may be necessary to push the scraper with a bulldozer during the loading phase. b) To transport soil, the apron is closed and the bowl is lifted. Scrapers usually transport soil at speeds of about 32 km/hr (20 mi/hr). c) Upon reaching the area to be filled, the operator again opens the apron, but lowers the bowl only slightly, leaving it 100–150 mm (4–6 in) above the ground surface. The ejector is moved forward, pushing the soil out and under the bowl, thus depositing a uniform thickness of soil (Wood, 1977).

Sometimes the ground is too hard to be excavated with a scraper or loader. This problem often can be overcome by first loosening it with a *ripper* attached to a tractor, as shown in Figure 6.13. This device consists of one or more teeth that are pressed into the ground, then pulled through to loosen it. The ripping operation can then be followed by excavating equipment.

Figure 6.13 A track-mounted tractor equipped with a ripper. This ripper has two teeth, others have as many as four or five, depending on the type of ground to be ripped and the power of the tractor. The teeth are hydraulically lowered into the ground, then pulled by the tractor.

When the ground is so hard that even rippers do not work, it may become necessary to use *blasting*. This consists of drilling strategically placed holes into the ground and packing them with an explosive. Then, the explosives are detonated, thus loosening the ground and allowing it to be excavated. Special chemicals also are available for use in areas where the noise and vibration from explosives is unacceptable. The contractor pours these liquid chemicals and a catalyst into the holes. The catalyst causes the chemical to solidify and expand, thus fracturing the rock.

Often the *excavatability* (ease of excavation) or *rippability* (ease of ripping) at a site is evident from a visual inspection, and the proper equipment and techniques may be selected accordingly. At questionable sites, measurements of the seismic wave velocity from a seismic refraction survey (see Chapter 3) can assist in selecting the proper equipment. Generally, soil and rock with velocities less than about 500 m/s (1600 ft/s) can be excavated without ripping. Higher velocities can be assessed using Figure 6.14.

Rippability also depends on other factors not reflected in the seismic wave velocity, such as the size and spacing of joints, ripper tooth penetration, presence of boulders, and operator technique.

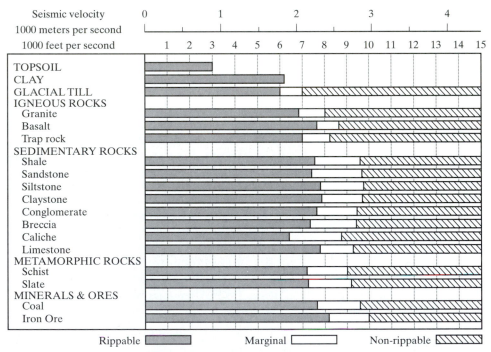

Figure 6.14 Ripping capabilities of a Caterpillar D9N tractor with a No. 9 ripper bar (Caterpillar, 1993). Conditions identified as "non-rippable" generally require blasting.

Because the excavatability of soil and rock can vary widely, even on a single job site, it becomes an important issue in computing payment to the contractor. If the excavation quantities are *unclassified*, the contractor receives the same unit price for all materials, while *classified* excavation sets different prices depending on the ease of excavation. Typically, classified excavation is divided into two categories: *common*, which includes soil, and *rock*. Rock excavation would have a much higher unit price. Unfortunately, there are many intermediate materials that could arguably fit into either classification, so this often becomes a point of contention between engineers and contractors, and has been the source of many lawsuits. To avoid such problems, specifications for classified excavation need to clearly define each category.

Transport and Placement

Although bulldozers and wheel loaders can transport soils for short distances (i.e., less than 100–150 m), they become very uneconomical with longer hauls. For these projects, it becomes necessary to use other equipment.

Scrapers are very efficient at moderate-length hauls, and many earthwork projects fall into this category. They can excavate, transport, and place the soil, as described above, but cannot be used to haul over public highways.

Dump trucks can be used instead of scrapers, especially when the soil is being excavated by loaders, as shown in Figure 6.15. Most dump trucks can travel over public highways, and they move faster than scrapers. However, this method requires more equipment and more operators.

When soil needs to be hauled for longer distances over the highway, contractors often use *wagons*. These are towed by semi tractors and are most cost-effective when the haul distance is long. Upon arriving at the fill site, the wagons are self-unloading, either through the bottom or the back. Large off-road wagons, such as the one in Figure 6.16, also are available.

Figure 6.15 A large dump truck. This one is too large to travel on highways, and is used only when all of its movement can occur off road. Smaller dump trucks also are available, and they can travel on highways. (Caterpillar Inc.)

When large quantities of soil need to be transported to a confined area, such as with earth dams, a system of *conveyor belts* sometimes is an efficient method. One such system is shown in Figure 6.17. Bull Shoals Dam in Arkansas was built using an exceptionally long series of conveyor belts that stretched 7 mi (11 km) from a borrow pit to the dam.

Figure 6.16 A very large off-road wagon. Upon reaching its destination, this wagon will unload the soil through doors on the bottom. Smaller versions of this design can be towed by semi tractors over the highway.

Figure 6.17 Use of conveyor belts to transport soil to the Seven Oaks Dam in California. These belts move soil about 3 km (2 mi) from the borrow site to the dam, then drop it at a convenient location. The loader places it in large dump trucks that haul it a short distance to the construction site.

The selection of equipment to transport soils depends on many factors, most notably the haul distance as described in Table 6.1.

TABLE 6.1 ECONOMICAL HAUL DISTANCES (Adapted from Caterpillar, 1993)

Equipment	Economical Haul Distance	
	(m)	(ft)
Bulldozer	< 100	< 300
Wheel loader	50–150	150–500
Scraper	300–2,500	1,000–8,000
Dump truck	350–6,500	1,100–21,000
Conveyor belt	30–11,000	100–36,000
Wagon	> 3,000	> 10,000

Once the soil arrives at the area to be filled, it must be laid out in thin horizontal *lifts*, typically about 200 mm thick. Each lift must be moisture-conditioned and compacted before the next lift is placed. Thus, the fill is constructed one lift at a time.

Moisture Conditioning

The soil must be at the proper moisture content before it is compacted. Soils that are too wet or too dry will not compact well. Usually the moisture content is not correct and needs to be adjusted accordingly. If the soil is too dry, this is usually done by spraying it with a water truck, as shown in Figure 6.18. A bulldozer or other equipment is then used to mix the soil so the water is uniformly distributed.

Figure 6.18 A water truck spraying a new lift of soil for a fill and preparing it to be compacted.

The grading contractor has a much more difficult problem when the soil is too wet. Mixing it with dryer soil is difficult, especially with clays, because it is very hard to achieve thorough mixing. The result often consists of alternating clumps of wet and dry soil instead of a smooth mixture. The most common technique is to spread the wet soil over a large area and allow the sun to dry it. This works well so long as a rainstorm does not occur!

Compaction

The next step is to compact the lift. *Compaction* is the use of equipment to compress soil into a smaller volume, thus increasing its dry unit weight and improving its engineering properties. The solids and water are virtually incompressible, so compaction produces a reduction in the volume of air, as shown in Figure 6.19.

Although many early fills were built without any special effort to compact them, some engineers recognized the importance of compaction as early as the nineteenth century. Animals were used as compaction "equipment" on some projects, including a team of 115 goats used to compact an earth dam near Santa Fe, New Mexico, in 1893 (Johnson and Sallberg, 1960). Heavy rollers also began to be used. Initially these rollers were pulled by teams of horses, but by 1920 the horses had been replaced with tractors. Further developments during the twentieth century enhanced the capabilities of compaction equipment. Today, a wide variety of effective compaction equipment is available.

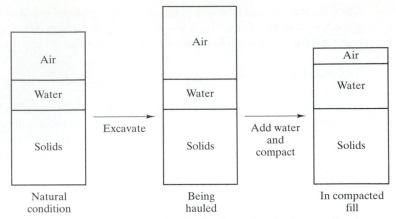

Figure 6.19 Three phase diagrams showing the changes in soil as it moves from its natural location to a compacted fill.

All of the equipment that drives over a fill, from pickup trucks to loaded scrapers, contributes to its compaction. However, we generally cannot rely only on this incidental compaction because:

- Most construction equipment is intentionally designed to have low contact pressures between the tires or tracks and the soil. This allows them to travel more quickly and easily through soft ground. For example, a Caterpillar 973 track loader has a contact pressure of only 83 kPa (12 lb/in^2). Such pressures are too low to produce the required compaction in normal-thickness lifts.
- Incidental traffic usually follows common routes, so their compactive effort is not uniformly distributed across the fill. Thus, some areas may receive sufficient compaction, while others receive virtually none.

Therefore, it is necessary to use specialized compaction equipment specifically designed for this task. Such equipment is much more efficient and effective. All compaction equipment uses one or more of the following four methods (Spann, 1986):

- **Pressure** — The contact pressure between the equipment and the ground is probably the most important factor in the resulting compaction of the underlying soils. A typical sheepsfoot roller has a contact pressure of about 3500 kPa (500 lb/in^2), which is far greater than the track-mounted equipment described earlier.
- **Impact** — Some equipment imparts a series of blows to the soil, such as by dropping a weight. This adds a dynamic component to the compactive effort.
- **Vibration** — Vibratory compaction equipment utilizes eccentric weights or some other device to induce strong vibrations into the soil, which can enhance its compaction. These vibrations typically have a frequency of 1000–3500 cycles per minute.
- **Manipulation** — Compaction equipment that imparts some shearing forces to the soil can also contribute to better compaction. This action is called *kneading* or

manipulation. However, excessive manipulation, such as in an overly wet fill, can be detrimental. When such fills are simply being moved around with no compaction occurring, we have a condition called *pumping.*

The proper selection of compaction equipment and methods depends on the type of soil, the size of the project, compaction requirements, required production rate, and other factors. No single device is the best choice for all situations. Figure 6.20 shows typical ranges of soil types for various types of compactors.

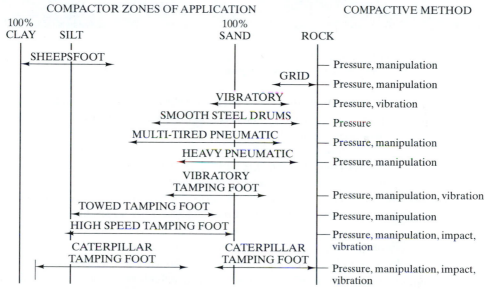

Figure 6.20 Soil types best suited for various kinds of compaction equipment (Adapted from Caterpillar, 1993).

One of the oldest and most common compaction machines is the *sheepsfoot roller*, shown in Figure 6.21. It consists of one or more rotating drums with numerous feet that concentrate its weight onto a small area, thus increasing the contact pressure to about 2000–5000 kPa (300–700 lb/in^2). The sheepsfoot roller was invented around 1906 by Walter Gillette, a contractor in Los Angeles. He had loosened the soil for a road construction project and left it in that condition at the end of a work day. Upon arriving at the job site the next morning, he discovered a flock of sheep had been driven over the road and had done an excellent job of compacting the soil. Based on this observation, he soon began fabricating the first sheepsfoot rollers, which were pulled across fills by teams of horses or mules (Southwest Builder and Contractor, 1936). By the 1930s they were being used extensively in construction of embankments.

Sheepsfoot rollers compact soil by pressure and manipulation. They can be used on a variety of soils, but work best in silts and clays. Most sheepsfoot rollers can accommodate soil lifts with loose thicknesses of about 200 mm (8 in).

a   b

Figure 6.21 Sheepsfoot rollers: a) Towed roller; b) Self-propelled roller. (Caterpillar Inc.)

Tamping foot rollers are very similar to sheepsfoot rollers, except they use larger feet with a correspondingly smaller contact pressure. They can be operated at a faster speed, but do not compact to as great a depth.

Pneumatic rollers (also known as *rubber-tire rollers*) are heavy units resting on several tires. The contact pressure is typically about 600 kPa (85 lb/in^2). Each tire is able to move up and down independently, so this device is good at finding small soft spots that rigid compaction equipment, such as sheepsfoot rollers, can miss. The tires also provide a kneading action that enhances compaction. These rollers can compact lifts with loose thicknesses of 250–300 mm (10–12 in).

Vibratory rollers, such as the one in Figure 6.22, are similar to sheepsfoot or tamping foot rollers, with the addition of a vibrating mechanism. Thus, they use pressure, manipulation, and vibration to compact the soil. Vibration is especially effective in sandy and gravelly soils. The heaviest of these rollers can accommodate loose lift thicknesses of up to 1 m (3 ft), and they provide some compactive effort to depths of about 2 m (7 ft).

Figure 6.22 Self-propelled vibratory roller.

Smooth steel-wheel rollers (don't call them "steam rollers"!), such as the one in Figure 6.23, leave a smooth compacted soil surface. The non-vibratory types are not well suited for compacting soil because the contact pressure is much lower than that of sheepsfoot rollers. However, they may be used to *proof roll* a subgrade just before paving (i.e., a final rolling to confirm compaction of the uppermost soils), and to compact the aggregate base course and asphalt pavement.

Figure 6.23 A smooth steel-wheel roller preparing to compact an asphalt pavement. (Caterpillar Inc.)

Fine Grading

Once the last lift has been placed and the fill is approximately at the final elevation, we say the *rough grading* is complete. Then the contractor begins *fine grading*, which consists of careful trimming and filling to produce the desired configuration. As the name implies, fine grading is more precise and therefore requires different equipment.

On highway subgrades and large building and parking lot pads, a *motor grader* (or *blade*) is often used, as shown in Figure 6.24. At more confined sites, a small loader equipped with a fine-grading accessory called a *gannon* might be used.

Figure 6.24 A motor grader fine grading a highway subgrade. (Caterpillar Inc.)

Fine grading slopes is more difficult, but it is nevertheless important both for aesthetics and to avoid future surficial stability problems. One method is to intentionally overfill the slope, then trim them back using a dozer equipped with a cutting arm called a *slope board*.

Small Backfills

Another type of earthwork, quite different from conventional grading, is the backfilling of small confined areas, such as retaining walls and small excavations. These areas are not large enough to accommodate the equipment described earlier, so it becomes necessary to use smaller equipment and more hand labor. Figure 6.25 shows examples of portable compaction equipment that may be used in confined spaces.

Figure 6.25 Using portable compaction equipment to compact soils in areas that are too confined to accommodate large equipment (Wacker Corporation).

Utility Trenches

Many civil engineering projects include installation of underground utilities, such as water pipes, electrical lines, sewer pipes, and so on. These are installed by digging a long trench, placing the pipe or conduit, and backfilling. The compaction of these backfills is important, especially because they often are beneath roadways. Unfortunately, many utility trenches are not properly compacted, and they eventually settle. This produces a condition that is both unsightly and hazardous.

Compaction can be difficult because of the narrow width, and because it must be accomplished without breaking the pipe. Often, the zone immediately around the pipe is filled with clean sand that is compacted by *jetting*, which consists simply of injecting water into the sand and relying on gravity and lubrication to compact the soil. The remainder of the trench is backfilled in lifts and compacted using a variety of equipment, such as the hoe-mounted sheepsfoot shown in Figure 6.26.

Figure 6.26 A hoe-mounted
sheepsfoot roller compacting
a utility trench backfill.

QUESTIONS AND PRACTICE PROBLEMS

6.1   You are preparing the earthmoving specifications for a project that will include a wide range
of materials, from loose soil to hard rock. Because of this variation, you plan to use a classified
excavation system so the contractor can give two unit prices, one for common excavation and
another for rock. However, to avoid disagreements on classification, you also need to develop
clear definitions for each type of excavation. What criteria might you use to differentiate
between common and rock excavation?

6.2   What are the four methods used by mechanical compaction equipment to densify soils? Which
of these methods are most effective in sands? In clays?

6.3   A contractor needs to excavate several thousand cubic yards of silty clay at one end of a site,
transport it to the other end, and place it as a compacted fill. The distance between the cut and
fill areas is about 3000 ft. The soils in the cut area can be excavated with little or no ripping.
However, they are very dry. Make a list of the construction equipment needed to complete this
project.

6.4   A developer plans to build a condominium complex on a site underlain by an old hydraulic fill.
What special geotechnical problems need to be considered at this site? Why?

6.3 SOIL COMPACTION STANDARDS AND ASSESSMENT

For geotechnical engineers, soil compaction is one of the most important parts of earthwork
construction. Compaction improves the engineering properties of the fill in many ways,
including:

  • Increased shear strength, which reduces the potential for slope stability problems, such
    as landslides, and enhances the fill's capacity for supporting loads, such as
    foundations.

- Decreased compressibility, which reduces the potential for excessive settlement.
- Decreased hydraulic conductivity, which inhibits the flow of water through the soil. This may be desirable or undesirable, depending on the situation.
- Decreased void ratio, which reduces the amount of water that can be held in the soil, and thus helps maintain the desirable strength properties.
- Increased erosion resistance, which helps maintain the ground surface in a serviceable condition.

Because these characteristics are so important, civil engineers include criteria for the placement and compaction of fills in the plans and specifications. The contractor is then obligated to satisfy these criteria. Although a typical specification might include many requirements, we will discuss only those that relate to compaction requirements.

Nearly all compaction specifications are based on achieving a certain dry unit weight, γ_d. Then, during construction, the geotechnical engineer usually has a staff of field engineers and technicians who measure the unit weight achieved in the field to verify the contractor's compliance with these specifications. However, it is important to recognize that dry unit weight itself is not particularly important. We use it only because it is an indicator of quality, is easy to measure, and correlates with the desirable engineering properties listed above. In other words, we want favorable strength, compressibility, hydraulic conductivity, void ratio, and erosion resistance properties, and know that we have attained them when the dry unit weight criteria has been met.

Proctor Compaction Test

During the 1930s, Mr. R.R. Proctor, an engineer with the City of Los Angeles, developed a method of assessing compacted fills that has since become a nearly universal standard (Proctor, 1933). This method consists of compacting the soil in the laboratory to obtain the *maximum dry unit weight*, $(\gamma_d)_{max}$, then requiring the contractor to achieve at least some specified percentage of this value in the field. The original version of the laboratory test is now known as the *standard Proctor test*, while a later revision is called the *modified Proctor test*. The dry unit weight achieved in the field is determined using several types of *field density tests*.

Standard Proctor Test

The standard Proctor test procedure is essentially as follows (see a laboratory manual or the ASTM standards book for the complete procedure):

1. Obtain a bulk sample of the soil to be used in the compacted fill and prepare it in a specified way.
2. Place some of the prepared soil into a standard 1/30 ft^3 (9.44×10^{-4} m^3) cylindrical steel mold[2] until it is about 40 percent full. This mold is shown in Figure 6.27.

---

[2] A larger mold is used for gravelly soils, but we will not consider it here.

3. Compact the soil by applying 25 blows from a special 5.5-lb (2.49-kg) hammer that drops from a height of 12 in (305 mm).
4. Place a second layer of the prepared soil into the mold until it is about 75 percent full and compact it using 25 blows from the standard hammer.
5. Place the third layer of the prepared soil into the mold and compact it in the same fashion. Thus, we have applied a total of 75 hammer blows.
6. Trim the sample so that its volume is exactly 1/30 ft³, then weigh it. The unit weight, γ, is thus:

$$\gamma = \frac{W_{ms} - W_m}{V_m} \tag{6.1}$$

where:

W_{ms} = weight of mold + soil
W_m = weight of mold
V_m = volume of mold = 1/30 ft³ = 9.44×10^{-4} m³

When using English units, γ is expressed in lb/ft³. With SI units, use kN/m³, which requires converting kg to kN (1 kN = 102.0 kg).

7. Perform a moisture content test on a representative portion of the compacted sample, then compute the dry unit weight, γ_d, using Equation 4.27.
8. Repeat steps 2–7 three or four times, each with the soil at a different moisture content.

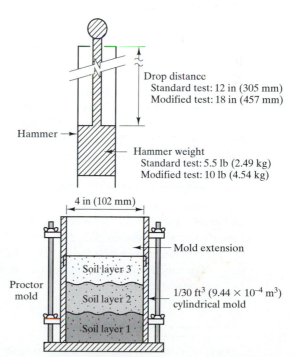

Drop distance
Standard test: 12 in (305 mm)
Modified test: 18 in (457 mm)

Hammer →

Hammer weight
Standard test: 5.5 lb (2.49 kg)
Modified test: 10 lb (4.54 kg)

4 in (102 mm)

Mold extension

Soil layer 3

Proctor mold

Soil layer 2

1/30 ft³ (9.44×10^{-4} m³) cylindrical mold

Soil layer 1

Figure 6.27 Mold and hammer for a Proctor compaction test. The standard test uses three layers, as shown, while the modified test uses five.

The test results are then plotted on a γ_d vs. w diagram as shown in Figure 6.28. The curve connecting the data points represents the dry unit weight achieved by compacting the soil at various moisture contents. Higher γ_d values indicate higher quality fill, so there is a certain moisture content, known as the *optimum moisture content*, w_o, that produces the greatest γ_d. The latter is called the *maximum dry unit weight*,[3] $(\gamma_d)_{max}$.

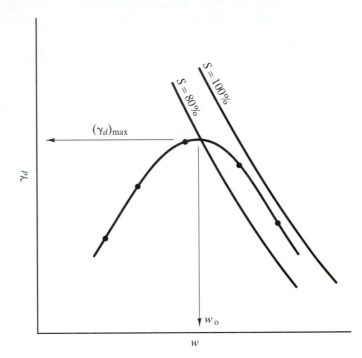

Figure 6.28 Results from a Proctor compaction test.

The mechanics behind the shape of this curve are very complex (Hilf, 1991). In a dry soil, we achieve better compaction by first adding water to raise its moisture content to near-optimum. This water provides lubrication, softens clay bonds, and reduces surface tension forces within the soil. However, if the soil is too wet, then there is little or no air left in the voids, and it thus becomes very difficult or impossible to compact. Such soils need to be dried before they are compacted. Usually, the peak of the compaction curve occurs at a degree of saturation, S, of about 80%.

Although most soils have compaction curves similar to that in Figure 6.28, clean sands and gravels (SP, GP, and some SW and GW soils) often do not. They typically have much flatter compaction curves (i.e., they are less sensitive to moisture content than clays) and these curves sometimes have two small peaks. Fortunately, these soils compact easily in the field, so quality assessments might rely more on observations of the contractor's procedures and less on test results.

---

[3] Many engineers use the term *maximum dry density* for $(\gamma_d)_{max}$, just as "density" is often used for "unit weight" as described in Chapter 4.

Before drawing the compaction curve, it is best to first draw two other curves: One representing $S = 100\%$ (sometimes called the *zero air voids curve*), and the other representing $S = 80\%$, as shown in Figure 6.28. These two curves can be developed using Equation 4.32, and are intended to help us draw the Proctor compaction curve. The $S = 100\%$ curve represents an upper limit for the Proctor data, for it is impossible to have $S > 100\%$. The $S = 80\%$ curve should go nearly through the peak in the Proctor curve. In addition, the right limb of the Proctor curve should be slightly to the left of $S = 100\%$, because compaction does not remove all of the air. Example 6.1 illustrates this technique.

Example 6.1

A Proctor compaction test has been performed on a soil sample. The test results were as follows:

	Data Point No.	1	2	3	4	5
	Mass of compacted soil + mold (kg)	3.762	3.921	4.034	4.091	4.040
Moisture Content	Mass of can (g)	20.11	21.24	19.81	20.30	20.99
	Mass of can + wet soil (g)	240.85	227.03	263.45	267.01	240.29
	Mass of can + dry soil (g)	231.32	212.65	241.14	238.81	209.33

The mass of the compaction mold was 2.031 kg. The moisture content tests were performed on small portions of the compacted soil samples.

a. Compute γ_d and w for each data point and plot these results
b. Develop and plot curves for $S = 80\%$ and $S = 100\%$ using a specific gravity of 2.69.
c. Using the data from steps a and b, draw the Proctor compaction curve and determine $(\gamma_d)_{max}$ and w_o.

Solution

a. Computations for data point no. 1:
    Using Equations 4.12 and 6.1,

$$\gamma = \frac{W_{ms} - W_m}{V_m}$$

$$= \frac{(M_{ms} - M_m)g}{V_m}$$

$$= \frac{(3.762\,\text{kg} - 2.031\,\text{kg})\,(9.81\,\text{m/s}^2)}{9.44\times10^{-4}\,\text{m}^3}\left(\frac{1\,\text{N}}{1000\,\text{kN}}\right)$$

$$= 17.99\,\text{kN/m}^3$$

Using Equation 4.3,

$$w = \frac{M_1 - M_2}{M_2 - M_c} \times 100\%$$

$$= \frac{240.85 \text{ g} - 231.32 \text{ g}}{231.32 \text{ g} - 20.11 \text{ g}} \times 100\%$$

$$= 4.5\%$$

Using Equation 4.27,

$$\gamma_d = \frac{\gamma}{1 + w} = \frac{17.99 \text{ kN/m}^3}{1 + 0.045} = 17.22 \text{ kN/m}^3$$

Thus, this data point plots at $w = 4.5\%$, $\gamma_d = 17.22$ kN/m³. Following the same procedure for the other data points gives:

Data Point No.	1	2	3	4	5
γ (kN/m³)	17.99	19.64	20.81	21.41	20.88
w	4.5%	7.5%	10.1%	12.9%	16.4%
γ_d (kN/m³)	17.22	18.27	18.90	18.96	17.94

b. Using Equation 4.32, select values of γ_d in the range of test data from step a and compute the corresponding values of w for $S = 80\%$ and 100%:

γ_d (kN/m³)	w	
	@ $S = 80\%$	@ $S = 100\%$
16.00	19.3%	24.1%
18.00	13.9%	17.3%
20.00	9.5%	11.9%

c. Plot the data points from part a and the curves from part b, then draw the compaction curve as shown in Figure 6.29.

The final results are:

$$(\gamma_d)_{max} = 19.0 \text{ kN/m}^3 \qquad \Leftarrow Answers$$
$$w_o = 11.8\%$$

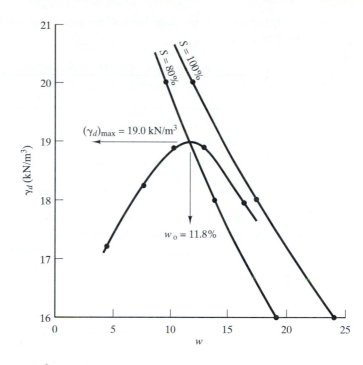

Figure 6.29 Laboratory test results for Example 6.1.

Relative Compaction

Once the maximum dry unit weight has been established for the soil being used in the compacted fill, we can express the degree of compaction achieved in the field by using the *relative compaction, C_R*:

$$C_R = \frac{\gamma_d}{(\gamma_d)_{max}} \times 100\% \qquad (6.2)$$

where:

γ_d = dry unit weight achieved in the field

$(\gamma_d)_{max}$ = maximum dry unit weight (from a Proctor compaction test)

Most earthwork specifications are written in terms of the relative compaction, and require the contractor to achieve at least a certain value of C_R. For example, if a certain soil has $(\gamma_d)_{max}$ = 120 lb/ft^3 and the project specifications require $C_R \geq 90\%$, then the contractor must compact the soil until $\gamma_d \geq 108$ lb/ft^3.

The minimum acceptable value of C_R listed in a project specification is a compromise between cost and quality. If a low value is specified, then the contractor can easily achieve the required compaction and, presumably, will perform the work for a low price. Unfortunately, the quality will be low. Conversely, a high specified value is more difficult to achieve and will cost more, but will produce a high-quality fill. Table 6.2 presents typical requirements.

TABLE 6.2 TYPICAL COMPACTION SPECIFICATIONS

Type of Project	Minimum Required Relative Compaction
Fills to support buildings or roadways	90%
Upper 150 mm of subgrade below roadways	95%
Aggregate base material below roadways	95%
Earth dams	100%

In Chapter 4 we defined the relative density, D_R, which is based on the minimum and maximum void ratio (Equation 4.20). This parameter is similar to C_R in that both reflect soil compaction and both are expressed as a percentage, but they are not numerically equal. In addition, some engineers and contractors incorrectly use the term "relative density" when they are really describing relative compaction. This is often a source of confusion, so it is important to be careful when using these terms.

Modified Proctor Test

During the 1940s and 1950s, geotechnical engineers found the standard Proctor test was no longer sufficient for airport and highway projects because fills were not providing adequate support for heavy trucks and aircraft. The U.S. Army Corps of Engineers addressed this problem by developing the *modified Proctor test*, which used greater levels of compaction and thus produced higher values of $(\gamma_d)_{max}$. The principal differences between the standard and modified tests are shown in Figure 6.30 and Table 6.3. This method was later adapted by the American Association of State Highway and Transportation Officials (AASHTO) and ASTM, and is now the most commonly used standard. The concurrent development of heavier and more efficient earthmoving and compaction equipment made it practical to implement this higher compaction standard.

TABLE 6.3 PRINCIPAL DIFFERENCES BETWEEN STANDARD AND MODIFIED PROCTOR COMPACTION TESTS

	Standard Proctor Test	Modified Proctor Test
Standards	ASTM D698 and AASHTO T-99	ASTM D1557 and AASHTO T-180
Hammer weight	5.5 lb (2.49 kg)	10.0 lb (4.54 kg)
Hammer drop height	12 in (305 mm)	18 in (457 mm)
Number of soil layers	3	5
Number of hammer blows per layer	25	25
Energy imparted from hammer for each sample	12,400 ft-lb$_f$/m^3 (600 kN-m/m3)	56,000 ft-lb$_f$/ft^3 (2,700 kN-m/m^3)

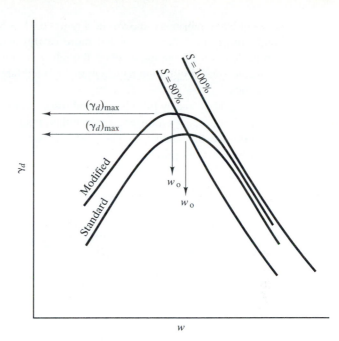

Figure 6.30 Comparison of standard Proctor test and modified Proctor test results on the same soil.

Figure 6.30 shows how the term "maximum" dry unit weight is misleading, because the standard and modified tests have two different "maximums." For each test, $(\gamma_d)_{max}$ is the greatest dry unit weight achieved for that level of compactive effort (i.e., for so many blows of a certain hammer). Neither of these values represents the largest possible dry density. Figure 6.30 also shows how the optimum moisture content for the modified test is slightly less than that for the standard test.

The γ_d achieved in the field will generally be less than $(\gamma_d)_{max}$ even if the moisture content is exactly equal to w_o because the contractor usually applies less compactive effort (i.e., so many passes of a certain kind of roller). This is acceptable so long as the project specifications permit relative compactions of less than 100%, which they usually do (see Table 6.2). However, sometimes the contractor applies large effort and exceeds $(\gamma_d)_{max}$, thus producing a relative compaction of more than 100%. This is especially likely on earth dams, where high levels of compaction are required.

Compaction Methods and Soil Fabric

Based on the Proctor compaction test results alone, it would appear that all fills should be compacted with a moisture content about equal to the optimum moisture content ($w \approx w_o$) because this method achieves the highest γ_d for the least compactive effort (i.e., for the lowest cost). This is what is done on most projects.

However, soils compacted slightly wet or dry of optimum also can achieve the required γ_d, through the use of additional compactive effort. Although this method is more expensive, it produces a fill with a different fabric, and possibly more favorable engineering properties, especially in clays (Lambe, 1958). When compacted dry of optimum, clays have a flocculated fabric as shown in Figure 6.31. Such soils have a higher hydraulic conductivity (i.e., they pass water more easily) and a greater shear strength than those compacted wet of optimum, even though γ_d is the same for both. Conversely, clays compacted wet of optimum have a more oriented fabric, which also affects the engineering properties.

The fabric in clay fills also depends on the compaction method. For example, pressure compaction produces a different fabric than manipulation compaction, even though both may produce the same γ_d. These effects are especially pronounced when the soil is compacted wet of optimum.

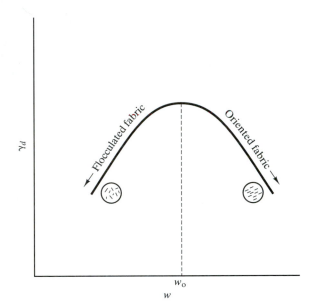

Figure 6.31 Effect of moisture content during compaction on soil fabric in clays.

Although these fabric effects are generally not considered in routine compacted fill projects, they can become important in critical projects, such as earth dams or very deep fills. For example, clays used to build the core of an earth dam might be placed wet of optimum to provide a low hydraulic conductivity. Such projects may include performing a series of laboratory tests on soils using various compaction moisture contents and methods, and developing compaction specifications based on the results of these tests.

Field Density Tests

The final link in assessing fill compaction by the Proctor method is to measure γ_d in the fill, thus enabling us to use Equation 6.2. Several methods have been developed to do so, and they are known as *field density tests*. All of these methods can be performed in the field, thus making it possible to immediately present the test results to the contractor. Such rapid feedback is important, because the contractor must rectify any inadequately compacted zones before they are buried by additional fill.

Sand Cone Test

One of the most common field density test methods is the *sand cone test* (ASTM D1556). The test procedure is essentially as follows (see ASTM for details):

1. Prepare a level surface in the fill and dig a cylindrical hole about 125 mm (5 in) in diameter and about 125 mm (5 in) deep. Save all of the soil that comes out of the hole and determine its weight, W.
2. Fill the sand cone apparatus, shown in Figure 6.32, with a special free-flowing SP sand similar to that found in an hourglass. Then determine the weight of the cone and the sand, W_1.
3. Place the sand cone over the hole, as shown in Figure 6.33. Then open the valve, and allow the sand to fill the hole and the cone.
4. Close the valve, remove the sand cone from the hole, and determine its new weight, W_2.

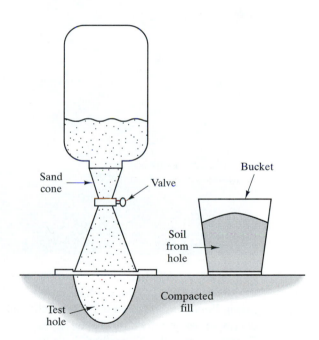

Figure 6.32 Use of a sand cone to measure the unit weight of a fill.

The volume of the hole, V, and unit weight of the fill, γ, are:

$$V = \frac{W_1 - W_2}{\gamma_{sand}} - V_{cone}$$

(6.3)

$$\gamma = \frac{W}{V}$$

(6.4)

where:
   V = volume of test hole
   W_1 = initial weight of sand cone apparatus
   W_2 = final weight of sand cone apparatus
   γ_{sand} = unit weight of sand used in sand cone
   V_{cone} = volume of sand cone below the valve
   γ = unit weight of fill
   W = weight of soil removed from test hole

Figure 6.33 A sand cone test being performed in the field.

Although a field engineer or soils technician can quickly measure γ in the field using the sand cone test, he or she still needs to measure the moisture content to obtain γ_d and C_R (using Equations 4.27 and 6.2). Normally this requires placing a soil sample in an oven overnight, as described in Chapter 4, but such a procedure would be unacceptably slow. Therefore, special rapid methods have been developed to measure the moisture content in the field. These include:

- "Stir-frying" the soil sample in a frying pan over a portable propane stove. This method is fast, and gives reasonably accurate results in clean sands and gravels, but is much hotter than a laboratory oven, and thus can produce erroneous results in clays and organic soils. The computed moisture content in such soils will usually be too high.

- Mixing the soil with calcium carbide in a special sealed container known as a *Speedy moisture tester*. The water in the soil reacts with the calcium carbide (CaC_2) to produce acetylene gas (C_2H_2), and this reaction continues until the water is depleted. A pressure gage measures the resulting pressure generated by the production of this gas, and it is experimentally calibrated to compute the moisture content. The results are not as accurate as a standard laboratory moisture test, but are usually sufficient for field density tests.

When carefully performed, the sand cone test is very accurate, and the equipment is inexpensive and durable.

Drive Cylinder Test

Another field density test method is the *drive cylinder test* (ASTM D2937). It consists of driving a thin-wall steel tube into the soil using a special drive head and a mallet as shown in Figure 6.34. The cylinder is then dug out of the fill using a shovel, the soil is trimmed smooth, and it is weighed. The unit weight of the fill is then computed based on this weight and the volume of the cylinder, and the moisture content is determined as discussed earlier.

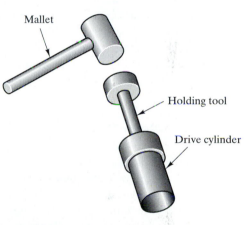

Figure 6.34 The drive cylinder test.

The drive cylinder test is much faster than the sand cone method and only slightly less precise. However, it is only suitable for fills with sufficient silt and clay to provide enough dry strength to keep the sample inside the cylinder. It is not satisfactory in clean sands, because they fall out too easily, or in gravelly soils.

Nuclear Density Test

A third type of field density test is the *nuclear density test* (ASTM D2922). It consists of a special device, shown in Figure 6.35, that emits gamma rays and detects how they travel through the soil. The amount of gamma rays received back into the device correlates with the unit weight of the soil. The nuclear density test also measures the moisture content of the soil in a similar way using alpha particles.

Both the unit weight and moisture content measurements depend on empirical correlations, which ultimately must be programmed into the device. This allows it to directly display both parameters on digital electronic readouts.

The nuclear test can encounter problems in fills with unusual chemistries, and needs regular calibration to maintain its accuracy. In addition, personnel must have special

radiation training before using this
equipment. Also, because the equipment
contains a source of radiation, it cannot
be shipped through normal channels,
such as in commercial aircraft or via
mail.

   In spite of its use of "hi-tech"
equipment, the nuclear method is slightly
less accurate than the sand cone. This is
because it is based on empirical
correlations with the transmission of
radiation, while the sand cone uses direct
measurements of weight and volume.
However, the nuclear test has sufficient
accuracy for compaction assessments of
normal fills, and is faster than the sand
cone. This saves time in the field, so its
chief attraction is an economic one.

Figure 6.35 Performing a nuclear density test in the field.

Therefore, it has generally become the preferred method for many geotechnical firms and
agencies.

Water Ring Test

Sometimes engineers wish to place fills that contain cobbles and small boulders. For
example, some earth dams contain zones made of such soils. Unfortunately, conventional
field density test methods are not applicable because of the large particle size. One way to
evaluate such fills is the *water ring test* shown in Figure 6.36. This test uses a much larger
sample (the ring is typically about 1800 mm in diameter) and uses water instead of sand to
measure the volume (ASTM D5030).

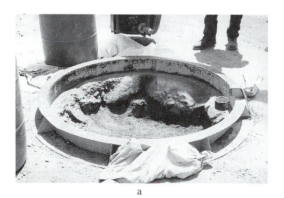

a   b

Figure 6.36 Use of a water ring test to assess compaction in an earth dam with cobbles and boulders: a) Hole has been
dug and excavated soils have been weighed; b) Hole has been lined with plastic and filled with water to measure its
volume.

Example 6.2

A sand cone test has been performed in a compacted fill made with the soil described in Example 6.1. The test results were as follows:

Initial mass of sand cone apparatus = 5.912 kg
Final mass of sand cone apparatus = 2.378 kg
Mass of soil recovered from hole = 2.883 kg
Moisture content of soil from hole = 7.0%
Density of sand = 1300 kg/m^3
Volume of cone below valve = 1.114×10^{-3} m^3

The project specifications require a relative compaction of at least 90%.

Compute γ_d and C_R and determine whether the project specifications have been met. If not, suggest a course of action.

Solution

$$(M_{sand})_{cone + hole} = 5.912 \text{ kg} - 2.378 \text{ kg} = 3.534 \text{ kg}$$

$$V_{cone + hole} = \frac{(M_{sand})_{cone + hole}}{\rho_{sand}} = \frac{3.534 \text{ kg}}{1300 \text{ m}^3} = 2.718 \times 10^{-3} \text{ m}^3$$

$$V_{hole} = V_{cone + hole} - V_{cone} = 2.718 \times 10^{-3} \text{ m}^3 - 1.114 \times 10^{-3} \text{ m}^3 = 1.604 \times 10^{-3} \text{ m}^3$$

$$W_{soil} = M_{soil} \, g = (2.883 \text{ kg})(9.81 \text{ m/s}^2)\left(\frac{1 \text{ kN}}{1000 \text{ N}}\right) = 2.828 \times 10^{-2} \text{ kN}$$

$$\gamma = \frac{W_{soil}}{V_{hole}} = \frac{2.828 \times 10^{-2} \text{ kN}}{1.604 \times 10^{-3} \text{ m}^3} = 17.63 \text{ kN/m}^3$$

$$\gamma_d = \frac{\gamma}{1 + w} = \frac{17.63 \text{ kN/m}^3}{1 + 0.070} = 16.5 \text{ kN/m}^3$$

$$C_R = \frac{\gamma_d}{(\gamma_d)_{max}} \times 100\% = \frac{16.5 \text{ kN/m}^3}{19.0 \text{ kN/m}^3} \times 100\% = \textbf{86.8\%} \quad \Leftarrow \textit{Answer}$$

Conclusion and Recommendation

The relative compaction is less than the required 90%, so the specifications have not been met. This may be at least partially due to the low moisture content, which is well below optimum. Suggest ripping the soil, adding water, mixing, and recompacting.

QUESTIONS AND PRACTICE PROBLEMS

6.5 How does the dry unit weight, γ_d differ from the unit weight γ? Why is the Proctor method of assessing soil compaction based on the dry unit weight and not the unit weight?

6.6 A compacted fill, currently under construction, will support a proposed supermarket. One of the field density tests in this fill found a unit weight of 121 lb/ft^3 and a moisture content of 12.5%. A Proctor compaction test has also been performed on the fill soils using the modified method (ASTM D1557), and it produced a maximum dry unit weight of 117 lb/ft^3 and an optimum moisture content of 13.0%. Compute the relative compaction and determine whether or not it satisfies the normal compaction specification for such projects.

6.7 A field density test has been conducted in a compacted fill that is currently under construction. The test results indicate a relative compaction of 101–103%, which made the construction manager believe the test must be incorrect. He believes it is impossible for the dry unit weight in the field to be greater than the maximum dry unit weight from the Proctor compaction test. Prepare a 200–400 word memo explaining why the relative compaction in the field could indeed be greater than 100%.

6.8 A certain soil has a dry unit weight of 18.0 kN/m^3 and a specific gravity of solids of 2.67. Compute the moisture contents that correspond to $S = 80\%$ and $S = 100\%$.

6.9 You are a geotechnical engineer and have just received the results from a Proctor compaction test. Upon examining the results you notice the data plots to the right of the zero air voids curve. Is this cause for concern? Why do you think the curve plots like this? Will you fire the laboratory staff or give them a raise? (Last question is optional!)

6.10 A Proctor compaction test has been performed on a soil that has $G_s = 2.68$. The test results were as follows. The weight of the empty mold was 5.06 lb.
    Plot the laboratory test results and the 80 and 100% saturation curves, then draw the Proctor compaction curve. Determine the maximum dry unit weight (lb/ft^3) and the optimum moisture content.

Point No.	Weight of Compacted Soil + Mold (lb)	Moisture Content Test Results		
		Mass of Can (g)	Mass of Can + Moist Soil (g)	Mass of Can + Dry Soil (g)
1	8.73	22.13	207.51	202.30
2	9.07	25.26	239.69	225.27
3	9.40	19.74	253.90	230.64
4	9.46	23.36	250.93	219.74
5	9.22	20.28	301.47	250.95

6.11 A Proctor compaction test has been performed on a soil that has $G_s = 2.70$. The test results were as follows:

Point No.	Mass of Compacted Soil + Mold (kg)	Moisture Content Test Results		
		Mass of Can (g)	Mass of Can + Moist Soil (g)	Mass of Can + Dry Soil (g)
1	3.673	22.11	205.74	196.33
2	3.798	23.85	194.20	180.54
3	3.927	19.74	196.24	177.92
4	3.983	20.03	187.43	165.71
5	3.932	21.99	199.59	171.11

The mass of the empty mold was 1.970 kg.

Plot the laboratory test results and the 80 and 100% saturation curves, then draw the Proctor compaction curve. Determine the maximum dry unit weight (kN/m^3) and the optimum moisture content.

6.12 A sand cone test has been performed in a recently compacted fill. The test results were as follows:

Initial weight of sand cone + sand = 13.51 lb
Final weight of sand cone + sand = 4.26 lb
Weight of sand to fill cone = 2.12 lb
Weight of soil from hole + bucket = 12.42 lb
Weight of bucket = 1.21 lb
Moisture content test
Mass of empty moisture content can = 23.11 g
Mass of moist soil + can = 273.93 g
Mass of oven-dried soil + can = 250.10 g

The sand used in the sand cone had a unit weight of 81.0 lb/ft^3, and the fill had a maximum dry unit weight of 121 lb/ft^3 and an optimum moisture content of 11.7%. Compute the relative compaction.

6.4 SUITABILITY OF SOILS FOR USE AS COMPACTED FILL

When imported soils are required, we often have the opportunity to choose between several borrow sites, each with a different soil. Our selection is based on their engineering properties, the cost of importing them, and other factors. Often we only require that the import soils be at least as good as those onsite, but sometimes other criteria might apply.

The following list summarizes the suitability of various soil types, and could be used to assist in selecting an import source.

Gravels (GW, GP, GM, GC)

Gravels make good fills that have high strength and low compressibility. These fills also retain their strength when wetted. GW and GP soils have a high hydraulic conductivity that allows water to drain quickly, which is especially important in highways. Base material below pavements is always a GW or GP. GM and GC soils have substantially lower hydraulic conductivity, and do not drain nearly as well. Gravels easily compact to a high unit weight over a wide range of moisture contents, but are prone to caving if they must later be excavated, such as for utility trenches.

Sands (SW, SP, SM, SC)

Sands also make good fills that have high strength and low compressibility. SW and SP soils retain virtually all of their strength when wetted; SM and SC lose some strength, but still remain good. If sands are too wet or too dry, they can easily be brought to the optimum moisture content, and they can be compacted over a wide range of moisture contents. However, SP and SW soils are problematic when exposed at the ground surface, because vehicles can become stuck and grades are difficult to maintain. SP and SW soils also are especially prone to caving into small excavations such as utility trenches. All sandy soils are susceptible to surface erosion unless protected, such as with vegetation. SM and SC soils are nearly ideal for most applications, so long as their moderate hydraulic conductivity is acceptable.

Low Plasticity Silts and Clays (ML and CL)

ML and CL soils are less desirable than SM or SC because they lose more strength when wetted and require more careful moisture control. They also are more difficult to dry if initial moisture content is far from the optimum, and are prone to frost heave problems in freezing climates. Nevertheless, they usually make acceptable fills.

High Plasticity Silts and Clays (MH and CH)

MH and CH soils are the best choice when very low hydraulic conductivity is required, such as in landfill caps. Otherwise they are difficult to handle and compact, especially when initial moisture content is above optimum. These soils expand when wetted, which can cause problems in pavements, lightly loaded foundations, flatwork concrete, and similar projects. They are a poor choice for retaining wall backfills because of their expansiveness, low wet strength, and low hydraulic conductivity.

Organic Soils (OL and OH)

Organic soils are very poor choices for compacted fills and are unsuitable for most applications. They are weak and compressible and difficult to compact. Never bring them in as an import soil (except for landscaping purposes). If OL or OH soils are being generated from onsite cuts, try to place them away from critical areas and consider hauling them away.

Peat (Pt)

This soil is extremely poor. It makes fills that are weak and compressible, and are unsuitable for support of buildings or pavements. However, peat is ideal for use in landscape areas because of its fertility.

6.5 EARTHWORK QUANTITY COMPUTATIONS

Grading Plans

When planning earthwork construction, civil engineers begin with a topographic map of the existing ground surface, then we develop a new topographic map showing the proposed ground surface. These two maps are usually superimposed to produce a *grading plan*, such as the one in Figure 6.37. When the proposed elevations are higher than the existing ones, a fill will be required, whereas when they are lower, a cut becomes necessary.

The design engineer also uses these two sets of elevations to compute the anticipated quantities of cut and fill. Such computations can be performed by hand, but they are more likely done by computer, and the results are always expressed in terms of volume, such as cubic yards or cubic meters.

If extra soil needs to be hauled to the site to produce the required grades, the project has a *net import*. Conversely, when extra soil is left over and needs to be hauled away, it has a *net export*. Finally, when neither import nor export is required, the earthwork is said to be *balanced*. We usually strive to design the proposed grades such that the earthwork is close to being balanced because importing and exporting soils can become very expensive.

Bulking and Shrinkage

Let us consider a grading project that consists of excavating soil from one area, transporting it, and placing it as compacted fill in another area, as illustrated in Figure 6.38. The excavation will occur in an area called the *bank*. At this particular site, each cubic yard of soil has a weight of 2700 lb (other sites would have different weights). The soil is loosened as it is removed and loaded into a dump truck, so the 2700 lb now occupies a volume of 1.25 yd^3. This is called the *loose* condition. Finally, it is placed and compacted and the volume shrinks to 0.80 yd^3; the *compacted* condition. These changes in volume are important to the contractor because they affect the equipment productivity, and to the design engineer because they affect the amount of fill produced by a given amount of cut.

The change in volume from the bank to the loose condition is called *bulking* or *swell* and depends on the unit weight in the bank, the soil type and moisture content, and other factors. Sands and gravels typically have about 10% bulking, while silts and clays usually have 30–40%. This means the contractor will need more equipment to transport a given volume of soil per day.

The net change in volume from the bank to compacted conditions is the *shrinkage factor*, $\Delta V/V_f$:

$$\frac{\Delta V}{V_f} = \left[\frac{(\bar{\gamma}_d)_f}{(\bar{\gamma}_d)_c} - 1 \right] \times 100\% \qquad (6.5)$$

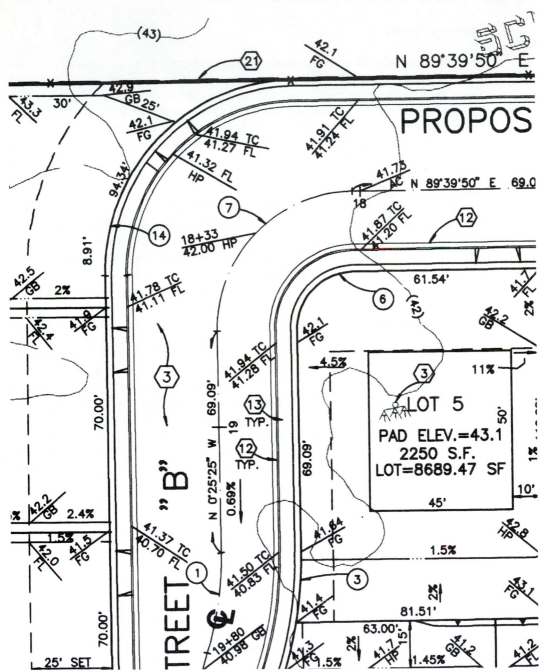

Figure 6.37 A small portion of a grading plan for a proposed residential tract. The contour lines represent the existing conditions, while the spot elevations represent the proposed conditions. Proposed lots are located along the left and bottom right portions of the drawing. Key to notation: FG = finish grade; FL = flow line; GB = grade break; HP = high point; TC = top of curb. Scale: 1 in = 30 ft. All elevations and dimensions are in feet (Algis Marciuska, Civil Engineering Department, Cal Poly University, Pomona).

where:

ΔV = change in volume during grading

V_f = volume of fill

$(\bar{\gamma}_d)_f$ = average dry unit weight of fill

$(\bar{\gamma}_d)_c$ = average dry unit weight of cut (i.e., the bank soil)

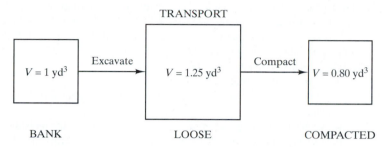

Figure 6.38 Changes in volume as soil is excavated, transported, and compacted. The numerical values are examples and would be different for each soil.

A positive value of $\Delta V/V_f$ indicates shrinkage (a net decrease in volume), while a negative value indicates swelling (a net increase). These are not the same as shrinkage or swelling of expansive soils, as discussed in Chapter 18. Most soils have a positive shrinkage factor because modern construction equipment usually compacts the soil to a dry unit weight greater than that in the bank.

If a civil engineer plans a project for 100,000 m³ of cut and 100,000 m³ of fill without accounting for shrinkage, he or she will probably have an unpleasant surprise during construction. If the soils at this site have 10% shrinkage, the 100,000 m³ removed from the cut area will produce only 90,910 m³ of fill, which means 9,090 m³ of import will suddenly be required to reach the proposed grades. This will come as a surprise during construction and will be even more expensive than usual because of the urgency of the situation, idle workers and equipment, and the lack of money in the project budget. Such situations have been the source of many lawsuits. However, if the design engineer had properly accounted for shrinkage, the proposed grades could have been set accordingly (perhaps with 104,000 m³ of cut and 94,550 m³ of fill), or at least the need for imported soils would have been known in advance.

The value of $(\gamma_d)_c$ may be obtained from unit weight measurements on undisturbed samples from borings made in the cut area, while $(\gamma_d)_f$ is obtained from Proctor compaction test results and project specifications for C_R. However, the project specifications quote a *minimum* acceptable value for C_R such as 90%, while Equation 6.5 requires the *average* value of $(\gamma_d)_f$. Grading contractors typically exceed the minimum requirements, so the shrinkage computations must be adjusted accordingly. Usually it is sufficient to compute $(\bar{\gamma}_d)_f$ using a $\bar{C}_R$ value 2% higher than the minimum required in the project specifications. For example, if the specifications require $C_R \geq 90\%$, then use $(\gamma_d)_f = 0.92 \, (\gamma_d)_{max}$.

Although Equation 6.5 is the most common definition for shrinkage factor, some engineers use $\Delta V/V_c$ expressed as a percentage, while others use the ratio V_f/V_c expressed in decimal form. This lack of consistency can be a source of confusion, and has lead to further lawsuits. Therefore, it is important to be careful when using this term. The best way seems to be to use both the term and the variable, (i.e., "the shrinkage factor, $\Delta V/V_f$, is about 5%").

Example 6.3

Twelve undisturbed soil samples were obtained from borings in a proposed cut area. These samples had an average γ_d of 108 lb/ft^3 and an average w of 9.1%. A Proctor compaction test performed on a representative bulk sample produced $(\gamma_d)_{max} = 124$ lb/ft^3 and $w_o = 12.8\%$. A proposed grading plan calls for 12,000 yd^3 of cut and 11,500 yd^3 of fill, and the specifications call for a relative compaction of at least 90%.

    a. Compute the shrinkage factor
    b. Compute the required quantity of import or export soils based on the unit weight of the cut.
    c. Compute the weight of import or export in tons using the moisture content of the cut.
    d. Compute the required quantity of water in gallons to bring the fill soils to the optimum moisture content.

Solution

a. Using an average relative compaction of 92%:

$$(\bar{\gamma}_d)_f = (\gamma_d)_{max} \, \bar{C}_R = (124 \text{ lb/ft}^3)(0.92) = 114 \text{ lb/ft}^3$$

$$\frac{\Delta V}{V_f} = \left[\frac{(\bar{\gamma}_d)_f}{(\bar{\gamma}_d)_c} - 1\right] \times 100\% = \left[\frac{114}{108} - 1\right] \times 100\% = 6\% \quad \textbf{(Shrinkage)} \quad \Leftarrow \textit{Answer}$$

b.

$$\Delta V = \frac{\Delta V}{V_f} V_f = (6\%)(11,500 \text{ yd}^3) = 690 \text{ yd}^3$$

Cut required to produce 11,500 yd^3 of fill = 11,500 yd^3 + 690 yd^3 = 12,190 yd^3

Required import = 12,190 yd^3 - 12,000 yd^3 = **190 yd^3** $\Leftarrow$ *Answer*

c.

$$W_s = V\gamma_d = (190 \text{yd}^3)\left(\frac{3\text{ft}}{1\text{yd}}\right)^3 (108 \text{lb/ft}^3)\left(\frac{1\text{T}}{2000 \text{lb}}\right) = 277 \text{ T}$$

$$W = W_s(1 + W) = (277 \text{ T})(1 + 0.091) = \textbf{302 T} \quad \Leftarrow \textit{Answer}$$

d.

$$(W_s)_{fill} = (\gamma_d)_{fill}(V_{fill}) = (114 \text{ lb/ft}^3)(11,500 \text{ yd}^3)\left(\frac{3\text{ft}}{1\text{yd}}\right)^3 = 35.4 \times 10^6 \text{ lb}$$

Based on Equation 4.1, the weight of water to be added is the required increase in the moisture content, Δw times the dry weight:

$$(W_w)_{added} = \Delta w \; W_s = (0.128 - 0.091)(35.4 \times 10^6 \text{ lb}) = 1.31 \times 10^6 \text{ lb}$$

Water has a unit weight of 8.34 lb/gal, so:

$$V_w = \frac{W_w}{\gamma_w} = \frac{1.31 \times 10^6 \text{ lb}}{8.34 \text{ lb/gal}} = \textbf{157,000 gal} \quad \Leftarrow Answer$$

Comments

If our estimate of the shrinkage factor was in error, perhaps because the undisturbed samples were not truly representative, and the true shrinkage factor was 10%, the required import volume becomes 650 yd^3 instead of the computed 190 yd^3. Such a situation would not be unusual in practice. Although this difference represents an 340 percent increase in the required import quantity, it is only 4 percent of the total fill volume. Therefore, it is important to keep in mind the potential effects of small errors in the computed shrinkage factor.

Sometimes the weight of fills cause significant settlement in the underlying natural soils. Some or all of this settlement may occur during construction, but much of it often occurs after construction. In either case, the earthwork quantity computations will be affected. We will discuss methods of predicting these settlements in Chapters 11 and 12.

6.6 LIGHTWEIGHT FILLS

Normal fills are very heavy. For example, a 2 m thick fill, which many people would hardly notice, has about the same weight as a six-story building. Usually this weight is not a major concern, but in some situations it can be problematic. For example,

- If the natural soils below the fill are soft, its weight will cause them to consolidate, thus producing large settlements at the ground surface. This can be especially problematic in fills leading to bridge abutments because the bridge (which is probably supported on pile foundations) does not settle with the abutment fill. See the example in Figure 11.2.
- The weight of a fill imposes new stresses on buried structures, such as pipelines or subway tunnels, and may overload them.
- The weight of a fill placed behind a retaining wall imposes stresses on the wall.
- The weight of a fill placed on a slope decreases the stability of that slope, possibly leading to a landslide.

When these issues are a concern, geotechnical engineers sometimes use special lightweight materials to build fills. These materials allow the fill to achieve the same elevation, without the adverse effects of excessive weight.

Various materials have been used as lightweight fills. When selecting the most appropriate material, the engineer needs to find the best compromise between cost and unit weight. For example, various cementitious materials with unit weights of 3.8–12.6 kN/m^3 (24–80 lb/ft^3) are available, and can be pumped into place. Automobile tires chopped into small pieces also have been used as lightweight fill material (Whetten, et al., 1997). Another lighter, but more expensive, option is to use expanded polystyrene (EPS) and extruded polystyrene (XPS), otherwise known as *geofoam* (Horvath, 1995; Negussey, 1997). These are essentially the same as the materials commonly called "styrofoam," except they are supplied in large blocks that are stacked on the ground as shown in Figure 6.39. The unit weight of geofoam used for geotechnical engineering purposes is only 1.25 lb/ft^3 (0.20 kN/m^3), yet it has sufficient strength and stiffness to support heavy external loads, such as those from vehicles.

For example, a 20 ft deep fill placed against a bridge abutment would induce an increase in total vertical compressive stress, $\Delta\sigma_z$, of about 2400 lb/ft^2 in the underlying soils. If these soils are soft, the resulting settlement could be substantial. However, the same fill made from geofoam would induce a $\Delta\sigma_z$ of only 25 lb/ft^2 plus the weight of the soil cover and pavement. The resulting settlement would be far less.

Geofoam also has other uses, such as providing thermal insulation, compressible inclusions (i.e., soft spots to protect underground structures from excessive stresses), vibration damping, and fluid transmission through intentional voids in the foam (Horvath, 1995).

Figure 6.39 Geofoam blocks being placed to produce a lightweight fill for a highway in Colorado. (BASF Corporation)

6.7 DEEP FILLS

Deep compacted fills have been used for many years in the construction of earth dams. However, during the last third of the twentieth century, deep fills also became increasingly common in other kinds of projects, such as highways and residential developments. Fill thicknesses of 30 m or more became quite common, and these fills were constructed using the same techniques that had been successful in other more conventional fills. Unfortunately, some of these deep fills have experienced excessive settlements, which in some cases have damaged buildings and other structures (Rogers, 1992c; Noorany and Stanley, 1994).

The settlement problems with these deep fills stem from the following characteristics and processes:

- The fills are very heavy, and thus induce large compressive stresses in the lower parts of the fill and in the underlying soils.
- Loose natural soils were usually removed before placement of the fills, but these removals often were not as extensive as they should have been.
- Although the fills were typically placed at a moisture content near optimum, they slowly became wetter due to the infiltration of water from rain, irrigation, broken utility pipelines, and other sources.
- The wetting of the lower portions of the fill and the underlying natural soils often caused them to compress, causing settlements that varied with the fill thickness. When such fills were placed in old canyons, the thickness often changed quickly over short distances, thus causing large differential settlements (i.e., differences in settlements) over short distances.
- If the fill was made of expansive soils, the wetting caused heaving in the upper zones, which sometimes offset the compression of the underlying fill, and other times aggravated the differential settlements.

Unfortunately, these processes typically required several years to develop, which was long after structures and other improvements had been constructed on the fill. As a result, a large number of lawsuits ensued. These problems have resulted in new methods of designing and constructing such projects, including the following:

- The customary requirement for 90% relative compaction may not be sufficient for the lower portions of deep fills. In some cases, higher standards may be required.
- Removals of loose natural soils beneath proposed fills must be more aggressive than previously thought.
- The use of highly expansive soils needs to be more carefully controlled.
- Designs must be based on the assumption that some wetting will occur during the lifetime of the fill.

These effects may be evaluated by conducting an appropriate laboratory testing program using the proposed fill soils, then performing special analyses to predict the long-term settlements (Noorany, Sweet, and Smith, 1992). Based on the results of such a testing and analysis program, we can develop a design to keep fill settlements within tolerable limits.

QUESTIONS AND PRACTICE PROBLEMS

6.13 A proposed building site requires 1,200 yd^3 of imported fill. A suitable borrow site has been located, and the soils there have a shrinkage factor of 13%. How many cubic yards of soil must be excavated from the borrow site?

6.14 A contractor needs to excavate 50,000 yd^3 of silty clay and haul it with Caterpillar 69C dump trucks. Each truck can carry 30.9 yd^3 of soil per load, and operates on a 15-minute cycle. The job must be completed in five working days with the trucks working two 8-hour shifts per day. Using a bulking factor of 30%, how many trucks will be required?

6.15 A proposed grading plan requires 223,120 m^3 of cut and 206,670 m^3 of fill. Laboratory tests on a series of undisturbed samples from the cut area produced the following results:

Sample No.	Dry Unit Weight (kN/m^3)	Moisture Content (%)
3-1	17.3	9.1
3-2	17.7	9.5
5-1	16.8	8.9
5-2	17.1	7.2
8-1	16.0	12.0

A Proctor compaction test on a representative bulk sample produced a maximum dry unit weight of 19.2 kN/m^3 and an optimum moisture content of 10.2%. The project specifications require a relative compaction of at least 90%.

   a. Determine the shrinkage factor and compute the required volume of import or export, if any. Use the dry unit weight of the cut when computing any import or export quantities.
   b. Determine the weight of any import or export soil, assuming it has a moisture content equal to the average moisture content in the cut. Express your answer in metric tons (1 metric ton = 1000 kg$_f$ = 1 Mg$_f$).

6.16 A 3.0 ft deep cut is to be made across an entire 2.5-acre site. The average unit weight of this soil is 118 lb/ft^3, and the average moisture content is 9.6%. It also has a Proctor maximum dry unit weight of 122 lb/ft^3 and an optimum moisture content of 11.1%. The excavated soil will be placed on a nearby site and compacted to an average relative compaction of 93%. Compute the volume of fill that will be produced, and express your answer in cubic yards.

6.17 A proposed highway is to pass through a hilly area and will require both cuts and fills. The horizontal alignment is fixed, but the vertical alignment can be adjusted within certain limits to make the earthwork balance. The design engineer has developed four trial vertical alignments, with A being the lowest one and D being the highest. The resulting earthwork requirements are as follows:

Trial Vertical Alignment	Cut Volume (m^3)	Fill Volume (m^3)
A	40,350	35,120
B	39,990	35,490
C	39,180	36,010
D	38,400	36,950

Using a shrinkage factor of 12%, determine which alignment would balance the earthwork. If none of the trial alignments work, then express your answer in terms of two of them (i.e., 20% of the way from C to D).

6.18 Make a xerox copy of the grading plan in Figure 6.37, then compare the existing and proposed grades. Using colored pencils, apply red shading to the fill areas and blue shading to the cut areas. Consider only the area south of the tract boundary (line 21). You may interpolate and extrapolate between and beyond the two contour lines (the full drawing, which is much larger, includes many more contour lines). Finally, locate the area that will receive the greatest depth of fill, and determine this depth.

SUMMARY

Major Points

1. Many civil engineering projects include making cuts and fills, and this work requires the active participation of geotechnical engineers.
2. Improperly designed or constructed fills often are problematic, but with proper design and construction, they provide reliable support for structures, highways, and other civil engineering works.
3. The advent of modern earthmoving equipment has made it possible to perform very large earthwork projects with high levels of reliability and efficiency at very low cost.
4. Design engineers need to be familiar with earthwork construction methods and equipment.
5. Proper compaction is very important. It is usually assessed using the Proctor compaction test, along with a series of field density tests.
6. When soils are excavated, transported, and placed as compacted fill, their volume usually changes. Contractors need to consider these changes when estimating their equipment needs and productivity, and design engineers need to do so when developing grading plans.
7. Sometimes it is useful to use lightweight materials, such as geofoam, to build fills. These materials produce less stress in the underlying soils, and thus can help control settlement problems.

Vocabulary

backhoe	field density test	pneumatic roller
blasting	fill	Proctor compaction test
borrow pit	fine grading	relative compaction
bulking	flocculated fabric	rippability
bulldozer	geofoam	ripper
clearing and grubbing	grading	rough grading
compacted fill	grading plan	sand cone test
compaction	hoe	scraper
conveyor belt	hydraulic fill	sheepsfoot roller
cut	loader	shrinkage
drive cylinder test	maximum dry unit weight	shrinkage factor
dump truck	modified Proctor test	smooth steel-wheel rollers
earthwork	moisture conditioning	standard Proctor test
embankment	motor grader	tamping foot roller
excavatability	nuclear density test	tractor
excavation	optimum moisture content	water ring test
excavator	oriented fabric	water truck

COMPREHENSIVE QUESTIONS AND PRACTICE PROBLEMS

6.19 A fill soil with a natural moisture content of 10% and an optimum moisture content of 14% is being used to construct a compacted fill. The contractor is placing this soil in 400 mm lifts, spraying the top with a water truck, and compacting it using a towed sheepsfoot roller. A soils technician has performed a series of field density tests in this fill and has found relative compaction values between 80 and 92%. The measured moisture contents ranged from 10 to 23%. The specifications require a relative compaction of at least 90%, so the fill is not acceptable. What is wrong with the contractor's methods, and what needs to be done to remedy the problem?

6.20 A contractor needs to import 100,000 yd^3 (compacted volume) of soil to build a small earth dam. Two methods of hauling this soil are being considered, as follows:

Method A
  Use model XL37 scrapers, which have a capacity of 20 yd^3. These scrapers will need to be pushed by a model BD12 bulldozer while they are loading soil at the borrow site, then can travel unassisted to the dam site and deposit the soil there. One BD12 will be required for every six scrapers, and the scrapers can work on a cycle time of 30 min. The labor and equipment costs are as follows:

XL37 scraper	$250 per hour
BD12 bulldozer	$200 per hour

Method B
  Use model 98F wheel loaders at the borrow site to load the soil into model 356 dump trucks. Each dump truck has a capacity of 11 yd^3, and one loader will be

required to service every five dump trucks The dump trucks will then haul the soil to the dam site and deposit it there, which will require a cycle time of 20 min. The labor and equipment costs are as follows:

98F wheel loader	$180 per hour
356 Dump truck	$175 per hour

Either method is acceptable, so the choice between them will be based solely on cost. The soil has a bulking factor of 30% and a shrinkage factor of 12%. The hauling needs to be completed within 20 working days, using one 8-hour shift per day.

Determine how many scrapers, bulldozers, loaders, and dump trucks will be needed to complete the hauling in the required time, then compute the cost of each method. Based on this computed cost, select the better method for this project.

Note: These hourly rates are not necessarily representative of the actual costs of this equipment, and are for illustrative purposes only.

6.21 The proposed grading at a project site will consist of 25,100 m^3 of cut and 23,300 m^3 of fill and will be a balanced earthwork job. The cut area has an average moisture content of 8.3%. The fill will be compacted to an average relative compaction of 93% based on a maximum dry unit weight of 18.3 kN/m^3 and an optimum moisture content of 12.9%. Compute the volume of water in kiloliters that will be required to bring these soils to the optimum moisture content.

6.22 A well-graded silty sand with a maximum dry unit weight of 19.7 kN/m^3 and an optimum moisture content of 11.0% is being used to build a compacted fill. Two field density tests have been taken in the recently completed fill, but one of these tests has produced results that are definitely incorrect. Test A indicated a relative compaction of 85% and a moisture content of 8.9%, while Test B indicated a relative compaction of 98% and a moisture content of 14.9%. Which test is definitely incorrect? Why?

7

Groundwater — Fundamentals

All streams flow into the sea, yet the sea is never full. To the place the streams come from, there they return again.

Ecclesiastes 1:7 (NIV)

By the ancients, man has been called the world in miniature; and certainly this name is well bestowed, because inasmuch as man is composed of earth, water, air, and fire, his body resembles that of the earth; and as man has in him bones, the supports and framework of his flesh, the world has its rocks, the supports of the earth; as man has in him a pool of blood in which the lungs rise and fall in breathing, so the body of the earth has its ocean tide which likewise rises and falls every six hours, as if the world breathed; as in that pool of blood veins have their origin, which ramify all over the human body, so likewise the ocean sea fills the body of the earth with infinite springs of water. The body of the earth lacks sinews, and this is because the sinews are made expressly for movements and the world being perpetually stable, no movement takes place, and no movement taking place, muscles are not necessary. But in all other points they are much alike . . . if the body of the earth were not like that of a man, it would be impossible that the waters of the sea — being so much lower than the mountains — could by their nature rise up to the summits of these mountains. Hence, it is to be believed that the same cause which keeps the blood at the top of the head in man keeps the water at the summits of the mountains.

Leonardo da Vinci (1452–1519) as quoted by Biswas (1970)

Karl Terzaghi once wrote "... in engineering practice, difficulties with soils are almost exclusively due not to the soils themselves, but to the water contained in their voids. On a planet without any water there would be no need for soil mechanics" (Terzaghi, 1939). The presence of water, or at least the potential for its presence, is a key aspect of most geotechnical analyses. Therefore, this is a topic that is worthy of careful study.

Specific water-related geotechnical issues include:

- The effect of water on the behavior and engineering properties of soil and rock
- The potential for water flowing into excavations
- The potential for pumping water through wells or other facilities
- The effect of water on the stability of excavations and embankments
- The resulting uplift forces on buried structures
- The potential for seepage-related failures, such as piping
- The potential for transport of hazardous chemicals along with the water

In this chapter we will explore the fundamental principles of subsurface water. Chapter 8 continues this discussion and applies these principles to practical engineering problems.

7.1 HYDROLOGY

Hydrology is the study of water movements across the earth. It includes assessments of rainfall intensities, stream flows, and lake water levels, known as *surface water hydrology*, as well as studies of underground water, known as *groundwater hydrology*. These various movements are part of the grand process called the *hydrologic cycle*.

The Hydrologic Cycle

The movement of water across the earth is ultimately driven by energy received from the sun. Thus, the hydrologic cycle begins with water rising into the sky from open bodies of water through the process of *evaporation*, as shown in Figure 7.1. This process also draws water out of the near-surface soil, which dries the soil. A related process, called *transpiration*, acts through plants and draws water out of the ground through their roots.

Water in the sky, which may be in the form of invisible water vapor or visible clouds, eventually falls to the earth as *precipitation* (rain, sleet, hail, and snow), much of which goes directly into the oceans. The precipitation that falls onto land and onto inland bodies of water becomes the source of virtually all surface water, much of which flows *overland* until it reaches streams or rivers, where it becomes *streamflow*.

A significant portion of the surface water soaks into the ground, either while it is flowing overland or after it has reached rivers or lakes. This *infiltration* recharges the groundwater. Water applied to the ground by irrigation also soaks in and can contribute to the groundwater. Although some of the infiltrated water remains near the ground surface, much of it penetrates down until it reaches the groundwater table, which is the fully saturated zone.

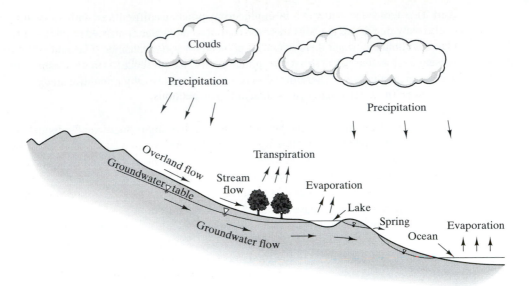

Figure 7.1 The hydrologic cycle.

These processes occur both in mountains and lowlands, so the groundwater eventually builds up and gains enough potential energy to begin flowing through the ground. Eventually this water reappears at the ground surface as *springs*, or it seeps directly into rivers or lakes. There it joins overland flows that eventually lead to *sinks* (low spots in the land) or the ocean, where the hydrologic cycle begins anew.

Groundwater Hydrology

Geotechnical engineers are mostly interested in the portions of the hydrologic cycle that occur underground. The term *subsurface water* encompasses all underground water, virtually all of which is located within the soil voids or rock fissures. A very small percentage of subsurface water is located in underground caverns, but this special case is not of much interest to geotechnical engineers.

We use various kinds information to describe and understand subsurface water. One of the most important is the *groundwater table* (also called the *phreatic surface*), which can be located by installing observation wells, as shown in Figure 3.23, and allowing the groundwater to seep into them until it reaches equilibrium. The water elevation inside these wells is, by definition, the groundwater table. Soil profiles represent the groundwater table as a line marked with a triangle, as shown in Figure 7.2. It also can be presented as a series of contour lines on a map. The groundwater table location is important, and finding it is one of the primary objectives of a site characterization program.

The groundwater table often changes with time, depending on the season of the year, recent patterns of rainfall, irrigation practices, pumping activities, and other factors. At some locations these fluctuations are relatively small (perhaps less than 1 m or 3 ft), while in other places the groundwater table elevation has changed by 10 m (30 ft) or more in only a year or two. Thus, the groundwater conditions encountered in an exploratory boring are not necessarily those we use for design. Often we need to use the observed conditions as a basis for estimating the worst-case conditions that are likely to occur during the project life.

Subsurface water may be divided into two sections:

- The portion below the groundwater table is called the *phreatic zone*. This water is subjected to a positive pressure as a result of the weight of the overlying water (and possibly due to other causes as well). Most subsurface water is in the phreatic zone.
- The portion above the groundwater table is called the *vadose zone*. This water has a negative pressure, and is held in place by capillary action and other forces present in the soil.

Technically, only the water in the phreatic zone is true *groundwater*. However, we often use the term "groundwater" to describe all subsurface water.

Some soils, such as sands and gravels, can transmit large quantities of groundwater. These are known as *aquifers*, and are good candidates for wells. Other soils, such as clays, transmit water very slowly. They are known as *aquicludes*. Intermediate soils, such as silty sand, pass water at a slow-to-moderate rate and are called *aquitards*. All three categories of soil might be present in a single soil profile, so the distribution and flow of groundwater can be quite complex. For example, a *perched groundwater* condition can occur when an aquiclude separates two aquifers, as shown in Figure 7.2. In this case there may be two or more groundwater tables.

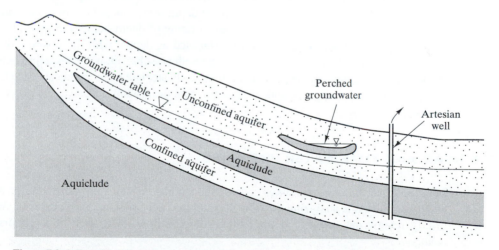

Figure 7.2 Soil profile showing complex nature of groundwater.

An *unconfined aquifer*, such as the upper aquifer in Figure 7.2, is one in which the bottom flow boundary is defined by an aquiclude, but the upper flow boundary (the groundwater table) is free to reach its own natural level. The groundwater occupies the lower portion of the aquifer, just as water in a kitchen pot occupies the lower part of the pot. The zone of soil through which the water flows is called the *flow regime*. If more groundwater arrived at the site, the groundwater table in an unconfined aquifer would rise accordingly.

Conversely, a *confined aquifer*, such as the lower aquifer in Figure 7.2, is one in which both the upper and lower flow boundaries are defined by aquicludes. This type of aquifer is similar to a pipe that is flowing full. Most confined aquifers also are *artesian*, which means the water at the top of the aquifer is under pressure. People often drill wells into such aquifers, because the water will rise up through the aquiclude without pumping. If the water pressure is high enough, *artesian wells* deliver water all the way to the ground surface without pumping.

Figure 7.2 shows an example of an artesian condition. Groundwater enters the confined aquifer from the left side of the cross-section, then travels down and to the right. By the time the water reaches the right side of the cross-section, it has developed an artesian head. In this case, someone has drilled a well through the overlying aquiclude and into the confined aquifer to take advantage of this artesian condition.

7.2 COORDINATE SYSTEM AND NOTATION

Our groundwater analyses will use the three-dimensional Cartesian coordinate system shown in Figure 7.3, where the x and y axes are in a horizontal plane and the z axis is vertical. For some analyses, geotechnical engineers express vertical dimensions in terms of elevation (z positive upward). However, it is usually more convenient to work in terms of depth (z positive downward). For example, we often speak of depth below the ground surface, depth below the groundwater table, or depth below the bottom of a foundation. In addition, boring logs are always presented in terms of depth below the ground surface. Therefore, in this book all z-values are expressed as depths with the positive direction downward, as shown in Figure 7.3. A z-value with no subscript indicates depth below the ground surface; z_w indicates depth below the groundwater table; and z_f indicates depth below a foundation or other applied load. The depth from the ground surface to the groundwater table is D_w.

7.3 HEAD AND PORE WATER PRESSURE

In a fluid mechanics or hydraulics course, you studied (or will study) the concept of *head* and its usefulness in analyzing the flow of water through pipes and open channels. We will briefly review this concept in the context of pipe flow, then apply it to groundwater analyses.

Consider the pipe shown in Figure 7.4. It has a cross-sectional area A, and contains water flowing from left to right at a velocity v. The *flow rate*, Q, is the quantity of water that passes through the pipe per unit of time:

$$Q = vA \qquad\qquad (7.1)$$

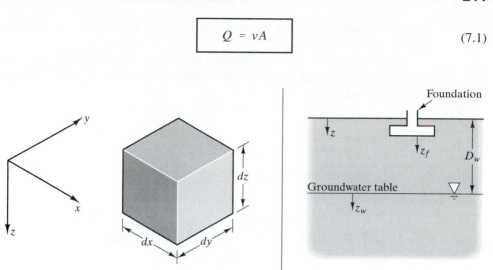

Figure 7.3 Coordinate system used in groundwater analyses, along with typical soil element.

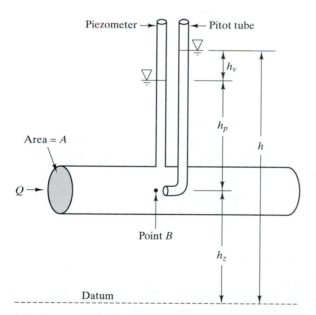

Figure 7.4 A pipe with a piezometer and a Pitot tube. These instruments measure the heads at Point B in the pipe.

This pipe also has a *piezometer*, which is simply a vertical tube with one end attached to the pipe and the other open to the atmosphere. Water from the pipe enters the piezometer and rises to the level shown. If the flow rate through the pipe remains constant, and the piezometer is sufficiently tall, the water level in the piezometer remains stationary and will not flow out of the top.

We also have installed a *Pitot tube*, named after its inventor, Henri DePitot (1695–1771), which is similar to a piezometer except the tip is pointed upstream and receives a ramming effect from the flowing water. The water level in the Pitot tube is thus slightly higher.

Finally, Figure 7.4 shows a horizontal datum elevation, which is the level from which elevations may be measured. This datum might be located at sea level, the laboratory floor, or some other suitable location.

Head

An element of groundwater, such as the one at Point B in Figure 7.4, contains energy in various forms, including:

- *Potential energy*, which is due to its elevation above the datum
- *Strain energy*, which is due to the pressure in the water
- *Kinetic energy*, which is due to its velocity

We could express these energies using Joules, BTUs, or some other suitable unit. However, it is more convenient to do so using the concept of *head*, which is energy divided by the acceleration of gravity, *g*. This method converts each form of energy to the equivalent potential energy and expresses it as the corresponding height. Thus, we express these three forms of energy as follows:

- The *elevation head*, h_z, is the difference in elevation between the datum and the point, as shown in Figure 7.4. It describes the potential energy at that point.
- The *pressure head*, h_p, is the difference in elevation between the point and the water level in a piezometer attached to the pipe. It describes the strain energy.
- The *velocity head*, h_v, is the difference in water elevations between the piezometer and the Pitot tube and describes the velocity head. It is related to the velocity, *v*, and acceleration due to gravity, *g*, as follows:

$$h_v = \frac{v^2}{2g} \tag{7.2}$$

The sum of these is the *total head*, *h*:

$$h = h_z + h_p + h_v \tag{7.3}$$

Equation 7.3 is called the *Bernoulli Equation*, and was named after the Swiss mathematician Daniel Bernoulli (1700–1782). It is one of the cornerstones of fluid mechanics and one of the most well-known equations in engineering, yet Bernoulli developed only part of the underlying theory and thus never wrote this equation. Later investigators completed the work and developed the equation as we now know it, but the credit has gone to Bernoulli.

The Bernoulli Equation is a convenient way to compare the energy at two points. For example, if the water at one point has an elevation head of 30 m, a pressure head of 10 m, and a velocity head of 5 m ($h = 30 + 10 + 5 = 45$ m), it has the same total head as the water at another point with an elevation head of 39 m, pressure head of 2 m, and velocity head of 4 m ($h = 39 + 2 + 4 = 45$ m).

Head Loss and Hydraulic Gradient

Figure 7.5 shows a pipe with piezometers and Pitot tubes at two points, A and B. As the water flows from A to B, some of its energy is lost due to friction, so the total head at B is less than at A. This difference is known as the *head loss, Δh*.

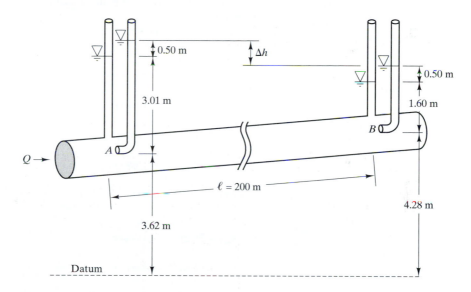

Figure 7.5 Head loss between two points in a pipe.

Water always flows from a point of high total head to a point of low total head. Thus, the water in this pipe must be flowing from Point A to Point B because the total head at A is greater than the total head at B. If the water was not flowing ($Q = 0$), then there would be no friction and no head loss, so the total heads at A and B would be equal.

The *hydraulic gradient, i,* is defined as:

$$i = -\frac{dh}{dl}$$

(7.4)

where:
i = hydraulic gradient
h = total head
l = distance the water travels

The total head decreases as water moves downstream (i.e., $dh < 0$, $dl > 0$), so i is always a positive number. In addition, both h and l are lengths, so i is unitless. A large hydraulic gradient reflects extensive friction, and thus usually means the water is flowing at a high velocity.

Example 7.1

Piezometers and Pitot tubes have been installed at Points A and B in the pipe shown in Figure 7.5. The water levels under steady-state flow are as shown. Determine the following:

The elevation, pressure, velocity, and total heads at Points A and B
The head loss between Points A and B
The hydraulic gradient between Points A and B

Solution

At Point A:
$$h_z = \textbf{3.62 m} \quad \Leftarrow Answer$$

$$h_p = \textbf{3.01 m} \quad \Leftarrow Answer$$

$$h_v = \textbf{0.50 m} \quad \Leftarrow Answer$$

$$h = h_z + h_p + h_v = 3.62 \text{ m} + 3.01 \text{ m} + 0.50 \text{ m} = \textbf{7.13 m} \quad \Leftarrow Answer$$

At Point B:
$$h_z = \textbf{4.28 m} \quad \Leftarrow Answer$$

$$h_p = \textbf{1.61 m} \quad \Leftarrow Answer$$

$$h_v = \textbf{0.50 m} \quad \Leftarrow Answer$$

$$h = h_z + h_p + h_v = 4.28 \text{ m} + 1.61 \text{ m} + 0.50 \text{ m} = \textbf{6.39 m} \quad \Leftarrow Answer$$

For pipe segment between A and B:
$$\Delta h = h_A - h_B = 7.13 \text{ m} - 6.39 \text{ m} = \textbf{0.74 m} \quad \Leftarrow Answer$$

$$i = \frac{\Delta h}{\Delta l} = \frac{0.74 \text{ m}}{200 \text{ m}} = \textbf{0.0037} \quad \Leftarrow Answer$$

Application to Soil and Rock

Although velocity head is important in pipe and open channel flow, the velocity of water flow in soil is much lower, so the velocity head is very small (less than 5 mm). Thus, we can neglect it for practical soil seepage problems. Equation 7.3 then reduces to:

$$h = h_z + h_p \qquad\qquad (7.5)$$

Thus, the total head, h, in soil also may be defined as the difference in elevation between the datum and the water surface in the piezometer. Also, in unconfined aquifers, i is equal to the slope of the groundwater table.

Field Instrumentation

Geotechnical engineers sometimes install piezometers in the ground to measure heads. Figure 7.6 shows a simple *open standpipe piezometer*, which is similar to the observation well shown in Figure 3.22 except the inside of the pipe is hydraulically connected only to one point in the soil. Observation wells are hydraulically open along nearly their entire length.

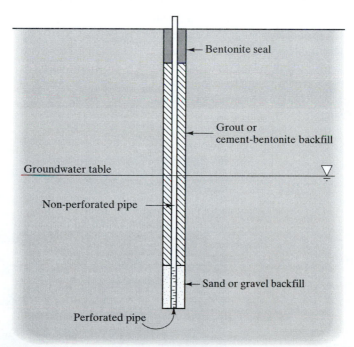

Figure 7.6 An open standpipe piezometer consists of a perforated pipe installed in a boring.

If the groundwater conditions consist of a simple groundwater table and the water is either stationary or flowing in a near-horizontal direction, piezometers and observation wells have virtually the same reading. However, if artesian or perched conditions are present, they would be quite different.

Both observation wells and open standpipe piezometers are read by simply determining the water elevation in the standpipe. This may be done by simply lowering a cloth tape measure with a weight on the end, or with an electronic water level indicator as shown in Figures 3.22 and 7.7.

Other types of piezometers also are available, such as the pneumatic piezometer shown in Figure 7.8. This device is read by applying nitrogen gas under pressure through a tube and matching the pressure to the pore water pressure. Pneumatic piezometers are more complex than open standpipe piezometers, and therefore less reliable, but they can be placed in difficult locations where open standpipes would be impossible to install. Similar units with electrical sensors also are available.

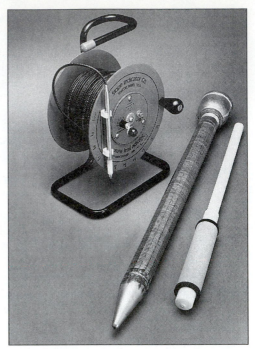

Figure 7.7 Filter tips for use on standpipe piezometers. The reel in the background is a probe to measure the water depth, as shown in Figure 3.22. (Slope Indicator Co.)

Figure 7.8 A pneumatic piezometer and readout unit. (Slope Indicator Co.)

Pore Water Pressure

The pressure in the water within the soil voids is known as the *pore water pressure*, *u*. This is what some engineers call gage pressure (i.e., it is the difference between the absolute water pressure and atmospheric pressure). For points below the groundwater table, the pore water pressure is:

$$u = \gamma_w h_p$$

(7.6)

where:

u = pore water pressure

h_p = pressure head

γ_w = unit weight of water = 62.4 lb/ft^3 = 9.81 kN/m^3

It is a simple matter to determine the pore water pressure at points where a piezometer is present. We simply determine the difference in elevation between the point and the water level in the piezometer (i.e., the pressure head), then use Equation 7.6. However, we usually do not have this luxury, and frequently need to compute *u* without the benefit of a piezometer.

To determine the pore water pressure at a point without a piezometer, we first need to determine if the pore water pressure is due solely to the force of gravity acting on the pore water. This is the case so long as the soil is not in the process of settling or shearing. We call this the *hydrostatic condition*, and the associated pore water pressure is the *hydrostatic pore water pressure*, u_h.

The next step is to determine if both of the following conditions also have been met:

- The aquifer is unconfined (i.e., the position of the groundwater table is not controlled by an overlying aquiclude)
- The groundwater is stationary or flowing in a direction within about 30° of the horizontal

If all these conditions have been met (which they often are), then the pressure head is simply the difference in elevation between the groundwater table and the point where the pressure head is to be computed. Therefore, the pore water pressure is:

$$u = u_h = \gamma_w z_w$$

(7.7)

where:

u = pore water pressure

u_h = hydrostatic pore water pressure

γ_w = unit weight of water = 62.4 lb/ft^3 = 9.81 kN/m^3

z_w = depth from the groundwater table to the point

For example, if the groundwater table is at elevation 215 ft and the point is at elevation 190 ft, then $u = \gamma_w z_w = (62.4 \text{ lb/ft}^3)(215\text{-}190 \text{ ft}) = 1600 \text{ lb/ft}^2$.

If the soil is not in a hydrostatic condition, then *excess pore water pressures* are present, as discussed in Chapters 11–13. If the water is flowing at a significant angle from the horizontal, or it is confined, then it is necessary to perform a two- or three-dimensional analysis as described later in this chapter.

Above the groundwater table, we normally consider the pore water pressure to be zero. In reality, surface tension effects between the water and the solid particles produce a negative pore water pressure above the groundwater table, and this negative pressure is sometimes called *soil suction*. Some advanced analyses sometimes consider soil suction, but they are beyond the scope of this book (see Fredlund and Rahardjo, 1993).

QUESTIONS AND PRACTICE PROBLEMS

7.1   Explain the difference between an aquifer, an aquiclude, and an aquitard, and the difference between confined flow and unconfined flow.

7.2   The water in a soil flows from Point K to Point L, a distance of 250 ft. Point K is at elevation 543 ft and Point L is at elevation 461 ft. Piezometers have been installed at both points, and their water levels are 23 ft and 74 ft, respectively, above the points. Compute the average hydraulic gradient between these two points.

7.3   Compute the pore water pressures at Points K and L in Problem 7.2.

7.4   The groundwater table in an unconfined aquifer is at a depth of 9.3 m below the ground surface. Assuming hydrostatic conditions are present, and the groundwater is virtually stationary, compute the pore water pressure at depths of 15.0 and 20.0 m below the ground surface.

7.5   An exploratory boring is being drilled. The soils encountered between the ground surface and a depth of 12 m have been dry, with no visible signs of groundwater. Then, at a depth of 12 m the soil became very wet. The boring continued to a depth of 15 m. An observation well was then installed in the boring. Within two days, the water in the observation well had risen to a depth of only 8 m below the ground surface. Explain the groundwater conditions that have been encountered.

7.4 GROUNDWATER FLOW CONDITIONS

Although evaluating pore water pressures is important, we often need to consider other groundwater characteristics as well. To do so, we need to identify certain conditions, as follows.

Laminar and Turbulent Flow

Sometimes water flows in a smooth orderly fashion, known as *laminar flow*. This flow pattern occurs when the velocity is low, and is similar to cars moving smoothly along an interstate highway. The other possibility is called *turbulent flow*, which means the water swirls as it moves. This happens when the velocity is high, and might be compared to an interstate highway filled with drivers who go too fast, weave back and forth, and occasionally do 360 degree turns. Turbulent flow consumes much more energy, and produces more head loss.

In most soils, the velocity is low, so the flow is laminar. This is important because many of our analyses are only valid for laminar flow. However, very coarse soils, such as clean, poorly-graded gravels, may have much higher velocities and thus have turbulent flow.

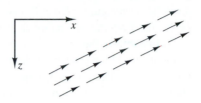

(a) One-dimensional flow

One-, Two-, and Three-Dimensional Flow

For analysis purposes, we need to distinguish between one-, two-, and three-dimensional flow conditions. A one-dimensional flow condition is one where the velocity vectors are all parallel and of equal magnitude, as shown in Figure 7.9. In other words, the water always moves parallel to some axis and through a constant cross-sectional area.

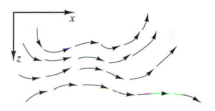

(b) Two-dimensional flow

Two-dimensional flow conditions are present when all of the velocity vectors are confined to a single plane, but vary in direction and magnitude within that plane. For example, the flow in natural soils beneath a concrete dam might be very close to a two-dimensional condition described along a vertical plane parallel to the river.

Three-dimensional flow is the most general condition. It exists when the velocity vectors vary in the x, y, and z directions. An example would be flow toward a water well.

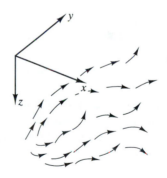

(c) Three-dimensional flow

Figure 7.9 One-, two-, and three-dimensional flow conditions.

Steady and Unsteady Flow

The term *steady-state condition* means a system has reached equilibrium. In the context of groundwater analyses, it means the flow pattern has been established and is not in the process of changing. We call this *steady flow* or *steady-state flow*. When this condition exists, the flow rate, Q, remains constant with time.

In contrast, the *unsteady condition* (also known as the *transient condition*) exists when something is in the process of changing. For seepage problems, *unsteady flow* (or *transient flow*) occurs when the pore water pressures, groundwater table location, flow rate, or other characteristics are changing, perhaps in response to a change in the applied head. In other words, steady flow does not vary with time, while unsteady flow does.

For example, consider a levee that protects a town from a nearby river. Some of the water in the river seeps through the levee as shown in Figure 7.10, forming a groundwater table. This is a steady-state condition. If the river rapidly rises, such as during a flood, the groundwater table inside the levee also rises. However, the groundwater inside the levee responds slowly, so some time will be required before achieving the new steady-state condition. During this transition period, the flow is unsteady.

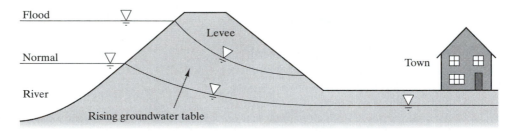

Figure 7.10 Flow through a levee adjacent to a river.

Analyses of unsteady flows are much more complex, and are generally beyond the scope of this chapter. However, we will study an important unsteady flow process called *consolidation* in Chapters 11 and 12.

7.5 ONE-DIMENSIONAL FLOW

One-dimensional flow is the easiest condition to understand. It is used in laboratory test equipment, and some field conditions may be idealized as one-dimensional flow problems.

Geotechnical engineers often need to predict the flow rate, Q, through a soil. We could use Newton's law of friction, combined with the Navier-Stokes equations of hydrodynamics to describe one-dimensional flow through soils. However, the resulting formulas are very complex and thus impractical for normal geotechnical engineering analyses. Therefore, engineers follow the simpler empirical method developed by the French engineer Darcy (1856).

Darcy performed experimental studies of the flow of water through sand and developed the following relationship, known as *Darcy's Law*:

$$\boxed{Q = kiA} \tag{7.8}$$

where:
 Q = flow rate
 k = hydraulic conductivity (also known as coefficient of permeability)
 i = hydraulic gradient
 A = area perpendicular to the flow direction

  The cross-sectional area, A, in Equation 7.8 includes both the voids and the solids, even though the groundwater flows only through the voids. For example, to evaluate groundwater flowing through a 2 m × 4 m zone of soil, we use $A = 8$ m^2, even though only a fraction of this area is voids. This definition simplifies the computations, and does not introduce any error in computations of Q so long as we use it consistently.

  Although Darcy's law was developed empirically, it has been found to be valid for a wide range of soil types, from clays through coarse sands. The primary exceptions are clean gravels, where its accuracy is diminished because of the turbulent flow, and possibly in clays with low hydraulic gradients, because the flow rate is so small.

Hydraulic Conductivity (Coefficient of Permeability)

The *hydraulic conductivity*, k, for a given liquid and soil depends on many factors, including:

   Soil properties
    • void size (depends on particle size, gradation, void ratio, and other factors)
    • soil structure
    • void continuity
    • particle shape and surface roughness
   Liquid properties
    • density
    • viscosity

  Most practical problems deal with clean water or water contaminated with small quantities of other substances. Any variations in density and viscosity are small and can usually be ignored. Thus, we normally think of the hydraulic conductivity as being dependant only on the soil.

  The most common unit of measurement for k is cm/s. However, many other units also are used, including ft/min, ft/yr, and even gallons/day/ft^2. Although k has units of length/time; the same as those used to describe velocity, it is not a measure of velocity. Table 7.1 presents typical values for different soil types.

  Notice the extremely wide range of k values in Table 7.1. For example, clays typically have a k that is 1,000,000 times smaller than that of sands. Thus, according to Equation 7.8,

Q also will be 1,000,000 times smaller. The low k in clays is due to the small particle size (and therefore small void size). It is not due to water being absorbed by the clay.

TABLE 7.1 TYPICAL VALUES OF HYDRAULIC CONDUCTIVITY, k, FOR SATURATED SOILS

Soil Description	Hydraulic Conductivity, k	
	(cm/s)	(ft/s)
Clean gravel	1 to 100	3×10^{-2} to 3
Sand-gravel mixtures	10^{-2} to 10	3×10^{-4} to 0.3
Clean coarse sand	10^{-2} to 1	3×10^{-4} to 3×10^{-2}
Fine sand	10^{-3} to 10^{-1}	3×10^{-5} to 3×10^{-3}
Silty sand	10^{-3} to 10^{-2}	3×10^{-5} to 3×10^{-4}
Clayey sand	10^{-4} to 10^{-2}	3×10^{-6} to 3×10^{-4}
Silt	10^{-8} to 10^{-3}	3×10^{-10} to 3×10^{-5}
Clay	10^{-10} to 10^{-6}	3×10^{-12} to 3×10^{-8}

Geotechnical engineers also use the term *coefficient of permeability* to describe k. Unfortunately, this term can generate some confusion because it has at least two definitions:

- The *coefficient of permeability*, k, in Darcy's Law describes the ease with which a certain liquid flows through a certain soil. It depends on both the soil and the liquid, has units of L/t, and is the same as *hydraulic conductivity*.
- The *intrinsic permeability* (also called the *specific permeability*) which depends only on the soil. It has units of L^2. This definition is used primarily by hydrogeologists and petroleum geologists.

Both parameters are frequently called "permeability" and the variables K and k have been used for both. In addition, the intrinsic permeability is sometimes measured with a special unit called Darcys, even though it is not the proper parameter for use in Equation 7.8. These inconsistencies in terminology can be a source of confusion. Whenever in doubt, check the units to determine which "permeability" is being used.

To avoid confusion, geotechnical engineers are gradually dropping the term "coefficient of permeability" and using "hydraulic conductivity" instead.

Hydraulic Conductivity Tests (Permeability Tests)

Various laboratory and in-situ tests are available to measure the hydraulic conductivity. However, the results from both types are often judged with some skepticism because we are not sure if the test samples are truly representative. Even small differences in the soil classification can make a big difference in k, as illustrated in Table 7.1. Thus, it is good to perform many tests and review the scatter in the results.

Another problem with laboratory tests is that the samples probably don't adequately represent small fissures, joints, sandy seams, and other characteristics in the field. In-situ tests are better in this regard.

Even carefully conducted tests on good samples typically have a precision on the order of ±50 percent or more. Therefore, test results are normally reported to only one significant figure (i.e., 5×10^{-4} cm/s). Even then, we need to recognize the true hydraulic conductivity in the field may be substantially different from the test values.

Constant-Head Test

The *constant head test* is a laboratory hydraulic conductivity test that applies a constant head of water to each end of a soil sample in a *permeameter* as shown in Figure 7.11. The

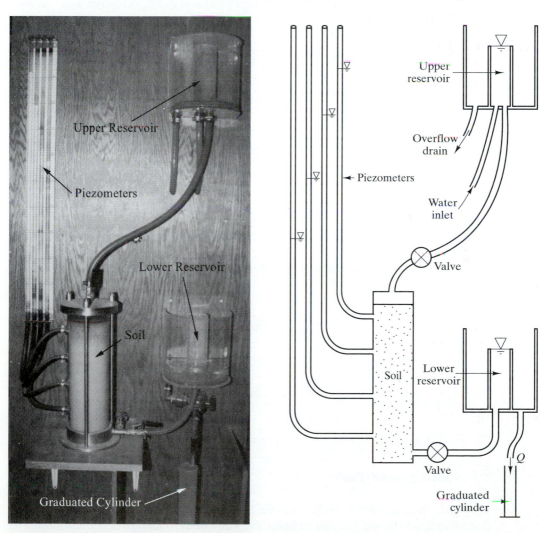

Figure 7.11 A constant head permeameter. The upper and lower reservoirs contain inner and outer chambers that maintain constant heads. The piezometers measure the head at four locations in the soil sample.

head on one end is greater than that on the other, so a flow is induced. We determine Q, i, and A from the test results, then compute k using Equation 7.8.

Some constant-head permeameters do not have piezometers, so we must compute the hydraulic gradient, i, by dividing the head loss between the two reservoirs (i.e., the difference in their water surface elevations) by the height of the sample. This method implicitly assumes the head losses in the tubes, valves, etc., is very small compared to that in the soil. This is generally a poor assumption. It is better to use permeameters with piezometer and compute i by dividing the difference in the total heads by the distance between the piezometer tips.

Example 7.2

The graduated cylinder in the right side of Figure 7.11 collects 892 ml of water in 112 seconds. The dimensions are:
- soil sample diameter = 18.0 cm
- elevation of water in upper piezometer = 181.0 cm
- elevation of water in lower piezometer = 116.6 cm
- the piezometer tips are spaced 16.7 cm on center

Compute the hydraulic conductivity, k.

Solution

$$Q = \frac{V}{t} = \left(\frac{892 \text{ ml}}{112 \text{ s}} \right) \left(\frac{1 \text{ cm}^3}{\text{ml}} \right) = 7.96 \frac{\text{cm}^3}{\text{s}}$$

$$A = \frac{\pi D^2}{4} = \frac{\pi \, 18.0^2}{4} = 254 \text{ cm}^2$$

Note how A includes both the voids and the solids, as defined earlier.

$$i = -\frac{\Delta h}{\Delta l} = \frac{116.6 \text{ cm} - 181.0 \text{ cm}}{3 \times 16.7 \text{ cm}} = 1.29$$

$$Q = kiA \quad \rightarrow \quad k = \frac{Q}{iA} = \frac{7.96 \text{ cm}^3/\text{s}}{(1.29)(254 \text{ cm}^2)} = 2 \times 10^{-2} \text{ cm/s} \quad \leftarrow \textit{Answer}$$

Falling-Head Test

The *falling-head test* also is conducted in the laboratory. It uses a standpipe on the upstream side as shown in Figure 7.12. In addition, the water in the standpipe is not replenished as it is in the constant-head reservoir. Thus, as the test progresses, the water level in the standpipe falls. This method is more suitable for soils with very low hydraulic conductivities, such as clays, where the flow rate is small and needs to be precisely measured.

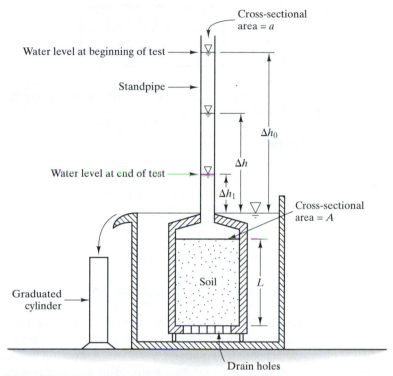

Figure 7.12 A falling-head permeameter.

The analysis of falling-head test results is more complex because the hydraulic gradient is not constant. This means the flow rate also is not constant (per Equation 7.8), so we must derive a new formula for k as follows:

Considering a head loss Δh_0 at the beginning of the test and Δh_1 after time t,

$$Q = kiA = k\frac{\Delta h}{L}A = -a\frac{d(\Delta h)}{dt}$$

$$\frac{kA}{L}\int_0^t dt = -a\int_{\Delta h_0}^{\Delta h_1} \frac{d(\Delta h)}{\Delta h}$$

$$\frac{kAt}{L} = -a \ln\left(\frac{\Delta h_1}{\Delta h_0}\right)$$

$$k = \frac{aL}{At} \ln\left(\frac{\Delta h_0}{\Delta h_1}\right) \qquad (7.9)$$

Notice how the volume of water is determined by the change in water level in the standpipe. The graduated cylinder is used only as a check.

In-Situ Tests

Hydraulic conductivity tests also may be performed in-situ, especially in sandy soils. These tests reflect the flow characteristics of a much larger volume of soil, and therefore should, at least in theory, produce more reliable results. Several kinds of in-situ hydraulic conductivity tests have been developed.

A *slug test* consists of installing a well through the aquifer, allowing it to reach equilibrium, then quickly adding or removing a "slug" of water. The natural flow of groundwater into or out of the well is then monitored until equilibrium is once again achieved and k is computed based on this data and theories of groundwater flow around wells.

Another method, which is more reliable but more expensive, consists of installing a pumping well and a series of two to four observation wells. These are then used to conduct a *pumping test*, which consists of pumping water out of the main well and monitoring the changes in water levels in the observation wells. This pumping continues until a steady-state condition is achieved, then k is computed based on the steady-state water levels and the principles of groundwater flow. The appropriate equations are presented in Chapter 8.

Hazen's Correlation

The hydraulic conductivity, k, is approximately proportional to the square of the pore diameter. In addition, the average pore diameter in clean sands is roughly proportional to D_{10}. Using this information, Hazen (1911) developed the following empirical relationship for loose, clean sands:

$$k = C D_{10}^2 \qquad (7.10)$$

where:

k = hydraulic conductivity (cm/s)

C = Hazen's coefficient = 0.8 to 1.2 (a value of 1.0 is commonly used)

D_{10} = Diameter at which 10 percent of the soil is finer (mm) (also known as the *effective size*)

Note: Be sure to use the stated units for k and D_{10}

Hazen's work was intended to be used in the design of sand filters for water purification, but can be used to estimate k in the ground. However, its applicability is limited to soils with 0.1 mm < D_{10} < 3 mm and a coefficient of uniformity, C_u < 5 (Kashef, 1986).

Example 7.3

A 3.2 m thick silty sand strata intersects one side of a reservoir as shown in Figure 7.13. This strata has a hydraulic conductivity of 4×10^{-2} cm/s and extends along the entire 1000 m length of the reservoir. An observation well has been installed in this strata as shown. Compute the seepage loss from the reservoir through this strata.

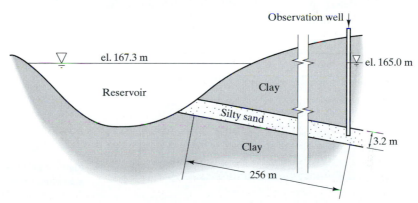

Figure 7.13 Cross-section through reservoir showing silty sand stratum. This cross-section is oriented parallel to the direction of flow.

Solution

The observation well indicates a water level above the top of the silty sand strata. Therefore, it is a confined aquifer.

$$i = -\frac{dh}{dl} = -\frac{165.0\,\text{m} - 167.3\,\text{m}}{256\,\text{m}} = 0.0090$$

$$A = (3.2\,\text{m})(1000\,\text{m}) = 3200\,\text{m}^2$$

$$k = (4\times10^{-2}\text{cm/s})\left(\frac{\text{m}}{100\,\text{cm}}\right)\left(\frac{3600\,\text{s}}{\text{hr}}\right)\left(\frac{24\,\text{hr}}{\text{d}}\right)\left(\frac{30\,\text{d}}{\text{mo}}\right) = 1000\,\text{m/mo}$$

$$Q = kiA = (1000\,\text{m/mo})(0.0090)(3200\,\text{m}^2) = \textbf{30,000\,m}^3\textbf{/mo} \qquad \leftarrow Answer$$

Flow Through Anisotropic Soils

Many natural soils, especially alluvial and lacustrine soils, contain thin horizontal stratifications that reflect their history of deposition. For example, there may be alternating layers of silt and clay, each only a few millimeters thick, as shown in Figure 7.14. The hydraulic conductivity in some layers is often much greater than in others, so groundwater flows horizontally much more easily than vertically. Such soils are said to be *anisotropic with respect to hydraulic conductivity*, so we need to determine two values of k: the horizontal hydraulic conductivity, k_x, and the vertical hydraulic conductivity, k_z.

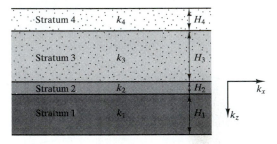

Figure 7.14 Flow of water through anisotropic soils.

Assuming the stratifications are horizontal, the horizontal and vertical hydraulic conductivities may be computed as follows:

$$k_x = \frac{\sum k_i H_i}{\sum H_i} \tag{7.11}$$

$$k_z = \frac{\sum H_i}{\sum \left(\dfrac{H_i}{k_i}\right)} \tag{7.12}$$

where:

k_x = horizontal hydraulic conductivity
k_z = vertical hydraulic conductivity
k_i = hydraulic conductivity of horizontal strata i
H_i = thickness of horizontal strata i

Many groundwater problems involve alluvial soils because they are often near rivers and often have shallow groundwater tables. Most alluvial soils have horizontal stratifications, so $k_x > k_z$. Varved clays also have $k_x > k_z$.

Example 7.4

A certain varved clay consists of alternating horizontal layers of silt and clay. The silt layers are 5 mm thick and have $k = 3 \times 10^{-4}$ cm/s; the clay layers are 20 mm thick and have $k = 6 \times 10^{-7}$ cm/s. Compute k_x and k_z.

Solution

$$k_x = \frac{\sum k_i H_i}{\sum H_i}$$

$$= \frac{(3 \times 10^{-4}\ \text{cm/s})(0.5\ \text{cm}) + (6 \times 10^{-7}\ \text{cm/s})(2\ \text{cm})}{0.5\ \text{cm} + 2\ \text{cm}}$$

$$= \mathbf{6 \times 10^{-5}\ cm/s} \quad \leftarrow Answer$$

$$k_z = \frac{\sum H_i}{\sum \left(\dfrac{H_i}{k_i} \right)}$$

$$= \frac{0.5\ \text{cm} + 2\ \text{cm}}{\dfrac{0.5\ \text{cm}}{3 \times 10^{-4}\ \text{cm/s}} + \dfrac{2\ \text{cm}}{6 \times 10^{-7}\ \text{cm/s}}}$$

$$= \mathbf{7 \times 10^{-7}\ cm/s} \quad \leftarrow Answer$$

Commentary

Water flowing horizontally moves through both types of soil in parallel. The ratio of thicknesses for the two strata is $20/5 = 4$, but the ratio of k values is $3 \times 10^{-4} / 6 \times 10^{-7} = 500$. Thus, the silt layers dominate the horizontal flow, even though they represent only 20% of the total cross-sectional area of flow. This is why the value of k_x is nearly equal to k of the silt. However, water flowing vertically must pass through all of the soil layers, and thus is controlled by the ones with the lowest hydraulic conductivity. Therefore, k_z is nearly equal to k of the clay.

Transmissivity

The *transmissivity*, T, of an aquifer (also called its *transmissibility*) is the product of the hydraulic conductivity and the saturated thickness of the aquifer, H_a:

$$T = kH_a \qquad (7.13)$$

For confined flow, H_a is the thickness of the aquifer.

Combining this formula with Darcy's Law (Equation 7.8) produces the flow rate through an aquifer of width L:

$$Q = TiL \qquad (7.14)$$

Rewriting to express Q as the flow per unit width of the aquifer gives:

$$q = \frac{Q}{L} = Ti \qquad (7.15)$$

where:
Q = flow rate through the aquifer
q = flow rate per unit width of the aquifer
T = transmissivity
i = hydraulic gradient
L = length of aquifer perpendicular to the direction of flow
H_a = saturated thickness of the aquifer

Example 7.5

Three observation wells have been installed in an unconfined aquifer, as shown in Figure 7.15. Steady-state hydrostatic conditions are present. The aquifer has a uniform thickness, and its bottom is 50 ft below the groundwater table. The hydraulic conductivity is 3×10^{-4} ft/s. Determine the hydraulic gradient, the direction of flow, the transmissivity, and the flow rate per unit width of the aquifer.

Solution

Evaluate the data on a horizontal plane by defining an x,y coordinate system as shown, then define i_x and i_y as the hydraulic gradients in the x and y directions, respectively. This allows us to write the following equations:

$$A \rightarrow B \qquad 8000\, i_x + 1800\, i_y = 35 - 22 = 13$$

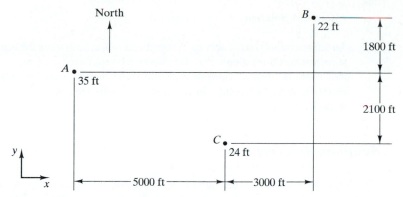

Figure 7.15 Plan view of observation wells A, B, and C, and elevations of observed water surfaces in the wells.

$$A \rightarrow C \qquad 5000\, i_x - 2100\, i_y = 35 - 24 = 11$$

Solving these two equations simultaneously gives $i_x = 0.00183$ and $i_y = -0.00089$. The resultant of these two components is:

$$i = \sqrt{i_x^2 + i_y^2} = \sqrt{0.00183^2 + (-0.00089)^2} = \mathbf{0.0020} \qquad \Leftarrow Answer$$

$$\theta = \tan^{-1}\left(\frac{0.00089}{0.00183}\right) = 26°$$

$$\text{Azimuth} = 90 + 26 = \mathbf{116}\,° \qquad \Leftarrow Answer$$

$$T = kh = (3\times10^{-4}\,\text{ft/s})(50\,\text{ft}) = \mathbf{1.5\times10^{-2}\,ft^2/s} \qquad \Leftarrow Answer$$

$$q = Ti = (1.5\times10^{-2}\,\text{ft}^2/\text{s})(0.0020\,\text{ft/ft}) = \mathbf{3\times10^{-5}}\,\frac{\text{ft}^3/\text{s}}{\text{ft}} \qquad \Leftarrow Answer$$

Alternative Solution

An alternative method of solving this problem would be to write the equation $z = Ax + By + C$ at each of the observation wells. This would produce three equations and three unknowns, and thus could be solved. Such an equation could then be used to compute the groundwater elevation at any x,y point. In addition, the coefficients A and B are equal to $-i_x$ and $-i_y$, respectively.

Seepage Velocity

The *seepage velocity*, v_s, is the rate of movement of an element of water through a soil. This rate is important in geoenvironmental problems because it helps us determine how quickly contaminants travel through the ground, as discussed in Chapter 9.

At first, it may seem that v_s could be computed by combining Equations 7.1 and 7.8, but this would be incorrect. Equation 7.1 describes flow through pipes, where A is the cross-sectional area of flow. However, Equation 7.8 describes flow through soils where A is the total cross-sectional area of the voids and the solids. Water in soil only flows through the voids, so these two A values are incompatible.

We reconcile this problem by accounting for the *effective porosity*, n_e, which is the percentage of A in Equation 7.8 that actually contributes to the flow. Thus,

$$v_s = \frac{ki}{n_e} \qquad (7.16)$$

where:
    v_s = seepage velocity
    k = hydraulic conductivity
    i = hydraulic gradient
    n_e = effective porosity

Although water flowing through a soil actually takes a circuitous route around the soil particles, the seepage velocity is based on the equivalent straight-line movement as shown in Figure 7.16. Even though this method produces a velocity that is smaller than the "true" velocity (when viewed on a microscopic scale), it is much more convenient for solving contaminant transport problems.

In sandy soils, n_e is equal to the porosity n as defined in Equation 4.16. However, clayey soils contain a static layer of water around the particles, so the actual flow area is less than the void area. Design values of n_e in clays are best determined using special laboratory tests (Kim, Edil, and Park, 1997). If no test data is available, the Environmental Protection Agency uses $n_e = 0.10$ in clays (Brumund, 1995).

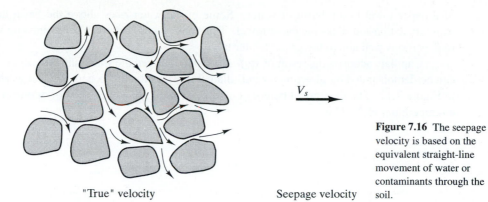

V_s

Figure 7.16 The seepage velocity is based on the equivalent straight-line movement of water or contaminants through the soil.

"True" velocity   Seepage velocity

Example 7.6

Using the data in Example 7.5 and a void ratio of 0.85, compute the seepage velocity through the aquifer.

Solution

The soil is sandy, so the effective porosity equals the porosity, n. Using Equation 4.17:

$$n = \frac{e}{1+e} = \frac{0.85}{1+0.85} = 46\%$$

$$v_s = \frac{ki}{n_e} = \frac{(3\times10^{-4}\ \text{ft/s})(0.0020)}{0.46}\left(\frac{3600\ \text{s}}{\text{hr}}\right) = 5\times10^{-3}\ \text{ft/hr} \qquad \leftarrow Answer$$

At this rate, about 120 years would be required for the water to travel one mile. This is very slow, due primarily to the low hydraulic gradient. At other sites, the seepage velocity can be much higher.

7.6 CAPILLARITY

Earlier in this chapter we defined the groundwater table as the elevation to which water would rise in an observation well. Soils below the groundwater table are saturated, and have a pore water pressure defined by Equation 7.7. For many engineering problems, this simple model is sufficient. However, the real behavior of soils is rarely so simple! One important aspect not addressed by this model is capillarity.

Capillarity (or *capillary action*) is the upward movement of a liquid into the vadose zone, which is above the level of zero hydrostatic pressure. This upward movement occurs in porous media or in very small tubes, and can be illustrated by gently lowering the edge of a paper towel into a basin of water. Some of the water rises above the basin due to capillary action and soaks the paper towel. The same process occurs in soils, drawing water to elevations well above the groundwater table.

Capillary action is the result of *surface tension* between the water and the media. This can be demonstrated by inserting a small-diameter glass tube into a pan of water as shown in Figure 7.17. The theoretical height of capillary rise, h_c, in a glass tube of diameter d, at a temperature of 20°C is:

$$h_c = \frac{0.03}{d} \tag{7.17}$$

where:
$\quad h_c$ = height of capillary rise (m)
$\quad d$ = diameter of glass tube (mm)

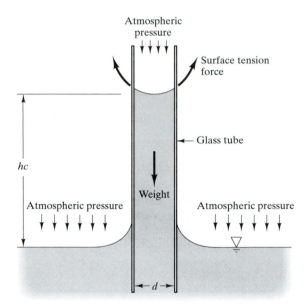

Figure 7.17 Capillary action demonstrated by a thin glass tube immersed in water.

Capillary rise in soils is more complex because soils contain an interconnected network of different-size pores. However, using $0.2 D_{10}$ as the equivalent d generally produces good results for sands and silts (Holtz and Kovacs, 1981). Thus, the height of capillary rise in these soils is approximately:

$$h_c = \frac{0.15}{D_{10}} \tag{7.18}$$

where:

h_c = height of capillary rise (m)

D_{10} = 10% grain diameter (mm)

The theoretical capillary rise in clays can be in excess of 100 m, but in reality it is much less.

When the near-surface soils are frozen, capillary rise can produce underground ice lenses, which result in frost heave. Chapter 18 discusses this important phenomena in more detail.

QUESTIONS AND PRACTICE PROBLEMS

7.6   A constant-head hydraulic conductivity test has been conducted on a 110 mm diameter, 270 mm tall fine sand sample in a permeameter similar to the one shown in Figure 7.11. The upper and lower reservoir elevations were 2010 mm and 1671 mm above the lab floor. The piezometers, whose tips are spaced 200 mm apart, had readings of 1809 and 1578 mm, and the graduated cylinder collected 910 ml of water in 25 min 15 s. Using the best available data, compute the hydraulic conductivity. Does the result seem reasonable? Why or why not?

7.7   May we use the Hazen correlation to estimate the hydraulic conductivity for soil C in Figure 4.13? Why or why not? If so, compute k.

7.8   Which of the following methods would be the better way to determine k for a clean sand? Why?

      a. Place a soil sample in a constant-head permeameter, conduct a hydraulic conductivity test, and compute k using Equation 7.8.

      b. Conduct a sieve analysis, determine D_{10}, and compute k using Equation 7.10.

7.9   A falling-head hydraulic conductivity test has been conducted on a clay sample in a permeameter similar to the one in Figure 7.12. The soil sample was 97 mm in diameter and 20 mm tall. The standpipe had an inside diameter of 6.0 mm. The water level in the bath surrounding the sample was 120 mm above the laboratory counter top and the water level in the standpipe fell from a height of 510 mm to 261 mm above the counter top in 46 hours 35 minutes. Compute the hydraulic conductivity. Does the result seem reasonable? Why or why not?

7.10   Derive Equations 7.11 and 7.12.

    Hint: For Equation 7.11, write Darcy's law for horizontal flow using the real soil ($k = k_1, k_2$ etc.), then write it again using the equivalent soil ($k = k_x$). Since Q is the same for both, you can solve for k_x.

7.11   A sandy soil with $k = 3 \times 10^{-2}$ cm/s contains a series of 5 mm thick horizontal silt layers spaced 300 mm on center. The silt layers have $k = 5 \times 10^{-6}$ cm/s. Compute k_x and k_z and the ratio k_x/k_z.

7.12   When drilling an exploratory boring through the soil described in Problem 7.11, how easy would it be to miss the silt layers? If we did miss them, how much effect would our ignorance have on computations of Q for water flowing vertically? Explain.

7.13 Three piezometers have been installed in the confined aquifer shown in Figure 7.18. The aquifer has a uniform thickness of 3.5 m and a hydraulic conductivity of 2×10^{-1} cm/s. Determine the hydraulic gradient, the direction of flow, the transmissivity, and the flow rate per unit width of the aquifer.

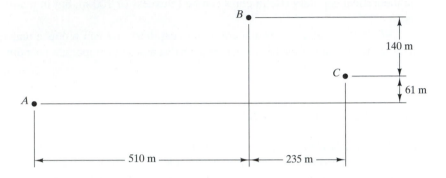

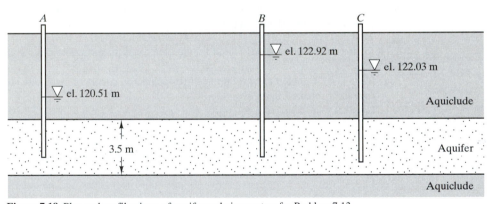

Figure 7.18 Plan and profile views of aquifer and piezometers for Problem 7.13.

7.14 A certain 20 m thick sandy aquifer has a transmissivity of 0.12 m²/s and a void ratio of 0.91. Groundwater is flowing through this aquifer with a hydraulic gradient of 0.0065. How much time would be required for water to travel 1 km through this aquifer?

7.15 The laboratory apparatus shown in Figure 7.19 maintains a constant head in both the upper and lower reservoirs. The soil sample is a silty sand (SM) with $k = 5 \times 10^{-3}$ cm/s and $w = 18.5\%$. Assume a reasonable value for G_s, then determine the time required for the plug of colored water to pass through the soil (i.e., from when the leading edge first enters the soil to when it begins to exit). Assume there is no diffusion (i.e., the colored water plug has the same volume when it exits as when it entered the soil). Also assume the colored water has the same unit weight and viscosity as plain water.

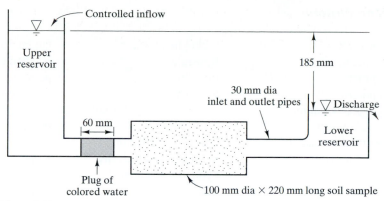

Figure 7.19 Laboratory apparatus for Problem 7.15.

7.16 Two small commercial buildings have been constructed at a site underlain by a sandy silt (ML) that has $D_{10} = 0.03$ mm. The groundwater table is at a depth of 6 ft. Both buildings have concrete slab-on-grade floors. In Building A, the slab was placed directly onto the natural soils, while Building B has a 4-inch layer of poorly- graded coarse gravel between the slab and the natural soils. Both buildings have vinyl floor coverings similar to those typically used in residential kitchens. Both buildings are now three years old.

Unfortunately, the tenant in Building A is having continual problems with the vinyl floors peeling up from the concrete slab. When the peeled sections are examined, moisture is always evident between the vinyl and the concrete. Curiously, the tenant in Building B has had no such problems, even though both buildings have the same floor covering. Could the problem in Building A be due to capillary action in the underlying soil? Explain why or why not. Also explain why Building B is not having any such problems.

SUMMARY

Major Points

1. The pores in a soil are interconnected, so the pore water can travel through it. Such movements of underground water are important to geotechnical engineers and others.
2. The energy in groundwater may be defined in terms of its head, h, which then may be used to evaluate water movements.
3. The hydraulic gradient, i, describes the head loss per unit distance of groundwater travel.
4. The pore water pressure is the gage pressure of the pore water at a given location.
5. The flow of water through soil is usually described using Darcy's Law.
6. The hydraulic conductivity is a factor in Darcy's Law that describes the ease with which water flows through a given soil when all other factors are fixed.
7. The seepage velocity describes the rate at which water flows through the ground. It is often used in contaminant transport analyses.
8. Water can rise well above the groundwater table through capillary action. One of the unfortunate consequences of this action is frost heave.

Vocabulary

anisotropic	falling head test	laminar flow
aquiclude	flow rate	perched groundwater
aquifer	flow regime	permeameter
aquitard	groundwater	phreatic zone
artesian condition	groundwater hydrology	piezometer
artesian well	groundwater table	pore water pressure
Bernoulli equation	Hazen's correlation	pressure head
capillary rise	hydraulic conductivity	seepage velocity
capillarity	hydraulic gradient	steady-state condition
confined aquifer	hydrologic cycle	total head
constant head test	hydrology	transmissivity
Darcy's law	hydrostatic condition	unconfined aquifer
effective porosity	hydrostatic pore water	vadose zone
elevation head	pressure	

COMPREHENSIVE QUESTIONS AND PRACTICE PROBLEMS

7.17 Apparatus A, shown in Figure 7.20, consists of a single 20 mm diameter pipe and is subjected to a head difference of 60 mm. Apparatus B consists of four 10 mm diameter pipes connected in parallel. How will the flow rate through A compare with that through B? Explain.

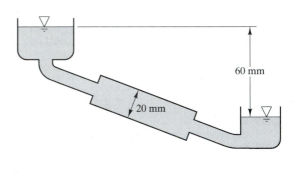

7.18 Based on your observations in Problem 7.17, explain why saturated clays have a significantly lower hydraulic conductivity than saturated sands, even though the total void areas per square foot of soil are about the same for both.

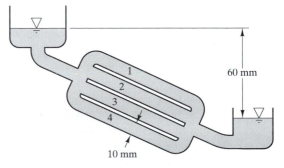

7.19 An engineer is searching for a suitable soil to cap a sanitary landfill. This soil must have a hydraulic conductivity no greater than 1×10^{-8} cm/s. A soil sample from a potential

Figure 7.20 Pipe networks for Problem 7.17.

borrow site has been tested in a falling head permeameter similar to the one in Figure 7.12. This sample was 120 mm in diameter and 32 mm tall. The standpipe had an inside diameter of 8.0 mm. Initially, the water in the standpipe was 503 mm above the water in the water bath surrounding the sample. Then, 8 hours 12 min later the water was 322 mm above the water in the water bath. Compute k and determine if this soil meets the specification.

7.20 An unlined irrigation canal is aligned parallel to a river, as shown in Figure 7.21. This cross-section continues for 4.25 miles. The soils are generally clays, but a 6 inch thick sand seam is present as shown. This sand has $k = 9 \times 10^{-2}$ cm/s. Compute the water loss from the canal to the river due to seepage through this sand layer and express your answer in acre-ft per month.

Note: One acre-foot is the amount of water that would cover one acre of ground to a depth of one foot, and thus equals 43,560 ft³.

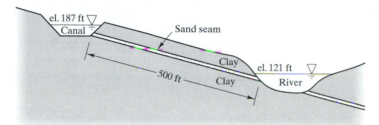

Figure 7.21 Cross-section for Problem 7.20.

7.21 The constant head permeameter shown in Figure 7.22 contains three different soils as shown. Their hydraulic conductivities are:

     Soil 1 — $k = 9$ cm/s
     Soil 2 — $k = 6 \times 10^{-2}$ cm/s
     Soil 3 — $k = 8 \times 10^{-3}$ cm/s

The four piezometer tips are spaced at 100 mm intervals, and the soil interfaces are exactly aligned with piezometer tips B and C. The total heads in piezometers A and D are 98.9 and 3.6 cm, respectively. Compute the total heads in piezometers B and C.

Figure 7.22 Constant head permeameter for Problem 7.21.

8

Groundwater — Applications

Boring — see Civil Engineers

Listing in the London telephone book

Civil Engineers are No Longer Boring

Headline in a London newspaper after the telephone company
agreed to revise its method of referring readers
to drilling and sampling companies

This chapter continues our discussion of groundwater, with more focus on applying the principles developed in Chapter 7 to practical engineering problems. Many of these problems require the analysis of two- and three-dimensional flow, so we will begin by expanding the analysis methods to accommodate these conditions.

8.1 TWO-DIMENSIONAL FLOW

Two-dimensional flow, as defined in Chapter 7, occurs when all of the velocity vectors are confined to a single plane. Many groundwater flow problems are very close to being two-dimensional, and may be analyzed as such. For example, groundwater flow into the long excavation shown in Figure 8.1 could be evaluated using a two-dimensional analysis in a vertical plane.

240

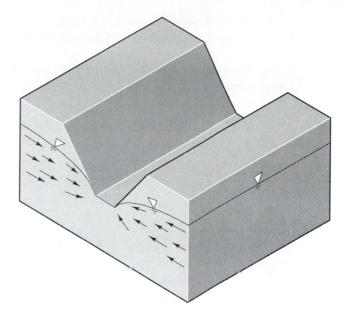

Figure 8.1 Example of two-dimensional flow analysis: a long excavation.

We will only consider two-dimensional analyses performed in vertical planes with the horizontal x axis oriented in the direction of flow and the vertical z axis increasing downward. In some situations, two-dimensional analyses also can be performed in horizontal planes, but these instances are beyond the scope of this book.

The LaPlace Equation

In general, Darcy's Law cannot be solved directly for two-dimensional flow because both i and A vary throughout the flow regime. Therefore, the analyses are more complex and need to incorporate a mathematical function called the *LaPlace Equation*.

We will begin our examination of the LaPlace Equation by considering a small element of soil in a vertical cross-section as shown in Figure 8.2, along with the following assumptions:

- Darcy's law is valid.
- The soil is completely saturated ($S = 100\%$).
- The size of the element remains constant (i.e., no expansion or contraction).

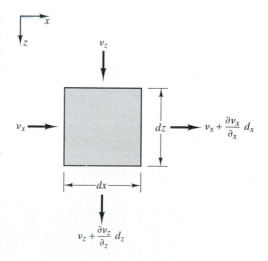

Figure 8.2 Element of soil for derivation of the LaPlace Equation. The arrows indicate water flowing into or out of the element.

- The soil is homogeneous (i.e., k is constant everywhere in the aquifer).
- The soil is isotropic (i.e., k is the same in all directions).

Using Equation 7.1, we will divide the flow of water into horizontal and vertical components, x and z. The total flows into and out of the element are then:

$$Q_{in} = (vAn_e)_{in} = Ln_e (v_x dz + v_z dx) \tag{8.1}$$

$$Q_{out} = (vAn_e)_{out} = Ln_e \left(\left(v_x + \frac{\partial v_x}{\partial x} dx \right) dz + \left(v_z + \frac{\partial v_z}{\partial z} dz \right) dx \right) \tag{8.2}$$

where v_x and v_z are the velocities in the x and z directions, respectively, L is the length of the element in the y direction, and n_e is the effective porosity.

The soil is saturated and its volume remains constant, so Q_{in} must equal Q_{out}:

$$\frac{\partial v_x}{\partial x} dx\, dz + \frac{\partial v_z}{\partial z} dz\, dx = 0 \tag{8.3}$$

which can be reduced to:

$$\frac{\partial v_x}{\partial x} + \frac{\partial v_z}{\partial z} = 0 \tag{8.4}$$

This means any change in velocity in the x direction must be offset by an equal and opposite change in the z direction.

Using Equation 7.16:

$$v = \frac{ki}{n_e} \tag{8.5}$$

$$v_x = -\frac{k}{n_e} \frac{\partial h}{\partial x} \tag{8.6}$$

$$v_z = -\frac{k}{n_e} \frac{\partial h}{\partial z} \tag{8.7}$$

Substituting Equations 8.6 and 8.7 into Equation 8.4 gives:

$$\frac{\partial^2 h}{\partial x^2} + \frac{\partial^2 h}{\partial z^2} = 0 \qquad (8.8)$$

This is the *LaPlace Equation* for two-dimensional flow. It describes the energy loss associated with flow through a medium, and is used to solve many kinds of flow problems, including those involving heat, electricity, and seepage.

For problems with simple boundary conditions, it is possible to derive analytical solutions to this equation. Unfortunately, the boundary conditions associated with most practical seepage problems are much too complex, so geotechnical engineers must rely on alternative solution methods, most notably:

- Flow nets
- Electrical analogy models
- Numerical solutions

We will discuss each of them.

Flow Net Solution

The flow net solution is a graphical method of solving the two-dimensional LaPlace Equation. This solution has been attributed to Forchheimer (1911) and others. We will first develop it for soils that are homogeneous and isotropic with respect to permeability. Then, we will consider the anisotropic case where $k_z \ne k_x$.

Theory

Flow nets are based on two mathematical functions: the *potential function*, ϕ, and the *flow function*, ψ (also known as the *stream function*). The potential function is defined as:

$$\phi = -kh + C \qquad (8.9)$$

where:
    ϕ = potential
    k = hydraulic conductivity
    h = total head
    C = a constant

Combining with Equations 8.6 and 8.7 gives:

$$\frac{\partial \phi}{\partial x} = -k\frac{\partial h}{\partial x} = v_x n_e \tag{8.10}$$

$$\frac{\partial \phi}{\partial z} = -k\frac{\partial h}{\partial z} = v_z n_e \tag{8.11}$$

Combining Equations 8.4, 8.10, and 8.11 would bring us back to Equation 8.8, which means our analysis is consistent with the LaPlace Equation.

We can draw a curve in the cross-section such that ϕ is constant everywhere along the curve. This is known as an *equipotential line* (even though it is a curve, not a line). Figure 8.3 shows a family of equipotential lines with $\phi = \phi_1, \phi_2, \phi_3,$ etc. According to Equation 8.9, the total head also is constant along each equipotential line. Thus, this family of curves is similar to the contour lines on a topographic map, except that we are using a vertical section (not a plan view) and the lines represent equal total heads (not equal elevations).

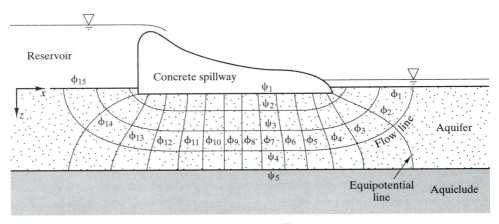

Figure 8.3 A sample flow net of seepage beneath a concrete spillway.

Combining Equations 8.10 and 8.11 gives:

$$d\phi = \frac{\partial \phi}{\partial x}dx + \frac{\partial \phi}{\partial z}dz$$
$$= n_e [v_x dx + v_z dz] \tag{8.12}$$

Along an equipotential line, ϕ is a constant, so $d\phi = 0$ and Equation 8.12 may be rewritten as:

$$\frac{dz}{dx} = -\frac{v_x}{v_z} \tag{8.13}$$

The slope of the equipotential line is dz/dx, and the direction of flow is v_z/v_x, so equipotential lines are always perpendicular to the direction of flow.

The flow function is the inverse of the potential function:

$$-\frac{\partial \psi}{\partial x} = -k\frac{\partial h}{\partial z} = v_z n_e \tag{8.14}$$

$$\frac{\partial \psi}{\partial z} = -k\frac{\partial h}{\partial x} = v_x n_e \tag{8.15}$$

We also can draw a family of curves in the cross-section such that ψ is constant everywhere along each curve. They are known as *flow lines*. Figure 8.3 also shows a family of flow lines with $\psi = \psi_1, \psi_2, \psi_3$, and so on. When presented together, these two families of curves (one set for ϕ and one for ψ) are known as a *flow net*.

Combining Equations 8.14 and 8.15 gives:

$$\begin{aligned} d\psi &= \frac{\partial \psi}{\partial x} dx + \frac{\partial \psi}{\partial z} dz \\ &= n_e[-v_z dx + v_x dz] \end{aligned} \tag{8.16}$$

Along a flow line, ψ is a constant, so $d\psi = 0$ and Equation 8.16 may be rewritten as:

$$\frac{dz}{dx} = \frac{v_z}{v_x} \tag{8.17}$$

Thus, the flow line is always parallel to the direction of flow and the zone between two flow lines as a *flow tube*. The flow rate, Q, through one such tube is:

$$\begin{aligned} Q_i &= vAn_e \\ &= Ln_e \int_{\psi_x}^{\psi_{x\Theta}} (-v_z dx + v_x dz) \\ &= Ln_e \int_{\psi_x}^{\psi_{x\Theta}} \left(\frac{\partial \psi}{\partial x} dx + \frac{\partial \psi}{\partial z} dz \right) \\ &= Ln_e (\psi_{i+1} - \psi_i) \end{aligned} \tag{8.18}$$

where L is the length of the flow zone perpendicular to the cross-section. In other words, the flow rate through a single flow tube is equal to the difference between the ψ values on each side of the tube multiplied by $L\,n_e$. In addition, Q_i is constant throughout the length of a flow tube.

If we draw N_F flow tubes ($N_F + 1$ flow lines) in such a way that each tube has the same Q_i, then the difference in ψ across each tube is $\Delta\psi/N_F$ and the total flow rate, Q, through the entire system is:

$$Q = N_F L n_e Q_i = N_F L n_e \frac{\Delta\psi}{N_F} = L n_e \Delta\psi \qquad (8.19)$$

We also will call the zone between two equipotential lines an *equipotential drop*, and draw the equipotential lines such that the difference in ϕ across each drop is $\Delta\phi/N_D$, where N_D is the number of equipotential drops.

Based on Equation 8.9, $\Delta\phi$ across the entire system is:

$$\Delta\phi = -k\,\Delta h \qquad (8.20)$$

where Δh is the total head loss from one side of the cross-section to the other.

Figure 8.4 shows the intersections of two flow lines and two equipotential lines. According to Equations 8.13 and 8.17, they must meet at right angles. The distances between these lines are a and b, as shown.

The velocity components, v_x and v_z, are then:

$$v_x = v\cos\alpha \qquad (8.21)$$

$$v_z = v\sin\alpha \qquad (8.22)$$

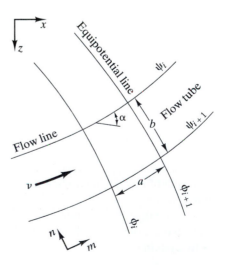

Figure 8.4 Intersection of flow lines and equipotential lines..

We also may write:

$$\frac{\partial \phi}{\partial m} = \frac{\partial \phi}{\partial x}\frac{dx}{dm} + \frac{\partial \phi}{\partial z}\frac{dz}{dm}$$

$$= v_x n_e \frac{dx}{dm} + v_z n_e \frac{dz}{dm}$$

$$= v n_e \cos\alpha \cos\alpha + v n_e \sin\alpha \sin\alpha$$

$$= v n_e (\cos^2\alpha + \sin^2\alpha)$$

$$= v n_e \qquad\qquad\qquad (8.23)$$

$$\frac{\partial \psi}{\partial n} = \frac{\partial \psi}{\partial x}\frac{dx}{dn} + \frac{\partial \phi}{\partial z}\frac{dz}{dn}$$

$$= -v_z n_e \frac{dx}{dn} + v_x n_e \frac{dz}{dn}$$

$$= -v n_e \sin\alpha (-\sin\alpha) + v n_e \cos\alpha \cos\alpha$$

$$= v n_e (\sin^2\alpha + \cos^2\alpha)$$

$$= v n_e \qquad\qquad\qquad (8.24)$$

Thus,

$$\frac{\partial \psi}{\partial n} = \frac{\partial \phi}{\partial m} \quad \rightarrow \quad \frac{b}{a} = \frac{\Delta\psi/N_F}{\Delta n} = \frac{\Delta\phi/N_D}{\Delta m} \quad \rightarrow \quad \frac{\Delta\psi}{\Delta\phi} = \frac{N_F}{N_D}\frac{b}{a} \qquad (8.25)$$

Finally, combining Equations 8.19, 8.20, and 8.25 produces:

$$\boxed{Q = k L \Delta h \frac{N_F}{N_D}\frac{b}{a}} \qquad (8.26)$$

where:

Q = flow rate

k = hydraulic conductivity

L = length of aquifer perpendicular to the cross-section

Δh = head loss through the flow net

N_F = number of flow tubes

N_D = number of equipotential drops

b/a = length-to-width ratio of pseudo-squares (see Figure 8.4)

Graphical Construction — Homogeneous Isotropic Soils

Equation 8.26 allows us to compute Q from a correctly drawn flow net. The next problem is knowing how to produce such a flow net. For flow problems where the soil is homogeneous and isotropic this may be done using a converging trial-and-error graphical technique as follows:

1. Draw a cross-section of the seepage zone to scale, with all of the fixed boundaries in ink. This drawing should include the ground surface, the limits of the pervious zone, the groundwater table and any free water surfaces (i.e., water levels in lakes or other bodies of water), and any other pertinent data. For confined aquifers, show both the upper and lower flow boundaries. For unconfined aquifers, show the lower flow boundary and the groundwater table. Be sure the vertical and horizontal scales are equal, and the cross-section is oriented parallel to the direction of flow. For example, when evaluating seepage under a dam, draw the cross-section perpendicular to its longitudinal axis. If the cross-section is symmetrical about some vertical axis, it is easier to draw only half of it, then double the Q computed from the flow net.

2. Select an integer value for N_F, the number of flow tubes. Larger values of N_F produce more precise flow nets, but also require more effort to finalize. Usually, N_F values of 4, 5, or 6 represent a good compromise between precision and effort.

3. Using a soft pencil, sketch in the approximate locations of the flow lines. You may wish to use the sample flow nets in Figures 8.3, 8.5, and 8.6 for guidance.

4. Using a soft pencil, sketch in the approximate locations of the equipotential lines. Keep in mind that the b/a ratio must be the same for each of the "rectangles" formed by the equipotential and flow lines and these lines must intersect at right angles. This can be done by using an integer value for N_D, spreading the equipotential lines accordingly, and measuring the resulting b/a. However, it is better to set $b/a = 1$ (so the "rectangles" are "squares") and keep drawing equipotential lines as needed. The proper number of equipotential lines will come out automatically as you draw them while maintaining right angles and the $b/a=1$. This method will not necessarily produce an integer value for N_D (for example, if each flow tube contains ten full "squares" and one half "square", then $N_D = 10.5$. You now have the first trial flow net.

5. Check the trial flow net against the following criteria:
   a. No two flow lines must ever intersect.
   b. No two equipotential lines must ever intersect.
   c. Flow lines and equipotential lines must always intersect at right angles.
   d. The b/a ratio must be the same for all of the "squares".
      Exception: Although we have chosen to set N_F equal to a whole number, N_D depends on the shape of the seepage zone and may not be a whole number. Thus, the last row of pseudo-squares may have a different b/a ratio.
   e. The curves are smooth.

6. Revise portions of the trial flow net as necessary to satisfy these criteria. Note that moving flow lines usually requires also moving nearby equipotential lines, and vice-versa. Continue revising until you obtain a satisfactory flow net.

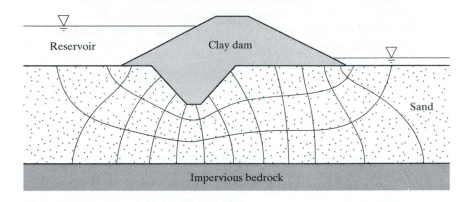

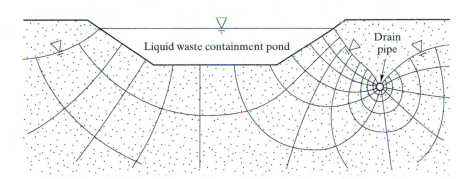

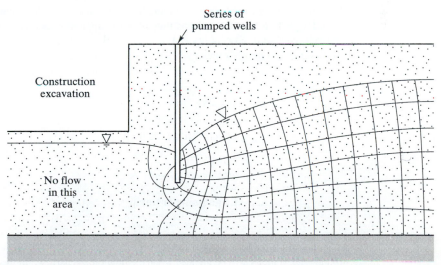

Figure 8.5 Sample flow nets.

This process could go on indefinitely, but you will quickly reach a point of diminishing returns where large amounts of additional effort result in only nominal improvements to the flow net. Keep in mind that the uncertainty in k is probably much greater than that in N_F and N_D, so there is no need to be overly meticulous with the flow net.

Harr (1962) presents a modification of the flow net method called the *method of fragments*. For some problems this method is quicker and easier than the conventional method. However, discussion of this method is beyond the scope of this book.

Example 8.1

A 6 ft diameter, 2000 ft long sewer drain pipe is to be installed 24 ft below the ground surface as shown in Figure 8.6. To build this pipe, a long trench must be excavated below the groundwater table as shown. Steel sheet piles (heavy, corrugated sheets of steel) will be installed to keep the sides from collapsing, and the trench will be dewatered with pumps. After the pipeline is installed, the pumps will be removed, the trench will be backfilled, and the sheet piles will be removed.

The dewatering pumps will be placed at 100 ft intervals along the trench. Assuming the sheet piles are perfectly watertight (probably not a good assumption!), draw a flow net and compute the Q for each pump.

Solution

1. The cross-section is symmetrical about a vertical axis through the center of the trench, so we will draw a flow net for the left half only, then double the computed Q.
2. The hydraulic conductivity in the clean sand strata is 1000 times higher than in the sandy silt, so the groundwater will move laterally into the sand much more easily than it will seep through the sandy silt. Therefore, the only significant head losses will be through the sandy silt, and we only need to draw the flow net through this strata.
3. The resulting flow net, developed by the converging trial-and-error method described above, is shown on the right side of Figure 8.6. We chose to use $N_F = 4$ and $a/b = 1$, and counted $N_D = 12.6$ from the finished flow net. The uppermost equipotential drop is not a full "square," so it only counts as 0.6.
4. The flow rate per pump is then:

$$Q = 2kL\Delta h \frac{N_F}{N_D}\frac{b}{a}$$

$$= 2(2\times10^{-6} \text{ ft/s})(100 \text{ ft})(20 \text{ ft})\left(\frac{4}{12.6}\right) \quad (1)$$

$$= 2.5\times10^{-3} \text{ ft}^3/\text{s}$$
$$\times 449 \text{ gal/min per ft}^3/\text{s}$$

$$= \textbf{1.1 gal/min per pump} \qquad \Leftarrow \textit{Answer}$$

5. According to Equation 7.1, the velocity is greatest when the area is small, so the groundwater velocity will be greatest when the flow tube is narrow. In this cross-section, the narrowest spot is near the tip of the sheet pile. Because the velocity here is high, there will be much more friction between the flowing water and the soil, so the hydraulic

gradient also will be high. Thus, the equipotential lines also are close together in this area.

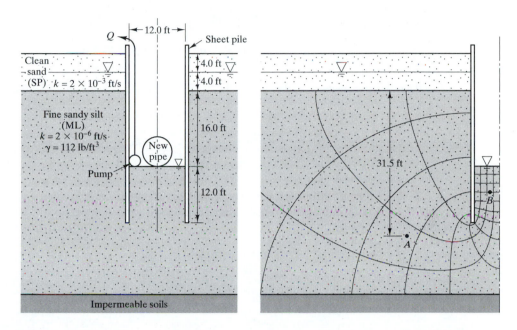

Figure 8.6 Cross-section of trench for Examples 8.1 and 8.2. The given data is on the left and the completed flow net is on the right.

Example 8.2

Using the flow net developed in Example 8.1, compute the pore water pressure at Point A. Assume steady-state hydrostatic conditions exist.

Solution

Set datum: Ground surface = Elevation 100.0 ft

$$(h_z)_A = 100.0 \text{ ft} - 4.0 \text{ ft} - 4.0 \text{ ft} - 31.5 \text{ ft} = 60.5 \text{ ft}$$

$$h \text{ at bottom of trench} = 100.0 - 24.0 = 76.0 \text{ ft}$$

From the flow net we determine there are 10.2 equipotential drops from the bottom of the trench to Point A. In addition, $N_D = 12.6$ and $\Delta h = 20.0$ ft. Therefore, 10.2/12.6 of the 20.0 ft head loss occurs between Point A and the bottom of the trench. Since the total head at the bottom of the trench is 76.0 ft, the total head at Point A is:

$$h_A = 76.0 \text{ ft} + 20.0 \text{ ft} \left(\frac{10.2}{12.6} \right) = 92.2 \text{ ft}$$

$$(h_p)_A = h_A - (h_z)_A = 92.2 \text{ ft} - 60.5 \text{ ft} = 31.7 \text{ ft}$$

Thus, if a piezometer were present at Point A, the water would rise to a level 31.7 ft above the point.

$$u = \gamma_w (h_p)_A = (62.4 \, \text{lb/ft}^3)(31.7 \, \text{ft}) = \mathbf{1980 \, lb/ft^2} \qquad \Leftarrow \textit{Answer}$$

The flow lines in a flow net also may be represented by injecting dye into a soil profile model, as shown in Figure 8.7. This method is not used for routine analyses, but can be useful for instructional purposes.

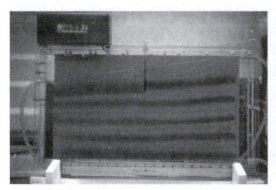 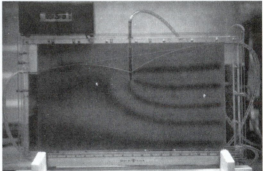

Figure 8.7 This laboratory model was constructed to illustrate groundwater flow. The left photograph shows steady-state conditions with the groundwater flowing from right to left. Dye has been injected along the right side of the model, and is being carried along by the flowing water, thus illustrating flow lines. In the right photograph, water is being pumped from the well. This alters the flow patterns, as reflected by the dye, even causing some of the water downstream of the well to reverse direction. The groundwater table in the vicinity of the well also has been drawn down. (National Ground Water Association, Westerville, OH)

Graphical Construction — Anisotropic Soils

In anisotropic soils, the coefficients of permeability are different in the horizontal and vertical directions. To draw a flow net in such soils, we first need to transform the cross-section by multiplying all vertical dimensions by $\sqrt{k_x/k_z}$ as shown in Figure 8.8. Then, we draw the flow net in the usual way on the transformed section and compute Q using Equation 8.26 with $k = \sqrt{k_x k_z}$.

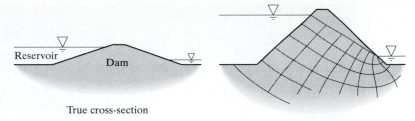

True cross-section

Transformed cross-section

Figure 8.8 Transformed section for an anisotropic soil. In this case, $k_x = 9\,k_z$, so the transformed section has been produced using a vertical exaggeration factor of 3.

Electrical Analogy Solution

The LaPlace Equation also describes many other kinds of flow problems, including electricity, heat, and magnetism. Thus, these flows are analogous to that of water through soil, as described in Table 8.1. We can take advantage of this similarity by constructing an *electrical analogy model* to analyze a seepage problem (Bardet, 1997).

Electrical analogy models, such as the one in Figures 8.9 and 8.10, use a pool of water or special conductive paper with the same shape as the seepage zone. We pass electricity through this pool or across the paper, thus producing the following analogies:

TABLE 8.1 ANALOGIES IN AN ELECTRICAL ANALOGY MODEL

Characteristic	Actual Seepage Problem	Electrical Analogy Model
Media	Soil	Water or conductive paper
Flowing substance	Water	Electricity
Resistance to flow	Hydraulic conductivity, k	Electrical conductivity, $\sigma = 1/R$
Quantity of flow per unit time	Flow rate, Q	Current, I
Potential	Total head, h	Voltage, E
Form of LaPlace Equation	$\dfrac{\partial^2 h}{\partial x^2} + \dfrac{\partial^2 h}{\partial z^2} = 0$	$\dfrac{\partial^2 E}{\partial x^2} + \dfrac{\partial^2 E}{\partial z^2} = 0$
Governing law	Darcy's Law ($Q = kiA$)	Ohm's Law ($I = E/R$)

The flow rate through this model can be determined from the current, as measured with an ammeter, and the total head at any point can be determined using a voltmeter and a movable probe, as shown.

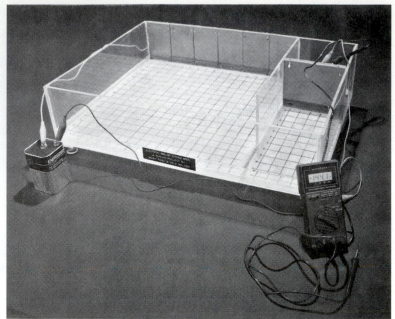

Figure 8.9 Electrical analogy model. The battery supplies current through the pool of water contained in this plexiglass tank.

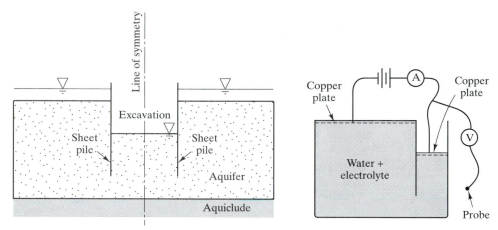

Figure 8.10 Schematic of electrical analogy model with corresponding cross-section. The A symbol represents an ammeter, and the V symbol represents a voltmeter. This is the same cross-section used in the model shown in Figure 8.9.

Electrical analogy models were popular before the days of powerful computers and sophisticated software. However, today their usefulness is limited to their instructional value. Practicing engineers and researchers are much more likely to use computer software as described in the next section.

Numerical Solutions

Most geotechnical engineering problems may be solved using *analytical solutions*. These are direct mathematical methods that find the desired answers. For example, Darcy's Law is an analytical solution to one-dimensional seepage problems. However, problems with more complex mathematical formulations or boundary conditions often have no analytical solution, and must be solved some other way. Two-dimensional seepage problems, as described by the LaPlace Equation, are in this category.

Although hand-drawn flow nets and electrical analogy solutions may be used to solve the LaPlace Equation, the widespread availability of digital computers has made numerical solutions the preferred method for many seepage problems.

Numerical solutions (also known as *numerical methods*) are mathematical techniques that solve complex problems by dividing them into small physical pieces and writing simpler equations that describe the functions within each piece and the relationships between the pieces. Several such techniques are available, and they are routinely used to solve a wide variety of engineering problems. The *finite element method* (*FEM*) is one of the most common numerical solutions, and it is the one most commonly used to solve seepage problems.

As its name implies, the finite element method divides the flow regime into a large number of discrete elements, as shown in Figure 8.11. A typical solution will have hundreds or thousands of such elements. We then assume the hydraulic gradients i_x and i_z are constant within each element, which allows us to directly apply Darcy's Law to describe the flow within an element. The next element has its own values of i_x and i_z, and the flow within that element also may be described using Darcy's Law. Finally, the flow rate, Q, exiting one side of an element is equal to the flow rate entering the side of the adjacent element. The elements in Figure 8.12 show these relationships.

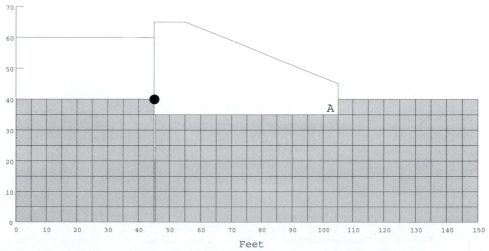

Figure 8.11 Finite element grid for numerical solution of seepage below a dam. The thin unshaded elements beneath the left edge of the dam represent an impervious cutoff (GEO-SLOPE International, Ltd.).

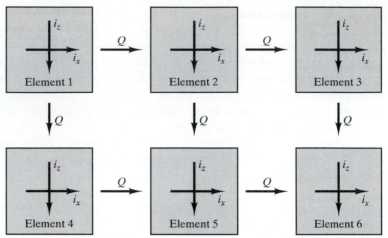

Figure 8.12 Relationships between adjacent elements in a finite element analysis. Each element has its own values of i_x and i_z, which are constant within that element. In addition, the Q exiting one side of an element is equal to the Q entering the side of the adjacent element.

This process generates thousands of simultaneous equations, and thus requires matrix algebra and a computer to solve. The final results represent an approximate solution whose precision depends on the number of elements and other factors. The output may be presented as a flow net, as shown in Figure 8.13, along with digital values of pore pressure in the center of each element, flow rate through the system, and other useful information.

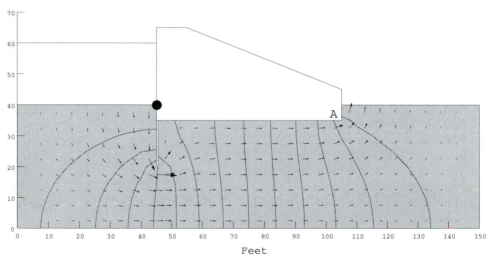

Feet

Figure 8.13 Typical output from a finite element seepage analysis. The curves are equipotentials and the arrows show the flow direction and velocity. Note how the flow goes around the impervious cutoff (GEO-SLOPE International, Ltd.).

Finite element analyses are attractive because of their analytical power and flexibility. They can consider problems with various complexities, including:

- Soil profiles that include strata with different k values
- Confined or unconfined flow
- Steady-state or transient conditions
- Complex boundary conditions
- Anisotropic soils

Commercially available software makes this method readily available for routine seepage problems.

QUESTIONS AND PRACTICE PROBLEMS

8.1 One of the factors in Equation 8.26 is N_F, yet when drawing a flow net we *assume* a value for this parameter. How can this formula produce correct results when one of the factors is assumed? (i.e., would assuming a higher N_F produce a higher computed value of Q?)

8.2 The flow net in Figure 8.14 is incorrect. Explain why.

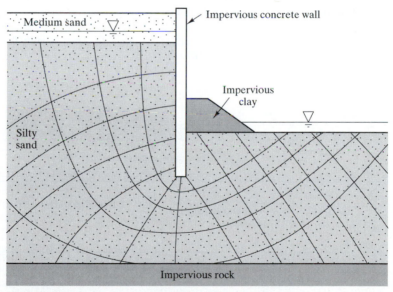

Figure 8.14 Trial flow net for Problem 8.2. Note: This flow net is not drawn correctly, and should not be used as an example!

8.3 Redraw the cross-section in Figure 8.15 to a scale of 1:500 (1 cm = 5 m), then draw a flow net that describes the seepage below this 150 m long concrete dam. Finally, compute the flow rate through this soil, expressed in liters per second, and the pore water pressure at Point A, which is at elevation 122.0 m.

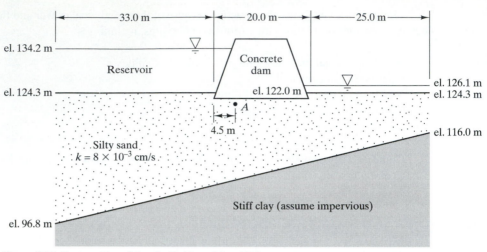

Figure 8.15 Cross-section of dam for Problems 8.3 and 8.4.

8.4   Using the flow net from Problem 8.3, develop a plot of seepage flow rate vs. the water elevation
in the reservoir. Consider reservoir elevations between 126.1 m and 135.0 m.

8.5   The earth dam shown in Figure 8.16 is to be built on a gravelly sand with silt and cobbles. This
dam will extend a distance of 850 ft perpendicular to the cross-section. To reduce the flow rate
through these soils, a concrete cutoff wall will be built as shown. Redraw this cross-section to
a scale of 1 in = 100 ft, draw a flow net, and compute Q. Then, identify the area in the flow net
that has the greatest hydraulic gradient.

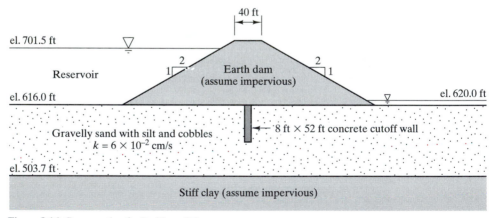

Figure 8.16 Cross-section for Problems 8.5.

8.2 THREE-DIMENSIONAL FLOW

The principles of three-dimensional flow are similar to those for two-dimensional flow. The
LaPlace Equation describes both, except now we must add a third term:

$$\frac{\partial^2 h}{\partial x^2} + \frac{\partial^2 h}{\partial y^2} + \frac{\partial^2 h}{\partial z^2} = 0$$

(8.27)

Many three-dimensional problems are very difficult to solve because of their complex geometries. Three-dimensional finite element analyses are a possibility, but they are not commonly used in engineering practice. A more likely solution would be to idealize the problem as a two-dimensional flow problem. However, one category of three-dimensional flow problems is easily solved: flow to wells.

Flow to Wells

Problems involving a single well are easily solvable because they are symmetric about a vertical axis passing through the well. The exact solution depends on the flow condition in the aquifer.

Confined Aquifers

A confined aquifer, as shown in Figure 8.17, is one that is sandwiched between two aquicludes. Thus, the upper and lower flow boundaries are fixed and the water flows through the entire depth of the aquifer.

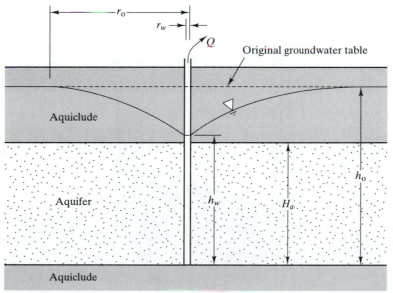

Figure 8.17 Well in a confined aquifer.

The flow rate, Q, produced by the well may be derived from Darcy's Law by considering a pie-shaped section of the aquifer as shown in Figure 8.18:

$$
\begin{aligned}
Q &= \int_0^{2\pi} kiA \\
&= \int_0^{2\pi} k \frac{dh}{dr} H_a r\, d\theta \\
&= \frac{2\pi k H_a (h_0 - h_w)}{\ln\left(\dfrac{r_0}{r_w}\right)}
\end{aligned}
\qquad (8.28)
$$

where:

Q = flow rate to well
k = hydraulic conductivity of aquifer
H_a = saturated thickness of aquifer
h_0 = total head in aquifer before pumping (datum = bottom of aquifer)
h_w = total head inside well casing during pumping (datum = bottom of aquifer)
r_0 = radius of influence
r_w = radius of well (includes casing and gravel-pack)

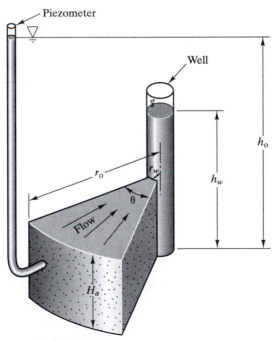

Figure 8.18 Pie-shaped section of a confined aquifer for the derivation of Equation 8.28.

The radius of influence, r_0, is the distance from the well to the farthest point of drawdown, and is a difficult parameter to assess. Fortunately, even approximate estimates are often sufficient because $\ln (r_0/r_w)$ is always very large, and it is not overly sensitive to errors in r_0. For example, changing r_0/r_w from 1000 to 5000 increases $\ln r_0/r_w$ by a factor of only 1.2.

Sichart and Kyrieleis (1930) presented the following empirical formula that gives an approximate value of r_0:

$$r_0 = 300 (h_0 - h_w) \sqrt{k} \qquad\qquad (8.29)$$

r_0, h_0, and h_w must be expressed in the same units, and k must be in cm/s.

Unconfined Aquifers

In an unconfined aquifer, as shown in Figure 8.19, the lower flow boundary is fixed, but the upper flow boundary is the groundwater table, which is free to seek its own level. The groundwater table is drawn down in the vicinity of the well, so the height of the flowing zone decreases as the water approaches the well.

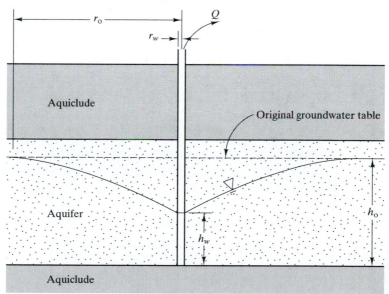

Figure 8.19 Well in an unconfined aquifer.

Using a derivation similar to that for confined aquifers, we can develop the following formula:

$$Q = \frac{\pi k (h_0^2 - h_w^2)}{\ln\left(\dfrac{r_o}{r_w}\right)} \qquad (8.30)$$

Mixed Aquifers

A mixed aquifer is one that was originally confined, but the portion near the well becomes unconfined because of the drawdown to the well as shown in Figure 8.20. This is the case when the water level inside the well casing is pumped to an elevation below the top of the confined aquifer.

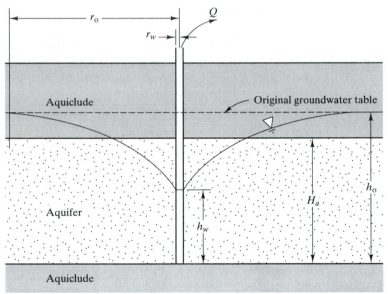

Figure 8.20 Well in a mixed aquifer.

Again, we can derive a formula for Q:

$$Q = \frac{\pi k (2H_a h_0 - H_a^2 - h_w^2)}{\ln\left(\dfrac{r_o}{r_w}\right)} \qquad (8.31)$$

Equations 8.28–8.31 are valid only when the well penetrates completely through the aquifer and is hydraulically open (i.e., has a well screen) throughout the aquifer. The flow rate to partially penetrating or partially open wells would be less (see Driscoll, 1986, or Powers, 1992).

All three well formulas may be used in anisotropic soils by using a transformed section as described earlier.

Example 8.3

A municipal water supply well is to be installed in the aquifer shown in Figure 8.21. This well will have an 8-in diameter casing in a 24-in diameter boring with a submersible pump that draws the water level down to a depth of 62 ft below the ground surface. Compute the maximum flow rate that could be produced by this well.

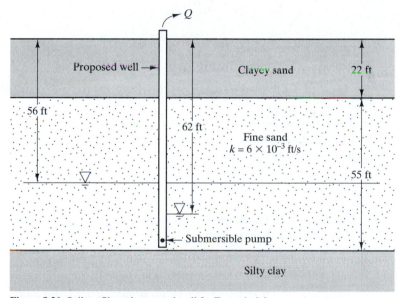

Figure 8.21 Soil profile and proposed well for Example 8.3.

Solution

This is an unconfined aquifer, so we will use Equation 8.30.

$$h_0 = 55 \text{ ft} + 22 \text{ ft} - 56 \text{ ft} = 21 \text{ ft}$$

$$h_w = 55 \text{ ft} + 22 \text{ ft} - 62 \text{ ft} = 15 \text{ ft}$$

$$k = (6 \times 10^{-3} \text{ ft/s}) \left(\frac{30.5 \text{ cm}}{1 \text{ ft}} \right) = 0.18 \text{ cm/s}$$

$$r_0 = 300(h_0 - h_w)\sqrt{k} = 300(21 \text{ ft} - 15 \text{ ft})\sqrt{0.18 \text{ cm/s}} = 760 \text{ ft}$$

Using a unit conversion factor of 449 gal/min per ft^3/s,

$$
\begin{aligned}
Q &= \frac{\pi k (h_0^2 - h_w^2)}{\ln\left(\dfrac{r_0}{r_w}\right)} \\[2mm]
&= \frac{\pi (6 \times 10^{-3} \text{ ft/s})[(21 \text{ ft})^2 - (15 \text{ ft})^2]}{\ln\left(\dfrac{760 \text{ ft}}{1 \text{ ft}}\right)} \\[2mm]
&= 0.61 \text{ ft}^3/\text{s} \\[2mm]
&= \textbf{276 gal/min} \qquad \leftarrow Answer
\end{aligned}
$$

Use of Pumped Wells to Conduct In-Situ Permeability Tests

Pumped wells that are already in place may be used to conduct in-situ permeability tests. This method should produce much more accurate values of k than obtained from laboratory tests because it mobilizes a much larger volume of soil. One way of doing this would be to use Equations 8.28–8.31 with a known Q and solve for k. Another, more precise, method involves installing two or more observation wells as shown in Figure 8.22 (Driscoll, 1986). These observation wells must be located at a distance less than r_0 from the pumped well. We then can rewrite Equations 8.28–8.31 as follows:

For confined aquifers,

$$k = \frac{Q \ln(r_1/r_2)}{2\pi H_a (h_1 - h_2)} \tag{8.32}$$

For unconfined aquifers,

$$k = \frac{Q \ln(r_1/r_2)}{\pi (h_1^2 - h_2^2)} \tag{8.33}$$

For mixed aquifers,

$$k = \frac{Q \ln(r_1/r_2)}{\pi (2 H_a h_1 - H_a^2 - h_2^2)} \tag{8.34}$$

where:

k = hydraulic conductivity of aquifer
Q = flow rate from pumped well
r_1 = radius from pumped well to farthest observation well (must be $< r_0$)
r_2 = radius from pumped well to nearest observation well
h_1 = total head in farthest observation well (datum = bottom of aquifer)
h_2 = total head in nearest observation well (datum = bottom of aquifer)
H_a = saturated thickness of aquifer

Again, these equations are valid only when the pumped well extends through the entire aquifer and is open to the groundwater through the entire aquifer.

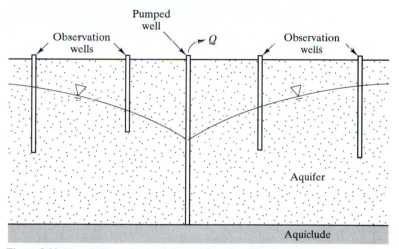

Figure 8.22 Use of observation wells near a pumped well to determine k in-situ.

This method of determining k is valid only after steady-state conditions have been achieved. This may be accomplished by continuously pumping from the pumped well and monitoring the water levels in the observation wells. When these water levels and the flow rate in the pumped well have both reached equilibrium, steady-state conditions have been achieved.

Example 8.4

A pumped well and two observation wells have been installed through a fine-to-medium sand as shown in Figure 8.23. A pump has been discharging water from the pumped well for a sufficient time to achieve steady-state conditions. The water levels in observation wells A and B were then observed to be 20 and 35 ft below the ground surface, respectively. Compute the hydraulic conductivity of the aquifer.

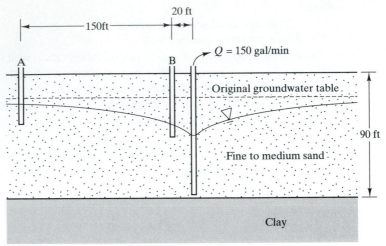

Figure 8.23 Cross-section for Example 8.4.

Solution

This is an unconfined aquifer, so use Equation 8.33.

$$Q = (150 \text{ gal/min}) \left(\frac{1 \text{ ft}^3}{7.48 \text{ gal}} \right) \left(\frac{1 \text{ min}}{60 \text{ s}} \right) = 0.33 \text{ ft}^3/\text{s}$$

$$h_1 = 90 - 20 = 70 \text{ ft}$$

$$h_2 = 90 - 35 = 55 \text{ ft}$$

$$
\begin{aligned}
k &= \frac{Q \ln(r_1/r_2)}{\pi (h_1^2 - h_2^2)} \\
&= \frac{0.33 \text{ ft}^3/\text{s} \ln(170 \text{ ft}/20 \text{ ft})}{\pi [(70 \text{ ft})^2 - (55 \text{ ft})^2]} \\
&= 1 \times 10^{-4} \text{ ft/s} \qquad \Leftarrow \textit{Answer}
\end{aligned}
$$

This answer seems reasonable, per Table 7.1.

QUESTIONS AND PRACTICE PROBLEMS

8.6   The proposed well shown in Figure 8.24 will be used to supply a municipal water system. Compute its pumping capacity with the groundwater level as shown, and express your answer in gallons per minute.

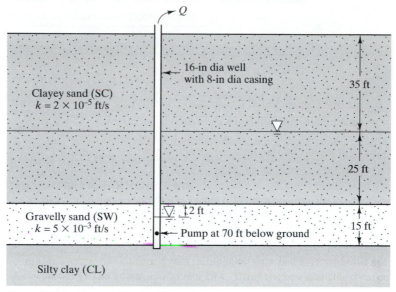

Figure 8.24 Proposed well for Problem 8.6.

8.7 After reaching steady-state conditions, the test well shown in Figure 8.25 is producing a flow rate of 17 liters per second. The aquifer is an alluvial soil with interbedded medium-to-coarse sand and silty sand.

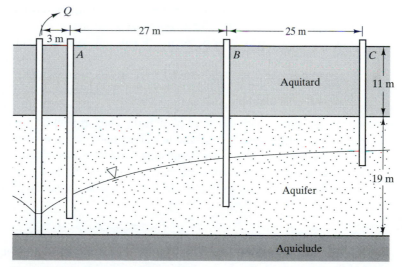

Figure 8.25 Cross-section for Problem 8.7.

The water depths in the observation wells are as follows:

Well	Water Depth From Ground Surface (m)	
	Before Pumping	During Pumping
Pumping	16.9	26.0
Observation A	16.9	23.5
Observation B	16.9	18.1
Observation C	16.9	16.9

Using the best available data, compute the hydraulic conductivity and transmissivity of the aquifer. Is the computed k value reasonable? Explain why or why not.

8.3 UPLIFT PRESSURES ON STRUCTURES

When buried structures extend below the groundwater table, they are subjected to uplift pressures from the pore water. These pressures are comparable to those acting on the hull of a ship. Uplift pressures are important, especially if the structure does not have much weight. For example, buried tanks or vaults might experience uplift pressures that exceed their weight, which would cause them to lift out of their intended position.

Engineers once thought full hydrostatic uplift pressures could occur only in sands, and only one-third to one-half of the full pressure would occur in clays (Terzaghi, 1936). Even Terzaghi once stated "we are not sure as to the extent to which hydrostatic uplift acts within a mass of plastic clay" (Terzaghi, 1929). Fortunately, he and others eventually realized full hydrostatic pore water pressures occur in all soils, so designs should be based on full hydrostatic uplift forces.

If the groundwater table is horizontal, or nearly so, then the uplift pressure at a point may simply be computed using Equation 7.7. This is true even though when viewed on a microscopic scale, only part of the submerged structure is in contact with the pore water (the remainder is in contact with the solid particles). Thus, the uplift force is the same as it would be if the structure was submerged to a comparable depth in a lake.

If the groundwater table is far from being horizontal, then significant head losses are occurring as the water flows from one side of the structure to the other. The concrete spillway in Figure 8.3 is an example. In this case it is necessary to use the following procedure:

1. Draw a flow net in the soil beneath the structure.
2. Using the equipotential lines, compute the total head at several points along the base of the structure.
3. Determine the elevation head at the points used in Step 2.
4. Using Equation 7.5 and the data from Steps 2 and 3, determine the pressure head at each point.

5. Using Equation 7.6, determine the pore water pressure at each point. This is the uplift pressure.
6. Develop a plot of uplift pressure across the structure.

Example 8.5

A sewage pump is to be located in the proposed underground vault shown in Figure 8.26. The site will be temporarily dewatered during construction, but afterwards the groundwater table will be allowed to return to its natural level. The highest probable level during the life of the project is as shown.

    The vault will be made of reinforced concrete and waterproofed so very little groundwater will enter. The vault and overlying soil has a mass of 30,000 kg, exclusive of the floor. A sump pump will eject any water that might accumulate inside.

    The weight of the vault and the overlying soil must be sufficient to provide a factor of safety of at least 1.5 against buoyant uplift (i.e. these weights must be at least 1.5 times the total uplift force). Write an equation for the buoyant uplift force, then compute the required base thickness, t. Neglect the weight of the pump and neglect any sliding friction between the sides of the vault and the soil.

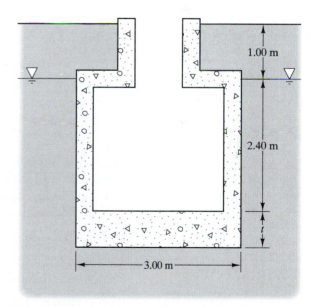

Figure 8.26 Cross-section for Example 8.5.

Solution

Uplift pressure

$$u = \gamma_w z_w = (9.81 \text{ kN/m}^3)(2.40 + t \text{ m}) = 23.5 + 9.81t \text{ kPa}$$

Uplift force

$$P = uA = (23.5 + 9.81t)(3^2) = 212 + 88.3t \text{ kN}$$

Weight

Use ρ concrete = 2400 kg/m^3

$$W = Mg$$
$$= \left[30,000 \text{ kg} + (2400 \text{ kg/m}^3)(3^2 \text{ m})t\right](9.81 \text{ m/s}^2)\left(\frac{1 \text{ kN}}{1000 \text{ N}}\right)$$
$$= 294 + 212t \text{ kN}$$

For factor of safety = 1.5

$$1.5(212 + 88.3t) = 294 + 212t$$
$$t = \textbf{200 mm} \qquad \leftarrow \textit{Answer}$$

Example 8.6

Compute the uplift pressures acting on the concrete spillway shown in Figure 8.27.

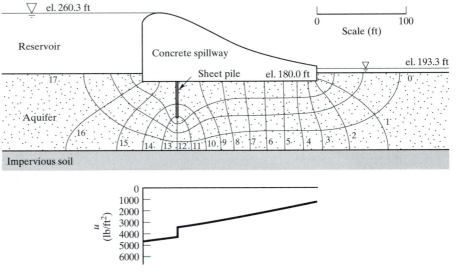

Figure 8.27 Cross-section for Example 8.6.

Solution

Using sea level as the datum:

h at upstream end = 260.3 ft

h at downstream end = 193.3 ft

$\Delta h = 260.3 - 193.3 = 67.0$ ft

From the flow net:

$N_D = 17.0$

$$\frac{\Delta h}{N_D} = \frac{67.0}{17.0} = 3.92 \text{ ft/equipotential drop}$$

Note: As discussed earlier, equipotential lines in a flow net are similar to contour lines in a topographic map. Therefore, $\Delta h/N_D$ is similar to the contour interval on a topographic map.

For each equipotential line, compute h based on h at the downstream end, the number of equipotential drops, and $\Delta h/N_D$. Then compute h_p and u along the bottom of the structure and plot the results.

Equipotential Line	h (ft)	Along bottom of structure		
		h_z (ft)	h_p (ft)	u (lb/ft^2)
0 (downstream)	193.3			
1	197.2			
2	201.2	180.0	21.2	1323
3	205.1	180.0	25.1	1566
4	209.1	180.0	29.1	1816
5	213.0	180.0	33.0	2059
6	216.9	180.0	36.9	2303
7	220.9	180.0	40.9	2552
8	224.8	180.0	44.8	2796
9	228.8	180.0	48.8	3045
10	232.7	180.0		
11	236.7	180.0		
12	240.6	180.0		
13	244.5	180.0		
14	248.5	180.0		
15	252.4	180.0	72.4	4518
16	256.4	180.0	76.4	4767
17	260.3			

These computed pressures are plotted on Figure 8.27. Note the drop in pressure at the sheet pile. This is due to the head losses as the groundwater flows around the sheet pile.

The weight of this structure is certainly much greater than the uplift force, so there is no danger of it rising out of position. However, when evaluating its ability to resist the horizontal hydrostatic force from the reservoir, we need to compute the normal force acting between the structure and the underlying soil and multiply it by the coefficient of friction. This force is equal to the weight of the structure less the uplift force. Thus, the uplift force reduces the horizontal sliding stability of the spillway.

8.4 GROUNDWATER CONTROL AND DEWATERING

Engineers and contractors often need to control groundwater flows, either temporarily or permanently. Temporary controls are necessary when performing underground construction below the groundwater table; permanent controls may be needed to keep groundwater from reaching sensitive areas. The various processes of removing groundwater are called *dewatering*.

Construction Dewatering

The most common type of temporary groundwater control is *construction dewatering*, which is performed to keep groundwater out of construction sites. Construction dewatering is in place only during the construction period, and is for the "convenience" of the contractor, so it is not shown on the design drawings or specifications. The contractor is responsible for designing the dewatering system, and normally does so by subcontracting this work to a dewatering specialist.

There are four basic categories of construction dewatering methods: open pumping, predrainage, cutoffs, and exclusion (Powers, 1992). The proper selection depends on many factors, including:

- The soil conditions, especially the hydraulic conductivity
- The size and depth of the construction excavation, especially the depth below the groundwater table
- The planned method of excavation and ground support
- The type and proximity of nearby structures
- The type of structure being built
- The planned schedule
- The presence and characteristics of groundwater contaminants (if any)

Figure 8.28 shows an excavation that has been temporarily dewatered to permit construction of two pipelines.

Open Pumping

Open pumping methods allow groundwater to enter the excavation, then direct it to low points known as *sumps* where it is pumped out. Figure 8.29 shows an open pumping system. This method is most appropriate in soils with a low to moderate hydraulic conductivity because these soils discharge controllable quantities of water into the excavation. When used in favorable conditions, open pumping is generally inexpensive and effective.

However, open pumping is inappropriate in soils with high hydraulic conductivity values, especially when the excavation extends to significant depths below the groundwater table. Such soils can produce very large flow rates that cannot be effectively collected or pumped. This method also is inappropriate when the excavation is potentially unstable as a result of boils, quicksand, or heave (as discussed in Chapter 13), or when the excavation is to performed by scrapers or other equipment that has traction problems in wet soil.

Figure 8.28 This 11 m (35 ft) deep excavation was required to construct these two water supply pipelines. The excavation extends 10 m (35 ft) below the groundwater table, and a river is located just beyond the excavation on the left side of the photograph. Therefore, a temporary dewatering system has been installed to draw down the groundwater table. This system includes a series of pumped wells, including the one in the right foreground (Photo courtesy of Foothill Engineering).

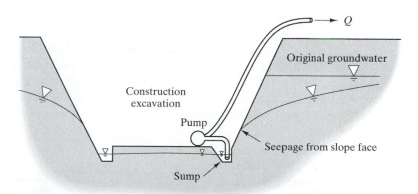

Figure 8.29 Cross-section through a construction excavation being dewatered by open pumping.

Predrainage

Predrainage encompasses several methods of intercepting groundwater before it reaches the excavation. Unlike open pumping, these methods allow the contractor to lower the groundwater table before excavation begins, which often solves equipment mobility problems. They also are suitable for a wider range of soil conditions.

Most predrainage systems use wells located a short distance outside the perimeter of the excavation, as shown in Figure 8.30. Several methods are available to extract water from the wells, including:

- *Pumped well systems* with a submersible pump in each well. This method can accommodate large flow rates, but requires a large investment in equipment. The excavation in Figure 8.28 used pumped wells.
- *Wellpoint systems* extracting the water by applying a vacuum to each well. This is much less expensive than installing individual pumps, but is limited to depths of no more than 5 to 6 m.
- *Ejector systems* using a nozzle and a venturi in each well to lift the water. These systems do not have the depth limitations of wellpoints, and often are less expensive than pumped well systems. However, they are inherently inefficient and thus quickly lose their cost advantage when the required flow rates are high.

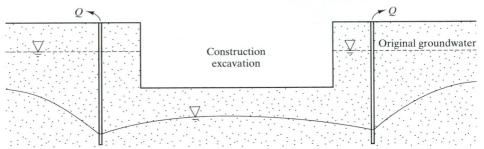

Figure 8.30 Cross-section of a construction excavation being dewatered by predrainage.

Cutoffs

Cutoffs (also known as *seepage barriers*) are intended to physically block the groundwater before it reaches the excavation, as shown in Figure 8.31. Although it is impossible to completely block the groundwater flow, cutoffs can be very effective and typically reduce the water inflow to a small fraction of what it otherwise would have been. The small amount of groundwater that does eventually reach the excavation can usually be removed by open pumping.

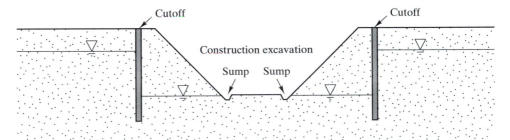

Figure 8.31 Use of a cutoff to block groundwater flow before it reaches a construction excavation.

Many methods and techniques have been used to cutoff groundwater. These include the following (Powers, 1992):

- *Steel sheet piles* are heavy corrugated steel sheets driven vertically into the ground. They are installed around the perimeter of a proposed excavation, then the soil inside is removed. The primary purpose of sheet piles is to provide structural support to keep the excavation from caving in. However, they also provide a partially effective groundwater cutoff. See Chapter 16 for more information on sheet piles.
- *Diaphragm walls* are constructed by excavating a trench (typically about 1 m wide) around the perimeter of the proposed excavation and filling it with concrete. To avoid caving, the trench is normally filled with bentonite slurry [1] as it is being excavated, and the concrete is placed using a tremie [2] extending through the slurry. These walls provide both structural support and groundwater control.
- *Slurry trenches* are constructed similar to diaphragm walls, except they are filled with low-permeability soils instead of concrete. They provide groundwater control, but do not provide structural support. Figure 8.32 shows a slurry trench under construction.

a   b

Figure 8.32 A slurry trench barrier under construction: a) An excavator is digging the 20 m (65 ft) deep trench through a slurry, b) then a bulldozer pushes in a clayey backfill.

- *Secant drilled shaft walls* are drilled shaft foundations (large diameter borings filled with concrete) placed on close centers, thus creating a continuous underground wall.
- *Tremie seals* are concrete barriers placed at the bottom of excavations to block water and resist hydrostatic pressures. They are typically used in combination with sheet piles or diaphragm walls, and are often placed underwater using a tremie.

---

[1] *Bentonite slurry* is a mixture of bentonite clay and water. When placed in a trench or drilled hole, it provides hydrostatic pressures that prevent caving.

[2] A *tremie* is a long tube used to place concrete. They are used when the concrete must be placed under water or through a bentonite slurry, because simply dropping the concrete through these materials would produce excessive segregation and result in a very poor product.

- *Permeation grouting* consists of injecting cement or special chemicals into the ground. This fills the voids and significantly reduces the hydraulic conductivity of the soil.
- *Ground freezing* consists of inserting refrigeration devices into the ground and freezing the groundwater. This method has been used successfully, and is very effective. It also is very expensive.

Exclusion

Exclusion methods use compressed air to prevent groundwater from entering a construction excavation, as shown in Figure 8.33. This method has been used in tunnel and foundation construction, because these excavations can be sealed and pressurized. The air pressure is maintained at a level approximately equal to the pore water pressure, thus preventing or at least greatly reducing seepage into the excavation. Workers, equipment, and excavated soil must pass through a decompression chamber located between the excavation and the atmosphere.

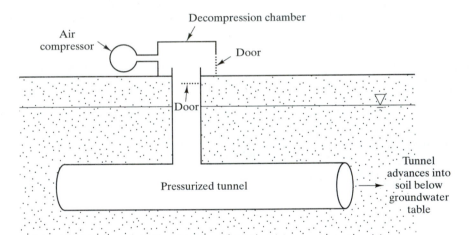

Figure 8.33 Tunnel construction under compressed air.

Permanent Dewatering

Permanent dewatering measures are necessary when groundwater must be controlled throughout the useful life of a civil engineering project. Some of the construction dewatering methods also may be used for permanent dewatering. However, economics and reliability concerns generally prevent the use of extensive long-term pumping. The specific methods of permanently controlling groundwater depend on the application (Cedergren, 1989). The following sections describe some common techniques and concerns for specific applications.

Underground Structures

Buildings with basements, tunnels, utility vaults, sewage pumping stations, and other structures need to be protected from flooding, especially when they extend below the groundwater table. Usually these structures have thick concrete walls that are reasonably watertight. In addition, the outside areas of these walls are usually coated with a waterproof material, and drain pipes are sometimes installed around the perimeter. However, in spite of these precautions, some water may enter the structure. Therefore, it is generally necessary to direct any such water to a sump and pump it out. Normally these sump pumps do not need to operate continuously.

Retaining Walls

If groundwater is allowed to build up behind retaining walls, the resulting hydrostatic pressures may be much greater than the lateral pressures due to the soil. Sometimes it becomes necessary to design for these pressures, but this can significantly increase the cost of the wall. The other option is to install drain pipes or weep holes (drainage holes near the bottom of the wall) to drain any groundwater, thus maintaining the groundwater table at a suitably low elevation.

Pavements

Pavements for highways, parking lots, airports, and other facilities are very sensitive to the accumulation of water in the underlying soils, and many pavement failures may be attributed to poor drainage of the subgrade soils. Water inevitably passes through joints, cracks, and other openings in pavements, and in the case of certain open-graded asphalt it may pass through the pavement itself. It is virtually impossible to stop water from reaching this zone. Therefore, engineers need to drain this water to a safe location.

Earth Slopes

Permanent drainage measures are often installed in earth slopes to control the groundwater level and thus enhance stability. This method is discussed in more detail in Chapter 14.

Earth Dams

Earth dams include extensive drainage facilities to control the groundwater flow through the dam. This enhances their stability.

8.5 SOIL MIGRATION AND FILTRATION

Geotechnical engineers often intentionally place highly pervious soils in key locations to capture and drain groundwater. Poorly graded coarse gravels are especially useful in this regard because they have very high hydraulic conductivities. However, a problem occurs when we place such soils adjacent to finer-grained soils (and nearly everything is finer than coarse gravel!) because the seepage forces push the finer soils into the gravel drainage layer, as shown in Figure 8.34. We call this process *soil migration*.

This migration has at least two detrimental results. First, the drainage layer becomes clogged and no longer functions properly. This clogging can occur whenever a coarse soil is downstream of a significantly finer soil and the seepage forces (i.e., hydraulic gradients) are large enough to cause soil migration. Once the subsurface drainage system ceases to work, failure often follows. Second, the migrating soils leaves voids in the upstream strata. In some cases, these voids can propagate for long distances, creating underground channels that lead to *piping failure* as shown in Figure 8.35. This is an important cause of failure in dams and levees, as discussed in Chapter 15.

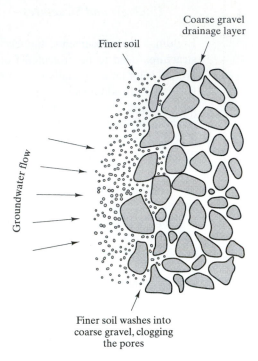

Figure 8.34 Migration of finer soils into a coarse drainage lager. The groundwater flow is washing the finer soils into the voids between the coarse gravel particles.

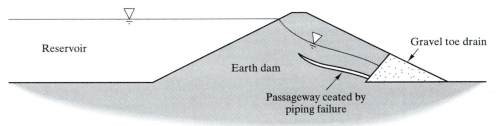

Figure 8.35 A piping failure in an earth dam.

Migration problems can be avoided by providing *filters* which are intended to pass water but retain potentially migrating soil. There are two principal types of filters: *graded soil filters* and *geosynthetic filters*.

Graded Soil Filters

Graded soil filters consist of one or more layers of carefully graded soil placed between the potentially migrating soil and the drain. Each filter zone is fine enough to prevent significant migration of the upstream soil, yet coarse enough not to migrate into the downstream soil. The filter also must have a sufficiently high hydraulic conductivity to effectively transmit water to the drain.

Sherard, et al. (1984a, 1984b, 1985, 1989) developed design criteria for graded soil filters based on the D_{15} size, as shown in Table 8.2, where:

D_{15} = the grain size at which 15% of the filter material is finer
d_{85} = the grain size at which 85% of the soil to be filtered is finer
A = the percentage of the soil to be filtered that passes the #200 sieve

For soil groups 1, 2, and 4, the values of d_{85} and A should be based only on the portion of the soil that passes the #4 sieve. In other words, if these soils contain plus #4 material, the grain-size curve needs to be reconstructed to what it would be if that material was not present. This adjustment is not necessary for soil group 3.

The shape of the filter grain-size curve should be similar to that of the soil being filtered (i.e., their slopes on the grain-size distribution chart should be about equal).

TABLE 8.2 GRADED SOIL FILTER DESIGN CRITERIA (Sherard, et al., 1984a, 1984b, 1985, 1989; SCS, 1986)

Soil Group No.	Soil Description	D_{15} Design Criteria
1	Fine silts and clays with >85% passing the #200 sieve	$D_{15} \leq 9d_{85}$, but not smaller than 0.2 mm
2	Silty and clayey sands and sandy silts and clays with 40–85% passing the #200 sieve	$D_{15} \leq 0.7$ mm
3	Silty and clayey sands and gravels with 15–39% passing the #200 sieve	$D_{15} \leq \left(\dfrac{40 - A}{25} \right) (4d_{85} - 0.7 \text{ mm}) + 0.7 \text{ mm}$
4	Silty and clayey sands and gravelly sands with ≤15% passing the #200 sieve	$D_{15} \leq 4d_{85}$

Other more detailed filter design criteria also are available (i.e., Reddi and Bonala, 1997).

Example 8.7

The design of a proposed earth dam is to include a gravel drain to control the groundwater inside the dam. This drain is to be protected with a graded soil filter zone. The grain-size distribution curves of the soils used to build the core of the dam and the gravel used to build the drain are shown in Figure 8.36. Determine the range of acceptable filter material.

Solution

Migration of core soils into filter:

91% passing #200 sieve, so this is soil type 1 per Table 8.2

$d_{85} = 0.056$ mm

$D_{15} \leq 9d_{85} \rightarrow D_{15} \leq (9)(0.056 \text{ mm}) \rightarrow D_{15} \leq 0.50$ mm

Therefore, the filter soils must have $D_{15} \leq 0.50$ mm, as shown by the mark on Figure 8.36.

Migration of filter soils into drain:

We use the same analysis, but now D represents the drain and d represents the filter. Based on the earlier results, the filter will be a type 4 soil

$D_{15} = 4.0$ mm

$D_{15} \leq 4d_{85} \rightarrow 4.0 \text{ mm} \leq (4)(d_{85}) \rightarrow 1.0 \leq d_{85}$

Therefore, the filter soils must have $D_{85} \geq 1.0$ mm, as shown by the mark on Figure 8.36.

Based on these two marks, the range of acceptable filter material is as shown in Figure 8.36.

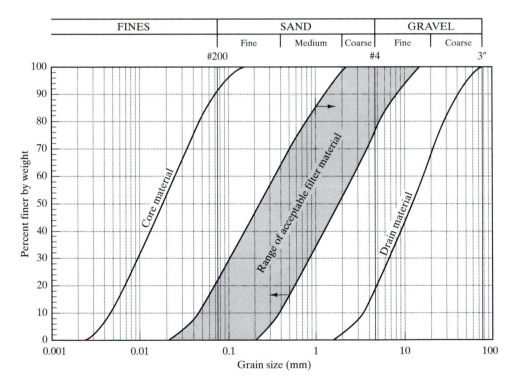

Figure 8.36 Grain-size distribution curves for Example 8.7.

Geosynthetic Filters

A wide variety of geosynthetic filter materials (often called *filter fabrics*) are now used as alternatives to graded soil filters. These are known as *geotextiles* (Koerner, 1998) and are supplied as rolls of fabric as shown in Figure 8.37.

Geotextile manufacturers provide various specifications for their materials to help the engineer select proper fabric for each application. One of these specifications is the *equivalent opening size* or *EOS* (also called *apparent opening size* or *AOS*), which is expressed either as O_{95} or as the equivalent sieve size. For example, a geotextile with EOS = AOS = #30 (O_{95} = 0.60 mm) has openings similar to those in a #30 sieve. Such a fabric would retain 95% of soil particles that have a diameter of 0.60 mm.

Carroll (1983) recommends selecting geotextiles for filtration based the following criterion:

$$O_{95} < (2 \text{ or } 3)(D_{85})_{soil}$$
(8.35)

Figure 8.37 A geotextile that could be used as a filter or for other purposes. This one is a Tensar® Vectra fabric (Courtesy Tensar Earth Technologies, Inc., Atlanta, Georgia, USA).

The geotextile filter fabric also must have sufficient *permittivity* to pass the required groundwater flow rate. This parameter is a measure of the fabric's ability to pass water, and is defined as:

$$\psi = \frac{k_n}{t}$$
(8.36)

Combining with Darcy's Law (Equation 7.8) gives:

$$\psi = \frac{\left(\dfrac{Q}{A}\right)}{\Delta h}$$
(8.37)

where:

ψ = permittivity of the geotextile
k_n = hydraulic conductivity for flows normal to the fabric face
t = thickness of the geotextile
Q = flow rate normal to fabric
A = area of fabric
Δh = head loss through fabric

The minimum required permittivity may be determined by assigning an allowable head loss and using Equation 8.37. Measured permittivity values and other physical properties for various geotextiles from one manufacturer are presented in Table 8.3.

TABLE 8.3 PHYSICAL PROPERTIES OF GEOSYNTHETIC FABRICS MANUFACTURED BY EVERGREEN TECHNOLOGIES, INC. (Courtesy of Evergreen Technologies, Inc., Atlanta, GA)

Product Designation	Thickness (mils)	Puncture Resistance (lb)	Apparent Opening Size (AOS)		Permittivity (s^{-1})
			O_{95} (mm)	Sieve Size	
TG 1000	155	170	0.150	100	0.6
TG 800	120	145	0.177	80	1.0
TG 750	100	115	0.177	80	1.1
TG 700	85	100	0.210	70	1.3
TG 650	75	95	0.210	70	1.6
TG 600	65	80	0.250	60	1.8
TG 550	55	70	0.300	50	1.9
TG 500	45	60	0.300	50	2.3

Evergreen is one of many manufacturers. Information on other products may be found in IFAI (1997) or by contacting the manufacturers.

Example 8.8

A sandy silt with $D_{85} = 0.10$ mm is to be drained by a perforated drainage pipe surrounded by a 3/4 inch gravel and a filter fabric as shown in Figure 8.38. The estimated flow rate to this drain will be 5 gal/min per lineal foot (half of which is from each side), and the head loss through the fabric must not exceed 0.1 ft. Select an appropriate filter fabric from Table 8.3.

Solution

Check maximum allowable apparent opening size (AOS).

$$O_{95} < (2 \text{ or } 3)\,(D_{85})_{\text{soil}}$$
$$< (2 \text{ or } 3)\,(0.10 \text{ mm})$$
$$< 0.2 - 0.3 \text{ mm}$$

Check minimum permittivity requirement.

$$Q = (5 \text{ gal/min}) \left(\frac{1 \text{ ft}^3/\text{s}}{449 \text{ gal/min}} \right) = 0.01 \text{ ft}^3/\text{s}$$

$$A = (0.5 \text{ ft} + 1.0 \text{ ft} + 0.5 \text{ ft}) (1.0 \text{ ft}) = 2.0 \text{ ft}^2$$

$$\frac{Q}{A} = \frac{0.01 \text{ ft}^3/\text{s}}{2.0 \text{ ft}^2} = 0.005 \text{ ft/s}$$

$$\text{minimum required } \psi = \frac{Q/A}{\Delta h} = \frac{0.005 \text{ ft/s}}{0.1 \text{ ft}} = 0.05 \text{ s}^{-1}$$

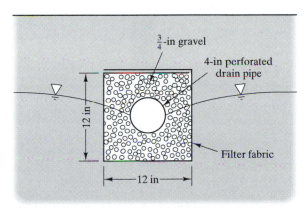

Figure 8.38 Proposed drain for Example 8.8.

Conclusion

Fabrics TG 750–TG 1000 definitely satisfy the AOS requirement, and the other fabrics marginally satisfy it. All of the fabrics satisfy the permittivity requirement by a wide margin. The costs of each are not given, so select the TG 1000 fabric, because it has the highest puncture strength, while still meeting the hydraulic requirements. However, if this fabric is too expensive or unavailable, the TG 800 or TG 750 also would be acceptable.

Occasionally, geotextile drains may be subject to soil clogging, which blocks the pores and prevents or severely restricts the flow of water. Although this is usually not a problem, the following conditions have been found to cause clogging (Koerner, 1998):

- Cohesionless sands and silts
- Gap-graded soils, and
- High hydraulic gradients

When all three of these conditions are present, it is better to use graded soil filters.

Clogging also can occur in soils where chemical precipitates tend to build up on the fabric, or when the groundwater contains a large concentration of biological matter (i.e., municipal landfill leachate).

QUESTIONS AND PRACTICE PROBLEMS

8.8   A proposed twenty-story office building with three levels of underground parking will be supported on a concrete mat foundation, as shown in Figure 8.39. The bottom of this mat will be 40 ft below the street, and its plan dimensions will be 200 ft × 150 ft. The groundwater table is currently at a depth of 25 ft below the ground surface, but could rise to only 13 ft below the ground surface during the life of the building. Compute the total hydrostatic uplift force to be used in the design.

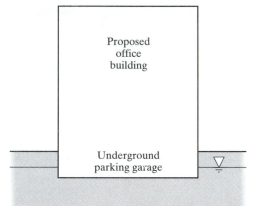

Figure 8.39 Proposed underground parking area for Problem 8.8.

8.9   A proposed levee is to be built using the soil described in Figure 8.40. This levee will include a toe drain similar to the one in Figure 8.35 to control the groundwater flow, and thus maintain adequate stability. This toe drain must be coarse enough to adequately collect and transmit the water, yet fine enough to provide sufficient filtration to prevent migration of the main levee soils. There will be no separate filter layer; the drain must act as the filter.

To maintain sufficient hydraulic conductivity, the drain must have no more than 3% passing the #200 sieve. In addition, to provide adequate filtration, it must meet the criteria described in this chapter. Determine the acceptable range of grain-size distribution for this material and plot it on a grain-size distribution curve.

8.10   The sheet pile in Example 8.6 was located near the upstream end of the spillway. Would the total hydrostatic uplift force acting on the structure change if the sheet pile was located near the downstream end? Explain. Which position would be best? Why?

8.11   An alternative design for the levee in Problem 8.9 uses a perforated pipe drain instead of the toe drain. This perforated pipe drain would be surrounded with gravel and wrapped with a filter fabric, similar to the one shown in Figure 8.38. The design flow rate into the drain is 80 gal/min per square foot, and the maximum acceptable head loss is 0.10 ft. Select an appropriate fabric from Table 8.3.

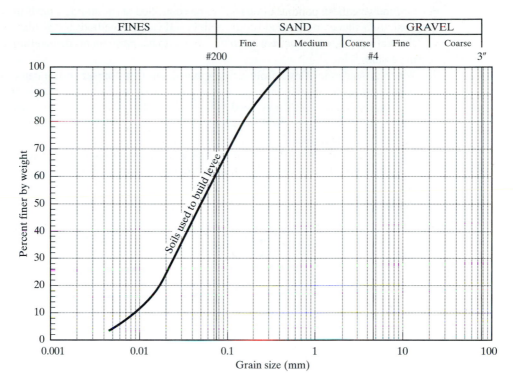

Figure 8.40 Grain-size distribution of proposed levee soils.

SUMMARY

Major Points

1. Two- and three-dimensional flow is governed by Darcy's Law, but its solution is more complex and requires use of the LaPlace Equation.
2. The LaPlace Equation for two-dimensional flow may be solved using a graphical solution called flow nets, by an electrical analogy, or by using numerical methods.
3. In general, three-dimensional flow is much more difficult to analyze, and generally requires numerical solutions. However, when certain symmetries are present, the problem simplifies and can sometimes be solved using simple equations. The flow of groundwater to a well is one such case.
4. Data collected from a pumped well along with two or more observation wells may be used to back-calculate the in-situ hydraulic conductivity. Such measurements are generally more precise than laboratory permeability tests because they encompass a much larger volume of soil.
5. Structures that extend below the groundwater table are subject to hydrostatic uplift pressures. These pressures may be computed and used to evaluate the stability of the structure.

6. Groundwater flow patterns often need to be controlled or modified. This is most often done to facilitate underground construction, and various methods are available. These are called *construction dewatering*. Sometimes permanent dewatering also is necessary.

7. When groundwater flows from a fine-grained soil to a coarser soil, it can cause the finer particles to migrate into the coarser zone. This migration can cause problems, such as piping, and thus may need to be controlled, especially in dams, levees, and other critical facilities. Soil migration also can clog drains, and thus needs to be controlled for that reason as well.

8. There are two ways to control soil migration: through the use of graded soil filters, or by installing filter fabrics. Both methods must satisfy certain design criteria.

Vocabulary

cutoffs	flow line	piping failure
dewatering	flow net	predrainage
electrical analogy	flow tube	soil migration
equipotential drop	geosynthetic filter	three-dimensional flow
equipotential line	graded soil filter	two-dimensional flow
equivalent opening size	LaPlace equation	uplift pressure
exclusion	numerical solution	
filtration	open pumping	

COMPREHENSIVE QUESTIONS AND PRACTICE PROBLEMS

8.12 Several years ago a state highway department built a highway across a shallow lake by placing the clayey fill shown in Figure 8.41. The top of this fill is above the high water level, thus protecting the highway from flooding.

It is now necessary to install a buried pipeline beneath the roadway. To install this pipe, the contractor plans to make a temporary excavation using steel sheet piles as shown. The contractor plans to use sump pumps at 50 ft intervals to maintain the water level at the bottom of the excavation. Once the pipe has been installed, the excavation will be backfilled and the pumps and sheet piles removed.

You are to perform the following tasks in connection with this project:
   a. Recognizing that the proposed cross-section is symmetrical, redraw half of it to a scale of 1 in = 10 ft and construct a flow net.
   b. Determine where the largest hydraulic gradient occurs and mark this spot on the cross-section.
   c. Using the flow net, compute the minimum required capacity for each pump, expressed in gallons per minute.
   d. Describe two methods of reducing the flow rate into the excavation (and thus the required pump size). Explain how each of them would reduce Q.

8.13 Using the flow net from Problem 8.3, compute the uplift pressures acting on the bottom of the dam in Figure 8.15 and develop a plot similar to the one shown in Figure 8.27.

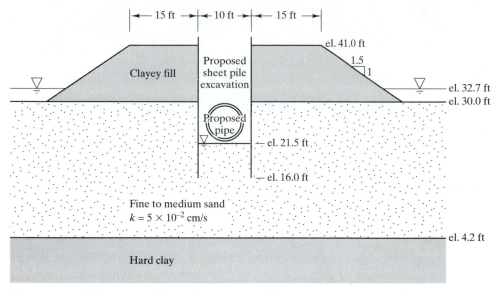

Figure 8.41 Cross-section for Problem 8.12.

8.14 A 30 m wide, 40 m long, 8 m deep construction excavation needs to be made in a silty clay (CL). The groundwater table is at a depth of 2 m. The sides of the excavation will be sloped at an angle of about 1 horizontal to 1 vertical, and no sensitive structures or other improvements are nearby. Suggest an appropriate method of construction dewatering for this site, and explain the reason for your choice. Include statements of any assumptions, if any.

8.15 After the analysis described in Example 8.3 was completed, the well was installed to the depth indicated in Figure 8.21. However, when the pump was installed, it produced a flow rate of only 102 gal/min. Is the difference between this value and the computed flow rate within the normal range of uncertainty for these kinds of analyses? Explain. What portion of the analysis usually introduces the greatest error?

8.16 Compute Q for the cross-section in Figure 8.15 using the following hydraulic conductivities for the silty sand: $k_x = 5 \times 10^{-2}$ cm/s, $k_z = 4 \times 10^{-3}$ cm/s.

8.17 By observing the groundwater drawdown in the vicinity of a proposed well, an engineer had determined the radius of influence, r_0. This engineer then used Equation 8.29 to compute k for the aquifer. Write a 200–300 word memo to this engineer, explaining why this is not a good method of computing k, then suggest a better method.

8.18 A below-ground swimming pool was built twenty years ago and, until recently, has been performing satisfactorily. About ten years ago it became necessary to temporarily drain the pool to clean out some algae. It was then refilled without any problems. However, the pool recently developed some cracks in its concrete shell, so it became necessary to drain it once again. Soon after it was drained, the pool rose out of the ground a distance of about 1 m. Provide a possible explanation for this behavior, and a possible explanation for why the pool did not rise out of the ground when it was drained the first time.

9

Geoenvironmental Engineering

> *Civil engineers protect public health by designing and building facilities that provide clean water and sanitation. No other profession, including medicine, has done more to reduce the spread of disease and save lives.*

Environmental engineering is the branch of civil engineering that deals with assessing, avoiding, and remediating environmental problems. Before 1970, these efforts were mostly limited to the design, construction, and operation of sewage treatment plants to protect surface waters from hazardous discharges. Since then, environmental engineering has grown to encompass a wide range of issues, including the identification and remediation of various sources of waste and other pollutants.

Geoenvironmental engineering deals with underground environmental problems, and thus is a blend of geotechnical and environmental engineering. Most of this work focuses on protecting groundwater aquifers from contamination, and involves assessing the kinds of materials being discharged, the processes by which they travel through the ground, and methods of remediating contaminated sites.

Problems with underground contamination began to receive widespread attention in the United States during the late 1970s. Two of the most noteworthy examples discovered thus far have been the Love Canal in Niagara Falls, New York and the Rocky Mountain Arsenal near Denver, Colorado.

288

Love Canal

In 1894, Mr. William T. Love, a land developer, began construction of a canal in what is now the City of Niagara Falls, New York. This canal was to be part of a hydroelectric project, but it was abandoned after construction had barely begun. The partially completed canal was about 1000 m long, 24 m wide, and 6 m deep, but it was never connected to a source of water and thus never used for its intended purpose.

In 1942, the Hooker Chemical Company began using the abandoned canal as a disposal site for chemical wastes from their manufacturing facilities. They eventually placed about 22,000 tons of waste products in the canal. The U.S. Army and others also may have dumped wastes into the canal. This indiscriminate dumping continued until 1953, when the site was covered with a thin layer of soil and sold to the Niagara Falls Board of Education for one dollar.

The School Board then built an elementary school directly over the site of the old canal. After the school was completed in 1955, other parties built new houses immediately adjacent to (but not directly over) the canal. By 1972, all of these houses were completed and occupied.

Unfortunately, some of the chemical wastes migrated through the soil and into the basements of nearby houses. Then, in the mid-1970s, heavy precipitation caused large amounts of water to infiltrate into the waste-filled canal, driving some of the contaminants up to the ground surface, where they traveled overland into backyards. By 1978, it became apparent that a major problem existed, so President Carter declared the former Love Canal a federal disaster area and approved emergency financial aid. Soon thereafter, all of the residents moved out and the school was closed.

Subsequent investigations have found a variety of hazardous chemicals at this site, including pesticides, the highly toxic dioxin, and radioactive cesium 137. Various health problems observed in the former residents of this neighborhood have been attributed to their exposure to these chemicals.

Fortunately, the natural soils surrounding the old canal generally have a low hydraulic conductivity, so the waste materials were not able to travel very far. Following an exhaustive study, the Environmental Protection Agency (EPA) determined the zone of underground contamination is limited to the immediate vicinity of the canal (Deegan, 1987). Although this zone included the basements of houses built immediately adjacent to the old canal, it would have been much worse if the soils were more permeable. Most of the wastes that escaped did so through overland flow.

By 1989, over $140 million had been spent cleaning the site and relocating the former residents. Remediation efforts have included the following (Mercer, et al., 1987):

- Removal, treatment, and disposal of some of the contaminated soil.
- Installation of French drains to collect contaminated groundwater, which is then processed in a special treatment facility
- Construction of a clay cap with a high density polyethylene (HDPE) synthetic membrane over the site to contain the wastes, reduce rainwater infiltration, and prevent direct human contact.
- Construction of impervious underground barriers to minimize migration of the wastes
- Sealing of underground utilities, which were found to act as conduits for the wastes

There are no plans to remove the waste materials from the old canal.

The Love Canal incident is often quoted as the classic case of haphazard and careless disposal of hazardous industrial wastes followed by inappropriate site development (Hartman, 1983). Although its physical effects were limited to a small area, its significance goes far beyond Niagara Falls because it was one of the primary motives for developing geoenvironmental legislation that affects the entire country.

Rocky Mountain Arsenal

In 1942, the U.S. Army began construction of the Rocky Mountain Arsenal at a site near Denver, Colorado. One of the purposes of this facility was to manufacture and store various gasses for use in chemical warfare, including mustard gas, lewisite, and chlorine gas (U.S. Army, 1996). Fortunately, nobody used chemical weapons during World War II, primarily because the American capabilities in this area were a strong deterrent to the Germans. The arsenal also manufactured other munitions, including incendiary bombs made from napalm. Then, during the early cold-war era of the 1950s, it began producing nerve gas, a deadly chemical warfare material, eventually becoming the largest producer in the free world. Finally, the Army used the facilities to produce rocket fuel.

By 1969, production of war materials at Rocky Mountain Arsenal had ended. Then, as a result of changing political and military priorities, the facilities were used to destroy many of the munitions that had been manufactured and stored there.

In addition to these military activities, part of the arsenal was leased to Shell Oil Company and its predecessor, who used it to produce agricultural pesticides. This occurred from 1952 to 1982.

These various activities produced large quantities of industrial waste, much of which was in liquid form. Initially it was discharged into unlined "evaporation" ponds, which was typical practice for industry at that time. Although some of this material evaporated (with resulting air pollution consequences), much of it soaked into the ground. In 1956, a new waste disposal pond was built with a 3/8-inch thick asphalt liner. Unfortunately, the liner eventually failed, thus releasing more contaminants into the ground.

The soils at this site are much more pervious than those at Love Canal, so they can transport larger quantities of contaminants for longer distances. Concerns about groundwater contamination began as early as 1959 when crop damage was noticed near the arsenal. It soon became clear that this contamination extended several miles beyond the arsenal property.

In 1974, the State of Colorado ordered the Army and Shell to "stop polluting ground and surface waters," which initiated an intensive investigation of the nature and extent of contamination. A number of lawsuits also were generated, and one of the first large-scale clean-up operations began (Civil Engineering, 1981).

Cleanup efforts have included (Mercer, 1987; U.S. Army, 1996):

- Removal and treatment of remaining materials in the ponds and discharge to safe locations

- Installation of wells and decontamination equipment to remove, treat, and reinject contaminated groundwater. Engineers plan to continue operating this equipment until the groundwater is acceptably clean, which is expected to require 15 to 30 years.
- Construction of underground barriers to retard further migration of contaminated groundwater beyond the arsenal property and to prevent backflow between the injection and withdrawl wells.

The Rocky Mountain Arsenal case is similar to Love Canal in that it was caused by what we would now consider indiscriminate disposal of hazardous wastes (although the methods used this site were typical for the period). However, because of the higher hydraulic conductivity of the soils, the contaminants traveled much farther and thus require a far more extensive clean-up effort.

Legislation

Although the Love Canal and Rocky Mountain Arsenal events are extreme examples, smaller-scale problems have been identified at tens of thousands of other sites in the United States alone. About 1,200 of these sites have problems significant enough to be included in the 1994 federal Superfund National Priorities List.

At some locations, contaminated groundwater poses a serious threat to wells used for domestic and agricultural water supply, and to rivers and lakes. About half of U.S. residents rely primarily on well water, and hundreds of these wells have been closed due to contamination. Therefore, government agencies, especially the federal government, have implemented legislation to identify and clean up existing geoenvironmental problems and avoid creating new ones. Key federal legislation in the United States has included the following:

- Safe Drinking Water Act (SDWA) of 1974
- Resource Conservation and Recovery Act (RCRA) of 1976
- Comprehensive Environmental Response, Compensation and Liability Act (CERCLA) of 1980
- Hazardous and Solid Waste Amendments (HSWA) of 1984
- Superfund Amendments and Reauthorization Act (SARA) of 1986

Many state and local laws also address these issues, and a body of case law has developed.

At the federal level, the Environmental Protection Agency has the primary responsibility for enforcing these laws. Similar agencies also are present at state and local levels.

Much of the work mandated by this legislation needs to be performed by geoenvironmental engineers. Thus, this field has grown to become a significant part of geotechnical and environmental engineering, and will continue to be important for the foreseeable future. Many other professions also are involved in various capacities.

9.1 TYPES OF UNDERGROUND CONTAMINATION

Biological Contamination

Biological contamination is the result of activities that produce excessive numbers of harmful microorganisms in the ground. These microorganisms include both bacteria and viruses, both of which are potential health hazards.

Fortunately, bacteria die within days of being discharged into the soil (Sinton, 1980). Groundwater flows very slowly through most soils, so live bacteria are normally present only within the immediate vicinity of the source. In addition, soil acts as a filter and traps bacteria before they travel very far. For example, in unsaturated medium sands, virtually all of the bacteria discharged by septic tank leach fields are filtered out within 10 m (Franks, 1993). Therefore, most groundwater is virtually free from bacterial contamination. The most notable exceptions are in karst aquifers where groundwater travels much more quickly and therefore carries bacteria much farther.

However, viruses are a much more difficult problem because of their longer life and smaller size. Live viruses have been observed up to 131 days after discharge (Yates and Yates, 1989), so they travel farther before dying. In addition, because of their smaller size, viruses are less likely to be filtered by the soil.

In the United States, septic tanks are the most common source of disease associated with consumption of groundwater (Yates and Yates, 1989). Usually this occurs because water supply wells are placed too close to septic tank leach fields. Plumbing codes typically require setback distances of 15 to 30 m (50-100 ft), which should be more than sufficient for removing bacteria, but much greater setbacks may be necessary in highly pervious soils to avoid virus contamination. It also is far better to place wells upstream of septic tanks.

Biological contamination is a much greater problem in surface waters, such as rivers and lakes, partially because these waters travel much faster and thus spread the contamination over much larger areas.

Chemical Contamination

Chemical contamination results from large concentrations of undesirable chemicals in the groundwater. These problems are generally more widespread and difficult than biological contamination because chemicals do not "die" and thus can travel for long distances. Contaminant plumes have been traced for several kilometers downstream of their source.

The presence of certain chemicals in groundwater can be injurious to people and the environment. Some are carcinogenic (they cause cancer), others induce different kinds of health problems. Table 9.1 lists the most common chemical contaminants detected in groundwater at hazardous waste sites. This ranking was generated by the Agency for Toxic Substances and Disease Registry using groundwater data from the National Priorities List of sites to be cleaned up under CERCLA. The ranking is based on the number of sites at which the substance was detected in groundwater.

TABLE 9.1 THE 25 MOST FREQUENTLY DETECTED GROUNDWATER CONTAMINANTS
AT HAZARDOUS WASTE SITES (NRC, 1994)

Rank	Compound	Common Sources
1	Trichloroethylene	Dry cleaning; metal degreasing
2	Lead	Gasoline (prior to 1975); mining; construction material (pipes); manufacturing
3	Tetrachloroethylene	Dry cleaning; metal degreasing
4	Benzene	Gasoline; manufacturing
5	Toluene	Gasoline; manufacturing
6	Chromium	Metal plating
7	Methylene chloride	Degreasing; solvents; paint removal
8	Zinc	Manufacturing; mining
9	1,1,1-Thrichloroethane	Metal and plastic cleaning
10	Arsenic	Mining; manufacturing
11	Chloroform	Solvents
12	1,1-Dichloroethane	Degreasing; solvents
13	1,2-Dichloroethane	Transformation product of 1,1,1-trichloroethane
14	Cadmium	Mining; plating
15	Manganese	Manufacturing; mining; occurs in nature as oxide
16	Copper	Manufacturing; mining
17	1,1-Dichloroethene	Manufacturing
18	Vinyl chloride	Plastic and record manufacturing
19	Barium	Manufacturing; energy production
20	1,2-Dichloroethane	Metal degreasing; paint removal
21	Ethylbenzene	Styrene and asphalt manufacturing; gasoline
22	Nickel	Manufacturing; mining
23	Di(2-ethylhexyl)phthalate	Plastics manufacturing
24	Xylenes	Solvents; gasoline
25	Phenol	Wood treating; medicines

Many common chemical contaminants are not readily soluble in water. These are
called *non-aqueous phase liquids* (*NAPLs*), and have been the subject of substantial research
because they are so common. There are two types: *Light NAPLs* (*LNAPLs*) have specific
gravities less than water, and thus float on top of the groundwater, while *dense NAPLs*
(*DNAPLs*) have specific gravities greater than water, and thus sink through groundwater.
Gasoline is an example of an LNAPL, while trichloroethylene is a DNAPL.

Hazardous Waste

The term *hazardous waste* is used to describe materials that pose significant threats to health or the environment, whether alone or in combination with other materials. This includes wastes with any of the following properties:

- Toxic
- Ignitable
- Explosive
- Corrosive
- Reactive

- Radioactive
- Infectious
- Irritating
- Sensitizing
- Bioaccumulative

Hazardous wastes are subject to special legal requirements, and are especially important in geoenvironmental investigations.

It is important to recognize that the potential hazard from a material depends not only on its presence, but also on its concentration. A hazardous material in very dilute solution may be virtually harmless, even though the same material in highly concentrated form might be highly toxic. This distinction is becoming increasingly important as instrumentation and test methods become more sophisticated. We are now able to detect very low concentrations of many chemicals, often well below that required to pose significant threats to health or the environment. Thus, the mere detection of a "hazardous" material in the ground does not necessarily indicate a significant problem.

9.2 SOURCES OF UNDERGROUND CONTAMINATION

Underground contaminants come from a variety of sources. *Point sources* are confined to a small area, such as a leaking underground storage tank or the site of an accidental spill, while *non-point sources* enter the ground over a large area, such as agricultural chemicals entering from a field.

Intentional Underground Disposal

For many years, liquid waste materials have been intentionally discharged into the ground as a means of disposal, as shown in Figure 9.1. Sometimes this was accomplished through open waste disposal ponds (also called "evaporation" ponds), such as the ones at the Rocky Mountain Arsenal (described earlier in this chapter). The Environmental Protection Agency has identified over 18,000 waste disposal ponds, many of which are unlined and near drinking water wells (Bedient, Rifai, and Newell, 1994).

Shallow injection wells also have been used, sometimes in close proximity to groundwater aquifers. These were very convenient for the user, but are potential sources of extensive underground contamination.

Deep injection wells that dispose of liquid contaminants well below groundwater aquifers also have been used. These wells are typically several thousand feet deep and, when properly used, might be an acceptable means of discharging certain kinds of waste. However, the continued use of such wells is a controversial topic.

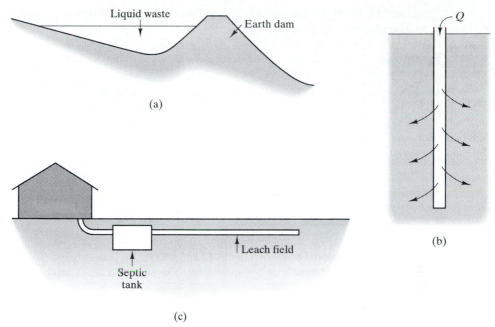

Figure 9.1 Intentional discharge of wastes into the ground: a) waste disposal ponds; b) shallow injection wells; c) septic tank sewage disposal systems.

Septic tank systems are the most common source of intentional underground disposal. They are used to treat and discharge residential, commercial, and industrial wastes in areas where sewers are not available. About 30 percent of the population in the United States relies on such systems.

The raw sewage enters the underground septic tank, where it is separated and digested by physical, chemical, and biological processes. The accumulated solids are periodically removed with a special pump and hauled away, while the liquids are discharged into a series of perforated pipes called a drain field (or leach field), where they soak into the ground. Septic tank "failures" normally refer to leach fields that are unable to discharge the required flow rate, which causes sewage to back up into the house.

In residential applications, contaminants in this discharge are primarily biological and can usually be controlled by using adequate setbacks from wells and other sensitive locations (see earlier discussion on biological contamination). Some chemical contaminants also may be present at some commercial sites (such as dry cleaners[1]) and in industrial applications.

Because of evaporation and transpiration, much of the effluent from leach lines often travels upward, not downward. Thus, the biological "contaminants" often become

[1] So called "dry" cleaning uses liquid solvents to clean clothing and fabrics. These solvents are hazardous, and should not be indiscriminately discharged into the ground.

"fertilizer" for lawns and trees, which inspired the title for Erma Bombeck's book *The Grass is Always Greener Over the Septic Tank.*[2]

Leakage

Many hazardous materials have entered the ground unintentionally through leakage in tanks, pipes, and other facilities, especially those that are buried underground. Such leaks can easily continue for many years without being detected, releasing large quantities of contaminants.

Underground storage tanks (USTs) have received the most attention in this context. There are an estimated 2.5 million USTs in the United States (OTA, 1984), many of which are located at gasoline service stations. The Environmental Protection Agency has estimated that 35 percent of USTs used for motor fuel storage have leaks (Bedient, Rifai, and Newell, 1994), so this is a widespread problem.

One of the most noteworthy examples of groundwater contamination from USTs is at the U.S. Coast Guard Station in Traverse City, Michigan. Leaking gasoline and jet fuel from USTs there have produced a 1 mile long, 500 ft wide plume that has contaminated about 100 municipal water wells.

Spills

Accidental and intentional spills onto the ground surface also have been a frequent source of underground contamination. Sources of these spills have included ruptures of above-ground tanks and pipes, discharges from derailed railroad cars, spills from vehicle and aircraft refueling facilities, accidents at chemical plants and refineries, and others.

In the past, clean-up efforts (if any) focused almost exclusively on materials remaining above ground, especially when they threatened to contaminate rivers or lakes. Until recently, portions that soaked into the ground were usually ignored.

Materials Applied to the Ground Surface

Pesticides, fertilizers, de-icing salts, and other materials are often applied to the ground surface for various purposes. They often are transported by surface water runoff, and can become sources of contamination in both surface water and groundwater.

Landfills

Solid waste materials are most often placed in landfills, which can range from uncontrolled open dumps to engineered sanitary landfills. Liquid *leachate* flowing out of these facilities can contaminate the groundwater. We will discuss the design of modern landfills later in this chapter.

---

[2] A geoenvironmental engineer would have called it *The Grass is Always Greener Over the Leach Field*, a much less marketable title.

9.3 FATE AND TRANSPORT OF UNDERGROUND CONTAMINANTS

Once a contaminant has escaped into the ground, it flows from pore to pore through the soil, sometimes traveling several miles. The manner and rate of transport depend on many factors, including:

- Whether the soil is saturated or unsaturated
- The type of soil
- The type of material flowing through the soil, especially its solubility in water and its specific gravity
- The velocity and direction of natural groundwater flow
- The rate of infiltration from the source

Advection

Contaminants travel with moving groundwater through a process called *advection*. The contaminants are simply "going along for the ride," so an advection analysis is simply an extension of the groundwater flow analyses discussed in Chapter 7.

Geoenvironmental engineers are especially interested in the velocity of flow due to advection, because this helps us understand how far the contaminants will travel in a given time. The *seepage velocity*, as discussed in Chapter 7, is:

$$v_s = \frac{k\,i}{n_e}$$

(9.1)

where:

v_s = seepage velocity
k = hydraulic conductivity
i = hydraulic gradient
n_e = effective porosity

Example 9.1

A chemical waste is being discharged into a shallow injection well. The surrounding soil is a fine to medium sand with a hydraulic conductivity of 4×10^{-3} ft/s and a void ratio of 0.91. The groundwater table is at a depth of 20 ft and the hydraulic gradient is 0.006. A municipal water well is located 2 miles downstream of the source. Considering only advection, how much time will be required for the contaminants to travel from the source to the well?

Solution

$$n = \frac{e}{1+e} \times 100\% = \frac{0.91}{1+0.91} \times 100\% = 48\%$$

The soil is sandy, so the effective porosity is equal to the porosity, n, as discussed in Chapter 7.

$$v_s = \frac{ki}{n_e} = \frac{(4\times10^{-3}\,\text{ft/s})(0.006)}{0.48} = 5\times10^{-5}\,\text{ft/s}$$

$$t = \frac{l}{v_s} = \frac{(2\,\text{mi})(5280\,\text{ft/mi})}{5\times10^{-5}\,\text{ft/s}} = 2\times10^8\,\text{s} = 7\,\text{yr} \quad \leftarrow \textbf{\textit{Answer}}$$

Seven years is a very short time, especially considering that injection wells such as this often are in service for decades.

Two-dimensional advection analyses could be performed by combining the technique of Example 9.1 with a flow net. The results could be expressed as shown in Figure 9.2.

Source

$\longrightarrow$ Direction of groundwater flow

Figure 9.2 Two-dimensional transport of contaminants from a continuous source considering only advection.

Dispersion

When water and contaminants flow through soil, the irregular shape of the pores and the particulate nature of the soil cause some of the contaminants to spread out over a wider area than predicted by advection alone. This spreading process is called *dispersion*. Figure 9.3 shows the spread of contaminants considering both advection and dispersion. These two processes dominate contaminant transport in highly permeable soils, such as sands, particularly when the hydraulic gradient also is high. This is because groundwater flows more quickly through such soils.

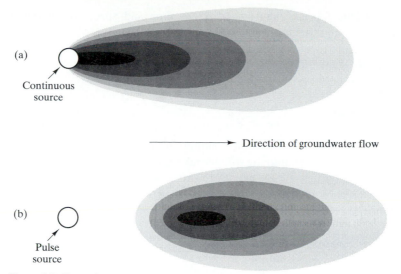

Direction of groundwater flow

Figure 9.3 Contaminant transport due to boty advection and dispersion: a) from a continuous source (compare with Figure 9.2); b) from a short-term "pulse" source, such as a spill.

The degree of dispersion may be defined by the *hydrodynamic dispersion coefficient*, D, which is then introduced into contaminant transport analyses (Bedient, et al., 1994). Engineers use *advection–dispersion analyses*, which combine both modes of transport, to assess contamination problems and to assist in the design of remediation methods.

Diffusion

When the concentration of a chemical in a liquid varies from place to place, the chemical naturally moves from the areas of high concentration to areas of low concentration through a process called *diffusion*. This process may be observed by placing a drop of food coloring in a cup of water. Initially the food coloring is concentrated in one small area, but it quickly diffuses throughout the cup. The same process occurs in groundwater.

Diffusion is described by *Fick's first law of diffusion*:

$$f = -D_d \frac{dC}{dl}$$

(9.2)

where:
f = mass flux (M/L^2/T)
D_d = diffusion coefficient (L^2/T)
C = concentration (M/L^3)
l = flow distance (L)

This formula may be solved in time and space to determine the contaminant concentration with time at any point.

In soils with a low hydraulic conductivity, such as clays, advection and dispersion are very slow and diffusion usually becomes the dominant process.

Sorption

Some waste chemicals form coatings around soil particles through a process called *adsorption*, or soak into the soil particles through *absorption*. In the field, it is difficult to distinguish which of these two is occurring, or in what proportions they occur, so we use the term *sorption* to describe their collective action. Factors that affect sorption include the following (Bedient, et al., 1994):

Contaminant characteristics
- Water solubility
- Polar-ionic character
- Octanol-water partition coefficient

Soil characteristics
- Mineralogy
- Hydraulic conductivity-porosity
- Texture
- Homogeneity
- Organic carbon content
- Surface charge
- Surface area

Sorption retards the flow of contaminants, thus producing plumes that move more slowly and have higher concentrations than would otherwise occur. Its effect may be estimated using various laboratory, field, or empirical methods.

Volatilization

Volatilization is the conversion of liquid materials into their gaseous phase. Materials are said to be "highly volatile" if this process occurs very quickly. For example, benzene is a highly volatile liquid. Conversely, materials that experience little or no volatilization are classified as "nonvolatile."

Volatilization occurs both above and below the groundwater table, and is an important process in the transport of certain contaminants. It also can be a source of problems, because the gasses can accumulate in undesirable places, such as underground utility vaults or basements, and cause explosions or health problems.

Chemical Reactions

A wide range of chemical reactions also can occur as contaminants travel through soil. Some of these reactions are between two contaminants traveling together, while others are between a contaminant and the soil. Water often is an agent in these reactions.

This is potentially one of the most complex aspects of contaminant transport analyses because so many different reactions are possible.

Biodegradation

Biodegradation occurs when waste materials are "digested" by microbes in the soil, thus converting them into new, less harmful materials. This is a very important process, and is the basis for certain remediation methods (see discussion of bioremediation later in this chapter).

Biodegradation is both a biological and a chemical process, and several conditions must be met for it to occur:

1. The waste materials must be carbon-based, such as petrochemicals.
2. The proper organisms must be present. They often occur naturally, or may be artificially introduced for the purpose of enhancing clean-up efforts.
3. An electron acceptor must be present so the biochemical reactions will proceed. Oxygen and carbon dioxide are two common electron acceptors.
4. Certain nutrients, such as nitrogen, phosphorous, and calcium, must be present
5. The temperature, pH, salinity, and other environmental conditions must be within acceptable limits.

Because of its use as a remediation method, biodegradation has been the object of extensive research.

Radioactive Decay

Radioactive isotopes occur in the ground, both naturally and in contaminants. These materials decay with time and form new materials. The rate is expressed as a *half-life* which is the time required for half of the material to decay. Although radioactive decay has no effect on the rate of contaminant movement, it does decrease the hazards associated with the contaminants as they flow away from the source.

One of the most important radioactive conditions in the ground is a natural one that even occurs in soils far from any source of human contamination. This process involves the decay of radon, which is produced by rocks high in uranium. High radon contents in drinking water can pose certain health problems. In addition, radon can enter homes though emanations from the soil and by diffusion from tap water with high radon contents.

9.4 GEOENVIRONMENTAL SITE CHARACTERIZATION

The work associated with identifying and characterizing potential geoenvironmental problems ranges from routine checks to detailed studies with extensive testing and monitoring. The appropriate level of work depends on the type and magnitude of the problems, if any, the potential threat to humans and the environment, and many other factors.

When commercial real estate in the United States is sold or refinanced, an *Environmental Site Assessment* (ESA) is usually performed. The purpose of these assessments is to determine whether the site is likely to be contaminated by hazardous materials as defined by applicable laws. These reviews are conducted by "environmental professionals" and are typically required by the buyer or the buyer's lender, who use this information to assist in their purchasing decisions. In addition, a favorable ESA report limits the buyer's liability to clean up pre-existing geoenvironmental problems discovered after the property is purchased. In some states, ESAs are required by law.

The most common type is called a *Phase I Environmental Site Assessment*. These studies typically include the following:

- Reviewing current and past uses of the property and adjacent properties
- Conducting a site reconnaissance
- Reviewing official records to identify known problems on this or nearby sites
- Interviewing owners and occupants

ASTM E1527 and E1528 describe suggested practices for performing Phase I ESAs.

Phase I ESAs do not involve "intrusive site investigations," which means no exploratory borings or trenches are used, and do not involve collecting samples or performing tests. They are intended to assess known, documented, or obvious problems, but can easily miss problems that are unknown and hidden. They also do not address other types of environmental hazards, such as asbestos or lead-based paint in building materials, nor do they address the costs or methods of remediating geoenvironmental problems.

If the Phase I study indicates hazardous wastes may be present, then it is generally followed by a Phase II study, which is a more extensive effort to identify potential problems. Phase II studies do include subsurface exploration, sampling and testing to assess the soil and groundwater, along with engineering analyses. These studies range widely in scope and detail, depending on the site conditions and extent of contamination.

Once geoenvironmental problems have been found, a remediation plan needs to be developed. A monitoring plan also may be necessary.

9.5 REMEDIATION METHODS

Once a contaminated site has been discovered and evaluated, we focus our attention on cleaning, or at least containing it. This may be a simple matter if the contaminated zone is small, but the difficulty and cost increase dramatically when large volumes of soil and groundwater are affected.

Remediation efforts normally are intended to achieve certain specified results. Complete recovery of all pollutants is rarely possible or practical, except when the zone of contamination is small. Normally, the remediation operation must only achieve a certain level of recovery to be considered successful.

Various standards have been developed for the quality of municipal drinking water, and these are often used as goals for groundwater cleanups. Alternatively, clean-up standards might be based on reducing certain contaminants such that the risk of developing health problems is below some specified level, such as a carcinogen risk of less than 10^{-4}. Other clean-up goals also have been proposed and used.

These groundwater quality standards can sometimes be attained quickly and at a reasonable cost, but often the time and/or cost required to achieve them becomes excessive. For example, there are some situations where remediation to drinking water standards would require half a century or more and cost tens of millions of dollars. Therefore, finding the optimum balance between the costs and benefits of remediation is a difficult and contentious issue. As the geoenvironmental industry matures, these issues should become more clear (NAS, 1994).

A wide variety of methods is available to clean up contaminated sites, and new technologies are being developed. The proper selection depends on the type of contaminants, the soil and site conditions, expected results, cost, and other factors.

Source Control

The most obvious aspect of reducing subsurface contamination is to stop the influx of contaminants from the source. For example, leaking underground storage tanks need to be repaired or replaced to prevent further contamination of the surrounding soil. Shallow injection wells and waste disposal ponds should be taken out of service and new methods found to dispose of the waste materials. However, these measures can be implemented only when we know the location of the source, which may be difficult to determine, especially if it is buried.

Excavation and Disposal

If only a small volume of soil has been contaminated, it may be practical to excavate and transport it to a safer location, such as a secure landfill. Although this method simply moves the problem to a new location, a well-designed landfill should provide much more protection for the surrounding environment. Often the excavated soil can be treated, such as by incineration or through other processes, to remove or neutralize the hazardous materials.

Another possibility is to mix non-hazardous contaminated soil with asphalt to form an asphaltic concrete pavement. For example, 2,700 tons of soil contaminated with diesel fuel at Fort Irwin, California were mixed with natural aggregates, water-based asphalt emulsion, and setting agents and placed as a 12 inch thick intermediate grade industrial pavement (CDE, 1996). The asphalt prevents significant quantities of the contaminants from leaching back into the soil.

The process of excavating and transporting contaminated soil can expose workers to hazardous materials, and can become a source of air pollution. These aspects need to be weighed against the potential benefits.

Excavation is generally not feasible when large volumes of soil have become contaminated, or when the contaminated zone extends below buildings or other obstructions.

Containment

Another option is to surround the contaminated soils with an impervious barrier to prevent the contaminants from traveling outside the containment zone. Containment is especially attractive when the cost or risk of removal is not acceptable. Methods of containment include constructing a *slurry trench wall*, a *grout curtain*, or *sheet piles*, as shown in Figure 9.4. Other containment methods also have been used.

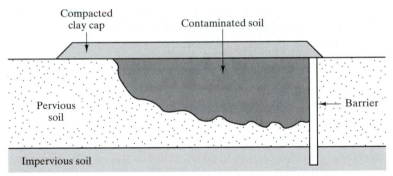

Figure 9.4 Use of containment barriers to block the flow of contaminants.

Slurry trench walls are built by excavating a trench while keeping it filled with a bentonite slurry (a combination of bentonite and water), as shown in Figure 8.32. The purpose of the slurry is to keep the sides of the trench from caving. Normally the excavation extends down to an impervious strata, which may be 20 m or more below the ground surface. Then, a clayey soil is pushed into the trench to fill it and displace most of the slurry. This mixture of clayey soil and the remaining bentonite forms the groundwater barrier. The process continues until the slurry trench wall has the desired length, which sometimes means placing it around the entire site.

Grout curtains are made by injecting cement grout into the soil to form an impervious barrier. Sheet piles are heavy, corrugated steel sheets that are driven into the ground to form a barrier.

The contaminated zone also may be covered with a compacted clay cap, which often includes one or more geosynthetic membranes. These caps are intended to reduce the infiltration of surface water and minimize the potential for human exposure to the waste. This was one of the containment measures used at Love Canal.

Containment systems can only be built around the sides and top of the contaminated zone. In some cases, the bottom might be naturally contained by impervious strata, but even then containment systems are not completely effective. As with any underground construction, uncertainties will always be present, and leaks can occur in unexpected places. Therefore, containment systems are often accompanied by other remediation measures, and almost always include monitoring systems.

Pump-and-Treat

Pump-and-treat remediation consists of extracting the contaminated groundwater, passing it through above-ground treatment facilities, then discharging it back into the ground through injection wells. This is one of the most commonly used remediation methods.

A wide variety of treatment methods is available, depending on the type and concentration of contaminants in the extracted water. Contaminants removed from the water are then hauled to a suitable disposal site.

Usually wells are used to extract the contaminated groundwater, as shown in Figure 9.5. Their placement and pumping rates are usually chosen so all of the contaminated groundwater flows toward the wells, thus increasing the recovery and reducing the potential for further expansion of the plume. These systems may be designed with the aid of numerical models that are available as computer software.

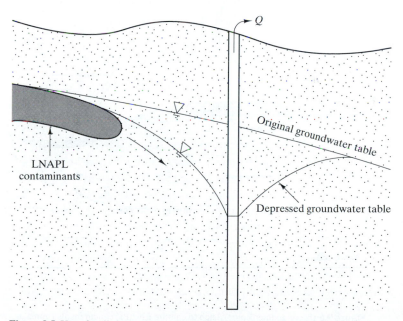

Figure 9.5 Use of wells in a pump-and-treat system.

In the case of LNAPL contamination, where the hazardous materials float on top of the groundwater, a trench might be used to capture the contaminant as shown in Figure 9.6.

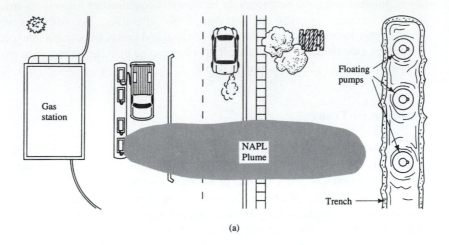

(a)

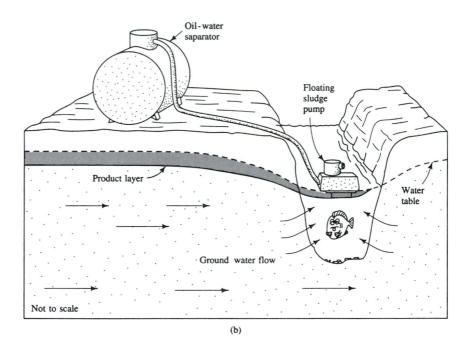

(b)

Figure 9.6 Use of an interceptor trench to capture LNAPL contaminants from a gasoline service station. Note how the LNAPL contaminants float on top of the groundwater table, thus making them much easier to capture (Fetter, 1993).

Bioremediation

Bioremediation is the engineered enhancement of natural biodegradation processes (Norris, et al., 1994). It is most frequently performed in-situ by providing favorable conditions for the biochemical processes. It also may be done ex-situ as a treatment method on excavated soils.

In-situ bioremediation systems usually consist of wells that inject nutrients, an electron acceptor, and possibly other substances to promote biodegradation. Pumps are installed in another set of wells to develop a hydraulic gradient across the site, thus distributing the injected materials, as shown in Figure 9.7.

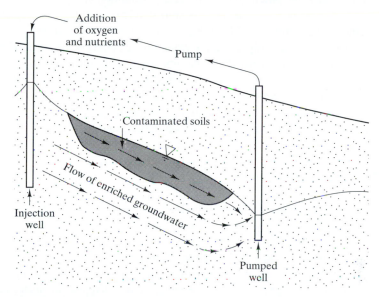

Figure 9.7 Typical bioremediation system.

Bioremediation is feasible only in soils with a sufficiently large hydraulic conductivity ($> 10^{-4}$ cm/s) to permit rapid delivery of the nutrients and oxygen. In some cases, such as halocarbons, biological processes can make matters worse instead of better by producing new substances that are even more hazardous than the original contaminant.

Soil Vapor Extraction

Unsaturated soils with volatile contaminants (i.e., those that quickly vaporize upon exposure to air) can sometimes be cleaned using a *soil vapor extraction system*. This method consists of applying a vacuum to a series of wells in an effort to draw large quantities of air through the soil. The effluent from these wells is then treated and discharged to the atmosphere.

This is a new technology that is still under development, so few design guidelines are yet available. It probably will be feasible only in soils with high air conductivities, which limits it to sandy soils with low moisture contents.

Soil vapor extraction systems may be enhanced through *in-situ air sparging*, which consists of injecting compressed air in the saturated soil below the contaminated zone (Reddy and Adams, 1996). The air rises through the contaminants, drawing them into the unsaturated zone above the groundwater table where they are captured by vapor extraction wells.

Intrinsic Remediation

In some cases, the potential benefits of a remediation scheme are so small and the cost of implementation is so high that it is best to simply do nothing and rely on natural processes to remediate the problem. This is known as *intrinsic remediation* (Hicks and Rizvi, 1996). This method is especially attractive when no wells, rivers, or other critical facilities are in immediate danger and when the contaminants are amenable to cleaning by natural processes.

Intrinsic remediation relies on natural biodegradation, reductions in concentration due to diffusion and dispersion, and other processes. These processes can be monitored using groundwater sampling wells. More aggressive techniques can then be introduced if the intrinsic remediation does not progress as anticipated.

9.6 SANITARY LANDFILLS

Although the identification and remediation of existing problems are important aspects of geoenvironmental engineering, we also need to design and build new facilities in ways that will avoid creating new environmental problems. Sanitary landfills are among the most common and most important of these facilities.

Sanitary landfills are facilities for disposing of solid wastes in a controlled fashion. They began to appear in the early twentieth century as replacements for the rat-infested open dumps that were being used to dispose of wastes. These new facilities are "sanitary" because the refuse is periodically covered with soil, thus forming a series of *cells*. Typically, each cell is covered with about 150 mm of soil at the end of each day. This greatly reduces problems with rats and vermin, and thus promotes public health. The completed landfill is then covered with more soil to form a *cap*. By 1960, sanitary landfills had become the preferred method of waste disposal in the United States.

Although sanitary landfills were a substantial improvement over earlier practices, more problems remained to be solved. Before 1975, most landfill sites were chosen with little or no regard for potential environmental impacts. As a result, many of them were built near lakes and rivers, or in close proximity to groundwater aquifers. In karst regions, old sinkholes were often favorite spots for landfills, even though these are usually important sources of groundwater recharge. Landfills placed in old quarrys, especially those formerly used to mine sand and gravel, had similar problems. As a result, contaminants often leached out of these landfills, causing contamination of ground and surface waters.

These problems became evident during the 1970s, thus beginning the era of *engineered sanitary landfills*. These modern facilities are carefully designed and operated to reduce the potential impact on the environment and to protect public health.

Solid Wastes

Solid wastes are inevitable by-products of human activities. Although conservation, recycling, and other efforts may be used to reduce waste quantities, we always will need to provide disposal facilities.

Sanitary landfills receive solid wastes from various sources, including the following:

- *Municipal solid waste (MSW)* consists of trash generated by residential and commercial sources. With no recycling, nearly half consists of paper and another one-third consists of organic wastes, including yard trimmings and food. The remainder includes glass, plastics, metals, and many other materials.
- *Industrial solid waste* is any waste material produced from industrial processes. These vary widely depending on the type of industry, and can include dangerous chemicals.
- *Agricultural solid waste* includes plant and animal materials generated on farms and in food processing plants.
- *Sewage sludge* is the solid waste produced by sewage treatment plants.
- *Combustion by-products* from the burning of coal include both ash from furnaces and flue gas desulphurization sludge from scrubbers. Some of these materials have productive uses, but others require disposal.
- *Mining solid waste* includes various by-products of mining operations. Coarse-grained mining wastes (sand-size or larger) are called *tailings*, while fine-grained wastes (silt and clay-size) are called *slimes*. In some cases, tailings may be used as construction aggregates, but slimes are almost universally worthless and may be contaminated with hazardous chemicals used in the mining process.

Site Selection

The selection of an appropriate site for a new sanitary landfill is probably the single most important task. It is much better to place new facilities at sites with favorable geotechnical conditions (i.e., those underlain by soil or rock with a low hydraulic conductivity and well above the groundwater table) rather than to choose a poor site (i.e., karst terrain, clean gravel, etc.) and attempt to protect it with special design features. In addition to the geotechnical concerns, landfill site selections also need to consider surface water and groundwater hydrology, the possibility of flooding, animal breeding areas, and so on.

If only technical concerns needed to be addressed, site selection would be a fairly straightforward process. In reality, nearly everyone wants refuse to be disposed of in a proper way, but virtually nobody wants a landfill nearby (an attitude summarized by NIMBY — not in my back yard). Even landfill proposals at remote sites can receive strong political opposition. For example, in 1996 a proposal to build a municipal solid waste landfill at a good site in a sparsely populated portion of the California desert was rejected because of exaggerated environmental impact claims and political ballyhooing.

Hauling costs, which once were among the most important considerations for municipal landfills, have since been overshadowed by these other concerns.

Facilities for disposal of especially hazardous materials, nuclear wastes, and other similar materials require much more intensive site selection efforts. However, the number of such facilities is small compared to those used for more conventional waste materials.

Decomposition

Landfills that receive primarily municipal solid waste (MSW) are subjected to extensive chemical, physical, and biological processes that we collectively call *decomposition*. These processes convert MSW into the following products (McBean, et al.,1995):

- Decomposed solid wastes
- New biomass
- Generated gasses
- Contaminants into solution (leachate)
- Heat

These processes are most active during the first twenty years, but they typically continue for several decades or more.

Leachate is especially important to geoenvironmental engineers. It consists of liquids brought in with the solid waste, liquids generated by decomposition, and water (which may have entered the landfill through inflow of groundwater or through infiltration of surface water). Leachate is the primary means of contaminant transport from the landfill to the surrounding ground, and thus needs to be controlled.

Liners, Covers, and Leachate Collection

Before refuse placement begins, the natural ground is covered with a *liner*, which is intended to prevent, or at least significantly retard, the flow of leachate into the ground. Figure 9.8 shows a liner being installed. Engineers often specify multiple liners to provide some level of redundancy. For example, a landfill liner system might include primary and secondary leachate collection zones, geosynthetic barriers, and leachate collection pipes all underlain by compacted clay. The leachate collection pipes in a liner system then lead to a treatment facility that removes or neutralizes the harmful chemicals.

When the landfill is completed, a final "impervious" *cover* (or *cap*) is placed over the site. It serves several purposes, including:

- Reducing the infiltration of surface water, thus reducing the production of leachate
- Controlling rats, vermin, and other pests
- Controlling gas emissions from the refuse
- Isolating people from direct contact with the refuse
- Reducing the potential for erosion and slumping
- Eliminating the potential for blowing debris
- Providing a pleasing appearance

Figure 9.8 A geomembrane being placed. The seams connecting each strip of geomemberane will be "welded" using special adhesives, thus forming a continuous liner that will serve as a virtually impervious barrier (Photo courtesy of Dr. Robert Koerner).

Gas Collection

Landfills generate large quantities of methane and other gasses as a result of decomposition of organic materials. Therefore, many landfills have been equipped with gas collection systems that recover a portion of this gas and direct it to productive use. The gas may be collected by applying a vacuum to a series of wells drilled through the refuse. These wells are connected by a series of pipes that direct the gas to a central location. Usually the gas is then burned to generate steam, which is used to produce electricity.

In addition to providing a useful product (electricity), gas collection systems also reduce the potential for migration of landfill gasses beyond the landfill property.

Closure and Post-Closure

Closure is the process of shutting down a landfill that is no longer accepting wastes. It includes such tasks as placing the final cover, installing surface drainage facilities, and establishing a vegetative cover. These need to be carefully designed and implemented to provide the required environmental protection.

Post-closure includes the long-term monitoring and maintenance of a closed landfill. Landfill settlements due to continued decomposition of the refuse are especially important, because these settlements can cause cracks and tears in covers, changes in surface drainage patterns, rupture of gas collection pipes, and other problems. Post-closure activities also include long-term groundwater monitoring programs to verify the required groundwater quality is being maintained.

SUMMARY

Major Points

1. Geoenvironmental engineering is a blend of geotechnical and environmental engineering, and deals with underground contamination.

2. Prior to the 1980s, wastes and hazardous materials were often intentionally or unintentionally discharged underground. This sometimes caused contamination of groundwater aquifers and other environmental problems.

3. As a result of various legislation, much of which was passed during the 1980s, underground discharge and underground disposal of wastes is now regulated. In addition, there now is an active effort to locate, assess, and clean up existing underground contamination.

4. Both biological and chemical contamination can occur underground, but chemical contamination is more often a significant problem.

5. Chemical wastes that are not readily soluble in water are called *non-aqueous phase liquids (NAPLs)*. Many chemical wastes, such as petrochemicals, are NAPLs. There are two types: *Light NAPLS* (LNAPLs) have specific gravities less than water, and thus float on top of the groundwater, while *dense NAPLs (DNAPLs)* have specific gravities greater than water, and thus sink through groundwater. Because of this behavior, DNAPLs are especially difficult to clean up.

6. Hazardous wastes are those waste products that pose significant threats to health or the environment.

7. Several processes control the fate and transport of contaminants through the ground. Advection and dispersion generally dominate in sandy soils, while diffusion generally dominates in clays.

8. Many remediation methods are available, and new methods continue to be developed. However, removal of all contaminants is practical only when the contaminated zone is small.

9. The increased emphasis on underground contamination also has resulted in stricter regulation and more careful design and operation of sanitary landfills. The various design features are intended to prevent contamination of the adjacent soils.

Vocabulary

absorption	environmental site	non-point sources
adsorption	assessment	point sources
advection	excavation and disposal	post-closure
biodegradation	fate and transport	pump-and-treat
biological contamination	gas collection	radioactive decay
bioremediation	geoenvironmental	sanitary landfill
chemical contamination	engineering	seepage velocity
closure	hazardous waste	septic tank
containment	in-situ air sparging	soil vapor extraction
covers	intrinsic remediation	solid waste
diffusion	leachate	sorption
dispersion	liner	source control
DNAPL	LNAPL	underground storage tank
engineered sanitary	municipal solid waste	volatilization
landfills	NAPL	

COMPREHENSIVE QUESTIONS AND PRACTICE PROBLEMS

9.1 What is the difference between diffusion and dispersion? Which process would be more important in the transport of contaminants through a gravelly soil? Why?

9.2 A Phase I Environmental Site Assessment at the site of a proposed office building has found that the property was previously occupied by a gasoline service station. The above-ground facilities were removed several years ago and the site is now a vacant lot, but the underground gasoline tanks remain. Further investigations found that one of these tanks had a leak, and about 10 yd^3 of the surrounding unsaturated silty clay soil has become contaminated. The groundwater table is at a depth of 50 ft below the ground surface, well below the contaminated zone. Suggest an appropriate method of remediating this problem.

9.3 What geotechnical characteristics should a site have for it to be considered a good candidate for a sanitary landfill? Why are these characteristics important?

9.4 An old waste disposal pond has introduced several contaminants into the groundwater at a certain location. The adjacent soils are sandy silts and silty sands, and chemical tests on samples from this site indicate a high mercury content. Should bioremediation be considered as a potential clean-up method? Why or why not?

9.5 The following import soils are available to build a cap over a sanitary landfill:
Import Site 1	ML
Import Site 2	SW
Import Site 3	CL
Import Site 4	SC

Which would probably be the best choice? Why?

9.6 An underground diesel fuel storage tank has a small leak that releases only 2 liters per day into the surrounding soil. This leak is far too small to be noticed by the owner. If the leak continues for ten years, how much diesel fuel will have entered the ground? Assuming the soil has a porosity of 45% and a degree of saturation of 60%, and assuming the diesel fuel fills two-thirds of the remaining voids, compute the total volume of soil contaminated by the leak. Also assume none of the fuel volatilizes. Would you consider this to be a significant problem? What additional information would you like to have to further assess it?

9.7 A shallow injection well in a suburban area has been used to dispose of liquid wastes for 35 years. The surrounding soils are fine-to-medium sands with $k = 2 \times 10^{-2}$cm/s and $e = 0.65$. The hydraulic gradient of the natural groundwater flow is 0.010. Considering advection only, how far downstream will the contaminant plume extend? Based on the results of this computation, is there cause for concern?

9.8 A small shopping center is to be built on a parcel of land in a karst area. It will require a new well to supply drinking water, and a septic tank/leach field system to dispose of sewage. Are there any special requirements or considerations that need to be addressed when determining the minimum acceptable separation between these two facilities and their relative proximity? Explain.

10

Stress

In soil mechanics the accuracy of computed results never exceeds that of a crude estimate, and the principle function of theory consists of teaching us what and how to observe in the field.

Karl Terzaghi, 1936

Virtually all civil engineering projects impart loads onto the ground that supports them, and these loads produce compressive, shear, and possibly even tensile stresses. For example, when we construct a building, its weight is transmitted to the ground through the foundations, thus inducing stresses in the underlying strata. These stresses might cause problems, such as shear failure or excessive settlement, and thus are important to geotechnical engineers.

Additional stresses exist in the ground due to the weight of the overlying soil and rock. These stresses also are important, and need to be considered in a wide range of geotechnical engineering problems. For example, the potential for slope stability problems, such as landslides, depends on the difference between these stresses and the strength of the soil or rock.

Many geotechnical problems depend on assessments of stresses in the ground, so this subject is worthy of careful study. Some of the methods we use to evaluate and describe these stresses are similar to those used with more conventional engineering materials, such as steel. However, because of the particulate nature of soils and the presence of water and/or air in the voids, we also need to introduce new concepts and methods.

This chapter discusses methods of analyzing stresses in the ground. Subsequent chapters, especially Chapters 11–14, 16, and 17, will apply these methods to specific geotechnical problems.

10.1 SIMPLIFYING ASSUMPTIONS

Soil and rock are very complex materials that cannot be modeled mathematically without introducing certain simplifying assumptions. We need to understand these assumptions before proceeding. For the purpose of computing stresses, we will assume the soil or rock has the following characteristics:

1. It is a *continuous material*, not a particulate, which means the actual transfer of stresses through the solid particles in a soil is very complex. The stresses at the particle contact points are very high, while elsewhere they are much smaller. The stresses in rock also are complex due to the presence of joints and fissures. Rather than attempting to deal with these complexities, we simply treat the ground as if it were a continuum, like steel or other familiar materials. We assume there are no cracks or open joints to block the flow of stress, and that the stresses are uniform, not a microscopic patchwork of stressed particles with very small contact areas. This assumption is reasonable so long as the dimensions in our problem are large compared to the particle size and the joint spacing. Later in this chapter, our discussion of effective stress will modify this assumption to allow consideration of the distribution of stresses between the solid particles and the pore water.

2. It is *homogeneous*, which means the relevant engineering properties are the same at all locations. In the context of stress analyses, we require only the modulus of elasticity, shear modulus, and Poisson's ratio (as defined later) to be constant. In other words, there are no "hard spots" and "soft spots," which can significantly alter the stress distribution. However, we will allow the unit weight to vary from place to place, and we will account for parts of the ground being above the groundwater table and parts below. This assumption virtually always introduces some error, since few soils are so homogeneous. Later in this chapter we will explore solutions that include two layers with different modulus values.

3. It is *isotropic*, which means the engineering properties (in this context, the two modulii and Poisson's ratio) are the same in all directions. Many soils nearly meet this criteria, but some materials, such as bedded sedimentary rock, do not.

4. It has *linear elastic stress–strain properties*, which means each increment of stress is associated with a corresponding increment of strain (i.e., the stress–strain curve is a straight line and has not reached a yield point). This means a load applied at one point will induce some increment of stress, even a small one, everywhere in the ground, and that there will be a smooth pattern of stress distribution. This is quite different from a *plastic* material, which is one that has exceeded its yield point, and thus has reached its maximum stress-carrying capacity. The distribution of stresses in plastic materials is therefore quite different. This assumption is satisfactory for deformation analyses, where the strains are small, but may not be acceptable for certain strength analyses where the stresses are nearly equal to the strength.

These assumptions need to remain in effect only until we have computed the appropriate stresses in the ground. The follow-up analyses do not necessarily need to adhere to the same assumptions. For example, the settlement computations in Chapter 11 will be based on a non-linear stress–strain relationship, and the strength analyses in Chapter 13 will consider the presence of cracks and joints.

10.2 MECHANICS OF MATERIALS REVIEW

Our simplifying assumptions have transformed soil and rock into something similar to standard engineering materials. Therefore, we will begin by reviewing principles of stress and strain that you learned in a mechanics of materials course, and discussing how we will apply them to geotechnical problems.

When conducting stress analyses, we will continue to use the coordinate system introduced in Chapter 7, where the x and y axes are horizontal, and the z axis is vertical, as shown in Figure 10.1. Usually we evaluate stresses acting on a small element of soil or rock such as the one shown. This particular element is aligned with the axes, but other orientations also are possible.

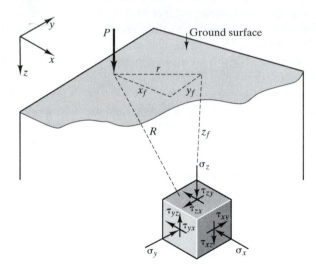

Figure 10.1 Element of soil in three-dimensional x-y-z space.

Stress

Each face of the element is subjected to a *normal stress*, σ, which could be either *tension* or *compression*. Most engineers use the following sign convention to differentiate between the two:

$$+ = \text{tension}$$
$$- = \text{compression}$$

However, geotechnical engineers like to be different. We use the opposite sign convention:

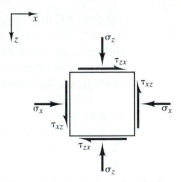

+ = compression
- = tension

Figure 10.2 A two-dimensional soil element aligned with the x and z axes.

Although it can be confusing at first, there is a good reason for using this unorthodox sign convention: We almost always deal with compression. This is because of the nature of soil stress problems and because soil has a very low tensile strength. Thus, our sign convention avoids the continual use of negative stresses, and eliminates the resulting mistakes that might occur.

When the element is aligned with the axes, as shown, the horizontal normal stresses (those acting on vertical planes) are σ_x and σ_y, and the vertical normal stress (which acts on a horizontal plane) is σ_z. Each face of the element also is subjected to a *shear stress*, τ, which we will divide into two perpendicular components as identified by two subscripts. For example, τ_{xz} is the component of shear stress in the x plane (i.e., the plane perpendicular to the x axis) acting in the z direction.

Although the normal stresses in the x, y, and z directions are independent of each other, the various shear stresses are not. To maintain static equilibrium, τ_{xz} and τ_{zx}, as shown in Figure 10.2, must be equal in magnitude and opposite in direction. The same relationship applies to the other pairs of shear stresses:

$$\tau_{xz} = -\tau_{zx} \tag{10.1}$$

$$\tau_{yz} = -\tau_{zy} \tag{10.2}$$

$$\tau_{xy} = -\tau_{yx} \tag{10.3}$$

Assigning the proper sign to shear stresses can be confusing, especially since engineers have proposed multiple definitions of "positive" shear stress. We will use the sign convention that defines positive shear stresses as those that cause the element to rotate clockwise, and negative shear stresses as those that cause it to rotate counterclockwise. Thus, τ_{zx} in Figure 10.2 is positive, τ_{xz} is negative.

Because we are treating the ground as if it were a continuous material, not a particulate, the area used to compute stresses is the total area of solids plus voids, not just the area of solids. For example, if a water tank has a total weight of 30,000 kN, and its base area is 300 m^2, then we say the vertical compressive stress in the soil immediately below the tank is 30,000 kN/300 m^2 = 100 kPa, even though the actual area of the soil solids is much less than 300 m². This means the real compressive stress within the soil particles is greater than 100 kPa, and the stresses at the particle contact points may be substantially greater. However, so long as we also define stress–strain properties and strengths in the same way, and the physical dimensions used in our analysis are large compared to the size of the individual soil particles, our computations will be essentially correct.

All stresses are expressed in units of force per area. When working with English measurements, geotechnical engineers normally use lb/ft², except in some laboratory tests where lb/in² is used (primarily because these are the units in pressure gages). With SI units, all soil stresses are expressed in kPa. Finally, geotechnical engineers in non-SI metric countries typically use kg_f/m^2, kg_f/cm^2 or bars (1 bar = 100 kPa).

Strain

When materials are subjected to a stress, they respond by deforming. Engineers call this deformation *strain*. Normal stresses produce normal strains and shear stresses produce shear strains.

Normal strain, ϵ, is the change in length divided by the initial length, as shown in Figure 10.3:

$$\epsilon = -\frac{dL}{L} \tag{10.4}$$

To be consistent with our sign convention for normal stresses, compressive strains are positive (i.e., the length becomes shorter, so $dL < 0$, which produces $\epsilon > 0$).

Shear strain, γ, is the angle of deformation shown in Figure 10.3 expressed in radians.

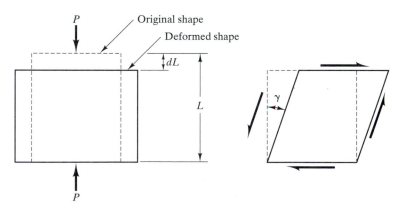

Figure 10.3 Definitions of normal and shear strain.

Modulus of Elasticity, Shear Modulus, and Poisson's Ratio

To perform deformation analyses, we need to define the relationships between stress and strain. In linear elastic materials, these relationships are expressed using three parameters, as follows: The *modulus of elasticity, E* (also known as *Young's modulus*), the *shear modulus, G* (also known as the *modulus of rigidity*), and *Poisson's ratio,* v.   The modulus of elasticity is the ratio of normal stress to normal strain:

$$E = \frac{\sigma}{\epsilon} \tag{10.5}$$

Thus, large values of E indicate a material that is very stiff and does not experience much deformation under an applied load, while low values indicate a soft material. The shear modulus has a similar definition:

$$G = \frac{\tau}{\gamma} \tag{10.6}$$

When a compressive load is applied to the element as shown in Figure 10.4, a compressive strain parallel to the stress, $\epsilon_\parallel$, occurs. In addition, if the element is unconfined, a tensile strain perpendicular to the load, $\epsilon_\perp$, also occurs. The ratio of these two strains is defined as *Poisson's ratio,* v:

$$v = -\frac{\epsilon_\perp}{\epsilon_\parallel} \tag{10.7}$$

The magnitude of v in elastic materials varies from 0 to 0.5. Those with v = 0.5 are said to be *incompressible* because the compression in the direction of the load is exactly matched by the expansion in the two perpendicular directions, resulting in no net volume change.

Although it is possible to measure v of soil or rock in the laboratory, geotechnical engineers usually rely on tabulated values such as those in Table 10.1. These values are sufficiently precise for nearly all geotechnical analyses.

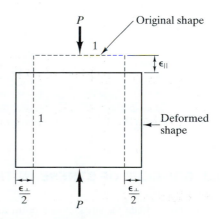

Figure 10.4 Deformation of an unconfined element and definition of Poisson's ratio.

TABLE 10.1 TYPICAL VALUES OF POISSON'S RATIO FOR SOILS AND ROCKS (Adapted from Kulhawy, et al., 1983)

Soil or Rock Type	Poisson's Ratio, ν
Saturated soil, undrained condition	0.50
Partially saturated clay	0.30–0.40
Dense sand, drained condition	0.30–0.40
Loose sand, drained condition	0.10–0.30
Sandstone	0.25–0.30
Granite	0.23–0.27

One-, Two-, and Three-Dimensional Analyses

Stresses propagate through soils in all three dimensions, so an analysis that keeps track of all the stresses identified in Figure 10.1 would be considered a *three-dimensional analysis*. Although such analyses are sometimes necessary, it often is possible to use more simplified methods. For example, a vertical cross-section through a long earth dam on a uniform soil deposit will be virtually constant along the entire length of the dam, as shown in Figure 10.5. Thus, we could reasonably evaluate such a problem using a *two-dimensional analysis* in a vertical plane oriented perpendicular to the dam axis.

Other scenarios can even be reduced to a *one-dimensional analysis*. For example, many problems require only the vertical stress, σ_z, due to the weight of the overlying ground. This requires only vertical dimensions and certain soil properties, making these one-dimensional analyses.

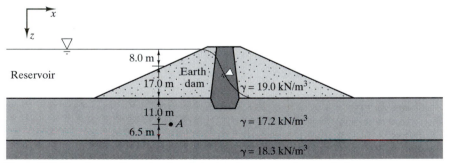

Figure 10.5 Use of a two-dimensional analysis to evaluate stresses in an earth dam.

10.3 SOURCES OF STRESS IN THE GROUND

To evaluate the stresses at a point in the ground, we need to know the locations, magnitudes, and directions of the forces that cause them. We will divide these sources into two broad categories:

- *Geostatic stresses* (sometimes called *body stresses*) are those that occur due to the weight of the soil above the point being evaluated. Geostatic stresses are naturally present in the ground. However, human activities, such as placing a fill or making an excavation, can cause them to change.
- *Induced stresses* are those caused by external loads, such as structural foundations, vehicles, or fluid in a storage tank. These are usually caused by human activities.

We will discuss each category separately, then combine them using superposition. Our discussions will be limited to static stresses. Dynamic stresses, such as those produced by earthquakes, explosions, or machine vibrations, are beyond the scope of this text (see Dowding, 1996 and Kramer, 1996).

10.4 GEOSTATIC STRESSES

Geostatic stresses are caused by gravity acting on the soil or rock, so the direct result is a vertical normal stress, σ_z. This stress has a significant impact on the engineering behavior of soil, and is one we frequently need to compute. This vertical normal stress indirectly produces horizontal normal stresses and shear stresses, which also are important to geotechnical engineers.

Vertical Stresses

To compute the geostatic σ_z at Point A in Figure 10.6, consider a column of soil that extends from the ground surface down to a point where we wish to compute σ_z. This column intercepts soil strata with unit weights γ_1, γ_2, and γ_3, so its weight is:

$$W = dx\,dy \sum \gamma H \qquad (10.8)$$

The geostatic vertical stress, σ_z at the bottom of the column is then:

$$\sigma_z = \frac{W}{A} = \frac{dx\,dy \sum \gamma H}{A} = \sum \gamma H \qquad (10.9)$$

where:

    W = weight of the column
    γ = unit weight of the soil strata
    H = thickness of the soil strata
    A = horizontal cross-sectional area of the column

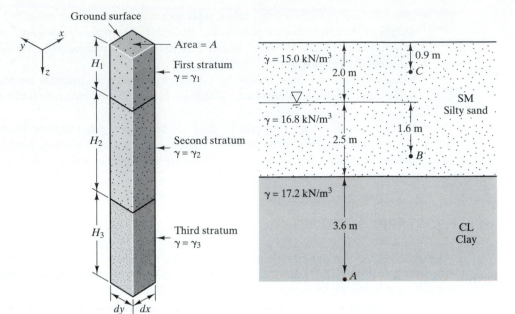

Figure 10.6 Imaginary column of soil to compute the geostatic σ_z.

Figure 10.7 Soil profile for Example 10.1.

Example 10.1

Compute σ_z at point A in Figure 10.7.

Solution

$$\sigma_z = \sum \gamma H$$
$$= (15.0 \, kN/m^3)(2.0 \, m) + (16.8 \, kN/m^3)(2.5 \, m) + (17.2 \, kN/m^3)(3.6 \, m)$$
$$= 134 \, kPa \qquad \Leftarrow Answer$$

Horizontal Stresses

The horizontal stresses, σ_x and σ_y, also are important for many engineering analyses. For example, the design of retaining walls depends on the horizontal stresses in the soil being retained. Some horizontal stresses are the direct result of applied external loads, such as the braking forces from the wheels of a large truck. However, most horizontal stresses are indirectly produced by vertical geostatic and induced stresses.

To understand how these indirect stresses are produced, consider the unconfined element of soil in Figure 10.4. When a vertical compressive stress is applied to this element, it induces both a vertical compressive strain and a horizontal tensile strains (per Equation 10.7). However, real soils in the field are not unconfined, and the adjacent elements of soil or rock also wish to expand, but in the opposite direction. These opposing forces may cancel each other (i.e., there may be no horizontal strain), or the horizontal strain may be much less than would occur in an unconfined sample. Either way, the result will be the formation of horizontal stresses in the ground.

You may recall solving a similar problem in a mechanics of materials course, where a metal bar was tightly fitted between two immovable barriers, then heated. Normally the bar would become longer due to thermal expansion, but the barriers prevented it from doing so. Therefore, instead of expanding, the bar developed new compressive stresses.

Horizontal stresses can be measured in-situ using the pressuremeter test (PMT), the dilatometer test (DMT) or other methods. Alternatively, it can be estimated from the vertical stress using the coefficient of lateral earth pressure, as discussed later in this chapter.

If the ground surface is horizontal, we normally assume the geostatic σ_x and σ_y are equal.

Shear Stresses

If the ground surface is horizontal, the geostatic shear stresses on horizontal and vertical planes are all equal to zero:

$$\tau_{xz} = \tau_{zx} = \tau_{yz} = \tau_{zy} = \tau_{xy} = \tau_{yx} = 0 \qquad (10.10)$$

However, shear stresses may be present on other planes. If the ground surface is inclined, then the geostatic shear stress conditions are more complex. The analysis of such stresses is beyond the scope of this text.

QUESTIONS AND PRACTICE PROBLEMS

10.1 Using the soil profile in Figure 10.7, develop a plot of σ_z vs. depth. Consider depths between 0 and 10 m.

10.2 A 0.500 ft × 0.500 ft × 0.500 ft cube of soil is subjected to a vertical compressive force of 500 lb. This force is being applied to the top of the cube. As a result of this force, the cube compresses to a height of 0.450 ft. Compute the vertical normal stress and the vertical normal strain.

10.5 INDUCED STRESSES

Civil engineering projects often introduce external loads onto the ground, thus producing induced stresses. These loads include structural foundations, vehicles, tanks, stockpiles, and many others. The resulting induced stresses are often significant, and can be the source of excessive settlement, shear failure, or other problems.

Boussinesq's Method

The French mathematician Joseph Boussinesq (1842–1929) developed a method of computing induced stresses in an infinite elastic half-space due to an applied external load (Boussinesq, 1885). The term *infinite elastic half-space* means the linear elastic material extends infinitely in all directions beneath a plane (which in our case is the ground surface). Boussinesq solved the problem where the *point load*, *P*, is perpendicular to this plane (in our case, this means the load is vertical), as shown in Figure 10.1. According to his solution, such a load will induce the following stresses at a point in the ground:

$$\sigma_x = \frac{P}{2\pi}\left[\frac{3x_f^2 z_f}{R^5} - (1-2v)\left(\frac{x_f^2 - y_f^2}{Rr^2(R+z_f)} + \frac{y_f^2 z_f}{R^3 r^2}\right)\right] \tag{10.11}$$

$$\sigma_y = \frac{P}{2\pi}\left[\frac{3y_f^2 z_f}{R^5} - (1-2v)\left(\frac{y_f^2 - x_f^2}{Rr^2(R+z_f)} + \frac{x_f^2 z_f}{R^3 r^2}\right)\right] \tag{10.12}$$

$$\sigma_z = \frac{3P z_f^3}{2\pi R^5} \tag{10.13}$$

$$\tau_{zx} = -\tau_{xz} = \frac{3P z_f^2 x_f}{2\pi R^5} \tag{10.14}$$

$$\tau_{xy} = -\tau_{yx} = \frac{P}{2\pi}\left[\frac{3x_f y_f z_f}{R^5} - (1-2v)\left(\frac{(2R+2)x_f y_f}{(R+z_f^2)R^3}\right)\right] \tag{10.15}$$

$$\tau_{zy} = -\tau_{yz} = \frac{3P z_f^2 y_f}{2\pi R^5} \tag{10.16}$$

$$R = \sqrt{x_f^2 + y_f^2 + z_f^2} \tag{10.17}$$

$$r = \sqrt{x_f^2 + y_f^2} \qquad (10.18)$$

The parameters x_f, y_f, and z_f in Equations 10.11 through 10.18 are the distances from the load to the point, as shown in Figure 10.1. The values of x_f and y_f may be either positive or negative, but z_f is always positive because the point is always below the load. When performing two-dimensional analyses in the x–z plane, we only need Equations 10.11, 10.13, and 10.14.

Example 10.2

The dimensions in Figure 10.1 are: $x_f = 10.0$ ft, $y_f = 0.0$ ft, and $z_f = 15.0$ ft. The load P is 132 k, and the soil is a partially saturated clay. Compute the induced σ_x, σ_z, and τ_{zx} in the soil element.

Solution

Per Table 10.1, use $\nu = 0.35$

$$
\begin{aligned}
R &= \sqrt{x_f^2 + y_f^2 + z_f^2} \\
&= \sqrt{10.0^2 + 0.0^2 + 15.0^2} \\
&= 18.0 \text{ ft}
\end{aligned}
$$

$$
\begin{aligned}
r &= \sqrt{x_f^2 + y_f^2} \\
&= \sqrt{10.0^2 + 0.0^2} \\
&= 10.0 \text{ ft}
\end{aligned}
$$

$$
\begin{aligned}
\sigma_x &= \frac{P}{2\pi}\left[\frac{3x_f^2 z_f}{R^5} - (1-2\nu)\left(\frac{x_f^2 - y_f^2}{Rr^2(R+z_f)} + \frac{y_f^2 z_f}{R^3 r^2}\right)\right] \\
&= \frac{132{,}000}{2\pi}\left[\frac{3(10.0)^2(15.0)}{18.0^5} - [1-2(0.35)]\left(\frac{10.0^2 - 0.0^2}{18.0(10.0)^2(18.0+15.0)} + \frac{0.0^2(15.0)}{18.0^3 10.0^2}\right)\right] \\
&= 39 \text{ lb/ft}^2 \qquad \Leftarrow \textit{Answer}
\end{aligned}
$$

$$
\begin{aligned}
\sigma_z &= \frac{3Pz_f^3}{2\pi R^5} \\
&= \frac{3(132{,}000)(15.0)^3}{2\pi 18.0^5} \\
&= 113 \text{ lb/ft}^2 \qquad \Leftarrow \textit{Answer}
\end{aligned}
$$

$$\tau_{zx} = \frac{3Pz_f^2 x_f}{2\pi R^5}$$

$$= \frac{3(132{,}000)(15.0)^2(10.0)}{2\pi(18.0)^5}$$

$$= 75 \text{ lb/ft}^2 \quad \leftarrow Answer$$

Application to Line Loads

Although Boussinesq developed formulas only for point loads, others have extended them to other loading conditions. The most simple extension is to a *line load*, which is a vertical load distributed evenly along a horizontal line. We express such loads using the parameter P/b, where P is the vertical load and b is a unit length along the line (i.e., $P/b = 100$ kN/m).

If we consider a line load of infinite length oriented parallel to the y axis, and integrate Equation 10.13 over its length, we obtain:

$$\sigma_z = \frac{2z_f^3 P/b}{\pi(x_f^2 + z_f^2)^2} \tag{10.19}$$

where x_f and z_f are the horizontal and vertical distances from line to the point at which σ_z is to be computed.

Application to Area Loads

The most common loading condition for geotechnical analyses is the *area load*, which is one distributed evenly across a horizontal area. Examples include spread footing foundations, tanks, wheel loads, stacked inventory in a warehouse, and small fills. We define the contact pressure between this load and the ground as the *bearing pressure*, q:

$$q = \frac{P}{A} \tag{10.20}$$

copy

where:
  q = bearing pressure
  P = applied vertical load
  A = area upon which the load acts

In the case of spread footing foundations, P must include both the column load and the weight of the foundation.

The induced stresses beneath the loaded area can be computed using an extension of the Boussinesq equations.

Analytic Solutions

Sometimes it is possible to integrate the Boussinesq equations over the area to produce new equations. Newmark (1935) used this method with Equation 10.13 to develop the following analytic solution for the vertical induced stress at a depth z_f beneath the corner of a loaded rectangle of width B and length L, as shown in Figure 10.8:

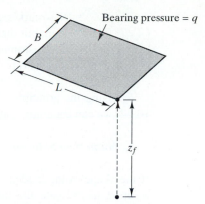

Bearing pressure = q

Figure 10.8 Newmark's solution for induced vertical stresses beneath the corner of a rectangular loaded area.

If $B^2 + L^2 + z_f^2 < B^2L^2/z_f^2$,

$$
\sigma_z = \frac{q}{4\pi}\left[\left(\frac{2BLz_f\sqrt{B^2+L^2+z_f^2}}{z_f^2(B^2+L^2+z_f^2)+B^2L^2}\right)\left(\frac{B^2+L^2+2z_f^2}{B^2+L^2+z_f^2}\right)\right.
$$
$$
\left. + \pi - \sin^{-1}\frac{2BLz_f\sqrt{B^2+L^2+z_f^2}}{z_f^2(B^2+L^2+z_f^2)+B^2L^2}\right]
$$

(10.21)

Otherwise,

$$
\sigma_z = \frac{q}{4\pi}\left[\left(\frac{2BLz_f\sqrt{B^2+L^2+z_f^2}}{z_f^2(B^2+L^2+z_f^2)+B^2L^2}\right)\left(\frac{B^2+L^2+2z_f^2}{B^2+L^2+z_f^2}\right)\right.
$$
$$
\left. + \sin^{-1}\frac{2BLz_f\sqrt{B^2+L^2+z_f^2}}{z_f^2(B^2+L^2+z_f^2)+B^2L^2}\right]
$$

(10.22)

where:
  σ_z = vertical induced stress at a point beneath the corner of the loaded rectangle
  B = width of loaded rectangle
  L = length of loaded rectangle
  z_f = vertical distance from loaded rectangle to the point (always > 0)
  q = bearing pressure on loaded rectangle

Notes:
   1. The $\sin^{-1}$ term must be expressed in radians.

2. Newmark's solution is often presented as a single equation with a $\tan^{-1}$ term, but that equation is incorrect when $B^2 + L^2 + z_f^2 < B^2 L^2 / z_f^2$.
3. It is customary to use B as the shorter dimension and L as the longer dimension.

Using the principle of superposition, as described later, and Equations 10.21 and 10.22, we can compute σ_z at any point beneath a rectangular loaded area.

Numerical Solutions

If the shape of the loaded area is too complex, it becomes necessary to use a numerical solution to compute the induced stresses. The term *numerical solution* (or *numerical method*) refers to a class of problem-solving methods that use a series of simplified equations assembled in a way that approximately models the actual system. Many engineering disciplines use these methods to develop solutions to problems that otherwise would be very difficult or impossible. The *finite element method* and the *finite difference method* are examples of numerical solutions. We discussed the finite element method and its application to seepage analyses in Chapter 8.

We can use a numerical method to solve Boussinesq stress problems as follows:

1. Divide the loaded area into hundreds or thousands of small elements.
2. Compute the total vertical load acting on each element and consider it to be a point load acting at the centroid of the element.
3. Apply Equations 10.11–10.16, as needed, to each element to compute the stress at the desired point in the ground due to the load on that element.
4. Using superposition, sum the stresses computed in Step 3 to find the net stress at the point.

The accuracy of this method increases as the number of elements increases. At least 1000 elements would generally be used for area loads, so this solution definitely requires a computer. Programs **STRESSP**, **STRESSL**, **STRESSR**, and **STRESSC**, which are included with this book, use this method to compute induced stresses. The use of these programs is explained later in this chapter.

Chart Solutions

Another option is to perform a series of computations using either analytic or numerical methods and express the results in non-dimensional charts. We then can use these charts to compute stresses in the soil. Figures 10.9 and 10.10 are two of the many such charts that may be developed. The curves in these charts, which connect points of equal induced stress, are sometimes called *pressure bulbs* or *stress bulbs*.

These charts are easy to use, and provide a visual sense of how the stresses are distributed. However, they do not have the flexibility or computational accuracy of a properly implemented numerical solution. Other charts also have been developed for more complex loading conditions (see U.S. Navy, 1982), but they generally provide only minimal help in visualizing the stress distributions, and serve only as computational aids. The ready availability of computer-based numerical solutions, such as the ones included in this book, has made them obsolete.

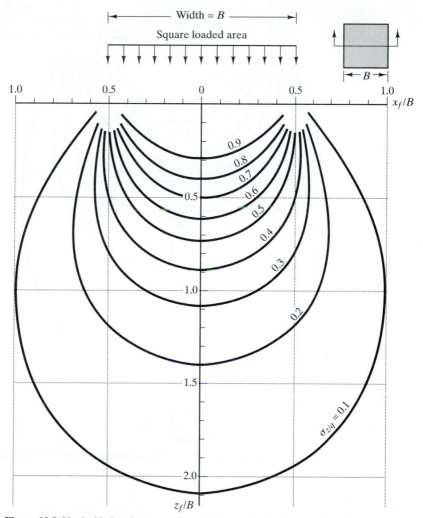

Figure 10.9 Vertical induced stress, σ_z beneath a square loaded area, per Boussinesq.

Example 10.3

Compute the induced σ_z at a depth of 10.0 m below the edge of a 25.0 m diameter water tank. The tank, its foundation, and its contents have a total mass of 6.1×10^6 kg, which is uniformly distributed across its base.

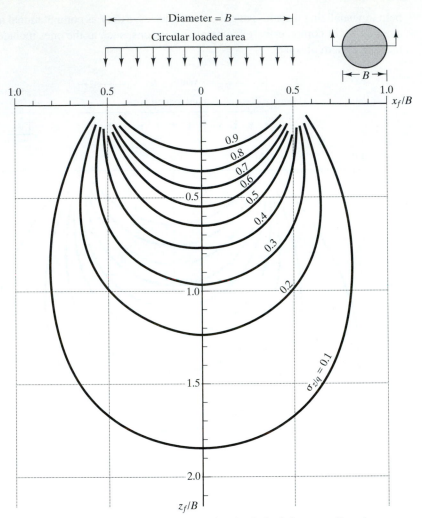

Figure 10.10 Vertical induced stress, σ_z beneath a circular loaded area, per Boussinesq.

Solution

$$W = Mg = (6.1 \times 10^6 \text{ kg})(9.81 \text{ m/s}^2) = 59{,}800 \text{ kN}$$

$$A = \pi r^2 = \pi (12.5 \text{ m})^2 = 491 \text{ m}^2$$

$$q = \frac{P}{A} = \frac{59,800 \text{ kN}}{491 \text{ m}^2} = 122 \text{ kPa}$$

$$\frac{z_f}{B} = \frac{10 \text{ m}}{25 \text{ m}} = 0.40$$

From Figure 10.10, at $z_f/B = 0.40$ beneath the edge of the tank ($x_f/B = 0.50$), $\sigma_z/q = 0.40$

$$\sigma_z = \left(\frac{\sigma_z}{q}\right) q = (0.40)(122 \text{ kPa}) = \textbf{49 kPa} \qquad \Leftarrow \textit{Answer}$$

Westergaard's Method

Westergaard (1938) solved the same problem Boussinesq addressed, but with slightly different assumptions. Instead of using a perfectly elastic material, he assumed one that contained closely spaced horizontal reinforcement members of infinitesimal thickness, such that the horizontal strain is zero at all points. This model may be a more precise representation of certain layered soils, such as varved clays.

Terzaghi (1943) presented the following formula for σ_z due to a vertical point load, P, based on Westergaard's method:

$$\sigma_z = \frac{PC}{2\pi z_f^2}\left[\frac{1}{C^2 + (r/z_f)^2}\right]^{1.5} \tag{10.23}$$

$$C = \sqrt{\frac{1-2\nu}{2(1-\nu)}} \tag{10.24}$$

The Westergaard solution produces σ_z values equal to or less than the Boussinesq values. As ν increases, the computed stress becomes smaller, eventually reaching zero at $\nu = 0.5$. Although some geotechnical engineers prefer Westergaard, at least for certain soil profiles, Boussinesq is more conservative, and probably more appropriate for most problems.

Approximate Methods

Sometimes it is useful to have simple approximate methods of computing stresses in soil. The widespread availability of computers has diminished the need for these methods, but they still are useful when a quick answer is needed, or when a computer is not available.

The following approximate formulas compute the induced vertical stress, σ_z, beneath the center of an area load.[1] They produce answers that are within 5 percent of the Boussinesq values, which is more than sufficient for virtually all practical problems.

For circular loaded areas (Poulos and Davis, 1974),

$$\sigma_z = q \left[1 - \left(\frac{1}{1 + \left(\frac{B}{2z_f} \right)^2} \right)^{1.50} \right] \tag{10.25}$$

For square loaded areas,

$$\sigma_z = q \left[1 - \left(\frac{1}{1 + \left(\frac{B}{2z_f} \right)^2} \right)^{1.76} \right] \tag{10.26}$$

For continuous loaded areas (also known as strip loads) of width B and infinite length,

$$\sigma_z = q \left[1 - \left(\frac{1}{1 + \left(\frac{B}{2z_f} \right)^{1.38}} \right)^{2.60} \right] \tag{10.27}$$

For rectangular loaded areas of width B and length L,

$$\sigma_z = q \left[1 - \left(\frac{1}{1 + \left(\frac{B}{2z_f} \right)^{1.38 + 0.62 B/L}} \right)^{2.60 - 0.84 B/L} \right] \tag{10.28}$$

---

[1] Equations 10.21 and 10.22 compute the induced vertical stress beneath the *corner* of the loaded area, while Equations 10.25–10.28 compute it beneath the *center* of the loaded area.

where:

σ_z = induced vertical stress beneath the center of a loaded area

q = bearing pressure

B = width or diameter of loaded area

L = length of loaded area

z_f = depth from bottom of loaded area to point

A commonly used approximate method is to draw surfaces inclined downward at a slope of 1 horizontal to 2 vertical from the edge of the loaded area, as shown in Figure 10.11. To compute the induced σ_z at a depth z_f below the loaded area, simply draw a horizontal plane at this depth, compute the area of this plane inside the inclined surfaces, and divide the total applied load by this area. The σ_z computed by this method is an estimate of the average σ_z across this area, and is most often used for approximate settlement computations.

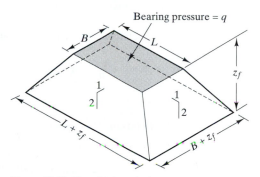

Figure 10.11 Use of 1:2 method to compute the average induced σ_z at a specified depth below a loaded area.

When applied to a rectangular loaded area of $B \times L$, the 1:2 method produces the following formula for the average induced vertical stress, $\bar{\sigma}_z$, at a depth z_f:

$$\bar{\sigma}_z = q\left(\frac{BL}{(B+z_f)(L+z_f)}\right) \qquad (10.29)$$

The primary advantage of this method is that Equation 10.29 can easily be derived from memory by simply applying the principles of geometry.

10.6 SUPERPOSITION

Since we have assumed the soil or rock is a linear elastic material, we can take advantage of the principle of superposition when computing σ and τ. This means problems that have multiple sources of stress may be evaluated by assessing each source separately, then adding the results. For example, if a certain point in the ground is subjected to geostatic stresses plus induced stresses from three different sources, we could perform four separate stress analyses (one for the geostatic and one for each of the induced stresses), then sum the results. This procedure greatly simplifies the analysis.

Example 10.4

The tank described in Example 10.3 is underlain by a soil that has a unit weight of 18.0 kN/m³.
Compute the pre-construction and post-construction σ_z at a depth of 10.0 m below the edge of
the tank.

Solution

Pre-construction condition:

Only the geostatic stresses are present.

$$\sigma_z = \sum \gamma H = (18.0 \text{ kN/m}^3)(10.0 \text{ m}) = \mathbf{180 \ kPa} \quad \leftarrow Answer$$

Post-construction condition:

Both geostatic and induced stresses are present.
Geostatic $\sigma_z = 180$ kPa
Induced $\sigma_z = 49$ kPa (from Example 10.3)

$$\sigma_z = 180 \text{ kPa} + 49 \text{ kPa} = \mathbf{229 \ kPa} \quad \leftarrow Answer$$

Example 10.5

The 1.0 m × 1.5 m footing shown in Figure 10.12 supports a vertical load of 475 kN. Compute
the magnitude of σ_z at Point A, considering both geostatic and induced stresses.

Solution

1. Geostatic stress

$$\sigma_z = \sum \gamma H = (17.0 \text{ kN/m}^3)(1.2 \text{ m}) = 20.4 \text{ kPa}$$

2. Induced stress
   Equations 10.21 and 10.22 compute σ_z below the corner of a loaded rectangle, but Point
   A is not beneath the corner. Therefore, it is necessary to compute the stress beneath a
   fictitious footing I+II, which is 1.0 m wide and 2.0 m long, and beneath a second
   fictitious footing II, which is 0.5 m wide and 1.0 m long. Both of these fictitious footings
   have a corner over Point A, and both have the same bearing pressure as the real footing.
   By superposition, the true σ_z at Point A is the difference between the σ_z values from these
   two footings.

$$q = \frac{P}{A} = \frac{475 \text{ kN}}{(1 \text{ m})(1.5 \text{ m})} = 317 \text{ kPa}$$

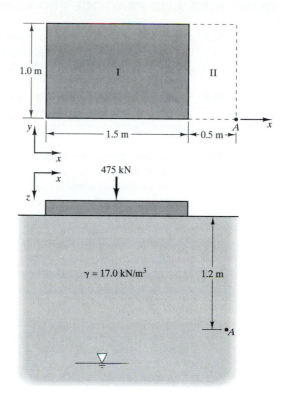

Figure 10.12 Plan and profile for Example 10.5.

Solving for Footing I+II

$B = 1.0, L = 2.0, z_f = 1.2$
$B^2 + L^2 + z_f^2 = 6.44, B^2 L^2 = 4$, Therefore, use Equation 10.22 → $\sigma_z = 57.6$ kPa

Solving for Footing II

$B = 0.5, L = 1.0, z_f = 1.2$
$B^2 + L^2 + z_f^2 = 2.69, B^2 L^2 = 0.25$, Therefore, use Equation 10.22 → $\sigma_z = 31.1$ kPa

By superposition, $\sigma_z = 57.6$ kPa - 31.1 kPa = 26.5 kPa

3. Combined results

$$\sigma_z = 20.4 \text{ kPa} + 26.5 \text{ kPa} = \textbf{46.9 kPa} \quad \leftarrow \textit{Answer}$$

QUESTIONS AND PRACTICE PROBLEMS

10.3 A vertical point load of 50.0 k acts upon the ground surface at coordinates $x = 100$ ft, $y = 150$ ft. Using a Poisson's ratio of 0.40, compute the induced σ_x, σ_z, and τ_{xz} at a point 3 ft below the ground surface at $x = 104$ ft, $y = 150$ ft.

10.4 A vertical line load of 75 kN/m acts upon the ground surface. Assuming this load extends for a very long distance in both directions, compute the induced σ_z at a point 1.5 m horizontal (measured perpendicular to the line) and 2.0 m below the line.

10.5 A dilatometer test (an in-situ test described in Chapter 3) has been conducted at a depth of 3.20 m in a soil that has a level ground surface and a unit weight of 19.2 kN/m³. According to this test, the geostatic σ_x at this point is 48 kPa. A proposed vertical point load of 1100 kN is to be applied to the ground surface at a point 1.10 m west of the test location. Using a Poisson's ratio of 0.37, compute the total σ_x, σ_z, and τ_{zx} at the test point after the load is applied.

10.6 A grain silo is supported on a 20.0 by 50.0 m mat foundation. The total weight of the silo and the mat is 180,000 kN. Using Newmark's method, compute the induced σ_z in the soil at a point 15.0 m below the center of the mat. Then repeat the computation using Equation 10.28. Compare the results from these two methods and comment on whether the difference is significant.

10.7 The circular water tank in Figure 10.13 imparts a bearing pressure of 3000 lb/ft² onto the soil below.

     a. Compute the geostatic σ_z at Point A. This is the stress that existed before the tank was built.
     b. Using Figure 10.10, compute the induced σ_z at Point A due to the weight of the tank.
     c. Combine the results from a and b to find the total σ_z at Point A after the tank is built.

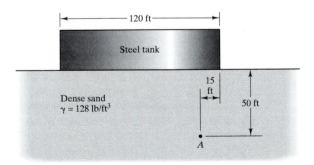

Figure 10.13 Water tank and soil profile for Problem 10.7.

10.7 EFFECTIVE STRESSES

The compressive stress, σ, computed using the techniques described thus far, is carried partially by the solid particles and partially by the pore water. Geotechnical engineers call it the *total stress* because it is the sum of the stresses carried by these two phases in the soil.

Although the total stress can be very useful, we gain even more insight by dividing it into two parts:

- The *effective stress*, σ', which is the portion carried by the solid particles, and
- The *pore water pressure*, u, which is the portion carried by the pore water. This is the same u we discussed in Chapters 7 and 8.

Karl Terzaghi was the first to recognize the importance of effective stress, and it has since become one of the most important concepts in geotechnical engineering.

Submerged Sphere Analogy

To understand the physics of soil particles under the groundwater table and the differences between total and effective stresses, let us consider the sphere resting on a scale as shown in Figure 10.14. It has a volume of 0.100 m³ as determined by measuring its diameter, and a weight of 2.60 kN as determined by the scale.

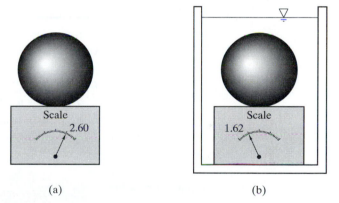

Figure 10.14 Submerged
sphere analogy.   (a)   (b)

Then, we take the scale and the sphere and place them into a tank of water, as shown. In this new environment, the sphere is subjected to a buoyancy force, F_B, equal to the weight of the displaced water:

$$F_B = V \gamma_w$$
$$= (0.100 \text{ m}^2)(9.8 \text{ kN/m}^2) \qquad (10.30)$$
$$= 0.98 \text{ kN}$$

The contact force between the sphere and the scale is thus reduced to:

$$F = 2.60 \text{ kN} - 0.98 \text{ kN}$$
$$= 1.62 \text{ kN} \qquad (10.31)$$

The weight of the sphere has not changed, but it is now being supported partially by the scale and partially by the water.

   The contact forces between soil particles above the groundwater table are similar to that between the dry sphere and the scale, while soils below the groundwater table are similar to the submerged sphere and scale. Buoyancy forces act on the soil solids the same way they act on the sphere. Therefore, the particle contact forces in an element of soil that is initially above the groundwater table will decrease if the groundwater rises above that element.

Vertical Effective Stress

From the submerged sphere analogy we see how σ in an element of soil below the groundwater table is distributed between the solid particles and the pore water. As discussed earlier, the portion carried by the solid particles is known as the *effective stress*, σ', while the portion carried by the pore water is equal to the *pore water pressure, u*.

   Under most unconfined hydrostatic conditions, we can compute u below the groundwater table using Equation 7.7. Then, we can compute the effective stress using:

$$\sigma' = \sigma - u \tag{10.32}$$

or

$$\sigma'_z = \sigma_z - u \tag{10.33}$$

where:

σ_z' = vertical effective stress
σ_z = vertical total stress
u = pore water pressure

If only geostatic stresses are present, we can combine Equations 10.9 and 10.33 to produce:

$$\sigma'_z = \sum \gamma H - u \tag{10.34}$$

Note how the distribution of force between the solids and water is *not* proportional to their respective cross-sectional areas.

   If both geostatic and induced stresses are present, then:

$$\sigma'_z = \sum \gamma H + \sum (\sigma_z)_{\text{induced}} - u \tag{10.35}$$

   The first two terms in Equation 10.35 are a restatement of the principle of superposition. Notice how we apply this principle to the total stresses, and then subtract the pore water pressure to find the effective stress. Do not attempt to combine effective stresses using superposition.

Example 10.6

Using the results from Example 10.1, compute σ_z' at Point A in Figure 10.7.

$$
\begin{aligned}
u &= \gamma_w z_w \\
&= (9.8\,\text{kN/m}^3)(6.1\,\text{m}) \\
&= 60\,\text{kPa}
\end{aligned}
$$

$$
\begin{aligned}
\sigma_z' &= \sigma_z - u \\
&= 134\,\text{kPa} - 60\,\text{kPa} \\
&= \textbf{74 kPa} \qquad \Leftarrow \textit{Answer}
\end{aligned}
$$

Commentary

A vertical compressive stress of 134 kPa is present at Point A, 74 kPa of this stress is being carried by the solid particles and 60 kPa by the pore water.

The principle of effective stress is the key to understanding many aspects of soil behavior. In the following chapters we will see how settlement and strength analyses are normally based on effective stresses, not total stresses.

Horizontal Effective Stress

The horizontal effective stresses, σ_x' and σ_y', are related to the horizontal total stresses as follows:

$$
\boxed{\sigma_x' = \sigma_x - u} \tag{10.36}
$$

$$
\boxed{\sigma_y' = \sigma_y - u} \tag{10.37}
$$

If multiple sources of stress need to be combined, do so by using superposition with the total stresses, then subtract the pore water pressure.

The ratio of the horizontal to vertical effective stresses is defined as the *coefficient of lateral earth pressure*, K. For geostatic stresses beneath a level ground surface, we normally assume K in the x direction is equal to that in the y direction:

$$K = \frac{\sigma'_x}{\sigma'_z} = \frac{\sigma'_y}{\sigma'_z}$$

(10.38)

However, if the ground surface is inclined, or if induced stresses are present, K may be different in the x and y directions. The value of K varies from about 0.3 to 3. We will discuss methods of evaluating it in Chapter 16.

Values of K normally reflect the existing or pre-construction condition. If a new load, such as that from a foundation, is to be applied, K will usually change because of the induced horizontal and vertical effective stresses. Depending on the type of load, K in the x direction also may be different than in the y direction. This can cause some confusion when solving stress problems, because the given K may only be used to evaluate the pre-construction stresses. Example 10.7 illustrates the proper way to solve such problems.

Example 10.7

A proposed vertical point load of 90.0 k is to be applied to the ground surface 3.0 ft south and 4.0 ft east of Point A in Figure 10.15. Compute all of the total and effective stresses acting on the vertical and horizontal planes at Point A. Consider both the geostatic and induced stresses, and use a coordinate system with the x and y axes oriented in the east and north directions, respectively.

Note: The x and y axes do not always need to be aligned with cardinal compass directions. Often it is more convenient to orient them parallel and perpendicular to a proposed structure or slope.

Solution

Geostatic stresses (initial condition):

$$\sigma_z = \sum \gamma H = (106 \text{lb/ft}^3)(3 \text{ ft}) + (110 \text{ lb/ft}^3)(1 \text{ ft}) + (112 \text{ lb/ft}^3)(2 \text{ ft}) = 652 \text{ lb/ft}^2$$

$$u = \gamma_w z_w = (62.4 \text{ lb/ft}^3)(2 \text{ ft}) = 125 \text{ lb/ft}^2$$

$$\sigma'_z = \sigma_z - u = 652 \text{ lb/ft}^2 - 125 \text{ lb/ft}^2 = 527 \text{ lb/ft}^2$$

$$\sigma'_x = \sigma'_y = K\sigma'_z = (0.68)(527 \text{ lb/ft}^2) = 358 \text{ lb/ft}^2$$

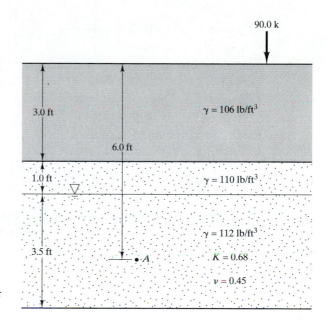

Figure 10.15 Cross-section for
Example 10.7.

$$\sigma_x = \sigma_y = \sigma'_x + u = 358 \text{ lb/ft}^2 + 125 \text{ lb/ft}^2 = 483 \text{ lb/ft}^2$$

Since the ground surface is horizontal:

$$\tau_{xz} = \tau_{zx} = \tau_{yz} = \tau_{zy} = \tau_{xy} = \tau_{yx} = 0$$

Induced stresses:

Using the Boussinesq equations:

$$x_f = -4 \text{ ft} \qquad y_f = 3 \text{ ft} \qquad z_f = 6 \text{ ft}$$

$$R = \sqrt{x_f^2 + y_f^2 + z_f^2} = \sqrt{(-4.0)^2 + 3.0^2 + 6.0^2} = 7.81 \text{ ft}$$

$$r = \sqrt{x_f^2 + y_f^2} = \sqrt{(-4.0)^2 + 3.0^2} = 5.0 \text{ ft}$$

$$\sigma_x = \frac{P}{2\pi}\left[\frac{3x_f^2 z_f}{R^5} - (1-2\nu)\left(\frac{x_f^2 - y_f^2}{Rr^2(R+z_f)} + \frac{y_f^2 z_f}{R^3 r^2}\right)\right]$$

$$= \frac{90{,}000}{2\pi}\left[\frac{3(-4.0)^2(6.0)}{7.81^5} - (1-2(0.45))\left(\frac{(-4.0)^2 - (3.0)^2}{(7.81)(5.0)^2(7.81+6.0)} + \frac{(3.0)^2(6.0)}{(7.81)^3(5.0)^2}\right)\right]$$

$$= 132 \text{ lb/ft}^2$$

$$\sigma_y = \frac{P}{2\pi}\left[\frac{3y_f^2 z_f}{R^5} - (1-2\nu)\left(\frac{y_f^2 - x_f^2}{Rr^2(R+z_f)} + \frac{x_f^2 z_f}{R^3 r^2}\right)\right]$$

$$= \frac{90{,}000}{2\pi}\left[\frac{3(3.0)^2(6.0)}{7.81^5} - (1-2(0.45))\left(\frac{(3.0)^2 - (-4.0)^2}{(7.81)(5.0)^2(7.81+6.0)} + \frac{(-4.0)^2(6.0)}{(7.81)^3(5.0)^2}\right)\right]$$

$$= 72 \text{ lb/ft}^2$$

$$\sigma_z = \frac{3P z_f^3}{2\pi R^5} = \frac{3(90{,}000)(6.0)^3}{2\pi(7.81)^5} = 319 \text{ lb/ft}^2$$

$$\tau_{zx} = -\tau_{xz} = \frac{3P z_f^2 x_f}{2\pi R^5} = -\tau_{xz} = \frac{3(90{,}000)(6.0)^2(-4.0)}{2\pi(7.81)^5} = -213 \text{ lb/ft}^2$$

$$\tau_{xy} = -\tau_{yx} = \frac{P}{2\pi}\left[\frac{3x_f y_f z_f}{R^5} - (1-2\nu)\left(\frac{(2R+2)x_f y_f}{(R+z_f^2)R^3}\right)\right]$$

$$= \frac{90{,}000}{2\pi}\left[\frac{3(-4.0)(3.0)(6.0)}{7.81^5} - (1-2(0.45))\left(\frac{(2(7.81)+2)(-4.0)(3.0)}{(7.81+6.0^2)(7.81)^3}\right)\right]$$

$$= -92 \text{ lb/ft}^2$$

$$\tau_{zy} = -\tau_{yz} = \frac{3P z_f^2 y_f}{2\pi R^5} = -\tau_{yz} = \frac{3(90{,}000)(6.0)^2(3.0)}{2\pi(7.81)^5} = 160 \text{ lb/ft}^2$$

Overall stresses (by superposition):

$$\sigma_x = 483 \text{ lb/ft}^2 + 132 \text{ lb/ft}^2 = \textbf{615 lb/ft}^2 \quad \Leftarrow Answer$$

$$\sigma_y = 483 \text{ lb/ft}^2 + 72 \text{ lb/ft}^2 = \textbf{555 lb/ft}^2 \quad \Leftarrow Answer$$

$$\sigma_z = 652 \text{ lb/ft}^2 + 319 \text{ lb/ft}^2 = \textbf{971 lb/ft}^2 \quad \Leftarrow Answer$$

$$\sigma_x' = \sigma_x - u = 615 \text{ lb/ft}^2 - 125 \text{ lb/ft}^2 = \textbf{490 lb/ft}^2 \quad \Leftarrow Answer$$

$$\sigma_y' = \sigma_y - u = 555 \text{ lb/ft}^2 - 125 \text{ lb/ft}^2 = \textbf{430 lb/ft}^2 \quad \Leftarrow Answer$$

$$\sigma_z' = \sigma_z - u = 971 \text{ lb/ft}^2 - 125 \text{ lb/ft}^2 = \textbf{846 lb/ft}^2 \quad \Leftarrow Answer$$

$$\tau_{zx} = -\tau_{xz} = \textbf{-213 lb/ft}^2 \quad \Leftarrow Answer$$

$$\tau_{xy} = -\tau_{yx} = \textbf{-92 lb/ft}^2 \quad \Leftarrow Answer$$

$$\tau_{zy} = -\tau_{yz} = \textbf{160 lb/ft}^2 \quad \Leftarrow Answer$$

Notes: 1. The K value given in the problem statement was applied only to the initial conditions, which in this case consisted of the geostatic stresses only. It would not be correct to apply this K to the $\sigma_z' = 846$ lb/ft² value, because it represents the proposed condition.
2. We compute the proposed stresses by first combining the geostatic and induced total stresses by superposition, then subtracting the pore water pressure.

Stresses Beneath Bodies of Water

Sometimes geotechnical engineers need to compute stresses in soils beneath bodies of water, such as lakes, rivers, or oceans. For example, this might be necessary while designing the foundation for a bridge, or when evaluating the stability of a slope that extends underwater.

Although this case appears confusing at first, it is really quite simple: The body of water makes equal contributions to total stress and pore water pressure, but no contribution to effective stress in the soil below. An easy way to remember is to simply think of the body of water as a "soil" with a unit weight equal to γ_w.

Stress Conditions with Negative Pore Water Pressures

Most geotechnical analyses assume the pore water pressure above the groundwater table is zero. Thus, according to Equation 10.32, the effective stress equals the total stress. However, this is a simplification of the truth. In reality, these soils generally have *negative pore water pressures* (also known as *soil suction*), which means the effective stress is greater than the total stress.

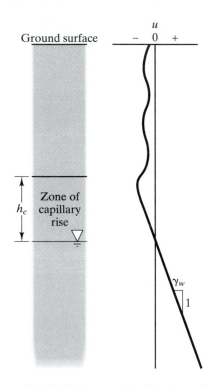

Within the zone of capillary rise, the pore water pressure may be computed using Equation 7.7, except z_w is now negative. This produces a negative u, as shown in Figure 10.16. In other words, the pore water in this zone may be visualized as a column of water held in tension by the capillary forces. Above the capillary zone, where the degree of saturation falls well below 100%, the pore water collects into small drops adjacent to the particle contact points as shown in Figure 10.17. Surface tension forces develop between this water and the particles, creating tensile forces (negative pore water pressures) in the water. The transition between these two zones is often poorly defined.

Negative pore water pressures can be measured in-situ or in the laboratory using a variety of techniques (Fredlund and Rahardjo, 1993), and these measurements have been used to develop more rational explanations of soil behavior above the groundwater table. These techniques are especially helpful in understanding soil strength issues, as discussed in Chapter 13.

Figure 10.16 Within the zone of capillary rise, a plot of pore water pressure vs. elevation is simply an extension of the plot below the groundwater table. Above the zone of capillary rise, the pore water accumulates near the particle contact points and surface tension forces develop between these pockets of pore water and the adjacent solid particles. These forces produce tensile stresses (negative pore water pressures) in the water. The magnitude of the negative pore water pressures in this zone depends on the soil type and other factors, and can be much greater (i.e., more negative) than those in the capillary zone.

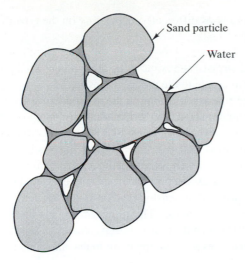

Sand particle

Water

Figure 10.17 Above the capillary zone, the pore water retreats to the particle contact points. Surface tension forces develop between this water and the solid particles, producing a negative pore water pressure.

QUESTIONS AND PRACTICE PROBLEMS

10.8 Develop a plot of σ_z, u, and σ_z' vs. depth for the soil profile in Figure 10.7. Consider depths from 0 to 10 m, assume hydrostatic conditions are present, and assume $u = 0$ above the groundwater table. Plot depth, z, on the vertical axis, with zero at the top and increasing downward. This method of plotting the data is easier to visualize, because depth on the plot is comparable to depth in a cross-section.

10.9 Compute the values of σ_x, σ_x', σ_z, σ_z', and τ_{zx} at Point B in Figure 10.7. The coefficient of lateral earth pressure in the SM soil is 0.60.

10.10 According to an in-situ soil suction measurement, the pore water pressure at Point C in Figure 10.7 is -5.0 kPa. Compute the vertical effective stress at this point.

10.11 A vertical point load P is to be applied to a level ground surface. The underlying soil has the following pre-construction characteristics:

      Groundwater table: 5.5 ft below the ground surface
      Unit weight above the groundwater table = 121 lb/ft^3
      Unit weight below the groundwater table = 124 lb/ft^3
      $K = 0.87$
      $\nu = 0.33$

The horizontal total stress, σ_x, at a point 8 ft below the ground surface and 3 ft east of the point of load application must not exceed 1000 lb/ft^2. Compute the maximum allowable value of P.

10.8 PROGRAMS STRESSP, STRESSL, STRESSR, AND STRESSC

A geotechnical analysis software package has been developed specifically for this book. This software package includes four programs that compute stresses in the ground. All of the programs compute the geostatic stresses using the techniques described in this chapter, but each program computes induced stresses differently, depending on the type of loading condition:

- Program **STRESSP** computes the stresses at a point in the ground beneath a point load. It uses Equations 10.11–10.16 to compute the induced stresses.
- Program **STRESSL** computes the stresses at a point in the ground beneath a line load. This line load may have any length and may be oriented in any compass direction. It divides the significant portions of the line load into 200 elements, replaces each element with an equivalent point load, computes the induced stresses from the point loads using the Boussinesq method, then adds them using superposition.
- Program **STRESSR** computes the stresses at a point in the ground beneath a rectangular load. This load may have any specified dimensions and be at any location, so long as the sides are aligned with the x and y axes. The program divides the significant portions of the rectangle into 2500 elements, replaces each element with an equivalent point load, computes the induced stresses from the point loads, and adds them using superposition.
- Program **STRESSC** computes the stresses at a point beneath a circular load of any diameter and location. It uses the same computational technique as **STRESSR**.

To use these programs, you must first download the software from the Prentice Hall website and install it onto a computer, as described in Appendix C. Then, select the desired program from the main menu and input the data using the following procedure:

1. Select the unit system (SI or English)
2. Fill in the soil profile table as necessary. If only one stratum is present, just enter its description (i.e., Silty clay) and its unit weight in the uppermost boxes and leave the remaining boxes empty. If additional strata are present, enter the appropriate depths to the strata interfaces and the descriptions and unit weights. Keep in mind the z coordinate increases with depth, as shown in Figure 10.1
3. Enter the depth from the ground surface to the groundwater table. If the groundwater is very deep, just enter a very large value.
4. Fill in the boxes that describe the applied load. You may use any origin for the x and y axes, but the z axis is always zero at the ground surface.
5. Fill in the boxes that describe the location and soil properties at the point where the stresses are to be computed.

Then, click on the CALCULATE button to perform the computations. The results are shown on the screen, and printouts may be obtained by clicking on the PRINT button.

These programs eliminate the tedium of solving complex stress analysis problems. They also provide a fast and easy means of conducting parametric studies.

Example 10.8

A proposed 3.0 m tall retaining wall with a level backfill, along with a proposed 350 kN vertical point load are shown in Figure 10.18. The groundwater table is very deep. Using program **STRESSP**, develop a plot of the earth pressure σ_x along the back of the wall.

Solution

To solve this problem place all of the dimensions and soil properties into Program **STRESSP**, then compute the stresses at a series of x, y, z points along the back of the wall. Figure 10.19 shows a **STRESSP** screen from one of these computations. The results are shown in Figure 10.18.

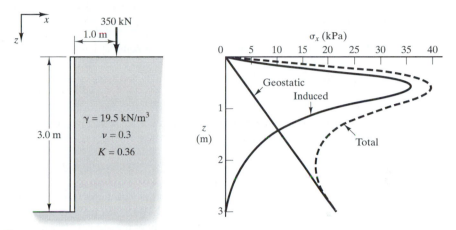

Figure 10.18 Cross-section and solution for Example 10.8.

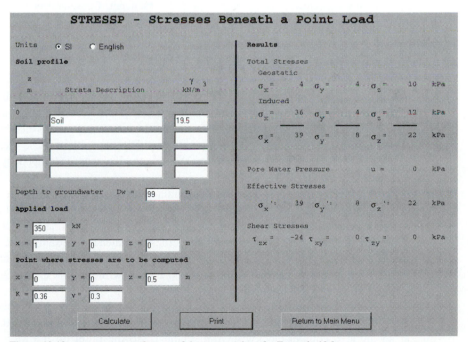

Figure 10.19 STRESSP screen for one of the computations for Example 10.8.

Chapter 16 discusses lateral earth pressures in much more detail, including methods of selecting K.

The wall is stiffer than the backfill soils, so the requirement for homogeneity described in Section 10.1 has technically not been satisfied. However, if the backfill is well compacted, this analysis procedure appears to provide sufficiently accurate results (Bowles, 1996).

QUESTIONS AND PRACTICE PROBLEMS

10.12 Use Program STRESSP to solve Example 10.7.

10.13 A 21.0 m diameter oil tank is to be built on a soil that has $\gamma = 18.4$ kN/m^3, $K = 0.60$, and $\nu = 0.40$. The tank and its contents have a total mass of 3.70×10^6 kg, and the bottom of the tank is flush with the ground surface. The groundwater table is at a great depth. Use program STRESSC to compute σ_x, σ_x', σ_y, σ_y', σ_z, σ_z', τ_{zx}, τ_{xy}, and τ_{zy} at the following two points: a) 8.0 m below the center of the tank, and b) 8.0 m below the east edge of the tank. Use a coordinate system with the origin at the center of the tank, x positive to the east, and y positive to the north.

10.14 Use program STRESSR to solve Example 10.5. Use $\nu = 0.30$ and $K = 0.65$.

10.15 A proposed 5 ft × 5 ft spread footing foundation will support an office building. The column load plus the weight of the foundation will be 80 k, and the bottom of the foundation will be 2 ft below the ground surface. The unit weight of the soil is 121 lb/ft^3, $\nu = 0.35$ and $K = 0.83$, and the groundwater table is at a great depth. Using program STRESSR, develop a plot of the induced σ_z below the center of this foundation vs. depth. Consider depths from the bottom of the footing to 15 ft below the bottom of the footing.

10.9 MOHR'S CIRCLE ANALYSES

So far we have considered only the stresses acting on vertical and horizontal planes. However, some problems may require computation of the stresses acting on other planes. For example, the reduction of data from certain laboratory tests, such as the triaxial compression test, requires such computations. We can obtain these stresses using a graphical representation called a *Mohr's circle*, which was developed by the German engineer Otto Mohr (1835–1918).

A Mohr's circle describes the two-dimensional stresses at a point in a material. It considers the stresses acting on each side of a two-dimensional element and plots them on a σ vs. τ diagram, as shown in Figure 10.20. Each point on the circle represents the normal and shear stresses acting on one side of an element oriented at a certain angle. For example, in this Mohr's circle Points A and B represent (σ_z, τ_{zx}) and (σ_x, τ_{xz}), which are the stresses acting on an element aligned with the x and z axes, while Points C and D represent the stresses on an element oriented as shown.

Notice how the angle between two points on a Mohr's circle, 2θ, is exactly twice the angle θ between the planes they represent. For example, Plane C in Figure 10.20 is oriented at an angle θ counter-clockwise from Plane E, so the point on the Mohr's circle that represents Plane C is at an angle 2θ counter-clockwise from the point that represents Plane E.

Principal Stresses

If the soil element is rotated to a certain angle, the shear stresses will be zero on all four sides. The planes on each side of this element are represented by Points E and F, and are known as *principal planes*. The stresses acting on them are known as *principal stresses*. The *major principal stress*, σ_1, also is the greatest normal stress that acts on any plane, while the *minor principal stress*, σ_3, is the smallest normal stress that acts on any plane. These two stresses act at right angles to each other. If we were conducting a three-dimensional analysis, there also would be an *intermediate principal stress*, σ_2, which acts at right angles to both σ_1 and σ_3. However, the vast majority of geotechnical analyses do not explicitly consider σ_2.

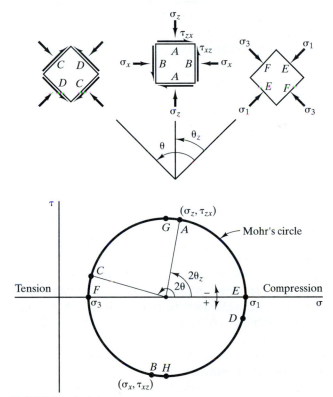

Figure 10.20 Mohr's circle in *x-z* space.

The magnitudes of σ_1 and σ_3 are:

$$\sigma_1 = \frac{\sigma_x + \sigma_z}{2} + \sqrt{\left[\frac{\sigma_x - \sigma_z}{2}\right]^2 + \tau_{zx}^2} \qquad (10.39)$$

$$\sigma_3 = \frac{\sigma_x + \sigma_z}{2} - \sqrt{\left[\frac{\sigma_x - \sigma_z}{2}\right]^2 + \tau_{zx}^2} \qquad (10.40)$$

The angle between σ_z and σ_1 is:

$$\theta_z = \frac{1}{2}\cos^{-1}\left(\frac{2\sigma_z - \sigma_1 - \sigma_3}{\sigma_1 - \sigma_3}\right) \qquad (10.41)$$

where:

σ_1 = major principal stress
σ_3 = minor principal stress
σ_x = horizontal stress
σ_z = vertical stress
τ_{zx} = shear stress acting on a horizontal plane
θ_z = angle between σ_z and σ_1

As discussed earlier, when the ground surface is level, the geostatic shear stresses on the vertical and horizontal planes are all zero. Therefore, these are the principal planes and the geostatic principal stresses act vertically and horizontally. If $\sigma_x < \sigma_z$, which is the most common case, then $\sigma_1 = \sigma_z$ and $\sigma_3 = \sigma_x$. Conversely, if $\sigma_x > \sigma_z$, then $\sigma_1 = \sigma_x$ and $\sigma_3 = \sigma_z$. However, the principal stresses due to induced loads can act in any direction. For example, Figure 10.21 shows the induced principal stresses beneath a circular loaded area.

Usually we are most interested in the Mohr's circle that represents the combined effects of both the geostatic and induced stresses. To develop such a circle, compute the geostatic σ_x and σ_z and the induced σ_x, σ_z, and τ_{zx}, then add them by superposition and use the combined values to develop the Mohr's circle. Do not attempt to combine σ_1 or σ_3 values by superposition.

Stresses on Other Planes

Once we have constructed the Mohr's circle that represents the stresses at a point in the soil, we can obtain the normal and shear stresses that act on any plane through that point. This could be done graphically by drawing the Mohr's circle to scale, but it is generally easier to use the circle simply as a graphical representation and perform the computations using the following equations:

$$\sigma = \frac{\sigma_1 + \sigma_3}{2} + \frac{\sigma_1 - \sigma_3}{2}\cos 2\theta \qquad (10.42)$$

$$\tau = \frac{\sigma_1 - \sigma_3}{2} \sin 2\theta \qquad\qquad (10.43)$$

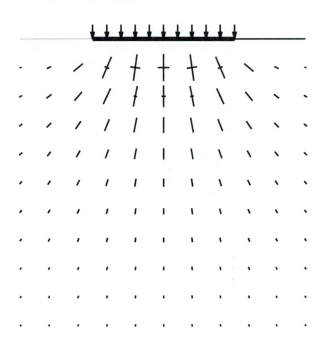

Figure 10.21 Principal stresses beneath a circular loaded area. The long and short lines indicate the major and minor principal stresses, respectively, and their lengths are proportional to their magnitudes. This diagram reflects only the induced stresses, and is based on $\nu = 0.4$.

The greatest shear stress, τ_{max}, occurs on the planes represented by points G and H. in Figure 10.20. These planes are oriented at $45°$ angles from the principal planes, and the shear stress acting on them is equal to the radius of the Mohr's circle:

$$\tau_{max} = \frac{\sigma_1 - \sigma_3}{2} \qquad\qquad (10.44)$$

where:
  σ = normal stress acting on a particular plane
  σ_1 = major principal stress
  σ_3 = minor principal stress
  τ = shear stress acting on a particular plane

τ_{max} = maximum shear stress acting on any plane

    θ = angle from σ_1 to σ (clockwise is positive; counterclockwise is negative), as shown in Figure 10.20

Use the following procedure to compute the stresses on a given plane:

1. Draw a soil element that is aligned with the x and z axes.
2. Compute σ_x, σ_z, τ_{xz}, and τ_{zx} and mark these stresses on the soil element.
3. Plot the points σ_x, τ_{xz}, and σ_z, τ_{zx} on a σ, τ diagram, then use these points to draw the Mohr's circle. The center of the circle is at $\sigma = (\sigma_x + \sigma_z)/2$.
4. Use Equations 10.39 and 10.40 to compute σ_1 and σ_3.
5. Use Equation 10.41 to compute the angle θ_z between σ_z and σ_1.
6. Compare the positions of the points on the Mohr's circle that represent σ_z and σ_1 to determine if θ_z extends clockwise or counter-clockwise from σ_z.
7. Draw another soil element that is rotated at an angle θ_z from the first soil element. The sides of this element are the principal planes. Mark the stresses σ_1 and σ_3 on this soil element. Since the sides of this element are the principal planes, the shear stresses are zero.
8. If the value of τ_{max} is required, it may be computed using Equation 10.44. It acts on planes oriented $\pm 45°$ from the principal planes.
9. Draw a third soil element with one of its sides oriented in the direction of the plane on which the stresses are to be computed. Mark the stresses σ and τ on this element.
10. Determine the angle θ between σ and σ_1, then locate the point on the Mohr's circle that represents the plane. This point is located at an angle 2θ from the point that represents σ_1. Be sure to follow the sign convention described earlier: angles measured clockwise from σ_1 are positive, while those measured counterclockwise are negative.
11. Use Equations 10.42 and 10.43 to compute σ and τ.

Example 10.9

The vertical and horizontal stresses at a certain point in a soil are as follows:

$$\sigma_x = 2100 \text{ lb/ft}^2$$
$$\sigma_z = 3000 \text{ lb/ft}^2$$
$$\tau_{zx} = -300 \text{ lb/ft}^2$$

  a. Determine the magnitudes and directions of the major and minor principal stresses.
  b. Determine the magnitude and directions of the maximum shear stress.
  c. Determine the normal and shear stresses acting on a plane inclined at $35°$ clockwise from the x axis.

Solution

  a.   See Figure 10.22.

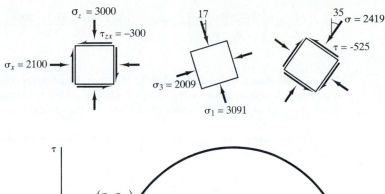

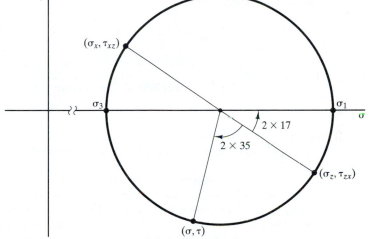

Figure 10.22 Mohr's circle and soil elements for Example 10.9.

$$\sigma_1 = \frac{\sigma_x + \sigma_z}{2} + \sqrt{\left[\frac{\sigma_x - \sigma_z}{2}\right]^2 + \tau_{xz}^2}$$

$$= \frac{2100 + 3000}{2} + \sqrt{\left[\frac{2100 - 3000}{2}\right]^2 + 300^2}$$

$$= \textbf{3091 lb/ft}^2 \quad \Leftarrow \textit{Answer}$$

$$\sigma_3 = \frac{\sigma_x + \sigma_z}{2} - \sqrt{\left[\frac{\sigma_x - \sigma_z}{2}\right]^2 + \tau_{xz}^2}$$

$$= \frac{2100 + 3000}{2} - \sqrt{\left[\frac{2100 - 3000}{2}\right]^2 + 300^2}$$

$$= \textbf{2009 lb/ft}^2 \quad \Leftarrow \textit{Answer}$$

$$\theta_z = \frac{1}{2}\cos^{-1}\left(\frac{2\sigma_z - \sigma_1 - \sigma_3}{\sigma_1 - \sigma_3}\right) = \frac{1}{2}\cos^{-1}\left(\frac{2(3000) - 3091 - 2009}{3091 - 2009}\right) = 17°$$

This equation does not give the sign for θ_z. However, based on this value of θ_z and the relative positions of σ_1 and σ_z on the Mohr's circle, we can determine the major principal stress acts at an angle of 17° counter-clockwise from the vertical.

b.   The maximum shear stress, τ_{max} occurs on the planes represented by the top and bottom of the Mohr's circle.

$$\tau_{max} = \frac{\sigma_1 - \sigma_3}{2}$$
$$= \frac{3091 - 2009}{2}$$
$$= \textbf{541 lb/ft}^2 \quad \leftarrow Answer$$

This stress acts on planes oriented ±45° from the principal planes. These planes are 17 + 45 = **62° counter-clockwise** and 17 - 45 = **28° clockwise from the vertical**

c.   The plane on which we wish to compute σ and τ is oriented at an angle of 35° clockwise from the horizontal. Therefore, the σ acting on this plane is oriented 35° clockwise from the vertical, as shown in the soil elements in Figure 10.22. Since σ_1 is oriented 17° counter-clockwise from the vertical, $\theta = -17 - 35 = -52°$. Note that θ has a negative value because σ is located clockwise from σ_1.

$$\sigma = \frac{\sigma_1 + \sigma_3}{2} + \frac{\sigma_1 - \sigma_3}{2}\cos 2\theta$$
$$= \frac{3091 + 2009}{2} + \frac{3091 - 2009}{2}\cos[2(-52)]$$
$$= \textbf{2419 lb/ft}^2 \quad \leftarrow Answer$$

$$\tau = \frac{\sigma_1 - \sigma_3}{2}\sin 2\theta$$
$$= \frac{3091 - 2009}{2}\sin[2(-52)]$$
$$= \textbf{-525 lb/ft}^2 \quad \leftarrow Answer$$

Mohr's Circles for Effective Stress

The Mohr's circle for effective stress is the same diameter as that for total stress, but it is offset horizontally by a distance equal to the pore water pressure, as shown in Figure 10.23. Notice how the shear stress on a given plane has the same value on both circles. This is because the pore water cannot carry a static shear stress, so the solid particles must carry all of the shear stress. In other words, the principle of effective stress applies only to normal stresses, not to shear stresses.

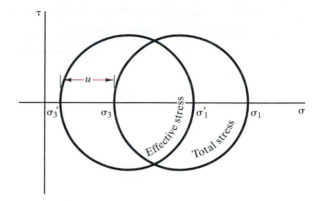

Figure 10.23 Mohr's circles for total and effective stresses.

QUESTIONS AND PRACTICE PROBLEMS

10.16 The major and minor principal stresses at a certain point in the ground are 450 and 200 kPa, respectively. Draw the Mohr's circle for this point, compute the maximum shear stress, τ_{max}, and indicate the points on the Mohr's circle that represent the planes on which τ_{max} acts.

10.17 The stresses at a certain point in the ground are $\sigma_x = 210$ kPa, $\sigma_z = 375$ kPa, and $\tau_{zx} = 75$ kPa. Draw the Mohr's circle for this point, then compute the following:

    a. The magnitudes and directions of the principal stresses.
    b. The magnitude and directions of the maximum shear stress.
    c. The normal and shear stresses acting on a plane inclined 55° clockwise from the horizontal.

10.18 The major principal stress at a certain point is 4800 lb/ft^2 and acts vertically. The minor principal stress is 3100 lb/ft^2. Draw the Mohr's circle for this point, then compute the normal and shear stresses acting on a plane inclined 26° counter-clockwise from the horizontal.

10.19 Use the "overall stresses" data in the x–z plane from Example 10.7 to perform the following computations:
    a. Draw the Mohr's circles for total and effective stresses for this point and identify the locations on the circle that represent the vertical and horizontal stresses.
    b. Compute σ_1, σ_3, σ_1', σ_3', and τ_{max}.
    c. Determine the angle between the major principal stress and the vertical, then prepare as a sketch showing the orientation of the major and minor principal stresses with respect to the vertical.
    d. Compute σ, σ', and τ that act on a plane inclined at an angle of 45° clockwise from the horizontal.

10.10 SEEPAGE FORCE

Thus far, the analyses in this chapter have considered only the case where the groundwater is nearly stationary. However, if the groundwater is moving, it imparts a drag force, called a *seepage force* on the solid particles:

$$j = i\,\gamma_w$$

(10.45)

where:

j = seepage force per unit volume of soil
i = hydraulic gradient
γ_w = unit weight of water

For example, if water is flowing through a certain soil with a hydraulic gradient of 0.15, the seepage force will be equal to $(0.15)(9.8 \text{ kN/m}^3) = 1.5 \text{ kN/m}^3$ and will act in the same direction the water is flowing.

The hydraulic gradient in soils is usually small enough that seepage forces may be ignored. However, seepage forces can be important if the water is flowing upward with a large i, as shown in Figure 10.24. This is because the seepage force now acts in the opposite direction of gravity, and thus reduces the effective stress. In this case, Equation 10.34 may be rewritten as:

$$\sigma'_z = \sum[(\gamma - j)H] - u$$

(10.46)

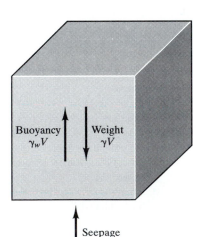

Figure 10.24 Forces acting on an element of soil when the seepage forces are acting vertically upward.

Thus, if the groundwater is flowing upward with a sufficiently large hydraulic gradient, the vertical effective stress can drop to zero. If this occurs in a sandy soil, the shear strength also drops to zero, thus producing a condition called *quicksand*. We will discuss this phenomenon in Chapter 13. If the seepage forces are so large that the computed vertical effective stress falls below zero, the soil experiences a tensile failure and moves upward. This phenomenon is called *heave*, and can be disastrous, as illustrated in Example 10.10.

Seepage forces also may be important in some slope stability problems, especially if the groundwater is flowing toward the slope face and the hydraulic gradient is high.

Example 10.10

The soil beneath the excavation in Figure 8.6 has a unit weight of 105 lb/ft^3. Evaluate the potential for heave in the soils immediately below the excavation.

Solution

The flow at Point B is upward, so the seepage force also acts upward. Each equipotential drop represents a head loss of $\Delta h/N_D = 20.0 \text{ ft}/12.6 = 1.59$ ft. The equipotential lines in the vicinity of Point B are about 2.0 ft apart. Therefore, the hydraulic gradient in the vicinity of Point B is:

$$i = -\frac{dh}{dl} = -\frac{-1.59}{2.0} = 0.80$$

$$j = i\gamma_w = (0.80)(62.4 \text{ lb/ft}^3) = 50 \text{ lb/ft}^3$$

The groundwater table inside the excavation is at the ground surface, so:

$$u = \gamma_w H = 62H$$

$$\begin{aligned} \sigma_z' &= \sum [(\gamma - j)H] - u \\ &= (105 - 50)H - 62H \\ &< 0 \end{aligned}$$

In this case, the seepage forces are sufficient to drop the computed vertical effective stress below zero. Since soil (especially sands) cannot sustain tensile stresses, this analysis indicates the seepage forces will result in an upward heave of the soils immediately below the excavation. This, in turn, would cause the sheet piles to move inward and the excavation to collapse. Such failures happen very suddenly, and have been the cause of serious injury and death, as well as significant property damage.

Problems with heave, such as the one described in Example 10.10, can be avoided by keeping the vertical effective stress well above zero. We accomplish this by maintaining the hydraulic gradient at acceptably low values (perhaps by extending the sheet piles or installing dewatering wells), or by covering the excavation with a highly pervious surcharge fill, such as gravel, which adds to the total stress, but does not contribute significantly to the seepage force.

10.11 STRESSES IN LAYERED STRATA

The beginning of this chapter included a list of several simplifying assumptions that have governed the remainder of our discussions. One of these stated the ground is homogeneous, which in this context meant the modulus of elasticity, E, shear modulus, G, and Poisson's ratio, v, are constants. Although this is an acceptable assumption for many soil profiles, sometimes we encounter conditions where some strata are significantly stiffer than others. Therefore, we need to understand the impact of such differences, and in some cases be able to quantify them.

One common condition consists of a soil layer underlain by a much stiffer bedrock ($E_1 < E_2$) as shown in Figure 10.25. In this case, there is less spreading of the load, so the induced stresses in the soil are greater than those computed by Boussinesq. Conversely, if we have a stiff layer underlain by a softer soil ($E_1 > E_2$), the load spreading is enhanced and the induced stresses are less than the Boussinesq values.

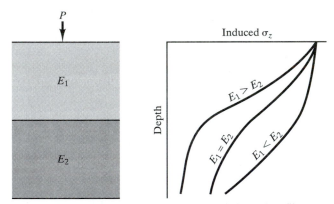

Figure 10.25 Distribution of induced σ_z with depth in layered profiles.

Sometimes we can use this behavior to our advantage, such as with highway pavements. The pavement and the underlying aggregate base course are much stiffer than the soils that support them, so they spread the wheel loads over a larger area of soil. This decreases the induced stresses, and thus enhances the soil's load-carrying capacity.

10.12 STRESS PATHS

Geotechnical engineers have discovered that soil behavior depends not only on the current stresses, but also on stresses that were present in the past. For example, a certain point in soil where $\sigma_z = 100$ kPa may have a coefficient of lateral earth pressure, K, of 0.65. But, if we increase σ_z to 500 kPa, then reduce it back to 100, the K value will become substantially higher than 0.65, even though σ_z is back to its original value. We can think of the soil as having a memory.

We can evaluate these effects by keeping track of the changes in stress with time through the use of *stress paths*. A study of these stress paths is beyond the scope of this book, but forms an important part of advanced studies in soil mechanics (see Holtz and Kovacs, 1981).

SUMMARY

Major Points

1. Soil and rock are much more complex than traditional engineering materials, so a true mathematical model to describe the propagation of stresses would be far too difficult to use in practice. Therefore, we treat soil and rock as if they were continuous, homogeneous, isotropic, linear elastic materials. The error introduced by these simplifying assumptions is acceptably small for most practical analyses.

2. Our analyses consider both normal stress, σ, and shear stress, τ, but we use a sign convention opposite that of most engineers: Compression is positive and tension is negative.

3. Although we can keep track of stresses in all three dimensions, thus performing a three-dimensional analysis, many problems can be simplified by using two-dimensional or even one-dimensional analyses.

4. Stresses in the ground come from two kinds of sources: geostatic stresses are those due to the weight of the ground itself, while induced stresses are due to external loads.

5. A series of formulas have been developed to compute induced stresses. In some cases they may be used directly. We also use simplified formulas, chart solutions, and numerical solutions.

6. We normally compute geostatic and induced stresses separately, then combine them using superposition.

7. Total stress is the compressive stress acting in a certain direction at a certain point. The portion of this stress that is carried by the solid particles is called the effective stress. The remainder is carried by the pore water, and is called the pore water pressure. The concept of effective stress is very important, and many geotechnical analyses use it.

8. The coefficient of lateral earth pressure, K, is the ratio of the effective horizontal stress to the effective vertical stress.

9. When evaluating stresses in ground below bodies of water, simply treat the water as if it were a "soil" with a unit weight equal to γ_w.

10. A Mohr's circle may be used to compute the stresses acting on planes other than the horizontal and vertical. The Mohr's circle for effective stress has the same diameter as that for total stress, but it is offset by a distance equal to the pore water pressure.

11. The major principal stress at a point in the ground is the maximum normal stress acting on any plane through that point. The minor principal stress is the smallest normal stress acting on any plane through that point.

12. When groundwater is flowing through a soil, it imparts a seepage force which can alter the effective stress. This is especially problematic when the flow is upward, because the effective stress is reduced, possibly producing problems with quicksand or heave.

Vocabulary

analytic solution	isotropic	principal stresses
area load	intermediate principal stress	seepage force
bearing pressure	line load	shear modulus
body stress	major principal stress	shear strain
Boussinesq's method	minor principal stress	shear stress
coefficient of lateral earth	modulus of elasticity	soil suction
pressure	modulus of rigidity	strain
compression	Mohr's circle	stress
continuous material	normal strain	stress bulb
effective stress	normal stress	stress paths
geostatic stress	numerical solution	superposition
heave	point load	tension
homogeneous	Poisson's ratio	total stress
incompressible	pore water pressure	Westergaard's method
induced stress	pressure bulb	Young's modulus
infinite elastic half-space	principal planes	

COMPREHENSIVE QUESTIONS AND PRACTICE PROBLEMS

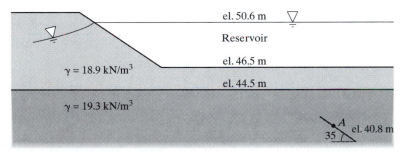

Figure 10.26 Soil profile for Problem 10.20.

10.20 Using $K = 0.61$ and assuming the major principal stress acts vertically, compute the following at Point A in Figure 10.26.

a. σ_x, σ_z, $\sigma_x{}'$, $\sigma_z{}'$, σ_1, σ_3, $\sigma_1{}'$, $\sigma_3{}'$, u
b. Mohr's circles for total and effective stresses
c. σ, σ', and τ on the plane shown in the figure

10.21 The data in the following table were obtained from three borings at a certain site. The ground surface is level, and the groundwater table is at a depth of 3.7 m below the ground surface.

Develop a representative one-dimensional design soil profile for this site, similar to the one in Figure 3.38. Then develop plots of total vertical stress, pore water pressure, and effective vertical stress vs. depth. All three plots should be superimposed on the same diagram, with the vertical axis (depth) increasing in the downward direction.

To develop the one-dimensional design soil profile, convert the information from the table into three boring logs. Then compare these logs, looking for similar soil types, and combine them into a single representative profile. Then use the γ_d and w values to compute the average unit weight for each strata. Keep in mind that computations of the total stress are based on the unit weight, γ, not the dry unit weight, γ_d.

Boring	Depth (m)	Soil Classification	Dry Unit Weight (kN/m³)	Moisture Content (%)
1	0.6	Medium sand (SP)	18.1	8.2
1	1.2	Fine to medium sand (SW)	17.9	8.0
1	2.1	Medium sand (SP)	18.7	8.9
1	2.7	Silty sand (SM)	18.4	10.3
1	3.3	Silty sand (SM)	18.5	11.0
1	4.3	Sandy gravel (GW)	19.6	12.0
1	5.2	Gravel (GP)	19.9	11.4
1	6.1	Sandy silt (ML)	17.1	19.5
1	6.7	Silty clay (CL)	16.5	21.7
1	7.6	Silty clay (CL)	16.3	22.0
2	0.9	Fine sand (SP)	17.6	7.5
2	1.5	Fine sand (SP)	17.4	9.1
2	2.1	Fine to medium sand (SW)	18.7	9.5
2	2.7	Fine sand (SP)	18.2	9.9
2	3.7	Sandy silt (ML)	17.6	11.9
2	4.9	Gravel (GP)	19.9	11.0
2	6.1	Silt (ML)	16.2	22.8
3	1.2	Fine to medium sand (SW)	18.2	8.0
3	2.7	Silty sand (SM)	17.8	8.1
3	4.3	Gravelly sand (SW)	19.2	13.4
3	5.8	Silty clay (CL)	16.0	23.2
3	7.3	Clay (CL)	15.4	25.9

10.22 A proposed spread footing foundation is to be built near an existing retaining wall as shown in Figure 10.27. It will be 4.0 ft wide, 4.0 ft long, embedded 1.5 ft into the ground, and carry a vertical load of 55.0 k. Using program **STRESSR**, develop plots of the induced contact pressure σ_x acting on the wall at the following locations:

- Immediately adjacent to the proposed foundation
- 3.0 ft away from the centerline of the proposed foundation, as measured perpendicular to this cross-section
- 6.0 ft away from the centerline of the proposed foundation, as measured perpendicular to this cross-section

Assume the weight of the foundation is equal to the weight of the soil excavated to build it, and thus does not contribute to the induced stresses.

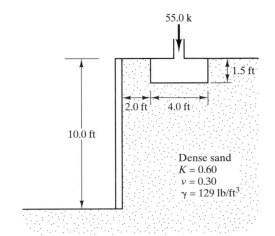

Figure 10.27 Cross-section of retaining wall for Problem 10.22.

10.23 When combining stresses from multiple sources, Equation 10.35 instructs us to combine the total stresses using superposition, then subtract the pore water pressure. Why would it be incorrect to compute the various effective stresses, then combine them by superposition?

10.24 An excavation similar to the one in Figure 8.6 has recently been constructed and dewatered. Unfortunately, this excavation is beginning to show signs of incipient heave and/or quicksand problems. An analysis similar to the one in Example 10.8 confirms that this is a potential problem.

   As an emergency measure, the contractor is proposing to remove the dewatering pumps, and fill the excavation with water. Evaluate this proposal and prepare a 200–300 word essay describing why this method would or would not provide temporary relief from the heave and quicksand problem.

10.25 A point load and a square area load are to be applied to the ground surface as shown in Figure 10.28. Using Programs **STRESSP** and **STRESSR** and associated hand computations, develop a plot of σ_z' vs. depth below Point A. This plot should contain two curves: one that represents the pre-construction condition (i.e., without the applied loads) and one that represents the post-construction condition. Then develop a similar plot for σ_x' vs. depth. Both plots should extend from the ground surface to the bottom of the fat clay stratum.

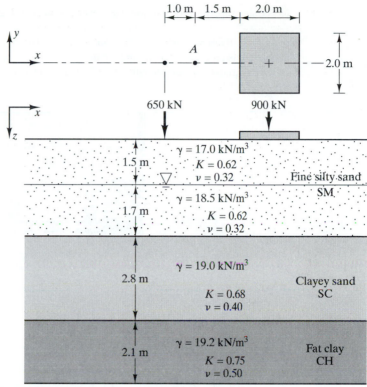

Figure 10.28 Plan and profile view for Problem 10.25.

10.26 A truck stop is to be built on a parcel of land adjacent to a major highway. During the planning stage of this project, the engineers found an existing 6 ft by 6 ft concrete box culvert under the proposed truck parking area, as shown in Figure 10.29. The project engineer is concerned that the weight of the parked trucks may overstress it, and has asked you to compute the vertical pressures acting on the top of the culvert. The results of your analyses will be provided to a structural engineer, who will then develop shear and moment diagrams and determine if the culvert can safely support the weight of the trucks.

   a. Compute the vertical pressure acting on the top of the culvert due to the weight of the overlying soil without any trucks. This is the same as the geostatic vertical stress at this depth, and represents the current condition. Use a unit weight of 120 lb/ft^3 and assume the groundwater table is at a depth of 45 ft.

   b. Using program **STRESSP**, compute the vertical pressure acting on the top of the culvert due to the wheel loads from a parked truck. This is the same as the induced vertical stress in the soil. Base your computations on two axles 48 inches apart, with the truck aligned parallel to the culvert. Perform all computations in the x-z plane of the first axle (i.e., $y = 0$ for the first axle, and $y = 48$ in for the second). Each axle carries a total vertical load of 18,000 lb, which is evenly divided among its four wheels. You may assume each wheel acts as a point load. Repeat this computation for various values of x along the top of the culvert, then present your results in the form of a pressure diagram.

c.   Using superposition, combine the results from parts a and b.

Note: The culvert is stiffer than the soil, so the Boussinesq solution gives an approximate solution to this problem. A more precise analysis would need to consider the ratio of modulii of elasticity in the soil and the culvert, and is beyond the scope of this book (see Poulos and Davis, 1974).

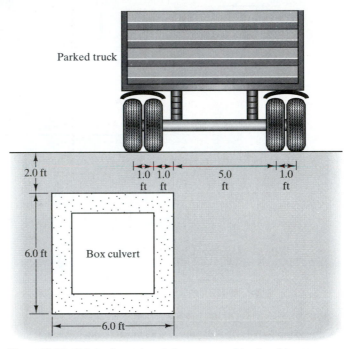

Figure 10.29 Existing box culvert below a proposed truck parking area.

10.27 The excavation shown in Figure 10.30 is to be made in a river. When the normal water level is present in the river, the hydraulic gradient at the bottom of the excavation is low enough to provide a sufficient margin of safety against heaving and quicksand. However, if the river rises to the design flood level, the hydraulic gradient will increase to 1.1, which will probably cause problems.

To provide sufficient protection against heave and quicksand, a gravel blanket is to be placed in the bottom of the excavation. This gravel, which has a unit weight of 20.2 kN/m³, will increase the effective stress in the underlying natural soils. However, because of its high hydraulic conductivity, the hydraulic gradient in the gravel will be very small, so the seepage force will be negligible. Thus, the gravel blanket will help protect the excavation against heave. The design requires a vertical effective stress of at least 25 kPa in the upper 3 m of soil. [2] Determine the minimum required thickness of the gravel blanket.

---

[2] This requirement is for illustrative purposes only, and is not necessarily an appropriate criteria for actual design problems.

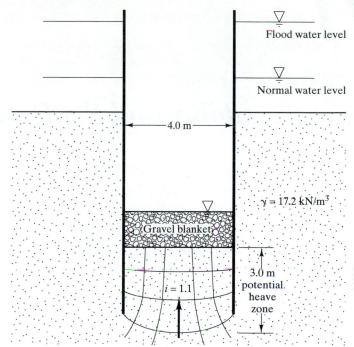

Figure 10.30 Cross-section for Problem 10.27

11

Compressibility and Settlement

Less than 10 years ago the Foundation Committee of a well-known engineering society decided, at one of its meetings, that the word "settlement" should be avoided in public discussions, because it might disturb the peace of mind of those who are to be served by the engineering profession.

Karl Terzaghi (1939)

Many civil engineering projects include placing loads onto the ground, which produce corresponding increases in the vertical effective stress, σ_z'. These increases are important because they induce vertical strains, ϵ_z, in the soil, and thus cause the ground surface to move downward. We call this downward movement *settlement*. When settlement occurs over a large area, it is sometimes called *subsidence*.

In a mechanics of materials course you learned that stresses in any material always produce strains. Therefore, whenever σ_z' increases, there always will be a corresponding settlement, δ. The issue facing a geotechnical engineer is not *if* settlements will occur, but rather the magnitude of these settlements and how they compare with tolerable limits.

This chapter discusses the various factors that influence settlement and presents methods of predicting its magnitude. Chapter 12 continues these discussions and addresses the rate of settlement, and Chapter 17 applies these methods to the design of structural foundations. Engineers use the results of these settlement analyses to design structures and other civil engineering projects. For example, if the analysis indicated the weight of a proposed building would cause excessive settlement in the soils below, the engineer may decide to place the building on pile foundations that penetrate through the soft compressible soils to deeper, harder strata.

Case Studies

Some of the most dramatic examples of soil settlement are found in Mexico City. Parts of the city are underlain by one of the most troublesome soils in any urban area of the world, a very soft lacustrine clay that was deposited in the former Lake Texcoco. Its engineering properties include (Hiriart and Marsal, 1969):

Mositure content, w	Average 281%, maximum 500%
Liquid limit, w_L	Average 289, maximum 500
Plastic limit, w_P	Average 85, maximum 150
Void ratio, e	Average 6.90

A comparison of these values with the typical ranges described in Tables 4.5 and 4.8 demonstrates that this is an extraordinary soil. For example, the very high void ratio indicates it contains nearly seven times as much water as solids! Another of its important properties is an extremely high compressibility.

As the city grew, municipal water demands increased and many wells were installed through this clay and into deeper water-bearing sand layers. These activities resulted in a significant drop in the groundwater levels which, as we will discuss later in this chapter, caused an increase in the effective stress. Because the clay is so compressible, and the stress increase was so large, the resulting settlements became a serious problem. Between 1898 and 1966, parts of the city settled 6 to 7 meters (Hiriart and Marsal, 1969)! At times, the rate of settlement has been as great as 1 mm/day. Fortunately, Mexican geotechnical engineers, most notably Dr. Nabor Carrillo, recognized the connection between groundwater withdrawal and settlement, and convinced government authorities to prohibit pumping in the central city area.

Figure 11.1 By 1950, the Palace of Fine Arts in Mexico City had settled about 3 m more than the surrounding streets.

In addition to the widespread settlements due to groundwater withdrawal, local settlements also have occurred beneath heavy structures and monuments. Their weight increased the stress in the underlying soil, causing it to settle. One example is the Palacio de las Bellas Artes (Palace of Fine Arts), shown in Figure 11.1. It was built between 1904 and 1934, and experienced large settlements even before it was completed. By 1950, the palace and the immediately surrounding grounds were about 3 m lower than the adjacent streets (Thornley et al., 1955). As a result, it has been necessary to build stairways from the street down to the building area.

As a result of these problems, geotechnical engineers in Mexico City have developed techniques for safely supporting large structures without the detrimental effects of excessive settlement. One of these, the 43-story Tower Latino Americana is discussed in *Foundation Design: Principles and Practices* (Coduto, 1999), the companion volume to this book. This building is across the street from the Palace of Fine Arts, and has been performing successfully since its completion in the mid-1950s.

The Tower of Pisa in Italy is another example of excessive settlement. In this case, one side has settled more than the other, a behavior we call *differential settlement,* which gives the tower its famous tilt. *Foundation Design: Principles and Practices* also explores this case study.

Settlement problems are not limited to buildings. For example, the highway bridge shown in Figure 11.2 is underlain by a soft clay deposit. This soil is not able to support the weight of the bridge, so pile foundations were installed through the clay into harder soils below and the bridge was built on the piles. These foundations protect it from large settlements.

It also was necessary to place fill adjacent to the bridge abutments so the roadway could reach the bridge deck. These fills are very heavy, so their weight increased σ_z' in the clay, causing it to settle. When this photograph was taken, about twelve years after the bridge was built, the fill had settled about 1 m, as shown by the sidewalk in the foreground.

Figure 11.2 The approach fills adjacent to this bridge in California have settled. However, the bridge, being supported on pile foundations, has not. Note the abrupt change in grade in the sidewalk, and the asphalt patch between the two signs. This photograph was taken about twelve years after the bridge was built.

11.1 PHYSICAL PROCESSES

The three most common physical processes that produce settlement in soils are:

- *Consolidation settlement* (also known as *primary consolidation settlement*), δ_c, occurs when a soil is subjected to an increase in σ_z', as discussed in the previous section, and the individual particles respond by rearranging into a tighter packing.

This process causes a decrease in the volume of the voids, V_v. If the soil is saturated ($S = 100\%$), this reduction in V_v can occur only if some of the pore water is squeezed out of the soil. All soils experience some consolidation when they are subjected to an increase in σ_z', and this is usually the most important source of settlement.

- *Secondary compression settlement*, δ_s, is due to particle reorientation, creep, and decomposition of organic materials, and does not require the expulsion of pore water. Secondary compression can be significant in highly plastic clays, organic soils, and sanitary landfills, but it is negligible in sands and gravels. Unlike consolidation settlement, secondary compression settlement is not due to changes in σ_z'.
- *Distortion settlement*, δ_d, results from lateral movements of the soil in response to changes in σ_z'. These movements occur when the load is confined to a small area, such as a structural foundation, or near the edges of large loaded areas, such as embankments.

The settlement at the ground surface, δ, is the sum of these three components:

$$\delta = \delta_c + \delta_s + \delta_d \qquad (11.1)$$

Other sources of settlement, such as that from underground mines, sinkholes, or tunnels, also can be important, but they are beyond the scope of our discussion.

11.2 CHANGES IN VERTICAL EFFECTIVE STRESS

Most settlement is due to changes in the vertical effective stress, so we will begin by examining these changes. The *initial vertical effective stress*, σ_{z0}', at a point in the soil is the value of σ_z' before the event that causes settlement occurs. The *final vertical effective stress*, σ_{zf}', is the value after the event has occurred and the settlement process is complete. Notice how settlement analyses are based on changes in effective stress, not total stress.

The value of σ_{z0}' may be computed using the techniques described in Chapter 10. Usually the initial condition consists of geostatic stresses only, and thus is evaluated using Equation 10.34.

The method of computing of σ_{zf}' depends on the kind of event that is causing the stresses to increase. The most common events are: placement of a fill, placement of an external load, and changes in the groundwater table elevation.

Stress Changes Due to Placement of a Fill

When a fill is placed on the ground, σ_z' in the underlying soil increases due to the weight of the fill. If the length and width of the fill are large compared to the depth of the point at which we wish to compute the stresses, and the point is beneath the central area of the fill, then we compute σ_{zf}' by simply adding another layer to the $\sum \gamma H$ of Equation 10.34. Therefore,

$$\sigma'_{zf} = \sigma'_{z0} + \gamma_{fill} H_{fill}$$ (11.2)

where:

σ'_{z0} = initial vertical effective stress
σ'_{zf} = final vertical effective stress
γ_{fill} = unit weight of the fill
H_{fill} = thickness of the fill

Unless stated otherwise, you may assume all of the fills described satisfy these criteria and that Equation 11.2 is valid.

If the width or length of the fill are less than about twice the depth to the point at which the stresses are to be computed, or if this point is near the edge of the fill, then we need to evaluate the fill as an area load using the techniques described in Chapter 10.

Stress Changes Due to Placement of an External Load

External loads, such as structural foundations, also produce increases in σ'_z. In this case, σ'_{zf} is:

$$\sigma'_{zf} = \sigma'_{z0} + (\sigma_z)_{induced}$$ (11.3)

Where $(\sigma_z)_{induced}$ is the induced vertical stress computed using the techniques described in Chapter 10. This computation may be performed by hand using the equations in Section 10.5, or with a computer by using program **STRESSP**, **STRESSL**, **STRESSR**, or **STRESSC**.

Stress Changes Due to Changes in the Groundwater Table Elevation

Sometimes natural events or construction activities produce changes in the groundwater table elevation. For example, pumping from wells causes a drop in the nearby groundwater table, as discussed in Chapter 8. When the groundwater table changes from one elevation to some lower elevation, the pore water pressure, u, in the underlying soils decreases and the vertical effective stress, σ'_z, increases. This is a more subtle process because there is no visible source of loading at the ground surface, yet it can be and has been the cause of significant settlements. For example, some of the settlement problems in Mexico City have been due to drops in the groundwater table because of excessive pumping from water supply wells.

In this case, it is usually easiest to compute σ_{zf}' using Equation 10.34 with the final groundwater position. When performing this computation, keep in mind that changes in the groundwater table elevation also may be accompanied by changes in the unit weight, γ. Soil that is now above the groundwater table will probably have a lower moisture content and therefore a lower unit weight than before. Thus, the zone of soil between the initial and final groundwater tables may have one unit weight for the σ_{z0}' computation, and another for the σ_{zf}' computation.

Stress Changes Due to Multiple Simultaneous Causes

Some civil engineering projects include multiple causes of settlement, each acting simultaneously. For example, a project might include both placement of a fill and construction of multiple structural foundations. In such cases, it may not be immediately clear how to compute σ_{zf}'. Whenever this kind of confusion arises, keep in mind that it is always possible to compute σ_{zf}' using Equation 10.35 with the post-construction condition.

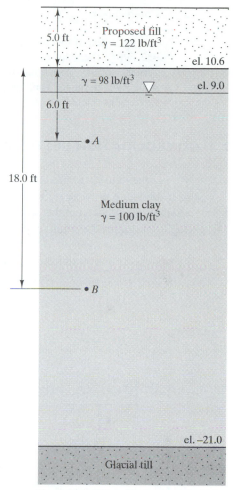

Figure 11.3 Soil profile for Example 11.1.

Example 11.1

A 5.0 ft thick fill is to be placed on a site underlain by medium clay, as shown in Figure 11.3. Compute σ_{z0}' and σ_{zf}' at Point A.

Solution

$$\sigma_{z0}' = \sum \gamma H - u$$
$$= (98 \text{ lb/ft}^3)(1.6 \text{ ft}) + (100 \text{ lb/ft}^3)(4.4 \text{ ft}) - (62.4 \text{ lb/ft}^3)(4.4 \text{ ft})$$
$$= \textbf{322 lb/ft}^2 \qquad \leftarrow Answer$$

$$\sigma_{zf}' = \sigma_{z0}' + \gamma_{fill} H_{fill}$$
$$= 322 \text{ lb/ft}^2 + (122 \text{ lb/ft}^3)(5.0 \text{ ft})$$
$$= \textbf{932 lb/ft}^2 \qquad \leftarrow Answer$$

Commentary

The placement of this fill will eventually cause σ_z' at Point A to increase from 322 lb/ft^2 to 932 lb/ft^2. The value of σ_z' at other depths in the natural soil also will increase, causing a vertical strain ϵ_z. As a result, the top of the natural ground will sink from elevation 10.6 ft to some lower elevation. Thus, the placement of a 5.0 ft thick fill will ultimately produce a ground surface that is less than 5.0 ft higher than the initial ground surface elevation.

11.3 CONSOLIDATION SETTLEMENT — PHYSICAL PROCESSES

We use the term *consolidation* to describe the pressing of soil particles into a tighter packing in response to an increase in effective stress, as shown in Figure 11.4. We assume the volume of solids remains constant (i.e., the compression of individual particles is negligible); only the volume of the voids changes. The resulting settlement is known as *consolidation settlement*, δ_c. This is the most important source of settlement in soils, and its analysis is one of the cornerstones of geotechnical engineering.

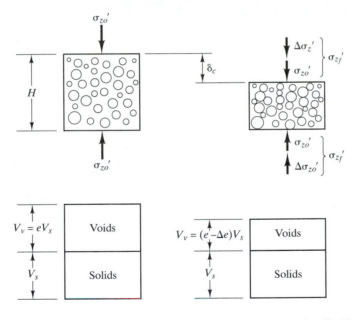

Figure 11.4 Consolidation of solid particles under the influence of an increasing vertical effective stress.

Consolidation analyses usually focus on saturated soils ($S = 100\%$), which means the voids are completely filled with water. Both the water and the solids are virtually incompressible, so consolidation can occur only as some of the water is squeezed out of the voids. We can demonstrate this process by taking a saturated kitchen sponge and squeezing it; the sponge compresses, but only as the water is pushed out. This relationship between consolidation and pore water flow was qualitatively recognized as early as 1809 when the British engineer Thomas Telford placed a 17 m deep surcharge fill over a soft clay "for the purpose of squeezing out the water and consolidating the mud" (Telford, 1830; Skempton, 1960). The American engineer William SooySmith also recognized that "slow progressive settlements result from the squeezing out of the water from the earth" (SooySmith, 1892).

The first laboratory soil consolidation tests appear to have been performed around 1910 in France by J. Frontard. He placed samples of clay in a metal container, applied a series of loads with a piston, and monitored the resulting settlement (Frontard, 1914). Although these tests provided some insight, the underlying processes were not yet understood. About the same time, the German engineer Forchheimer developed a crude mathematical model of consolidation (Forchheimer, 1914), but it was not very accurate and did not recognize important aspects of the problem.

Karl Terzaghi, who was one of Forchheimer's former students, made the major breakthrough. He was teaching at a college in Istanbul, and began studying the soil consolidation problem. This work, which he conducted between 1919 and 1923, produced the first clear recognition of the principle of effective stress, which led the way to understanding the consolidation process. Terzaghi's *theory of consolidation* (Terzaghi, 1921, 1923a, 1923b, 1924, 1925a, and 1925b) is now recognized as one of the major milestones of geotechnical engineering. Although this theory includes several simplifications, it has been verified and is considered to be a good representation of the field processes. We will study it in this chapter and in Chapter 12.

Piston and Spring Analogy

To understand the physical process of consolidation and its relationship to the flow of pore water, let us consider the mechanical piston and spring analogy shown in Figure 11.5a. This device consists of a piston and spring located inside a cylinder. The cylinder is filled with water and small drain holes are present in the piston. All of this represents an element of soil at some depth in the ground, with the spring representing the soil solids, the water representing the pore water, and the holes representing the soil voids through which the pore water must flow.

We will begin with the piston in static equilibrium under a certain vertical load, P, as shown in Figure 11.5a. The assembly is submerged in a tank, so the water is subjected to a hydrostatic pressure that represents the hydrostatic pore water pressure, u_h, in the soil (see Equation 7.7). In addition, the water pressures on the top and bottom are equal, so the applied load on the piston is carried entirely by the spring. This load divided by the cross-sectional area of the cylinder represents the initial vertical effective stress, σ_{z0}'.

Then, at time $= t_0$, we apply an additional load ΔP to the piston, as shown in Figure 11.5b. This represents the additional total vertical stress $\Delta\sigma_z$ in a soil, such as that induced by a new fill. It causes a very small downward movement of the piston, but this movement is resisted by both the spring and the water. The water is much stiffer than the spring, so it carries virtually all of this additional load and the water pressure increases. This additional pressure is known as *excess pore water pressure*, u_e. Thus, the water pressure, u, inside the cylinder now equals $u_h + u_e$.

The water pressure (and the total head) inside the cylinder is now greater than that outside, so some of the water begins to flow through the holes. These holes are very small, so the flow rate through them also is small, but eventually a certain quantity of water passes through. This allows the piston to move farther down, thus compressing the spring and relieving some of the load from the water. This process represents the gradual transfer of stress from the pore water to the soil solids. Note the relationship between compression of the spring and dissipation of the excess pore water. Understanding this relationship is a key part of this problem.

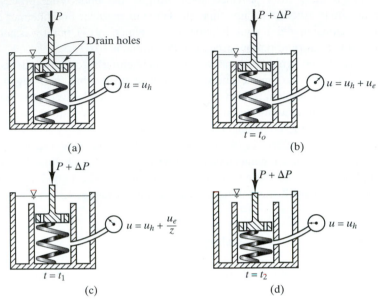

Figure 11.5 Piston and spring analogy.

At time = t_1, as shown in Figure 11.5c, half of $\Delta\sigma_z$ has been transferred to the soil solids and half is still being carried by excess pore water pressure. The process continues until the spring has compressed sufficiently to accommodate the original effective stress plus the additional stress as shown in Figure 11.5d (time = t_2). The excess pore water pressure is now zero, so flow through the holes ceases. We have returned to static equilibrium, but the piston is in a lower position than before. This change in position represents the vertical strain in that element of soil in the field.

Processes in the Field

The initial buildup of excess pore water pressures in soils is more complex than the piston and spring analogy because it depends on changes in both the vertical and horizontal total stresses, $\Delta\sigma_z$ and $\Delta\sigma_x$, and on certain empirical coefficients known as Skempton's pore pressure parameters A and B (Skempton, 1954). However, we will simplify the problem by assuming the excess pore water pressure, u_e, immediately after loading is equal to $\Delta\sigma_z$.

This increase in pore water pressure produces a hydraulic gradient in the soil, causing some of the pore water to flow away. As each increment of water is discharged, the solid particles consolidate and begin to carry part of the new load, just as the spring compressed in our analogy. Thus, $\Delta\sigma_z$ is gradually transferred from the pore water to the soil solids, and the vertical effective stress, $\Delta\sigma_z'$ rises. Eventually, all of the new load is carried by the solids, the pore water pressure returns to its hydrostatic value, and the flow of pore water ceases.

This transfer of load from water to solids is one of the most important processes in geotechnical engineering.

Example 11.2

The element of soil at point A in Figure 11.6 is initially subjected to the following stresses:

$$\sigma_{z0} = \sum \gamma H$$
$$= (18.7 \text{ kN/m}^3)(1.0 \text{ m}) + (19.0 \text{ kN/m}^3)(2.0 \text{ m}) + (16.5 \text{ kN/m}^3)(4.8 \text{ m})$$
$$= 136 \text{ kPa}$$

$$u = \gamma_w z_w$$
$$= (9.8 \text{ kN/m}^3)(6.8 \text{ m})$$
$$= 67 \text{ kPa}$$

$$\sigma_{z0}' = \sigma_{z0} - u$$
$$= 136 \text{ kPa} - 67 \text{ kPa}$$
$$= 69 \text{ kPa}$$

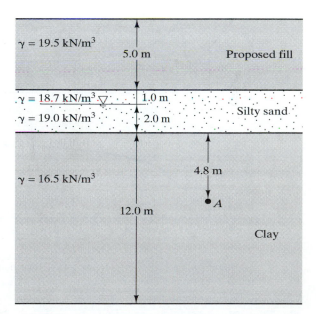

Figure 11.6 Soil profile for Example 11.2.

These conditions are illustrated on the left side of the plots in Figure 11.7. Then, we place a 5.00 m deep fill that has a unit weight of 19.5 kN/m³. This increases the vertical total stress to:

$$\sigma_{zf} = \sigma_{z0} + \gamma_{fill} H_{fill} = 136 \text{ kPa} + (19.5 \text{ kN/m}^3)(5.0 \text{ m}) = 234 \text{ kPa}$$

Notice the jumps in these curves in Figure 11.7. Initially, the applied load is carried entirely by the pore water, so the pore water pressure becomes:

$$u = u_h + u_e = u_h + \gamma_{fill} H_{fill} = 67 \text{ kPa} + (19.5 \text{ kN/m}^3)(5.0 \text{ m}) = 165 \text{ kPa}$$

but the vertical effective stress remains unchanged at:

$$\sigma_z' = \sigma_z - u$$
$$= 234 \text{ kPa} - 165 \text{ kPa}$$
$$= 69 \text{ kPa}$$

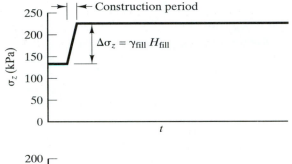

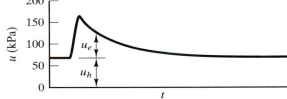

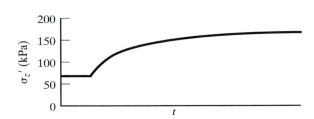

Figure 11.7 Stresses and pore water pressure at Point A in Example 11.2. σ_z steadily increases during the construction period, then remains constant. Initially it causes an equal increase in u, but as the excess pore water drains, the load is gradually transferred to the soil solids, causing σ_z' to increase.

As some of the pore water drains away, this element consolidates and $\Delta\sigma_z$ is gradually transferred from the pore water to the solids. After a sufficiently long time, $u_e = 0$ and the consolidation is complete. Then:

$$u = u_h + u_e = 67 \text{ kPa} + 0 \text{ kPa} = 67 \text{ kPa}$$

$$\sigma'_z = \sigma_z - u$$
$$= 234 \text{ kPa} - 67 \text{ kPa}$$
$$= 167 \text{ kPa}$$

Commentary

This example illustrates the process of consolidation and how it is intimately tied to the build-up and decay of excess pore water pressures. It also illustrates why this process could not be properly understood until Terzaghi developed the principle of effective stress.

QUESTIONS AND PRACTICE PROBLEMS

11.1 The current σ'_z at a certain point in a saturated clay is 181 kPa. This soil is to be covered with a 2.5 m thick fill that will have a unit weight of 19.3 kN/m^3. What will be the value of σ'_z at this point immediately after the fill is placed (i.e., before any consolidation has occurred)? What will it be after the consolidation settlement is completed?

11.2 A 4.0 m thick fill with a unit weight of 20.1 kN/m^3 is to be placed on the soil profile shown in Figure 10.7. Develop plots of σ'_{z0} and σ'_{zf} vs. depth. The plot should extend from the original ground surface to a depth of 10.0 m.

11.3 The groundwater table at a certain site was at a depth of 10 ft below the ground surface, and the vertical effective stress at a point 30 ft below the ground surface was 2200 lb/ft^2. Then a series of wells were installed, which caused the groundwater table to drop to a depth of 25 ft below the ground surface. Assuming the unit weight of the soil above and below the groundwater table are equal, compute the new σ'_z at this point.

11.4 A 1.00 m^3 element of soil is located below the groundwater table. When a new compressive load was applied, this element consolidated, producing a vertical strain, ϵ_z, of 8.5%. The horizontal strain was zero. Compute the volume of water squeezed out of this soil during consolidation and express your answer in liters.

11.4 CONSOLIDATION (OEDOMETER) TESTS

To predict consolidation settlement in a soil, we need to know its stress–strain properties (i.e, the relationship between σ'_z and ϵ_z). This normally involves bringing a soil sample to the laboratory, subjecting it to a series of loads, and measuring the corresponding settlements. This test is essentially the same as those conducted by Frontard in 1910, but now we have the benefit of understanding the physical processes, and thus can more effectively interpret the results. The test is known as a *consolidation test* (also known as an *oedometer test*), and is conducted in a *consolidometer* (or *oedometer*) as shown in Figure 11.8.

We are mostly interested in the engineering properties of natural soils as they exist in the field, so consolidation tests are usually performed on high-quality "undisturbed" samples. It is fairly simple to obtain these samples in soft to medium clays, and the test results are reliable. However, it is virtually impossible to obtain high-quality undisturbed samples in uncemented sands, so we use empirical correlations or in-situ tests instead of consolidation tests to assess the stress–strain properties as discussed in Section 11.6.

It also is important for samples that were saturated in the field to remain so during storage and testing. If the sample is allowed to dry, a process we call *desiccation*, negative pore water pressures will develop and may cause irreversible changes in the soil.

Sometimes engineers need to evaluate the consolidation characteristics of proposed compacted fills, and do so by performing consolidation tests on samples that have been remolded and compacted in the laboratory. These tests are usually less critical because well-compacted fills generally have a low compressibility.

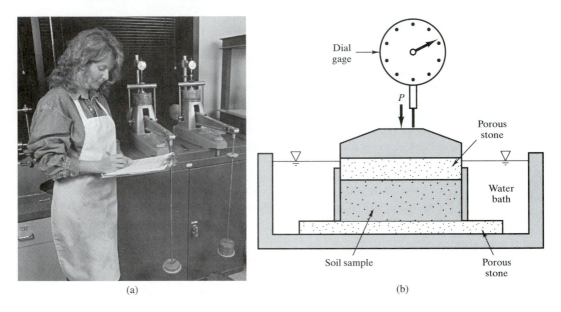

(a)   (b)

Figure 11.8 a) Performing consolidation tests in the laboratory. The two consolidometers use the weights in the foreground to load the samples; b) cross-section of a consolidometer.

Test Procedure

The soil sample, which has the shape of an upright cylinder, is placed in the consolidometer, and surrounded by a brass or stainless steel ring. The purpose of this ring is to maintain zero horizontal strain, thus producing *one-dimensional consolidation*. *Porous stones* are placed above and below the sample. These stones are manufactured products that are strong enough to carry the applied loads, yet porous enough to allow water to pass through freely.

The sample, rings, and porous stones are submerged in a water bath. This keeps the soil saturated, thus simulating the worst-case conditions in the field. A dial gage (or comparable electronic device) is placed above the sample to measure its compression as the test progresses.

The test begins by applying a vertical normal load, P. It produces a vertical effective stress of:

$$\sigma_z' = \frac{P}{A} - u \tag{11.4}$$

where:

σ_z' = vertical effective stress
P = applied load
A = cross-sectional area of soil sample
u = pore water pressure inside soil sample

The water bath barely covers the sample, so the pore water pressure is very small compared to the vertical stress and thus may be ignored. Thus:

$$\boxed{\sigma_z' = \frac{P}{A}} \tag{11.5}$$

Then the sample is allowed to consolidate. While conducting his early tests, Frontard noted "one of the most interesting facts which have been revealed is the great length of time required for the escape of the excess water." During this period some of the water is being squeezed out of the voids, and must pass through the soil to reach the porous stones. Because we are normally testing clayey soils, the hydraulic conductivity is low and the water flows slowly. Thus, several hours or more may be required for the sample to consolidate. We determine when the consolidation is complete by monitoring the dial gage. The vertical strain, ϵ_z, upon completion of consolidation is:

$$\boxed{\epsilon_z = \frac{\text{change in dial gage reading}}{\text{initial height of sample}}} \tag{11.6}$$

The strain is expressed using the sign convention defined in Chapter 10, where positive strain indicates compression. We now have one (σ_z', ϵ_z) data point.

The next step is to increase the normal load to some higher value and allow the soil to consolidate again, thus obtaining a second (σ_z', ϵ_z) data point. This process continues until we have obtained curve ABC in Figure 11.9. The stress–strain curve in soil is decidedly non-linear when shown in an arithmetic plot. However, when presented on a semilogarithmic plot as shown, the data is much easier to interpret. Finally, we incrementally unload the sample and allow it to rebound, thus producing curve CD. Curve AB is known as the *recompression curve*, BC is the *virgin curve*, and CD is the *rebound curve*.

Methods of Presenting Consolidation Test Results

Geotechnical testing laboratories use two different methods of presenting consolidation test results: a *strain plot* or a *void ratio plot*. The test results that arrive on a geotechnical engineer's desk could be presented in either or both forms, so it is important to recognize the difference, and be able to use both methods.

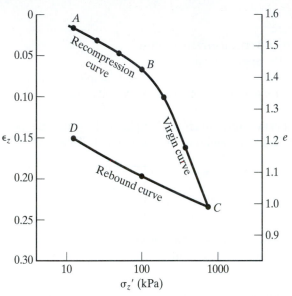

Figure 11.9 Results of laboratory consolidation test. The initial void ratio, e_0, is 1.60.

The first method uses a plot of ϵ_z vs. log σ_z', and thus is a direct representation of the data obtained in the laboratory. The horizontal and left axes of Figure 11.9 use this method. A strain plot is the most straightforward approach, because the purpose of a consolidation test is to measure the stress–strain properties of the soil.

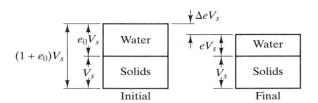

Figure 11.10 Phase diagram for derivation of Equations 11.7–11.9.

The second method presents the data as a plot of void ratio, e, vs. log σ_z' as shown in the horizontal and right axes in Figure 11.9. This was the method Terzaghi used, presumably because it emphasizes the reduction in void size that occurs during consolidation. To compute the void ratio at various stages of the test, we need to develop an equation that relates void ratio with strain. Using Figure 11.10:

$$e V_s = e_0 V_s - \Delta e V_s \tag{11.7}$$

$$\epsilon_z = \frac{\Delta e \, V_s}{(1 + e_0) V_s} \tag{11.8}$$

Combining Equations 11.7 and 11.8 gives:

$$
\begin{aligned}
e &= e_0 - \Delta e \\
&= e_0 - \epsilon_z (1 + e_0) \\
&= 1 + e_0 - \epsilon_z (1 + e_0) - 1 \\
&= (1 - \epsilon_z)(1 + e_0) - 1
\end{aligned} \tag{11.9}
$$

where:

  e = void ratio

  e_0 = initial void ratio (i.e. the void ratio at the beginning of the test)

  ϵ_z = vertical strain

  Δe = change in void ratio during test = $e_0 - e$

  V_s = volume of solids

The initial void ratio, e_0, is usually computed from the moisture content using Equation 4.26.

As Figure 11.9 illustrates, these two methods are just different ways of expressing the same data. Both methods produce the same computed settlements.

Plastic and Elastic Deformations

All materials deform when subjected to an applied load. If all of this deformation is retained when the load is released, it is said to have experienced *plastic deformation*. Conversely, if the material returns to its original size and shape when the load is released, it is said to have experienced *elastic deformation*. For example, we could illustrate plastic deformation by bending a copper wire, and elastic deformation by bending a rubber hose. The copper will retain nearly all of its deformation, while the rubber will not.

Soil exhibits both plastic and elastic deformations, which is why we see two slopes in the loading and unloading curves in Figure 11.9. To understand this behavior, let us consider soil element A in the profile shown in Figure 11.11, and consolidation data for this element, also shown in Figure 11.11.

In Figure 11.11a, the element of soil has recently been deposited and is described by Point 1 on the plot. The effective stress is low and the void ratio is high. Then, additional deposition occurs and the element becomes progressively buried by the newly deposited soil. The effective stress increases and the void ratio decreases (i.e., consolidation occurs). At this stage, both plastic and elastic deformations are occurring. Point 2 in Figure 11.11b describes the conditions that existed just prior to our drilling and sampling effort.

Then, we drill an exploratory boring, obtain an undisturbed sample from Element A, and take it to the laboratory. This process removes the overburden stress, so the effective stress drops and the sample expands slightly (i.e., the void ratio increases) as described by Point 3 in Figure 11.11c. This expansion reflects the elastic portion of the compression that occurred naturally in the field. Although there has been some elastic rebound, most of the compression was plastic, so the unloading Curve 2-3 is much flatter than the loading Curve 1-2.

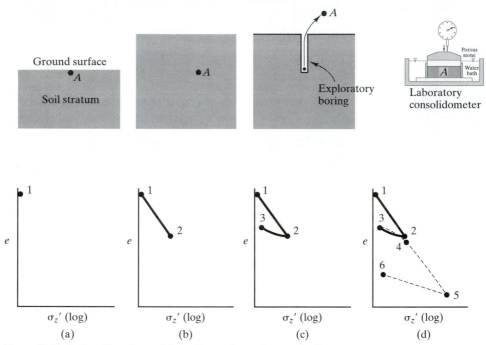

Figure 11.11 Soil profile and consolidation history for an element of soil.

Once the sample has been installed in the laboratory consolidometer, we once again load it and produce Curve 3-4-5 as shown in Figure 11.11d. The initial part, Curve 3-4, has already been defined as the reloading curve. It is nearly parallel to Curve 1-2 and reflects elastic compression only. The effective stress is less than the maximum past effective stress, so no new plastic deformation occurs. However, when the curve reaches Point 4, its slope suddenly changes, and Curve 4-5 reflects new plastic deformations, which occur only when the effective stress is higher than ever before. This is the virgin curve defined earlier. Finally, we unload the sample in the lab and form Curve 5-6, the decompression curve. It is nearly parallel to Curves 2-3 and 3-4, and reflects the elastic component only.

Thus, soils behave one way if the vertical effective stress is less than the past maximum, and another if it is increasing beyond that maximum.

Preconsolidation Stress

The point where the slope of the consolidation curve changes (Point B in Figure 11.9 or Point 4 in Figure 11.11) is an important event in the consolidation process. The stress at this point is called the *preconsolidation stress*, σ_c'. It is the greatest vertical effective stress the soil has ever experienced. The value of σ_c' is sometimes greater than σ_{z0}' at the sample location, which means the soil was once subjected to a higher stress. We will discuss this in more detail in Section 11.5.

The preconsolidation stress obtained from the consolidation test represents only the conditions at the point where the sample was obtained. If the sample had been taken at a different elevation, the preconsolidation stress would change accordingly.

Adjustments to Laboratory Consolidation Data

Consolidation tests are very sensitive to sample disturbance. Very high-quality samples produce distinct consolidation curves as shown in Figure 11.12. However, less than ideal sampling and handling techniques, drying during storage, and other effects can alter the sample and make the test results more obscure and difficult to interpret. It is especially difficult to obtain σ_c' from poor-quality samples because the transition between the recompression and virgin curves becomes much more rounded. Thus, it is best to be very careful with samples intended for consolidation tests.

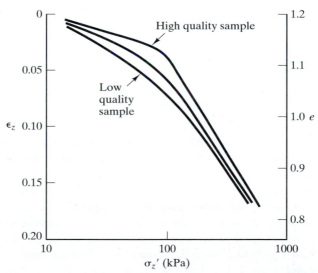

Figure 11.12 Effect of sample disturbance on consolidation test results.

Casagrande (1936) and Schmertmann (1955) developed methods of adjusting laboratory consolidation test results in an attempt to compensate for nominal sample disturbance effects. Both methods were developed primarily for soft clays, and often are more difficult to implement in stiffer soils.

The Casagrande procedure determines the preconsolidation stress, σ_c', from laboratory data. Implement this method as follows, and as illustrated in Figure 11.13:

1. Locate the point of minimum radius on the consolidation curve (Point A).
2. Draw a horizontal line from Point A.
3. Draw a line tangent to the laboratory curve at Point A.
4. Bisect the angle formed by the lines from Steps 2 and 3.
5. Extend the straight portion of the virgin curve upward until it intersects the line formed in Step 4. This identifies Point B, which is the preconsolidation stress, σ_c'.

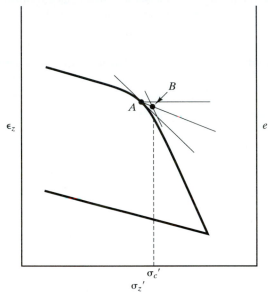

Figure 11.13 Casagrande's method of finding the preconsolidation stress.

Sample disturbance also affects the slope of the curves, so the Schmertmann procedure is an attempt to reconstruct the field consolidation curve (as illustrated in Figure 11.14). This procedure is performed as follows:

1. Determine σ_c' using the Casagrande procedure.
2. Compute the initial vertical effective stress, σ_{z0}', at the sample depth. This is the vertical effective stress prior to placement of the proposed load.
3. Draw a horizontal line at $e = e_0$ (or $\epsilon_z = 0$) from the vertical axis to σ_{z0}'. This locates Point C.
4. Beginning at Point C, draw a line parallel to the rebound curve. Continue to the right until reaching σ_c'. This forms Point D. In some cases, $\sigma_c' \approx \sigma_{z0}'$, so this step becomes unnecessary.
5. Extend the virgin curve downward to $e = 0.42\, e_0$ thus locating Point E. If no void ratio data is included on the consolidation plot, locate Point E at $\epsilon_z = 0.42$, which is

the same as $e = 0.42\, e_0$ when $e_0 = 2$ and sufficiently close for other initial void ratios (i.e., locating Point E more precisely has very little impact on the results of Schmertman's construction).

6. Draw a line connecting points D and E. This is the reconstructed virgin curve.

The final result of the Casagrande and Schmertman constructions is a bilinear function when plotted on a semilogarithmic diagram.

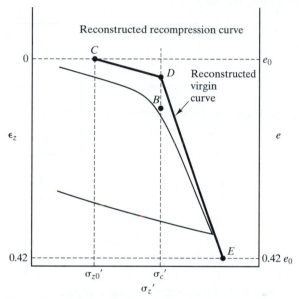

Figure 11.14 Schmertmann's method of adjusting consolidation test results. If void ratio data is available, then locate Point E at a void ratio of $0.42\, e_0$. If only strain data is available (i.e., no void ratios are given), then locate Point E at $\epsilon_z = 0.42$. If both void ratio and strain data are given, use the void ratio data to locate Point E, even though this point may not correspond to $\epsilon_z = 0.42$.

Soil Compressibility

The slopes on the consolidation plot reflect the *compressibility* of the soil. Steep slopes mean a given increase in σ_z' will cause a large strain (or a large change in void ratio), so such soils are said to be *highly compressible*. Conversely, shallow slopes indicate the same increase in σ_z' will produce less strain, so the soil is *slightly compressible*. Although we could use graphical constructions on these plots to determine the strain that corresponds to a certain increase in effective stress, it is much easier to do so mathematically, as follows:

The slope of the virgin curve is defined as the *compression index*, C_c:

$$C_c = -\frac{de}{d \log \sigma_z'}$$

(11.10)

There is a potential point of confusion here, because geotechnical engineers also use the variable C_c to represent the coefficient of curvature, as defined in Equation 4.36. However, these are two entirely separate parameters. The compression index is a measure of the compressibility, while the coefficient of curvature describes the shape of the grain-size distribution curve.

The reconstructed virgin curve is a straight line (on a semilogarithmic e vs. log σ_z' plot), so we can obtain a numerical value for C_c by selecting any two points, a and b, on this line and rewriting Equation 11.10 as:

$$C_c = \frac{e_a - e_b}{(\log \sigma_z')_b - (\log \sigma_z')_a} \tag{11.11}$$

Alternatively, if the data is plotted only in ϵ_z - σ_z' form (i.e., no void ratio data is given), then the slope of the virgin curve is:

$$\frac{C_c}{1 + e_0} = \frac{(\epsilon_z)_b - (\epsilon_z)_a}{(\log \sigma_z')_b - (\log \sigma_z')_a} \tag{11.12}$$

where the parameter $C_c/(1+e_0)$ is called the *compression ratio*.

If the reconstructed virgin curve is sufficiently long, it is convenient to select Points a and b such that $\log (\sigma_z')_b = 10 \log (\sigma_z')_a$. This makes the denominator of Equations 11.11 and 11.12 equal to 1, which simplifies the computation. This also demonstrates that C_c could be defined as the reduction in void ratio per tenfold increase (one log-cycle) in effective stress, as shown in Figure 11.15. Likewise, $C_c/(1+e_0)$ is the strain per tenfold increase in effective stress.

In theory, the recompression and rebound curves have nearly equal slopes, but the rebound curve is more reliable because it is less sensitive to sample disturbance effects. This slope, which we call the *recompression index*, C_r, is defined in the same way as C_c and can be found using Equation 11.13 with Points c and d on the decompression curve:

$$C_r = \frac{e_c - e_d}{(\log \sigma_z')_d - (\log \sigma_z')_c} \tag{11.13}$$

If the data is plotted on a strain diagram, then the slope is $C_r/(1+e_0)$, which is the *recompression ratio*:

$$\frac{C_r}{1 + e_0} = \frac{(\epsilon_z)_d - (\epsilon_z)_c}{(\log \sigma_z')_d - (\log \sigma_z')_c} \tag{11.14}$$

Kulhawy and Mayne (1990) compared C_c and C_r values obtained from laboratory consolidation tests and from a theoretical soil called "modified cam clay" with the plasticity index, I_p, and found the following empirical correlations:

$$C_c = \frac{I_p}{74} \qquad (11.15)$$

$$C_r = \frac{I_p}{370} \qquad (11.16)$$

Most soils probably have C_c and C_r values within about ±50 percent of those predicted by Equations 11.15 and 11.16. These equations are useful for checking the reasonableness of laboratory test results and for performing preliminary analyses. However, final designs normally require actual laboratory tests on samples from the project site.

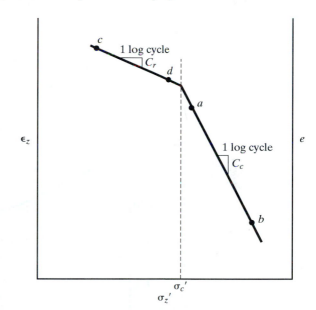

Figure 11.15 The slopes of consolidation curves on a semilogarithmic e vs. σ_z' plot are C_c and C_r. The break in slope occurs at the preconsolidation stress, σ_c'.

C_r values in saturated clays from conventional consolidation tests are typically about twice the true C_r in the field (Fox, 1995). This difference is due to the expansion of air bubbles in the pore water when the soil is unloaded during sampling and storage. This error is acceptable for most projects because the laboratory C_r is low enough that it does not produce large computed settlements (most consolidation settlement is due to C_c) and because it is conservative. However, when more precise measurements of C_r are needed, the consolidation tests can be performed in a special *backpressure consolidometer* that overcomes this problem.

Example 11.3

A consolidation test has been performed on a sample of soil obtained from Point A in Figure 11.6. The test results are shown in Figure 11.9. Compute σ_c' using Casagrande's method, then adjust the test results using Schmertmann's method. Finally, compute C_c and C_r

Solution

Stresses at sample depth:

From Example 11.2:

$$\sigma_{z0}' = 69 \text{ kPa}$$

From the Casagrande construction (Figure 11.16):

$$\sigma_c' = \textbf{140 kPa} \qquad \Leftarrow \textbf{\textit{Answer}}$$

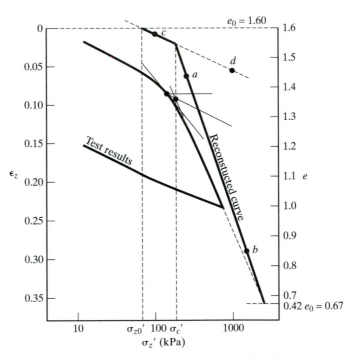

Figure 11.16 Adjusted consolidation data from Example 11.3.

Slopes of the reconstructed lines (Figure 11.16):

$$C_c = \frac{e_a - e_b}{(\log \sigma'_z)_b - (\log \sigma'_z)_a}$$

$$= \frac{1.43 - 0.84}{\log 1580 \text{ lb/ft}^2 - \log 250 \text{ lb/ft}^2}$$

$$= \mathbf{0.74} \qquad \Leftarrow Answer$$

$$C_r = \frac{e_c - e_d}{(\log \sigma'_z)_d - (\log \sigma'_z)_c}$$

$$= \frac{1.58 - 1.46}{\log 1000 \text{ lb/ft}^2 - \log 100 \text{ lb/ft}^2}$$

$$= \frac{0.12}{1}$$

$$= \mathbf{0.12} \qquad \Leftarrow Answer$$

QUESTIONS AND PRACTICE PROBLEMS

11.5 A consolidation test is being performed on a 3.50 in diameter saturated soil sample that had an initial height of 0.750 in and an initial moisture content of 38.8%.
   a. Using $G_s = 2.69$, compute the initial void ratio, e_0.
   b. At a certain stage of the test, the normal load P was 300 lb. After the consolidation at this load was completed, the sample height was 0.690 in. Compute σ'_z (expressed in lb/ft^2), ϵ_z, and e.

11.6 A consolidation test on a sample of clay produced the following data:

σ'_z (kPa)	8	16	32	64	128	256	512	1024	16
ϵ_z	0.032	0.041	0.051	0.069	0.109	0.173	0.240	0.301	0.220

The initial void ratio was 1.21, and σ'_{z0} at the sample depth was 40 kPa.

   a. Plot this data on a semilogarithmic diagram similar to that in Figure 11.9.
   b. Using Casagrande's method, find σ'_c.
   c. Using Schmertmann's method, adjust the test results.
   d. Determine C_c and C_r.
   e. The soil has a plasticity index of 23. Based on Kulhawy and Mayne's correlations, do the consolidation test results seem reasonable?

11.5 CONSOLIDATION STATUS IN THE FIELD

Normally Consolidated, Overconsolidated, and Underconsolidated Soils

When performing consolidation analyses, we need to compare the preconsolidation stress, σ'_c, with the initial vertical effective stress, σ'_{z0}. The former is determined from laboratory test data, as described earlier, while the latter is determined using Equation 10.34 with the original field conditions (i.e., without the new load) and the original hydrostatic pore water pressures (i.e., Equation 7.7). Both values must be determined at the same depth, which normally is the depth of the sample on which the consolidation test was performed. Once these values have been determined, we need to assess which of the following three conditions exist in the field:

- If $\sigma_{z0}' \approx \sigma_c'$, then the vertical effective stress in the field has never been higher than its current magnitude. This condition is known as being *normally consolidated* (NC). For example, this might be the case at the bottom of a lake, where sediments brought in by a river have slowly accumulated. In theory these two values should be exactly equal. However, in the "real world" both are subject to error due to sample disturbance and other factors, so the values obtained from our site characterization program will rarely be exactly equal, even if the soil is truly normally consolidated. Therefore, in order to avoid misclassifying the soil, we will consider it to be normally consolidated if σ_{z0}' and σ_c' are equal within about ± 20 percent.

- If $\sigma_{z0}' < \sigma_c'$, then the vertical effective stress in the field was once higher than its current magnitude. This condition is known as being *overconsolidated* (OC) or *preconsolidated*. There are many processes that can cause a soil to become overconsolidated, including (Brumund, et al., 1976):

    - Extensive erosion or excavation such that the ground surface elevation is now much lower than it once was.
    - Surcharge loading from a glacier, which has since melted.
    - Surcharge loading from a structure, such as a storage tank, which has since been removed.
    - Increases in the pore water pressure, such as from a rising groundwater table.
    - Desiccation (drying) due to evaporation, plant roots, and other processes, which produces negative pore water pressures in the soil (Stark and Duncan, 1991).
    - Chemical changes in the soil, such as the accumulation of cementing agents.
    - Aging effects.

The term *overconsolidated* can be misleading because it implies there has been excessive consolidation. Although there are a few situations, such as cut slopes, where heavily overconsolidated soils are less desirable, overconsolidation is almost always a good thing.

- If $\sigma_{z0}' > \sigma_c'$, the soil is said to be *underconsolidated*, which means the soil is still in the process of consolidating under a previously applied load. We will not be dealing with this case.

Table 11.1 gives a classification of soil compressibility based on $C_c/(1+e_0)$ for normally consolidated soils or $C_r/(1+e_0)$ for overconsolidated soils.

TABLE 11.1 CLASSIFICATION OF SOIL COMPRESSIBILITY

$\dfrac{C_c}{1+e_0}$ or $\dfrac{C_r}{1+e_0}$	Classification
0–0.05	Very slightly compressible
0.05–0.10	Slightly compressible
0.10–0.20	Moderately compressible
0.20–0.35	Highly compressible
> 0.35	Very highly compressible

For soils that are normally consolidated, base the classification on $C_c/(1+e_0)$. For soils that are overconsolidated, base it on $C_r/(1+e_0)$.

Example 11.4

Using the consolidation test results developed in Example 11.3, determine whether the soil at point A in Figure 11.6 is normally consolidated or overconsolidated. The proposed fill has not yet been placed.

Solution

At sample depth:

$$\sigma_{z0}' = 69 \text{ kPa} \quad \text{Per Example 11.2}$$

$$\sigma_c' = 140 \text{ kPa} \quad \text{Per Example 11.3}$$

$\sigma_{z0}' < \sigma_c'$ by more than 20 percent, so **the soil is overconsolidated** $\Leftarrow$ *Answer*

Overconsolidation Margin and Overconsolidation Ratio

The σ_c' values from the laboratory only represent the preconsolidation stress at the sample depth. However, we sometimes need to compute σ_c' at other depths (i.e., in a soil strata with the same geologic origin). To do so, compute the *overconsolidation margin*, σ_m', using σ_{z0}' at the sample depth and the following equation:

$$\boxed{\sigma_m' = \sigma_c' - \sigma_{z0}'} \tag{11.17}$$

Table 11.2 presents typical values of σ_m'.

The overconsolidation margin should be approximately constant throughout in a stratum with common geologic origins. Therefore, we can compute the preconsolidation stress at other depths in that stratum by using Equation 11.17 with σ_{z0}' at the desired depth. Another useful parameter is the *overconsolidation ratio* or *OCR*:

$$OCR = \frac{\sigma_c'}{\sigma_{z0}'}$$

(11.18)

Unlike the overconsolidation margin, the OCR varies as a function of depth, and therefore cannot be used to compute σ_c' at other depths in a stratum. For normally consolidated soils, OCR = 1.

TABLE 11.2 TYPICAL RANGES OF OVERCONSOLIDATION MARGINS

Overconsolidation Margin, σ_m'		Classification
(kPa)	(lb/ft^2)	
0	0	Normally consolidated
0 - 100	0 - 2000	Slightly overconsolidated
100 - 400	2000 - 8000	Moderately overconsolidated
> 400	> 8000	Heavily overconsolidated

11.6 COMPRESSIBILITY OF SANDS AND GRAVELS

The principles of consolidation apply to all soils, but the consolidation test described in Section 11.4 and the methods of assessing consolidation status in the field, as described in Section 11.5, are primarily applicable to clays and silts. It is very difficult to perform reliable consolidation tests on most sands because they are more prone to sample disturbance, and this disturbance has a significant effect on the test results. Clean sands are especially troublesome. Gravels have similar sample disturbance problems, plus their large grain size would require very large samples and a very large consolidometer.

Fortunately, sands and gravels subjected to static loads are much less compressible than silts and clays, so it often is sufficient to use estimated values of C_c or C_r in lieu of laboratory tests. For sands, these estimates can be based on the data gathered by Burmister (1962) as interpreted in Table 11.3. He performed a series of consolidation tests on samples reconstituted to various relative densities. Engineers can estimate the in-situ relative density using the methods described in Chapter 4, then select an appropriate $C_c/(1+e_0)$ from this table. Note that all of these values are "very slightly compressible" as defined in Table 11.1.

TABLE 11.3 TYPICAL CONSOLIDATION PROPERTIES OF SATURATED NORMALLY CONSOLIDATED SANDY SOILS AT VARIOUS RELATIVE DENSITIES (Adapted from Burmister, 1962)

Soil Type	$C_c / (1+e_0)$					
	$D_r = 0\%$	$D_r = 20\%$	$D_r = 40\%$	$D_r = 60\%$	$D_r = 80\%$	$D_r = 100\%$
Medium to coarse sand, some fine gravel (SW)	-	-	0.005	-	-	-
Medium to coarse sand (SW/SP)	0.010	0.008	0.006	0.005	0.003	0.002
Fine to coarse sand (SW)	0.011	0.009	0.007	0.005	0.003	0.002
Fine to medium sand (SW/SP)	0.013	0.010	0.008	0.006	0.004	0.003
Fine sand (SP)	0.015	0.013	0.010	0.008	0.005	0.003
Fine sand with trace fine to coarse silt (SP-SM)	-	-	0.011	-	-	-
Find sand with little fine to coarse silt (SM)	0.017	0.014	0.012	0.009	0.006	0.003
Fine sand with some fine to coarse silt (SM)	-	-	0.014	-	-	-

For saturated overconsolidated sands, $C_r / (1+e_0)$ is typically about one-third of the values listed in Table 11.3, which makes such soils nearly incompressible. Compacted fills can be considered to be overconsolidated, as can soils that have clear geologic evidence of preloading, such as glacial tills. Therefore, many settlement analyses simply consider the compressibility of such soils to be zero. If it is unclear whether a soil is normally consolidated or overconsolidated, it is conservative to assume it is normally consolidated.

Very few consolidation tests have been performed on gravelly soils, but the compressibility of these soils is probably equal to or less than those for sand, as listed in Table 11.3.

Another characteristic of sands and gravels is their high hydraulic conductivity, which means any excess pore water drains very quickly. Thus, the rate of consolidation is very fast, and typically occurs nearly as fast as the load is applied. Thus, if the load is due to a fill, the consolidation of these soils may have little practical significance.

However, there are at least two cases where consolidation of coarse-grained soils can be very important and needs more careful consideration:

1. **Loose sandy soils subjected to dynamic loads, such as those from an earthquake.** They can experience very large and irregular settlements that can cause serious damage. Kramer (1996) discusses methods of evaluating this problem.

2. **Sandy or gravelly soils that support shallow foundations.** Structural foundations are often very sensitive to settlement, so we often conduct more precise assessments of compressibility. These are usually done using in-situ tests, such as the SPT or CPT, and often expressed in terms of a modulus of elasticity, E, instead of C_c or C_r. Special analyses based on in-situ test results are available to predict such settlements, as discussed in Chapter 17.

QUESTIONS AND PRACTICE PROBLEMS

11.7 A consolidation test has been performed on a soil sample obtained from Point B in Figure 11.19. The measured preconsolidation stress was 88 kPa. Determine whether the soil is normally consolidated or overconsolidated, then compute the overconsolidation margin and overconsolidation ratio at Point B.

Note: These computations are based on the initial conditions, and thus should not include the weight of the proposed fill.

11.8 A saturated, normally consolidated, 1000-year-old fine-to-medium sand has an SPT $N_{60}= 12$ at a depth where the vertical effective stress is about 1000 lb/ft^2 and $D_{50}= 0.5$ mm. Using the techniques described in Chapters 3 and 4, determine the relative density of this soil, then estimate $C_c / (1+e_0)$ based on Table 11.3.

11.9 A consolidation test has been performed on a sample obtained from Point A in Figure 11.3. The measured preconsolidation stress was 1500 lb/ft^2.
   a. Determine if the soil is normally consolidated or overconsolidated
   b. Compute the overconsolidation margin and the overconsolidation ratio
   c. Compute σ_c' at Point B

11.7 CONSOLIDATION SETTLEMENT PREDICTIONS

The purpose of performing consolidation tests is to define the stress–strain properties of the soil and thus allow us to predict consolidation settlements in the field. We perform this computation by projecting the laboratory test results (as contained in the parameters C_c, C_r, e_0, and σ_c') back to the field conditions. For simplicity, the discussions of consolidation settlement predictions in this chapter consider only the case of one-dimensional consolidation, and we will be computing only the ultimate consolidation settlement.

One-dimensional consolidation means only vertical strains occur in the soil (i.e., $\epsilon_x = \epsilon_y = 0$). We can reasonably assume this is the case when at least one of the following conditions exist (Fox, 1995):

1. The width of the loaded area is at least four times the thickness of the compressible strata.
2. The depth to the top of the compressible strata is at least twice the width of the loaded area.

or 3. The compressible strata lie between stiffer soil strata whose presence tends to reduce the magnitude of horizontal strains.

In this context, "compressible strata" refers to strata that have a C_c or C_r large enough to contribute significantly to the settlement.

The most common one-dimensional consolidation problems are those that evaluate settlement due to the placement of a long and wide fill or due to the widespread lowering of the groundwater table. Many other problems, such as foundations, also may be idealized as being one-dimensional.

The *ultimate consolidation settlement*, $(\delta_c)_{ult}$, is the value of δ_c after all of the excess pore water pressures have dissipated, which may require many years or even decades. Chapter 12 explores this topic in more detail, and presents methods of developing time-settlement curves.

Normally Consolidated Soils ($\sigma_{z0}' \approx \sigma_c'$)

If $\sigma_{z0}' \approx \sigma_c'$, the soil is, by definition, normally consolidated. Thus, the initial and final conditions are as shown in Figure 11.17, and the compressibility is defined by C_c, the slope of the virgin curve.

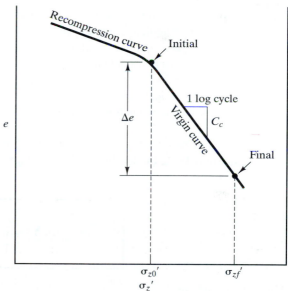

Figure 11.17 Consolidation of normally consolidated soils.

Rewriting Equation 11.10 gives:

$$de = -C_c \, d\log\sigma_z' \tag{11.19}$$

$$\Delta e = -C_c \log\left(\frac{\sigma_{zf}'}{\sigma_{z0}'}\right) \tag{11.20}$$

Combining with Equation 11.8 gives the vertical strain in the element of soil, ϵ_z:

$$\epsilon_z = -\frac{\Delta e}{1 + e_0}$$

$$= \frac{C_c}{1 + e_0} \log\left(\frac{\sigma'_{zf}}{\sigma'_{z0}}\right) \tag{11.21}$$

Integrating over the depth of the soil gives the consolidation settlement at the ground surface, δ_c:

$$(\delta_c)_{ult} = \int \epsilon_z \, dz$$

$$= \int \frac{C_c}{1 + e_0} \log\left(\frac{\sigma'_{zf}}{\sigma'_{z0}}\right) dz \tag{11.22}$$

where:

$(\delta_c)_{ult}$ = ultimate consolidation settlement at the ground surface

ϵ_z = vertical normal strain

C_c = compression index

e_0 = initial void ratio

σ'_{z0} = initial vertical effective stress

σ'_{zf} = final vertical effective stress

z = depth below the ground surface

For nearly all practical problems, geotechnical engineers evaluate the integral in Equation 11.22 by dividing the soil into n finite layers, computing δ_c for each layer, and summing:

$$(\delta_c)_{ult} = \sum \frac{C_c}{1 + e_0} H \log\left(\frac{\sigma'_{zf}}{\sigma'_{z0}}\right) \tag{11.23}$$

where:

H = thickness of the soil layer

When using Equation 11.23, compute σ'_{z0} and σ'_{zf} at the midpoints of each layer.

Overconsolidated Soils — Case I $(\sigma'_{z0} < \sigma'_{zf} \leq \sigma'_c)$

If both σ'_{z0} and σ'_{zf} do not exceed σ'_c, the entire consolidation process occurs on the recompression curve as shown in Figure 11.18. The analysis is thus identical to that for normally consolidated soils except we use the recompression index, C_r, instead of the compression index, C_c:

$$(\delta_c)_{ult} = \sum \frac{C_r}{1 + e_0} H \log \left(\frac{\sigma'_{zf}}{\sigma'_{z0}} \right)$$

(11.24)

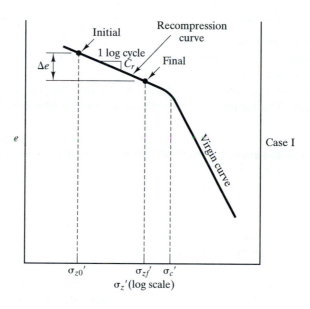

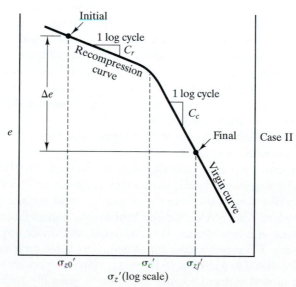

Figure 11.18 Consolidation of overconsolidated soils.

Overconsolidated Soils — Case II ($\sigma_{z0}' < \sigma_c' < \sigma_{zf}'$)

If the consolidation process begins on the recompression curve and ends on the virgin curve, as shown in Figure 11.18, then the analysis must consider both C_c and C_r:

$$(\delta_c)_{ult} = \sum \left[\frac{C_r}{1 + e_0} H \log\left(\frac{\sigma_c'}{\sigma_{z0}'}\right) + \frac{C_c}{1 + e_0} H \log\left(\frac{\sigma_{zf}'}{\sigma_c'}\right) \right] \qquad (11.25)$$

This condition is quite common, because many soils that might appear to be normally consolidated from a geologic analysis actually have a small amount of overconsolidation (Mesri, Lo, and Feng, 1994).

When using Equation 11.25, σ_{z0}', σ_c', and σ_{zf}' must be computed at the midpoint of each layer. This means σ_c' will need to be computed using Equation 11.17.

Ultimate Consolidation Settlement Analysis Procedure

Use the following procedure to compute $(\delta_c)_{ult}$:

1. Beginning at the original ground surface, divide the soil profile into strata, where each stratum consists of a single soil type with common geologic origins. For example, one stratum may consist of a dense sand, while another might be a soft-to-medium clay. Continue downward with this process until you have passed through all of the compressible strata (i.e., until you reach bedrock or some very hard soil). For each stratum, identify the unit weight, γ. Note: Boring logs usually report the dry unit weight, γ_d, and moisture content, w, but we can compute γ from this data using Equation 4.27. Also define the location of the groundwater table.
2. Each clay or silt stratum must have results from at least one consolidation test (or at least estimates of these results). Using the techniques described in Section 11.4, determine if each stratum is normally consolidated or overconsolidated, then assign values for $C_c/(1+e_0)$ and/or $C_r/(1+e_0)$. For each overconsolidated stratum, compute σ_m' using Equation 11.17 and assume it is constant throughout that stratum. For normally consolidated soils, set $\sigma_m' = 0$.
3. For each sand or gravel stratum, assign a value for $C_c/(1+e_0)$ or $C_r/(1+e_0)$ based on the information in Section 11.5.
4. For any very hard stratum, such as bedrock or glacial till, that is virtually incompressible compared to the other strata, assign $C_c = C_r = 0$.
5. Working downward from the original ground surface (i.e., do not consider any proposed fills), divide the soil profile into horizontal layers. Begin a new layer whenever a new stratum is encountered, and divide any thick strata into multiple layers. When performing computations by hand, each strata should have layers no more than 2 to 5 m (5 to 15 ft) thick. Thinner layers are especially appropriate near the ground surface, because the strain is generally larger there. Computer-based

computations can use much thinner layers throughout the entire depth, and achieve slightly more precise results.

6. Tabulate the following parameters at the midpoint of each layer:

For normally consolidated strata:

σ_{z0}'

σ_{zf}'

$C_c / (1+e_0)$

H

For overconsolidated strata:

σ_{z0}'

σ_{zf}'

$\sigma_c' = \sigma_{z0}' + \sigma_m'$

$C_c / (1+e_0)$

$C_r / (1+e_0)$

H

It is not necessary to record these parameters in incompressible strata.

Normally we compute σ_{z0}' and σ_{zf}' using the hydrostatic pore water pressures (Equation 7.7) with no significant seepage force, and this is the only case we will consider in this book. However, if preexisting excess pore water pressures or significant seepage forces are present, they should be evaluated. Sometimes this may require the installation of piezometers to obtain accurate information on the in-situ pore water pressures.

7. Using Equation 11.23, 11.24, or 11.25, compute the consolidation settlement for each layer, then sum to find $(\delta_c)_{ult}$. Note that each layer will not necessarily use the same equation. If σ_c' is only slightly greater than σ_{z0}' (perhaps less than 20 percent greater), then it may not be clear if the soil is truly overconsolidated, or if the difference is only an apparent overconsolidation due to uncertainties in assessing these two values. In such cases, it is acceptable to use either Equation 11.23 (normally consolidated) or 11.25 (overconsolidated case II).

Example 11.5

A 3.0 m deep compacted fill is to be placed over the soil profile shown in Figure 11.19. A consolidation test on a sample from point A produced the following results:

$C_c = 0.40$

$C_r = 0.08$

$e_0 = 1.10$

$\sigma_c' = 70.0$ kPa

This sample is representative of the entire soft clay stratum. Compute the ultimate consolidation settlement due to the weight of this fill.

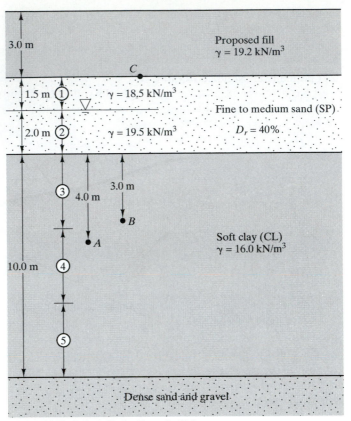

Figure 11.19 Soil profile for Example 11.5.

Solution

Using Equation 11.2:

$$\sigma'_{zf} = \sigma'_{z0} + \gamma_{fill} H_{fill}$$
$$= \sigma'_{z0} + (19.2 \text{ kN/m}^3)(3.0 \text{ m})$$
$$= \sigma'_{z0} + 57.6 \text{ kPa}$$

Compute the initial vertical stress at sample location, using Equation 10.34:

$$\sigma'_{z0} = \sum \gamma H - u$$
$$= (18.5 \text{ kN/m}^3)(1.5 \text{ m}) + (19.5 \text{ kN/m}^3)(2.0 \text{ m})$$
$$+ (16.0 \text{ kN/m}^3)(4.0 \text{ m}) - (9.8 \text{ kN/m}^3)(6.0 \text{ m})$$
$$= 72.0 \text{ kPa}$$

$$\frac{C_c}{1 + e_0} = \frac{0.40}{1 + 1.10} = 0.190$$

At the sample $\sigma_c' \approx \sigma_{z0}'$ $\rightarrow$ $\therefore$ clay is normally consolidated

If the soil at the sample depth is normally consolidated, and the sample is truly representative, then the entire stratum is normally consolidated.

Assume the sand also is normally consolidated, which is conservative. For the sand strata, use $C_c/(1+e_0) = 0.008$, per Table 11.3.

| Layer | H (m) | At midpoint of layer | | | Eqn. | $(\delta_c)_{ult}$ (mm) |
		σ_{z0}' (kPa) Eqn 10.35	σ_{zf}' (kPa)	$\dfrac{C_c}{1+e_0}$		
1	1.5	13.9	71.5	0.008	11.23	8
2	2.0	37.4	95.0	0.008	11.23	6
3	3.0	56.4	114.0	0.19	11.23	174
4	3.0	75.0	132.6	0.19	11.23	141
5	4.0	96.7	154.3	0.19	11.23	154
					$(\delta_c)_{ult} =$	483

Round off to:

$$(\delta_c)_{ult} = 480 \text{ mm} \qquad \leftarrow Answer$$

Notice how we have used the same analysis for soils above and below the groundwater table, and both are based on saturated $C_c/(1+e_0)$ values. This is conservative (although in this case, very slightly so) because the soils above the groundwater table are probably less compressible. Section 11.9 discusses unsaturated soils in more detail.

Example 11.6

An 8.5 m deep compacted fill is to be placed over the soil profile shown in Figure 11.20. Consolidation tests on samples from points A and B produced the following results:

	Sample A	Sample B
C_c	0.25	0.20
C_r	0.08	0.06
e_0	0.66	0.45
σ_c'	101 kPa	510 kPa

Compute the ultimate consolidation settlement due to the weight of this fill.

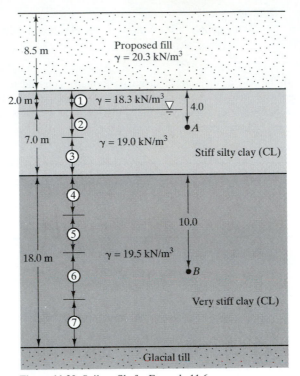

Figure 11.20 Soil profile for Example 11.6.

Solution

Using Equation 11.2:

$$\sigma'_{zf} = \sigma'_{z0} + \gamma_{fill} H_{fill}$$
$$= \sigma'_{z0} + (20.3 \text{ kN/m}^3)(8.5 \text{ m})$$
$$= \sigma'_{z0} + 172.6 \text{ kPa}$$

Applying Equation 10.34 at sample A:

$$\sigma'_{z0} = \sum \gamma H - u$$
$$= (18.3 \text{ kN/m}^3)(2.0 \text{ m}) + (19.0 \text{ kN/m}^3)(2.0 \text{ m}) - (9.8 \text{ kN/m}^3)(2.0 \text{ m})$$
$$= 55.0 \text{ kPa}$$

$$\sigma'_{zf} = \sigma'_{z0} + 172.6 \text{ kPa}$$
$$= 55.0 \text{ kPa} + 172.6 \text{ kPa}$$
$$= 227.6 \text{ kPa}$$

$$\sigma'_{z0} < \sigma'_{c} \le \sigma'_{zf} \quad \therefore \text{ overconsolidated case II}$$

$$\sigma'_m = \sigma'_c - \sigma'_z = 101 - 55 = 46 \text{ kPa}$$

Therefore, σ'_c at any depth in the stiff silty clay stratum is equal to $\sigma'_z + 46$ kPa.

At sample B:

$$\sigma'_{z0} = \sum \gamma H - u$$
$$= (18.3 \text{ kN/m}^3)(2.0 \text{ m}) + (19.0 \text{ kN/m}^3)(7.0 \text{ m})$$
$$+ (19.5 \text{ kN/m}^3)(10.0 \text{ m}) - (9.8 \text{ kN/m}^3)(17.0 \text{ m})$$
$$= 198.0 \text{ kPa}$$

$$\sigma'_{zf} = \sigma'_{z0} + \gamma_f H_f$$
$$= 198.0 \text{ kPa} + 172.6 \text{ kPa}$$
$$= 370.6 \text{ kPa}$$

$$\sigma'_{z0} < \sigma'_c \text{ and } \sigma'_{zf} \le \sigma'_c \quad \therefore \text{ overconsolidated case I}$$

Layer	H (m)	At midpoint of layer			$\frac{C_r}{1 + e_0}$	$\frac{C_c}{1 + e_0}$	Eqn.	$(\delta_c)_{ult}$ (mm)
		σ'_{z0} (kPa) Eqn. 10.34	σ'_c (kPa) Eqn. 11.17	σ'_{zf} (kPa)				
1	2.0	18.3	64.3	190.9	0.05	0.15	11.25	196
2	3.0	50.4	96.4	223.0	0.05	0.15	11.25	206
3	4.0	82.6	128.6	255.2	0.05	0.15	11.25	217
4	4.0	120.4	-	293.0	0.04	0.14	11.24	62
5	4.0	159.2	-	331.8	0.04	0.14	11.24	51
6	5.0	202.8	-	375.4	0.04	0.14	11.24	53
7	5.0	251.4	-	424.0	0.04	0.14	11.24	45
						$(\delta_c)_{ult} =$		830

$$(\delta_c)_{ult} = \textbf{830 mm} \quad \Leftarrow \textit{Answer}$$

Notice how most of the compression occurs in the upper stratum, which is overconsolidated case II (i.e., some of the compression occurs along the virgin curve). The lower stratum, which is overconsolidated case II, has much less compression even though it is twice as thick because it is overconsolidated case I and all of the compression occurs on the recompression curve.

Example 11.7

The groundwater table in the soil profile shown in Figure 11.21 is currently at the elevation labeled "initial." A proposed dewatering project will cause it to drop to the elevation labeled "final." Compute the resulting ultimate consolidation settlement.

A consolidation test performed on a sample from point A produced the following results:

$$C_c / (1 + e_0) = 0.14$$
$$C_r / (1 + e_0) = 0.06$$
$$\sigma_c' = 3000 \text{ lb/ft}^2$$

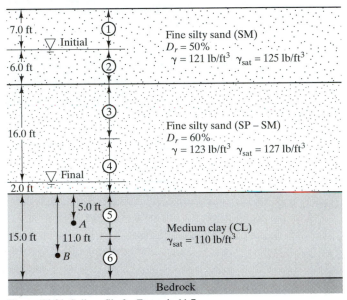

Figure 11.21 Soil profile for Example 11.7.

Solution

At sample A

$$\sigma_{z0}' = \sum \gamma H - u$$
$$= (121 \text{ lb/ft}^3)(7.0 \text{ ft}) + (125 \text{ lb/ft}^3)(6.0 \text{ ft}) + (127 \text{ lb/ft}^3)(18.0 \text{ ft})$$
$$+ (110 \text{ lb/ft}^3)(5.0 \text{ ft}) - (62.4 \text{ lb/ft}^3)(29.0 \text{ ft})$$
$$= 2623 \text{ lb/ft}^2$$

$$\sigma_{z0}' \approx \sigma_c' \qquad \therefore \text{ normally consolidated}$$

Layer	H (ft)	At midpoint of layer		$\frac{C_c}{1+e_0}$	Eqn.	$(\delta_c)_{ult}$ (in)
		σ_{z0}' (lb/ft²)	σ_{zf}' (lb/ft²)			
1	7.0	423	423	0.011	11.23	0.0
2	6.0	1035	1210	0.011	11.23	0.1
3	9.0	1513	2126	0.009	11.23	0.1
4	9.0	2095	3233	0.009	11.23	0.2
5	7.0	2552	3837	0.14	11.23	2.1
6	8.0	2909	4194	0.14	11.23	2.1
					$(\delta_c)_{ult} =$	4.6

$$(\delta_c)_{ult} = \textbf{4.6 in} \quad \Leftarrow \textit{Answer}$$

Notice how most of the compression occurs in the normally consolidated clay, even though it remains below the groundwater table. The cause of settlement is an increase in effective stress, not drying, and the clay is most susceptible to this increase because it has the highest $C_c/(1 + e_0)$.

Example 11.8

After the settlement due to the fill described in Example 11.5 is completed, a 20 m diameter, 10 m tall cylindrical steel water tank is to be built. The bottom of the tank will be at the top of the fill, and it will have an empty mass of 300,000 kg. Ultimately, the water inside will be 9.5 m deep. Compute the ultimate consolidation settlement beneath the center of this tank due to the weight of the tank and its contents. Assume the new fill is overconsolidated with $C_r/(1+e_0) = 0.002$.

Strategy

The settlement due to the fill is now complete, so the values of σ_{zf}' from the solution of Example 11.5 are now the initial stresses, σ_{z0}'. We will compute new σ_{zf}' values using $\Delta\sigma_z$ from Equation 10.25. Note how the increase in total stress, $\Delta\sigma_z$, due to the weight of the tank diminishes with depth. This is because the loads from the tank are distributed over a much smaller area compared to the wide fills of previous examples.

Solution

Compute weights:

$$W_{tank} = Mg$$
$$= (300,000\,\text{kg})(9.8\,\text{m/s}^2)\left(\frac{1\,\text{kN}}{1000\,\text{N}}\right)$$
$$= 2900\,\text{kN}$$

$$W_{water} = V_{tank}\,\gamma_w$$

$$= \frac{\pi B^2 H}{4}\,\gamma_w$$

$$= \frac{\pi\,(20.0\text{ m})^2\,(9.5\text{ m})}{4}\,(9.8\text{ kN/m}^3)$$

$$= 29{,}200\text{ kN}$$

The weight of the water is much greater than that of the empty tank, so it is reasonable for us to assume the bearing pressure q is constant across the bottom of the tank.

$$q = \frac{W}{A}$$

$$= \frac{2900\text{ kN} + 29{,}200\text{ kN}}{\pi\,(20.0\text{ m})^2\,/4}$$

$$= 102\text{ kPa}$$

			At midpoint of layer						
Layer	H (m)	σ_{z0}' (kPa) Eqn. 10.34	z_f (m)	$(\sigma_z)_{induced}$ (kPa) Eqn. 10.25	σ_{zf}' (kPa) Eqn. 11.3	$\dfrac{C_c}{1+e_0}$	$\dfrac{C_r}{1+e_0}$	Eqn.	$(\delta_c)_{ult}$ (mm)
1	3.0	28.8	1.5	101.7	130.5		0.002	11.24	4
2	1.5	71.5	3.7	97.7	169.2	0.008		11.23	4
3	2.0	95.0	5.5	90.6	185.6	0.008		11.23	5
4	3.0	114.0	8.0	77.1	191.1	0.19		11.23	128
5	3.0	132.6	11.0	60.7	193.3	0.19		11.23	93
6	4.0	154.3	14.5	45.1	199.4	0.19		11.23	85
								$(\delta_c)_{ult} =$	319

Round off to:

$$(\delta_c)_{ult} = 320\text{ mm} \qquad \Leftarrow \textit{Answer}$$

Commentary

1. If the tank were built immediately after the fill was placed, then σ_{z0}' would be the same as in Example 11.5, and everything else would remain unchanged. Such a solution would illustrate the use of superposition of stresses.
2. The values of $(\sigma_z)_{induced}$ beneath the edge of the tank are less than those beneath the center (see Figure 10.10). Thus, the consolidation settlement also will be less and the bottom of the tank will settle into a dish shape. The difference between these two settlements is called *differential settlement*. We will discuss differential settlements in more detail in Chapter 17.

Program FILLSETT

Program **FILLSETT**, which is in geotechnical analysis software package that accompanies this book, computes the ultimate consolidation settlement due to the weight of fills. It uses the analysis method described in this chapter, except that it divides each soil stratum into 50 layers, thus obtaining a slightly more precise solution.

To use this program, you must first download the geotechnical analysis software package from the Prentice Hall web site and install it onto a computer. See Appendix C for computer system requirements, downloading information, and installation instructions. Then select **FILLSETT** from the main menu.

Once the **FILLSETT** screen appears, select the units of measurement and enter the requested data. If fewer than five soil strata are present, then enter only the data for those strata and leave the lower fields empty.

Finally, click on the **CALCULATE** button to perform the computation. The results are shown on the screen, and printouts may be obtained by clicking on the **PRINT** button.

Example 11.9

Compute the ultimate consolidation settlement due to the weight of the fill shown in Figure 11.22.

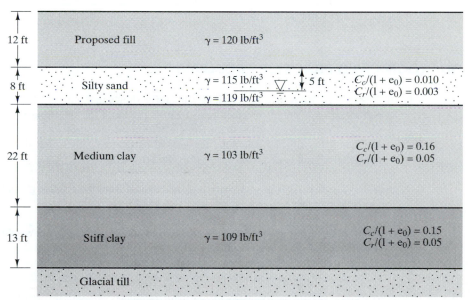

Figure 11.22 Cross-section for Example 11.9.

Solution

Using program **FILLSETT** — see screen capture in Figure 11.23

Final result: $(\delta_c)_{ult} = 1.078$ **ft   (round off to 1.1 ft)** $\Leftarrow$ *Answer*

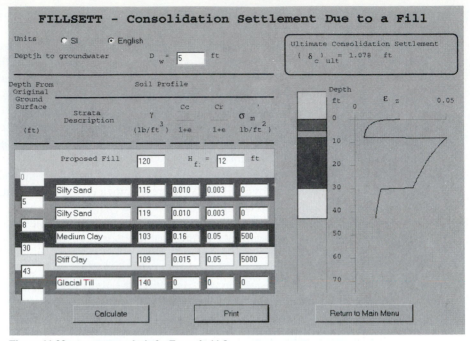

Figure 11.23 FILLSETT analysis for Example 11.9.

QUESTIONS AND PRACTICE PROBLEMS

11.10 A 5.0 ft thick fill is to be placed on the soil profile shown in Figure 11.3. A consolidation test performed on a sample obtained from Point B produced the following results: $C_c = 0.27$, $C_r = 0.10$, $e_0 = 1.09$, $\sigma_c' = 760$ lb/ft². Compute the ultimate consolidation settlement due to the weight of this fill and determine the ground surface elevation after the consolidation is complete. Check your answers using program **FILLSETT**.

Note: The first layer in your analysis should extend from the original ground surface to the groundwater table.

11.11 A 4.0 m thick fill is to be made of a soil with a Proctor maximum dry unit weight of 19.4 kN/m³ and an optimum moisture content of 13.0%. This fill will be compacted at optimum moisture content to an average relative compaction of 92%. The underlying soils are as shown in Figure 11.24. Consolidation tests were performed at Points A and B, with the following results:

Sample	C_c	C_r	e_0	σ_c' (kPa)
A	0.59	0.19	1.90	75
B	0.37	0.14	1.21	100

The silty sand is normally consolidated. Using hand computations, determine the ultimate consolidation settlement due to the weight of this fill. Then, check your answer using program **FILLSETT**.

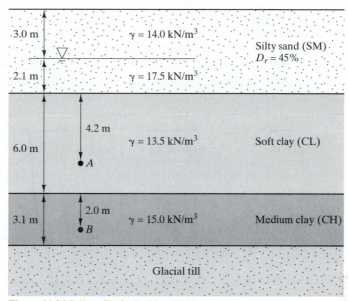

Figure 11.24 Soil profile for Problem 11.11.

11.12 The owner of the land shown in the profile in Example 11.5 has decided not to build the proposed fill. Instead, the land will be used for farming. To provide irrigation water, a series of shallow wells will be drilled into the sand, and these wells will cause the groundwater table to drop to the bottom of the sand layer (i.e., 2.0 m below its current position). Compute the ultimate consolidation settlement due to this drop in groundwater. Do you think such a settlement will adversely affect the farming?

11.13 A certain site is underlain by the soil profile shown in Figure 11.21 with the groundwater table at the location labeled "initial." The groundwater table will remain at this location, but a 20.0 ft deep fill with a unit weight of 119 lb/ft^3 is to be placed. The only consolidation data available is from a test conducted on a sample from Point B. The test results are as follows: $e_0 = 1.22$, $C_c = 0.31$, $C_r = 0.09$, $\sigma_c' = 3800$ lb/ft^2. Using hand computations, determine the ultimate consolidation settlement due to the weight of this proposed fill. Assume the sands are normally consolidated. Then, check your answer using program **FILLSETT**.

11.14 Using the data in Example 11.8, compute the consolidation settlement at the edge of the tank. Then compute the differential settlement, which is the difference between the settlement at the center and the edge.

Hint: Compute $(\sigma_z)_{induced}$ using Figure 10.10 or program **STRESSC**.

11.8 CRUSTS

Soft fine-grained soil deposits, such as those often found in wetlands, frequently have a thin crust near the ground surface, as shown in Figure 11.25. These crusts are typically less than 2 m (7 ft) thick, and are formed when the upper soils temporarily dry out. This drying process is called *dessication* and causes these soils to become overconsolidated. Thus, profiles that contain crusts have less settlement than identical profiles without crusts.

The presence of crusts has a significant impact on settlement computations, even if they are much thinner than the underlying compressible soils. In addition, variations in the crust thickness across a site can be a significant source of differential settlement. Thus, site characterization studies need to carefully evaluate the thickness and compressibility of crusts.

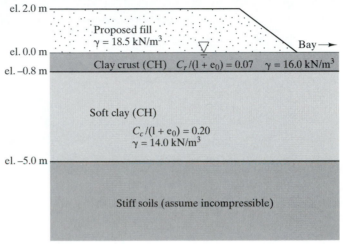

Figure 11.25 Typical crust near the ground surface in an otherwise normally consolidated clay.

11.9 SETTLEMENT OF UNSATURATED SOILS

Thus far we have treated unsaturated soils using the techniques developed for saturated soils, except we have set $u = 0$ (e.g., Example 11.5). However, some unsaturated soils are prone to other kinds of settlement problems, especially if they become wetted sometime after construction.

One of these problems occurs in certain kinds of clay that are known as *expansive soils*. These clays expand when they become wetted, and contract when dried. Another kind of problematic soil is called a *collapsible soil*, which compresses when it is wetted. Both types are discussed in Chapter 18.

Although the soft saturated soils generally have the worst problems with settlement, expansive and collapsible soils also can be problematic, especially in arid and semi-arid climates.

11.10 SECONDARY COMPRESSION SETTLEMENT

Once the excess pore water pressures have dissipated, consolidation settlement ceases. However, some soils continue to settle anyway. This additional settlement is due to *secondary compression* and occurs under a constant effective stress. We don't fully understand the physical basis for secondary compression, but it appears to be due to particle rearrangement, creep, and the decomposition of organics. Highly plastic clays, organic soils, and sanitary landfills are most likely to have significant secondary compression. However, secondary compression is negligible in sands and gravels.

The *secondary compression index*, C_α, defines the rate of secondary compression. It can be defined either in terms of either void ratio or strain:

$$C_\alpha = -\frac{de}{d\log t}$$ (11.26)

$$\frac{C_\alpha}{1 + e_p} = \frac{d\epsilon_z}{d\log t}$$ (11.27)

where:

C_α = secondary compression index

e = void ratio

e_p = void ratio at end of consolidation settlement (can use $e_p = e_0$ without introducing much error)

ϵ_z = vertical strain

t = time

Design values are normally determined while conducting a laboratory consolidation test. The consolidation settlement occurs very rapidly in the lab (because of the short drainage distance), so it is not difficult to maintain one or more of the load increments beyond the completion of consolidation settlement. The change in void ratio after this point can be plotted against log time to determine C_α.

Another way of developing design values of C_α is to rely on empirical data that relates it to the compression index, C_c. This data is summarized in Table 11.4.

TABLE 11.4 EMPIRICAL CORRELATION BETWEEN C_α
AND C_c (Terzaghi, Peck, and Mesri, 1996)

Material	C_α/C_c
Granular soils, including rockfill	0.02 ± 0.01
Shale and mudstone	0.03 ± 0.01
Inorganic clays and silts	0.04 ± 0.01
Organic clays and silts	0.05 ± 0.01
Peat and muskeg	0.06 ± 0.01

The settlement due to secondary compression is:

$$\delta_s = \frac{C_\alpha}{1 + e_p} H \log\left(\frac{t}{t_p}\right)$$ (11.28)

where:

δ_s = secondary compression settlement

H = thickness of compressible strata

t = time after application of load

t_p = time required to complete consolidation settlement (in theory this is infinite, but for practical problems we can assume it occurs when 95 percent of the consolidation in the field is complete.

We assume the secondary compression settlement begins at time t_p.

Usually secondary compression settlement is much smaller than consolidation settlement, and thus is not a major consideration. However, in some situations, it can be very important. For example, the consolidation settlement in sanitary landfills is typically complete within a few years, while the secondary compression settlement continues for many decades. Secondary compression settlements on the order of 1 percent of the refuse thickness per year have been measured in a 10-year-old landfill (Coduto and Huitric, 1990).

Significant structures are rarely built on soils that have the potential for significant secondary compression. However, highways and other transportation facilities are sometimes built on such soils.

Example 11.10

The soft clay described in Example 11.5 has $C_\alpha/(1+e_p) = 0.018$. Assuming the consolidation settlement will be 95 percent complete 40 years after the fill is placed, compute the secondary compression settlement that will occur over the next 30 years.

Solution

$$\delta_s = \frac{C_\alpha}{1 + e_p} H \log\left(\frac{t}{t_p}\right)$$

$$= (0.018)(10,000\,\text{mm})\log\left(\frac{40\,\text{yr} + 30\,\text{yr}}{40\,\text{yr}}\right)$$

$$= \mathbf{40\ mm}$$

This is approximately one-tenth of the consolidation settlement of 480 mm, as computed in Example 11.5.

A complete example including both consolidation and secondary compression settlements is included in Chapter 12.

11.11 DISTORTION SETTLEMENT

When heavy loads are applied over a small area, the soil can deform laterally, as shown in Figure 11.26. Similar lateral deformations also can occur near the perimeter of larger loaded areas. These deformations produce additional settlement at the ground surface, which we call *distortion settlement*.

Distortion settlement is generally much smaller than consolidation settlement, and can usually be ignored. However, it is sometimes considered in the design of spread footing foundations, as discussed in *Foundation Design: Principles and Practices* (Coduto, 1999).

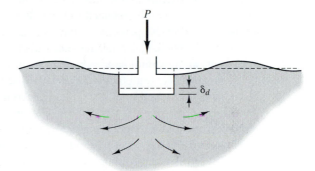

Figure 11.26 Distortion settlement beneath a small loaded area.

11.12 HEAVE DUE TO UNLOADING

Our discussions thus far have considered only settlement of soils in response to an increased load. Another possibility is heave (negative settlement) due to a decreased load, such as an excavation. In this case, $\sigma_{zf}' < \sigma_{z0}'$. The soil will heave according to the rebound curve in Figure 11.9, which has a slope of C_r, so we compute the heave using Equation 11.24. Because the soil is expanding, the excess pore water pressure is negative, causing pore water to be drawn into the voids. This is the opposite of the process described in Figure 11.7.

There are other processes that also can cause heave in soils. The most notable one is the swelling of expansive clays, which is discussed in Chapter 18.

11.13 ACCURACY OF SETTLEMENT PREDICTIONS

As with all other geotechnical analyses, settlement predictions are subject to many errors. These include:

- Differences between the soil profile used in the analysis and the real soil profile, especially the proper identification of crusts
- Differences between the engineering properties of the soil samples and the average properties of the strata they represent (i.e., are they truly representative?)
- Sample disturbance

- Errors introduced due to testing techniques in the laboratory
- Errors in assessing $\sigma_c{}'$
- The assumption that consolidation in the field is one-dimensional (i.e., there is no horizontal strain)
- Differences between Terzaghi's theory of consolidation and the real behavior of soils in the field

The compression index, C_c, and the recompression index, C_r, vary widely within soil deposits, even those that appear to be fairly uniform. Kulhawy, Roth, and Grigoriu (1991) reported coefficients of variation in C_c values of 26 to 52 percent. This means C_c values from a single randomly obtained soil sample would have only a 30 to 56 percent probability of being within 20 percent of the true C_c of the stratum. This uncertainty can be significantly reduced by testing more than one sample from each stratum, but it still represents an important source of error in our analyses.

Fortunately, settlement analyses consist of a summation for multiple strata, which introduces an averaging effect on test uncertainties. Even so, the error in consolidation settlement predictions is typically on the order of ±25 to 50 percent, even with careful sampling and testing. Analyses of secondary compression settlement are even less accurate, having errors on the order of about ±75 percent (Fox, 1995). We need to consider these potential errors when setting allowable settlement values, and incorporate an appropriate factor of safety in these allowable values. These margins of error also underscore the usefulness of monitoring the actual settlements in the field, comparing them to the predicted settlements, and, if necessary, modifying the designs accordingly.

QUESTIONS AND PRACTICE PROBLEMS

11.15 Using the data from Example 11.10, develop a plot of secondary compression settlement vs. time for the period 40 to 100 years after completion of the fill. Is the rate of secondary compression settlement increasing or decreasing with time?

11.16 Point C in Figure 11.19 was originally at elevation 12.00 m, but it dropped to elevation 11.52 m as a result of the consolidation settlement described in Example 11.5. Now that the consolidation is complete, the fill is to be removed. Compute the new elevation of Point C after the natural soils rebound in response to the fill removal. Ignore any secondary compression settlement.

11.17 A cross-section through a tidal mud flat area is shown in Figure 11.25. This site is adjacent to a bay, is subject to varying water levels according to tides, and is occasionally submerged when heavy runoff from nearby rivers raises the elevation of the water in the bay. For analysis purposes, use a groundwater table at the ground surface, as shown. A crust has formed in the upper 0.8 m of soil due to dessication (drying) and is stiffer than the underlying soil. This crust is overconsolidated case I, and the soils below are normally consolidated. The proposed fill is required to protect the site from future flooding, and thus permit construction of a commercial development. Using program **FILLSETT**, determine the ultimate consolidation settlement due to the weight of this proposed fill. Then, consider the possibility that the crust was not recognized in the site characterization program, and perform another **FILLSETT** analysis using

$C_c/(1 + e_0) = 0.20$ and $\gamma = 14.0$ kN/m^3 for the entire 5.0 m of clay. Compare the results of these two analyses and comment on the importance of recognizing the presence of crusts.

SUMMARY

Major Points

1. Settlement can be caused by several different physical processes. We have considered three: consolidation settlement, δ_c, secondary compression settlement, δ_s, and distortion settlement, δ_d.

2. Consolidation is usually the most important type of settlement. It occurs when the vertical effective stress increases from an initial value of σ_{z0}' to a final value of σ_{zf}'. This change causes the solid particles to move into a tighter packing, which results in a vertical strain and a corresponding settlement.

3. If the soil is saturated, which is the case in most consolidation analyses, the applied load is first carried by the pore water. This causes a temporary increase in the pore water pressure. This increase is called an excess pore water pressure, u_e. The presence of this pressure induces a hydraulic gradient in the soil, forcing some pore water to flow out of the voids, thus relieving the excess pore water pressures. After some period, which may be years or decades, $u_e \rightarrow 0$, the applied load is transferred to the solid particles, and $\sigma_{z0}' \rightarrow \sigma_{zf}'$.

4. We measure the stress–strain properties of a soil by conducting consolidation tests in the laboratory on undisturbed samples. The test results are expressed in the following parameters:

   σ_c = preconsolidation stress
   e_0 = initial void ratio
   C_c = compression index
   C_r = recompression index

   The preconsolidation stress is the greatest vertical effective stress the soil has ever experienced at the point where the sample was obtained. The parameters C_r and C_c define the slope of the consolidation curve at stresses less than and greater than σ_c', respectively.

5. Using the parameters from the consolidation test, the changes in effective stress in the field, and other data, we can compute the consolidation settlement.

6. The greatest consolidation settlements occur in soft clays. Sandy and gravelly soils are usually much less compressible. In addition, it is nearly impossible to obtain sufficiently undisturbed samples to conduct reliable consolidation tests on sands and gravels, so the compressiblity of these soils is determined by empirical correlations or by in-situ tests.

7. Normally consolidated soils are those that have never experienced a vertical effective stress significantly greater than the present value of σ_z'. Conversely, overconsolidated soils are those that have experienced higher stresses.

8. Normally consolidated soils often have an overconsolidated crust near the ground surface. It is important to recognize the presence of these crusts in the site characterization program.

9. Secondary compression settlement is the result of particle rearrangement, creep, decomposition of organic materials, and other processes. It produces settlement even though σ_z' remains constant, and this settlement continues at an ever-decreasing rate with time.

10. Distortion settlement is due to the horizontal movements of soil, and occurs primarily when the loaded area is small and the bearing pressure is high, such as structural foundations.

11. If the soil is unloaded, σ_z' decreases and $\sigma_{zf}' < \sigma_{z0}'$, so the soil heaves instead of settles.

12. Settlement predictions are subject to several sources of error. Even careful predictions of consolidation settlement typically have a precision on the order of ±25 to 50 percent. Predictions of secondary compression settlement typically are even less accurate, with errors on the order of ±75 percent.

Vocabulary

collapsible soil
compressibility
compression index
compression ratio
consolidation
consolidation settlement
consolidation test
consolidometer
crust
desiccation
differential settlement
distortion settlement
elastic deformation
excess pore water pressure

expansive soil
heave
normally consolidated
oedometer
one-dimensional
    consolidation
overconsolidated
overconsolidation margin
overconsolidation ratio
plastic deformation
porous stones
preconsolidated
preconsolidation stress
rebound curve

recompression curve
recompression index
recompression ratio
secondary compression
    index
secondary compression
    settlement
settlement
subsidence
theory of consolidation
ultimate consolidation
    settlement
virgin curve

COMPREHENSIVE QUESTIONS AND PRACTICE PROBLEMS

11.18 Explain the difference between normally consolidated soil and overconsolidated soils, and give examples of geologic conditions that would form each type.

11.19 What types of natural soils are best suited for testing in a consolidometer? Why? Which are not well suited? Why?

11.20 According to the results from a consolidation test, the preconsolidation stress for a certain soil sample is 850 lb/ft². The in-situ vertical effective stress at the sample location is 797 lb/ft², and the proposed load will cause σ_z to increase by 500 lb/ft². Which equation should be used to compute the consolidation settlement, 11.23, 11.24, or 11.25? Why?

11.21 A 3.0 m thick fill with a unit weight of 18.1 kN/m³ is to be placed on the soil profile shown in Figure 11.24. Consolidation test results at Points A and B are as stated in Problem 11.11, except that σ_c' at point B is now 200 kPa. Using Equation 11.21 with C_c or C_r as appropriate, develop a plot of vertical strain, ϵ_z vs. depth from the original ground surface to the top of the glacial till. How does this curve vary within a given soil stratum? Why? Does it suddenly change at the strata interfaces? Why?

11.22 Considering the variation of strain with depth, as found in Problem 11.21, does a 1 m thick layer near the top of a stratum contribute more or less to the consolidation settlement than a 1 m thick stratum near the bottom? Explain. Does this finding support the statement in Section 11.8 that "The presence of crusts has a significant impact on settlement computations, even if they are much thinner than the underlying compressible soils?" Explain.

11.23 A shopping center is to be built on a site adjacent to a tidal mud flat. The ground surface elevation is +0.2 m, and the groundwater table is at the ground surface. The underlying soils consist of 7.3 m of medium clay with $C_c/(1+e_0) = 0.18$, $C_r/(1+e_0) = 0.06$, $\sigma_m' = 0$, and $\gamma = 15.1$ kN/m³. The clay stratum is underlain by relatively incompressible stiff soils.

In order to provide sufficient flood control protection, a fill must be placed on this site before the shopping center is built, thus maintaining the entire site above the highest flood level. This fill will have a unit weight of 19.0 kN/m³. According to a hydrologic study, the fill must be thick enough so that the ground surface elevation is at least +1.8 m after all of the consolidation settlement is complete. Using program **FILLSETT**, determine the required ground surface elevation at the end of construction. Assume all of the settlement occurs after construction.

11.24 A consolidation test has been performed on a sample of lodgement till from a region that was once covered with a glacier. The current vertical effective stress at the sample location is 1800 lb/ft² and the measured preconsolidation stress is 32,500 lb/ft².

a. Assuming the glacier was in place long enough for complete consolidation to occur, and assuming the ground surface and groundwater table elevations have remained unchanged, compute the maximum thickness of the glacier. The specific gravity of glacial ice is about 0.87.

b. Although glacial ice was present for a very long time, it also extended over very large areas, so the required drainage distance for the excess pore water was very long. As a result, all of the excess pore water pressures may not have dissipated (Chung and Finno, 1992). Therefore, our assumption that complete consolidation occurred may not be accurate. If so, would our computed thickness be too large or too small? Explain.

11.25 A highway is to be built across a wetlands with the soil profile shown in Figure 11.27 below. These wetlands are subject to flooding, so a fill must be placed to keep the pavement above the highest flood level. According to a hydrologic analysis, the roadway must be at elevation 7.0 ft or higher to satisfy this requirement. Sandy fill material that has a compacted unit weight of 122 lb/ft³ is available from a nearby borrow site.

A subsurface exploration program has been completed at this site, and laboratory tests have been performed. The results of this program are tabulated below:

Depth (ft)	Dry unit weight (lb/ft³)	Moisture content (%)	$C_c/(1+e_0)$	$C_r/(1+e_0)$	σ_c' (lb/ft²)
2.0	95	28.6	0.13	0.06	3000
7.5	89	33.0	0.16	0.06	550
13.0	92	30.5	0.12	0.05	850
24.0	93	29.9	0.14	0.07	4800

All depths are measured from the original ground surface.

The natural soils will settle under the weight of the proposed fill. Approximately 25 years will pass before this settlement is complete. Therefore, the road must be built at an elevation higher than 7.0 ft so that after the settlement is complete it is at 7.0 ft. The pavement thickness is 0.5 ft, so the top of the fill must remain at or above elevation 6.5 ft.

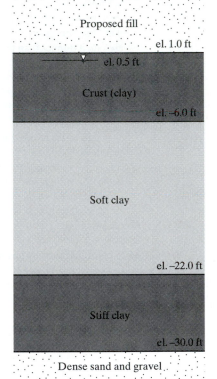

a. Using program **FILLSETT,** determine the required elevation of the roadway immediately after construction. Assume that no settlement occurs during construction, and the pavement has the same unit weight as the fill.

   Hint: As the fill thickness becomes greater, the settlement increases. Thus, this problem requires a trial-and-error solution. You will need to estimate the required fill thickness, then compute the settlement and final roadway elevation. Try to have one trial that produces a road elevation that is too high, and another that produces one too low. Then interpolate to find the required fill thickness.

   Note 1: As the fill settles, the lower portion will become submerged below the groundwater table, so σ_{zf}' will be less than predicted by Equation 11.2. However, we will ignore this effect.

Figure 11.27 Cross-section for Problem 11.25.

   Note 2: Laboratory tests have been performed on two samples of the soft clay. Combine these two sets of test results, then assign γ, $C_c/(1+e_0)$, $C_r/(1+e_0)$, and σ_m' values that apply to the entire stratum.

   Note 3: We do not have any unit weight data for the portion of the crust above the groundwater table. Therefore, assume it is the same as that below the groundwater table. In this case, this assumption should introduce very little error.

b. Is our assumption regarding the submergence of the fill (per note 1 in part a) conservative or unconservative? Is this a reasonable assumption? Explain.

11.26 An engineer has suggested an alternative design for the proposed highway in Problem 11.25. This design consists of using geofoam for the lower part of the fill, as shown in Figure 6.39. It will be covered with at least 1.0 ft of soil to provide a buffer between the pavement and the geofoam. Compute the minimum required geofoam thickness so that the roadway will always be at elevation 7.0 or higher.

Hint: The geofoam is an extra "hidden" layer between the fill and the natural ground surface. If we assume the geofoam has $\gamma = 0$, then it can be ignored when using program **FILLSETT**.

11.27 A series of prefabricated dual-bore steel tubes similar to the one in Figure 11.28 are to be installed in an underwater trench to form a tunnel. The trench will be in seawater, which has a unit weight of 64.0 lb/ft^3. The tubes will be floated into position, and sunk into place by temporarily flooding the interior. Then, non-structural concrete will be placed into chambers along the tube to act as ballast, and the inside will be pumped dry. The completed tube will be 80 ft wide, 300 ft long, and 40 ft tall, and weigh 32,000 tons exclusive of buoyant forces. Finally, the tube will be covered with soil, producing the cross-section shown in Figure 11.29.

a. The interior of the tube will be dewatered after the concrete is placed, but before the trench is backfilled. Once this is done, will the tube remain at the bottom of the trench, or will it float up to the water surface?

b. After the trench is backfilled, what will be the net $\Delta\sigma_z$ in the soft clay? Assume $\Delta\sigma_z$ is constant with depth.

c. Using the final cross-section, compute the ultimate consolidation settlement or heave of the tube due to $\Delta\sigma_z'$ in the soft clay. Assume no heave occurs during construction.

d. The weakest parts of the completed tunnel will be the connections between the tube sections. In order to avoid excessive flexural stresses at these connections, the structural engineer has specified a maximum allowable differential settlement or differential heave of 5 in along the length of the tube (the term "allowable" indicates this value already includes a factor of safety). An evaluation of the soil profile suggests the differential settlement or heave will be no more than 50 percent of the total. Has the structural engineer's criteria been met?

Figure 11.28 This prefabricated tunnel section is part of the Central Artery Project in Boston. It was floated into position, then sunk to the bottom of the bay (Photograph by Peter Vanderwarker, courtesy of the Central Artery/Tunnel Project).

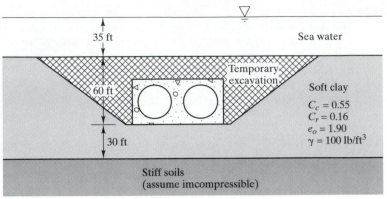

Figure 11.29 Final cross-section for underwater tunnel as described in Problem 11.27.

11.28 A proposed building is to have three levels of underground parking, as shown in Figure 11.30. To construct this building, it will be necessary to make a 10.0 m excavation, which will need to be temporarily dewatered. The natural and dewatered groundwater tables are as shown, and the medium clay is normally consolidated. The chief geotechnical engineer is concerned that this dewatering operation may cause excessive differential settlements in the adjacent building and has asked you to compute the anticipated differential settlement across the width of this building. Assume the wall is perfectly rigid, and thus does not contribute to any settlement problems, and that the maximum allowable differential settlement from one side of the building to the opposite side is 50 mm. Neglect any loss in σ_z' below the existing building due to the removal of soil from the excavation. Discuss the implications of your answer.

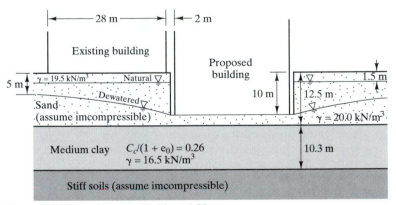

Figure 11.30 Cross-section for Problem 11.28.

11.29 The Palacio de las Bellas Artes in Mexico City, shown in Figure 11.1, is an interesting example of large consolidation settlement. It is supported on a 1.8 to 3.0 m thick mat foundation which is approximately 65 m wide and 115 m long. The average bearing pressure between the bottom of this mat and the supporting soil is 115 kPa (Ledesma, 1936).

   The soil conditions beneath the palace are too complex to describe in detail here. However, we can conduct an approximate analysis using the following simplified profile:

Depth (m)	Description	γ (kN/m^3)	$\dfrac{C_c}{1 + e_0}$
0 - 5	Sandy fill	17.5	0
5 - 45	Normally consolidated soft clays	11.5	0.53
> 45	Stiff soils		0

For our simplified analysis, use a groundwater table at a depth of 5 m, and assume the bottom of the mat is at the original ground surface. In addition, assume the fill has been in pace for a very long time, so the consolidation settlement due to the weight of the fill is complete.

Divide the soft clay zone beneath the center of the building into five layers of equal thickness. Then, compute $(\sigma_z)_{induced}$ at the midpoint of each layer using the methods described in Chapter 10. Finally, compute the ultimate consolidation settlement beneath the center of the palace due to its own weight.

11.30 Develop a spreadsheet that can compute one-dimensional consolidation settlement due to the weight of a fill. This spreadsheet should be able to accommodate a fill of any unit weight and thickness, underlain by multiple compressible soil strata. It also should be able to accommodate both normally consolidated and overconsolidated soils. Since the computer does all of the computations, the spreadsheet should use at least 50 layers. Once the spreadsheet is completed, use it to solve Examples 11.5 and 11.6. Submit printouts of both analyses.

11.31 A fill is to be placed at a proposed construction site, and you need to determine the ultimate consolidation settlement due to its weight. Write a 200–300 word essay describing the kinds of field exploration, soil sampling, and laboratory testing you will need to perform to generate the information needed for this analysis. Your essay should describe specific things that need to be done, and what information will be gained from each activity.

12

Rate of Consolidation

Ut tensio sic vis
(as to stretch, so the force)

Robert Hooke's 1678 description of the relationship
between stress and strain, now known as Hooke's Law

When static loads are applied to structural members, such as beams or columns, the resulting deformations occur virtually as fast as the loads are applied. For example, when floor loads are applied to a steel beam, we assume all of the resulting deflection occurs immediately. However, deformations in soil sometimes occur much more slowly, especially in saturated clays. Many years, or even decades, may be required for the full settlement to occur in a soil, so geotechnical engineers often need to evaluate both the magnitude and the rate of consolidation settlement. Therefore, this chapter extends the analyses performed in Chapter 11 and develops the ability to produce settlement vs. time plots.

12.1 TERZAGHI'S THEORY OF CONSOLIDATION

Karl Terzaghi's most significant contribution to geotechnical engineering was his *theory of consolidation*, which he developed in Istanbul between 1919 and 1923 (Terzaghi, 1921, 1923a, 1923b, 1924, 1925a, and 1925b). Although others had studied the consolidation problem and made useful contributions, it was Terzaghi's work that properly identified and quantified the underlying physical processes. During this time he identified the principle of effective stress, which became the key to understanding the consolidation process. Terzaghi's academic training as a mechanical engineer was very useful, because the

422

processes that control consolidation are very similar to certain thermodynamic processes. In fact, he was teaching thermodynamics while conducting his consolidation experiments, which probably helped inspire the new theory.

Review of the Consolidation Process

The consolidation process, as discussed in Chapter 11, begins when the placement of a fill or some other load produces an increase in the vertical total stress, $\Delta\sigma_z$. Initially, this increase is carried entirely by the pore water, thus producing an excess pore water pressure, u_e. In one-dimensional consolidation analyses, the initial value of u_e equals $\Delta\sigma_z$. Thus, the vertical effective stress, σ_z', immediately after loading is unchanged from its original value, σ_{z0}'.

The excess pore water pressure produces a localized increase in the total head, and thus induces a hydraulic gradient. Therefore, some of the pore water begins to flow away from the zone that is being loaded. This flow causes the excess pore water pressure to slowly dissipate, the vertical effective stress to increase, and the soil to consolidate. After sufficient time has elapsed, $u_e \to 0$, $\sigma_z' \to \sigma_{zf}'$, and the consolidation settlement, $\delta_c \to (\delta_c)_{ult}$. Terzaghi's theory of consolidation quantifies this process.

It is important to recognize this theory is not simply an empirical description of settlement data obtained in the field; it is a rational method based on a physical model of the consolidation process. This is an important distinction, because it illustrates the difference between organized empiricism and the development of more fundamental understandings of soil behavior.

The various soil parameters needed to implement the theory of consolidation are normally obtained from a site characterization program, including laboratory consolidation tests, and thus are subject to many sources of error (i.e., are the samples truly representative?, what are the effects of soil disturbance?, and so on). Therefore, it does not give exact answers. However, the validity of this theory has been confirmed, and it is the basis for nearly all time–settlement computations.

Assumptions

To keep the computational process from becoming too complex, the theory of consolidation is based on certain simplifying assumptions regarding the compressible stratum:

1. The soil is homogeneous ($C_c/(1+e_0)$, $C_r/(1+e_0)$ and k are constant throughout).
2. The soil is saturated ($S = 100\%$).
3. The settlement is due entirely to changes in the void ratio, and these changes occur only as some of the pore water is squeezed out of the voids (i.e., the individual solid particles and the water are incompressible).
4. Darcy's Law (Equation 7.8) is valid.
5. The applied load causes an instantaneous increase in vertical total stress, $\Delta\sigma_z$. Afterwards, the vertical total stress, σ_z, at all points remains constant with time.

6. Immediately after loading, the excess pore water pressure, u_e, is constant with depth, and equal to $\Delta\sigma_z$. This is generally true when the load is due to an extensive fill, but not when it is from a smaller loaded area, such as a foundation.
7. The *coefficient of consolidation*, c_v, as defined below, is constant throughout the soil, and remains constant with time. In normally consolidated soils, c_v is:

$$c_v = \left(\frac{2.30\, \sigma_z'\, k}{\gamma_w} \right) \left(\frac{1+e}{C_c} \right) \tag{12.1}$$

where:
     c_v = coefficient of consolidation
     k = hydraulic conductivity
     e = void ratio
     C_c = compression index
     γ_w = unit weight of water
     σ_z' = vertical effective stress

For overconsolidated soils, substitute C_r for C_c in Equation 12.1.
8. The consolidation process is one-dimensional, as discussed below.

Some of these assumptions, such as number 4, are very realistic. Others, such as number 7, are only approximately correct, and are intended to simplify the analysis. Some of these assumptions may be modified, with corresponding changes in the solution to Terzaghi's theory, but these enhancements are beyond the scope of this book. For most practical problems, the error introduced by these assumptions is acceptable. Section 12.5, later in this chapter, discusses the sources and probable magnitudes of error in consolidation analyses.

One-Dimensional Consolidation

Terzaghi's theory assumes the excess pore water flows only vertically, either up or down, and consolidation occurs only in the vertical direction. In other words, there is no horizontal drainage and no horizontal strain. This condition is called *one-dimensional consolidation*, and is shown in Figure 12.1.

One of the important parameters in one-dimensional consolidation analyses is the *length of the longest drainage path*, H_{dr}. This is the longest distance any molecule of excess pore water must travel to move out of the consolidating soil. There are two possibilities, as shown in Figure 12.2:

- If the strata above and below are much more permeable than the consolidating soil (i.e., they have a much greater hydraulic conductivity, k), then the excess pore water will drain both up and down. This condition is known as *double drainage* and H_{dr} is equal to half the thickness of the consolidating strata.

- If the stratum below is less permeable, such as bedrock, then all of the excess pore water must travel up, a condition known as *single drainage*. In this case, H_{dr} equals the thickness of the consolidating strata.

In both cases, H_{dr} is measured in a straight line, even though the actual flow path is a circuitous one that winds around the individual soil particles. We do it this way to be consistent with the definition of Darcy's Law.

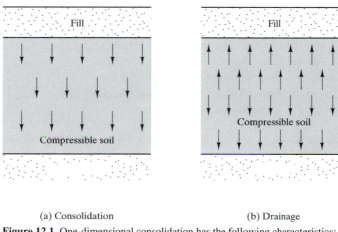

(a) Consolidation   (b) Drainage

Figure 12.1 One-dimensional consolidation has the following characteristics: a) the consolidation settlement is assumed to occur only in the vertical direction; and b) the excess pore water is assumed to escape only by flowing vertically.

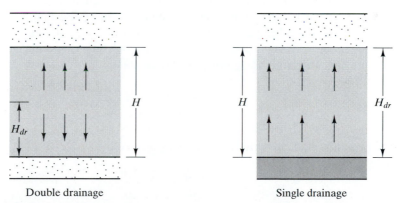

Double drainage   Single drainage

Figure 12.2 Computation of the length of the longest drainage path, H_{dr}, for one-dimensional consolidation problems.

The value of H_{dr} has a significant effect on the time required to complete the consolidation process. All else being equal, this time is proportional to H_{dr}^2. Thus, if a 6 m thick stratum requires 10 years to fully consolidate, a 12 m stratum of the same soil (double the thickness) would require 40 years (four times as long).

Derivation of the One-Dimensional Consolidation Equation

In Section 11.3 we discussed how the application of the additional total stress $\Delta\sigma_z$ onto the ground induces an excess pore water pressure, u_e. Immediately after the application of this load, which is assumed to occur instantaneously, u_e is constant with depth (according to Assumption 6). The rate of consolidation depends on the dissipation of these excess pore water pressures and the corresponding transfer of stress to the solid particles. Therefore, we can derive the governing equation by first examining the dissipation of excess pore water pressure in a typical soil element as shown in Figure 12.3. Then we will examine the consolidation that occurs in this element as the stresses are transferred. Finally we will combine these two processes into one equation.

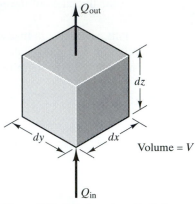

Figure 12.3 Soil element used to derive Terzaghi's theory of consolidation.

We will assume the excess pore water flows upward through the element. At the top of the element, the hydraulic gradient is:

$$i_z = \frac{dh}{dz} \tag{12.2}$$

where z is the depth to the top of the element. Unlike Equation 7.4, this equation does not have a negative sign because the water is flowing in the $-z$ direction (upward).

The elevation head at the top of the element remains constant with time. Only the pressure head, h_p, changes, and this change is due only to changes in the excess pore water pressure, u_e. Therefore, using Equation 7.6 we obtain:

$$i_z = \frac{dh_p}{dz} = \frac{1}{\gamma_w}\frac{du_e}{dz} \tag{12.3}$$

The hydraulic gradient varies with depth, as defined by:

$$\frac{di}{dz} = \frac{1}{\gamma_w}\frac{d^2u_e}{dz^2} \tag{12.4}$$

Therefore, the hydraulic gradient at the bottom of the element is:

$$i_{z+dz} = i_z + \frac{di}{dz}dz$$

$$= \frac{1}{\gamma_w}\left(\frac{du_e}{dz} + \frac{d^2u_e}{dz^2}dz\right) \tag{12.5}$$

Using Darcy's Law (per Assumption 4):

$$Q = \frac{dV}{dt} = k\,i\,A \quad \Rightarrow \quad dV = k\,i\,A\,dt \tag{12.6}$$

$$dV_{in} = k\,\frac{1}{\gamma_w}\left(\frac{du_e}{dz} + \frac{d^2u_e}{dz^2}\,dz\right) dx\,dy\,dt \tag{12.7}$$

$$dV_{out} = k\,\frac{1}{\gamma_w}\,\frac{du_e}{dz}\,dx\,dy\,dt \tag{12.8}$$

$$\begin{aligned} dV &= dV_{in} - dV_{out} \\ &= \frac{k}{\gamma_w}\,\frac{d^2u_e}{dz^2}\,dx\,dy\,dz\,dt \end{aligned} \tag{12.9}$$

Next, we will consider the relationship between volume changes and excess pore water pressures inside the sample. According to Equation 11.10:

$$C_c = -\frac{de}{d\log\sigma'_z} \tag{12.10}$$

which may be rewritten as:

$$de = -\frac{C_c}{2.30\,\sigma'_z}\,d\sigma'_z \tag{12.11}$$

Changes in effective stress are due solely to changes in excess pore water pressure, so:

$$d\sigma'_z = -du_e \tag{12.12}$$

$$de = \frac{C_c}{2.30\,\sigma'_z}\,du_e \tag{12.13}$$

An extension of Equation 11.8 gives the following formula for the vertical strain, ϵ_z:

$$d\epsilon_z = -\frac{de}{1+e} \tag{12.14}$$

The change in volume is then:

$$
\begin{aligned}
dV &= -d\epsilon_z A \, dz \\
&= \frac{de}{1+e} dx \, dy \, dz \\
&= \frac{C_c}{(2.30 \, \sigma_z')(1+e)} dx \, dy \, dz \, du_e
\end{aligned} \tag{12.15}
$$

Finally, we combine Equations 12.1, 12.9, and 12.15. The excess pore water pressure, u_e, now varies with both depth z and time t, so we have a partial differential equation:

$$\frac{\partial u_e}{\partial t} = c_v \frac{\partial^2 u_e}{\partial z^2} \tag{12.16}$$

where:
  u_e = excess pore water pressure
  t = time
  c_v = coefficient of consolidation (a constant, per Assumption 7)
  z = vertical distance below the ground surface

Equation 12.16 is called the *one-dimensional consolidation equation*. It is similar to Fick's Law of Thermal Diffusion, which mechanical engineers use to analyze the flow of heat through a material and the resulting temperature changes.

Solution of the One-Dimensional Consolidation Equation

The solution of Equation 12.16 requires the establishment of two boundary conditions for z and one initial condition for u_e. For the single drainage condition with z_t = depth to the top of the compressible stratum, and H = the thickness of the compressible stratum, they are as follows:

1. At $z = z_t$, $u_e = 0$ at $t = 0$ (the excess pore water pressure is zero at the top of the compressible stratum). This appears to be in violation of Assumption 6, and more correctly describes the conditions immediately after $t = 0$. This initial dissipation of excess pore water pressures sets up a hydraulic gradient that permits the process to continue.

2. At $z = z_t + H$, $i = du_e/dz = 0$ (the hydraulic gradient is zero at the bottom of the compressible stratum). This is because we are considering the single drainage condition.

3. At $t = 0$, $u_e = \Delta\sigma_z$ (immediately after placement of the load, the applied vertical stress is carried entirely by the excess pore water pressure, is equal to the change in total stress, $\Delta\sigma_z$, and is constant with depth). This is a restatement of Assumption 6.

For the double drainage case, simply add a mirror image of the drainage model to the bottom half of the compressible stratum.

An analytic solution based on these boundary conditions produces the following infinite series formula for u_e at any point in the compressible strata (Means and Parcher, 1963):

$$u_e = \Delta\sigma_z \sum_{N=0}^{\infty} \left(\frac{4}{(2N+1)\,\pi} \, \sin\left[\frac{(2N+1)\,\pi}{2} \, \frac{z_{dr}}{H_{dr}} \right] e^{-\left[\frac{(2N+1)^2 \pi^2}{4} T_v \right]} \right) \qquad (12.17)$$

The parameter T_v is known as the *time factor*:

$$T_v = \frac{c_v\,t}{H_{dr}^2} \qquad (12.18)$$

where:

   u_e = excess pore water pressure

   $\Delta\sigma_z$ = change in vertical total stress due to applied load = $\gamma_{fill}\,H_{fill}$

   z_{dr} = vertical distance from point to nearest drainage boundary

   H_{dr} = length of longest drainage path (for single drainage, H_{dr} = thickness of the compressible stratum; for double drainage, H_{dr} = half the thickness of the compressible stratum)

   e = base of natural logarithms = 2.7183

   c_v = coefficient of consolidation

   t = time since application of load

   T_v = time factor

The sin term in Equation 12.17 must be in radians. The summation has a value of 1 at $t = 0$, and a value of 0 at $t = \infty$, which means u_e has an initial value equal to $\Delta\sigma_z$ and a final value of 0.

Each increment of N in Equation 12.17 produces a progressively smaller change in the summation. Thus, the summation needs to continue only until this incremental change becomes negligible. Frequently this occurs at an N less than 10, although sometimes many more increments are needed.

Application of the One-Dimensional Consolidation Equation

Equation 12.17 describes the excess pore water pressure, u_e, produced in a soil subjected to an instantaneous increase in total stress $\Delta\sigma_z$. Immediately after this load is applied, $u_e = \Delta\sigma_z$, then it gradually diminishes with time, eventually becoming equal to zero. At the end of this process, the groundwater will have returned to the hydrostatic condition. The time required for this process and the applicability of this equation to practical problems depend on many factors, including the soil type.

Clays and Silts

The excess pore water pressures dissipate only as some of the pore water flows away from the zone of soil that is being loaded. Clays and silts have a low hydraulic conductivity, k, so water flows very slowly through these soils, and a long time is required to return to the hydrostatic condition. The theory of consolidation reflects this through the use of a low coefficient of consolidation, c_v (per Equation 12.1). Using these low c_v values in Equations 12.17 and 12.18 demonstrates that years or even decades will be required to fully dissipate the excess pore water pressures and return to the hydrostatic condition.

Our assumption that the load is applied instantaneously is not too far from the truth, because the duration of construction is probably very short compared to the time required to dissipate the excess pore water pressures. Therefore, the analyses described in this chapter are applicable to these soils. We will use these methods to compute the dissipation of excess pore water pressures and thus develop plots of consolidation settlement vs. time.

Sands and Gravels

The hydraulic conductivity, k, of sands and gravels is much greater than that of clays and silts, so their time–settlement behavior is correspondingly different. According to Table 7.1, k in sands is typically about 1,000,000 times greater than that in clays, and according to Equation 12.1, c_v is proportional to k. If we place these high c_v values in Equations 12.17 and 12.18 it becomes clear that the excess pore water pressures dissipate very quickly, perhaps in a few minutes or less. This is much faster than the rate of construction, so the consolidation settlement occurs virtually as fast as the load is applied.

Therefore, it is not necessary to conduct rate of consolidation analyses in sandy and gravelly soils. We simply compute the ultimate consolidation settlement, $(\delta_c)_{ult}$, using the methods described in Chapter 11 and assume it occurs as quickly as the load is applied.

Example 12.1

Consider the soft clay strata in Example 11.5. According to a laboratory consolidation test, $c_v = 0.0021$ m²/day. Compute the hydrostatic, excess, and total pore water pressures at Point B, 2000 days after placement of the fill.

Solution

$$u_h = \gamma_w z_w = (9.8 \text{ kPa})(5.0 \text{ m}) = 49.0 \text{ kPa}$$

The soils above and below the clay are much more permeable than the clay. Therefore, use double drainage.

$$H_{dr} = \frac{10.0 \text{ m}}{2} = 5.0 \text{ m}$$

$$z_{dr} = 3.0 \text{ m}$$

Using Equations 12.17 and 12.18:

N	u_e (kPa)
0	39.2
1	0.2
2	0.0
Sum	39.4

This time, only three increments of N are required.

$$u = u_h + u_e = 49.0 \text{ kPa} + 39.4 \text{ kPa} = 88.4 \text{ kPa}$$

$$u_h = \textbf{49.0 kPa} \quad \Leftarrow \textit{Answer}$$
$$u_e = \textbf{39.4 kPa} \quad \Leftarrow \textit{Answer}$$
$$u = \textbf{88.4 kPa} \quad \Leftarrow \textit{Answer}$$

To develop a complete plot of pore pressure vs. depth, the computations in Example 12.1 would need to be repeated at many different depths. This would be tedious to do by hand, but easy to do with a computer.

When a computer is not available, the excess pore water pressures also may be computed using Figure 12.4, which presents plots of $u_e/\Delta\sigma_z$ for various values of T_v. It was developed from Equation 12.17. Notice how the consolidation process (i.e., the dissipation of excess pore water pressures) occurs very quickly at the top and bottom because the excess pore water pressures drain most easily there. However, the process is much slower in the center because it is farther from the drainage boundaries.

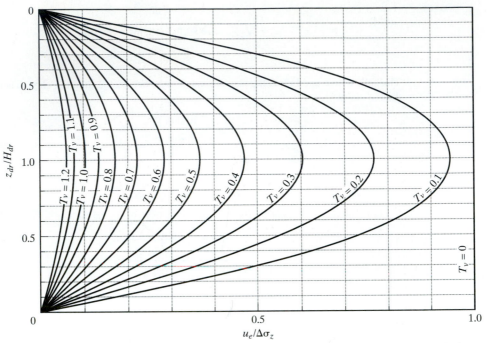

Figure 12.4 $u_e/\Delta\sigma_z$ for various values of T_v with double drainage. For the single drainage case, use only the upper half of this diagram.

Example 12.2

A fill is to be placed on the soil profile shown in Figure 12.5. Using the curves in Figure 12.4, develop a plot of u_h, u_e, and u vs. depth at $t = 10$ yr after placement of the fill.

Solution

The hydraulic conductivities of the SM and ML strata are much greater than that of the CH strata, so the double drainage condition exists.

$$H_{dr} = \frac{10.0 \text{ m}}{2} = 5.0 \text{ m}$$

$$T_v = \frac{c_v t}{H_{dr}^2} = \frac{(3\times 10^{-3} \text{ m}^2/\text{d})(10\,\text{yr})(365\,\text{d/yr})}{(5.0\,\text{m})^2} = 0.438$$

$$\Delta\sigma_z = \gamma_{fill} H_{fill} = (19.7 \text{ kN/m}^3)(3.5 \text{ m}) = 68.9 \text{ kPa}$$

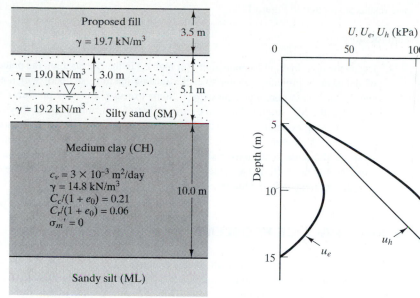

Figure 12.5 Soil profile and results for Example 12.2.

Depth from original ground surface (m)	Soil	$u_h = \gamma_w z_w$ (kPa)	z_{dr} (m)	z_{dr}/H_{dr}	$u_e/\Delta\sigma_z$ from Fig. 12.4	$u_e = 68.9\, u_e/\Delta\sigma_z$ (kPa)	$u = u_h + u_e$ (kPa)
0		0	-	-	-	-	0
1		0	-	-	-	-	0
2	SM	0	-	-	-	-	0
3		0	-	-	-	-	0
4		9.8	-	-	-	-	9.8
5	—	19.6	0	0	0	0	19.6
6		29.4	1.0	0.20	0.13	9.0	38.4
7		39.2	2.0	0.40	0.25	17.2	56.4
8		49.0	3.0	0.60	0.35	24.1	73.1
9		58.8	4.0	0.80	0.41	28.2	87.0
10	CH	68.6	5.0	1.00	0.44	30.3	98.9
11		78.4	4.0	0.80	0.41	28.2	106.6
12		88.2	3.0	0.60	0.35	24.1	112.3
13		98.0	2.0	0.40	0.25	17.2	115.2
14		107.8	1.0	0.20	0.13	9.0	116.8
15	—	117.6	0	0	0	0	117.6
16	ML	127.4	-	-	-	-	127.4
17		137.2	-	-	-	-	137.2

The results are plotted in Figure 12.5.

Commentary

1. Due to the high hydraulic conductivity (or high c_v) in the SM stratum, the potential rate of drainage is probably greater than the rate of loading, so there will not be any significant excess pore water pressures. Some excess pore water pressures might be present in the ML stratum during the early stages of consolidation, but they will have dissipated long before $t = 10$ years. Therefore, excess pore water pressures are present only in the CH stratum.
2. If single drainage conditions had been present in the CH strata, then the z_{dr}/H_{dr} values would range from 0 at the top of the stratum to 1 at the bottom, and the u_e values would be correspondingly higher.

QUESTIONS AND PRACTICE PROBLEMS

12.1 A 12.0 m thick saturated clay stratum with double drainage is to be subjected to a $\Delta\sigma_z$ of 75 kPa. The coefficient of consolidation in this soil is 3.5×10^{-3} m²/day. Using Equation 12.17, compute the excess pore water pressure at a point 2.7 m above the bottom of this stratum 10 years after placement of the load.

12.2 Solve Question 12.1 using Figure 12.4.

12.3 A 20 ft thick fill with a unit weight of 120 lb/ft³ is to be placed on the soil profile shown in Figure 12.6. Assuming the fill is placed instantaneously, use the curves in Figure 12.4 to develop a plot of u_e vs. depth at $t = 1.5$ yr. Plot depth on the vertical axis, increasing downward, and consider depths from the original ground surface to the bottom of the CL stratum.

12.4 Use Equation 12.17 to compute the hydrostatic, excess, and total pore water pressures at Point F in Figure 12.6 at $t = 1, 2, 4, 8$, and 16 years after placement of the fill. Then use this data to develop a plot of u_h, u_e, and u at this point vs. time. All three curves should be on the same diagram, with time on the horizontal axis.

12.5 Using the soil profile in Figure 12.6, develop a spreadsheet (Excel, Lotus 1-2-3, etc.) that solves Equation 12.17 at 1.0 ft depth intervals through the entire soft clay stratum. Use summations for $N = 0$ to 8. Then use this spreadsheet to develop a curve of excess pore water pressure vs. depth at $t = 6$ yr after construction. Submit a printout of the spreadsheet, and a plot of the excess pore water pressure curve.

Note for those who may wish to develop spreadsheets or other software for more general solutions: The natural exponent term in Equation 12.17 may cause difficulties for some programming languages when they attempt to take e to a large negative power. However, these difficulties appear to occur only when N has risen to values beyond those necessary for the summation. Therefore, avoid such difficulties by terminating the summation whenever the exponent term generates an error, or when the increment of N produces a negligible change in the summation.

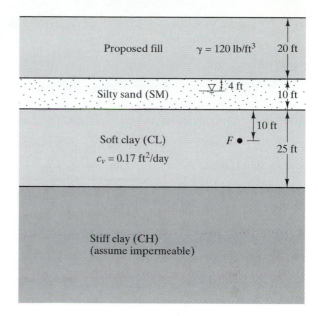

Figure 12.6 Soil profile for Problems 12.3–12.5.

12.2 CONSOLIDATION SETTLEMENT VS. TIME COMPUTATIONS

Now that we have Equation 12.17 and are able to compute excess pore water pressures as a function of depth and time, we also can compute consolidation settlement, δ_c, as a function of the time since loading, t. Plots of anticipated settlement vs. time are very valuable to geotechnical engineers because they help us plan appropriate mitigative measures. For example, if the weight of a proposed fill will produce a certain amount of settlement, but this settlement will be virtually complete before construction of any buildings begins, then its impact on the buildings will be minimal. However, if the settlement will continue for many years after construction of the buildings, then it may be necessary to provide some different type of foundation or some other measures to avoid damaging the buildings.

To compute the consolidation settlement, δ_c, at a particular time, we need to know the vertical effective stress, σ_z'. At the beginning of consolidation ($t = 0$, $\delta_c = 0$), $\sigma_z' = \sigma_{z0}'$; at the end of consolidation ($t = \infty$, $\delta_c = (\delta_c)_{ult}$), $\sigma_z' = \sigma_{zf}'$. Between these times, σ_z' may be computed using the following equation:

$$\sigma_z' = \sigma_{zf}' - u_e \qquad (12.19)$$

where:

σ_z' = vertical effective stress at any time in the consolidation process
σ_{zf}' = vertical effective stress at the end of consolidation
u_e = excess pore water pressure

The vertical effective stress, σ_z', varies with depth, so we must compute the consolidation settlement at time t using the following revised versions of Equations 11.23, 11.24, and 11.25:

For normally consolidated soils ($\sigma_{z0}' \approx \sigma_c'$):

$$\delta_c = \sum \frac{C_c}{1 + e_0} H \log\left(\frac{\sigma_z'}{\sigma_{z0}'}\right) \tag{12.20}$$

For overconsolidated soils — case I ($\sigma_{z0}' < \sigma_z' \leq \sigma_c'$):

$$\delta_c = \sum \frac{C_r}{1 + e_0} H \log\left(\frac{\sigma_z'}{\sigma_{z0}'}\right) \tag{12.21}$$

For overconsolidated soils — case II ($\sigma_{z0}' < \sigma_c' < \sigma_z'$):

$$\delta_c = \sum \left[\frac{C_r}{1 + e_0} H \log\left(\frac{\sigma_c'}{\sigma_{z0}'}\right) + \frac{C_c}{1 + e_0} H \log\left(\frac{\sigma_z'}{\sigma_c'}\right) \right] \tag{12.22}$$

where:
δ_c = consolidation settlement at time t
C_c = compression index
C_r = recompression index
e_0 = initial void ratio
σ_{z0}' = initial vertical effective stress
σ_z' = vertical effective stress at time t
σ_c' = preconsolidation stress
H = thickness of soil strata

Equations 12.20–12.22 need to be selected carefully, because the appropriate choice may vary with both depth and time. For example, a given point in the soil may be overconsolidated case I during the early stages of consolidation, then change to overconsolidated case II when σ_z' reaches σ_c'.

It also is helpful to define a new parameter, the *degree of consolidation*, U, which is the percentage of the ultimate consolidation settlement that has occurred at a certain time after loading:

$$U = \frac{\delta_c}{(\delta_c)_{ult}} \times 100\%$$ (12.23)

where:

U = degree of consolidation (percent)

δ_c = consolidation settlement

$(\delta_c)_{ult}$ = ultimate consolidation settlement

We will consider two ways to develop time–settlement curves: one that explicitly considers the dissipation of pore water pressures but requires a computer, and another simplified method that may be solved by hand.

Computer-Based Solution

The analysis described thus far may be implemented using a computer as follows:

1. Divide the compressible stratum into horizontal layers.
2. Using Equation 7.7, compute the hydrostatic pore water pressure, u_h, at the midpoint of each layer.
3. Using the pore water pressure from Step 2, compute the initial vertical effective stress, σ_{z0}', at the midpoint of each layer.
4. Select an appropriate time, t, after placement of the load.
5. Using Equation 12.17, compute the excess pore water pressure, u_e, at the midpoint of each layer.
6. Add the values obtained from Steps 2 and 5 to find the pore water pressure, u, at the midpoint of each layer ($u = u_h + u_e$).
7. Using the pore water pressure from Step 6, compute the vertical effective stress, σ_z', at the midpoint of each layer. This is the σ_z' at time t.
8. Using Equation 12.20, 12.21, or 12.22, as appropriate, compute the consolidation settlement for each layer and sum. This is δ_c at time t. If Equation 12.22 is to be used, it will be necessary to compute σ_c' at the midpoint of each layer using Equation 11.17.
9. Repeat Steps 4 through 8 using new values of t until an acceptable δ_c vs. t plot has been obtained.

This is a type of numerical solution, and its precision depends on the number of layers used in the computations. At least fifty layers are typically necessary to develop reasonable plots.

Program SETTRATE

Program **SETTRATE**, which is part of the geotechnical analysis package that accompanies this book, uses this method to compute the consolidation settlement as a function of time.

It considers only consolidation settlements due to the weight of a wide fill. The program also is limited to soil profiles where the groundwater table is located above the top of the compressible stratum. To use this program, you must first install the geotechnical analysis software package onto a computer. See Appendix C for computer system requirements and installation instructions.

To run the program, select **SETTRATE** from the main menu, then type in the requested information. Once all of the information is entered, click on the **CALCULATE** button to begin a short animation. The time–settlement curve progresses from $t = 0$ to the time specified by the user. Simultaneously, plots of u_e, σ_z', and ϵ_z vs. depth show how these functions vary with time. The animation speed may be adjusted by selecting slow, medium, or fast in the upper left portion of the screen. At the end of the animation, δ_c, T_v, and U are displayed in the upper right portion of the screen.

This program has two purposes. The first is to permit faster solution of time–settlement problems. The second and perhaps more important purpose is to help the reader visualize how the various parameters change with time and depth. This is the purpose of the animation. The plots also show the maximum value of each parameter.

To obtain a time-settlement plot, enter a series of values for the time since loading and record the corresponding calculated values of δ_c. If sandy strata also are present, compute their ultimate consolidation settlement by hand, and add it to the δ_c values from **SETTRATE**.

When finished, a printout may be obtained by clicking on the **PRINT** button. Then, exit by clicking on the **EXIT TO MAIN MENU** button.

Example 12.3

A fill is to be placed over the soil profile shown in Figure 12.7. Using program **SETTRATE**, develop a plot of consolidation settlement vs. time. Both the SP and CL/ML strata are normally consolidated.

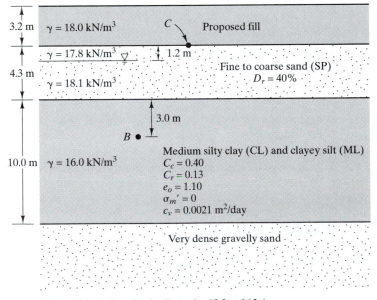

Figure 12.7 Soil profile for Examples 12.3 and 12.4.

Solution

Silty clay/clayey silt stratum:

> Compute using program **SETTRATE**. See Figure 12.8 for typical screen. The results are plotted in the table on the next page.

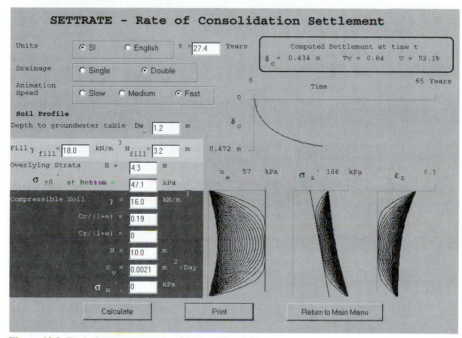

Figure 12.8 Typical **SETTRATE** screen for Example 12.3.

Sand stratum:

> The consolidation settlement in the sand stratum will occur as quickly as the load is applied. Therefore, we only need to compute the ultimate consolidation settlement.

> At midheight in the sand stratum:

$$\sigma'_{z0} = \sum \gamma H - u$$
$$= (17.8 \text{ kN/m}^3)(1.2 \text{ m}) + (18.1 \text{ kN/m}^3)(0.95 \text{ m}) - (9.8 \text{ kN/m}^3)(0.95 \text{ m})$$
$$= 29.2 \text{ kPa}$$

$$\sigma'_{zf} = \sigma'_{z0} + \gamma_{fill} H_{fill}$$
$$= 29.2 \text{ kPa} + (18.0 \text{ kN/m}^3)(3.2 \text{ m})$$
$$= 86.8 \text{ kPa}$$

Per Table 11.4, $C_c/(1+e_0) = 0.007$

$$(\delta_c)_{ult} = \sum \frac{C_c}{1+e_0} H \log\left(\frac{\sigma'_{zf}}{\sigma'_{z0}}\right)$$

$$= (0.007)(4.3 \text{ m}) \log\left(\frac{86.8 \text{ kPa}}{29.2 \text{ kPa}}\right)$$

$$= 14 \text{ mm}$$

Gravelly sand stratum:

This soil is classified as very dense, so we can assume its compressibility is negligible compared to that of the other strata.

The results of these computations are shown in the following table, and the time-settlement curve is plotted in Figure 12.9.

Time		Settlement (mm)		
(days)	(years)	Sand Strata	Clay Strata	Total
2		14	7	21
5		14	12	26
10		14	18	32
20		14	26	40
50		14	41	55
100		14	58	72
200		14	81	95
500	1.4	14	127	141
1000	2.7	14	178	192
1400	3.8	14	209	223
2000	5.5	14	247	261
3200	8.8	14	304	318
5000	13.7	14	361	375
7100	19.5	14	402	416
10000	27.4	14	434	448
14000	38.4	14	455	469
20000	54.8	14	466	480
32000	87.7	14	471	485

Commentary

Although most of the settlement occurs in the first 25 years, the consolidation process continues for at least another 25 years.

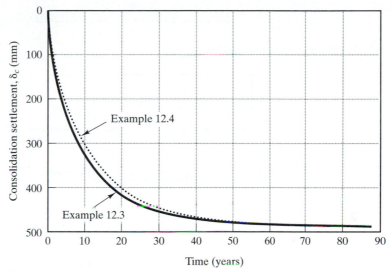

Figure 12.9 Time-settlement plot for Examples 12.3 and 12.4.

Simplified Solution

The second method of computing the consolidation settlement introduces an additional simplifying assumption: For the purpose of computing the dissipation of u_e, we will assume the vertical strain, ϵ_z, due to consolidation in the compressible strata is proportional to the vertical effective stress, σ_z'. In other words, the stress–strain curve is linear. This is clearly not true — it is a logarithmic relationship, as discussed in Chapter 11. However, this assumption simplifies the computations in a significant way because the vertical strain becomes proportional to the drop in excess pore water pressure ($\epsilon_z \propto -u_e$). Thus, a certain drop in u_e at one depth in the compressible stratum produces the same strain as an equal drop in u_e at another depth.

This assumption can be confusing in that it applies only to the settlement rate computation. The value of $(\delta_c)_{ult}$ remains unchanged, and is still based on the non-linear Equations 11.23–11.25. The only difference in the results obtained from this simplified method and the more precise computer-based solution is the shape of the time–settlement curve.

With this new simplifying assumption, U becomes equal to half the area to the right of each T_v curve in Figure 12.4. Therefore, we can develop a unique relationship between U and T_v, as shown in Figure 12.10. This relationship also may be represented by the following fitted equations (adapted from Terzaghi, 1943):

For $T_v \leq 0.217$ ($U \leq 52.6\%$):

$$U = \sqrt{\frac{4T_v}{\pi}} \times 100\% \tag{12.24}$$

For $T_v > 0.217$ ($U > 52.6\%$):

$$U = \left[1 - 10^{-\left(\frac{0.085 + T_v}{0.933}\right)}\right] \times 100\% \tag{12.25}$$

where:
   U = degree of consolidation (percent)
   T_v = time factor

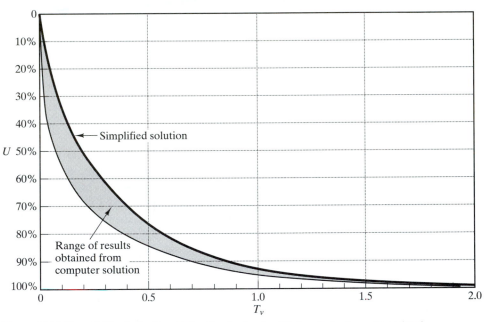

Figure 12.10 The solid line is the U vs. T_v function for the simplified analysis of one-dimensional consolidation. The shaded area represents the range of values obtained from the more precise computer solution described earlier.

Figure 12.10 also shows the range of U vs. T_v values obtained from the more precise computer solution described earlier. The difference between these two methods is often small, but it can be quite significant, especially during the early stages of consolidation.

Example 12.4

Solve Example 12.3 using Figure 12.10.

Solution

According to program **FILLSETT**, or from a hand analysis similar to the ones in Chapter 11, the ultimate consolidation settlement, $(\delta_c)_{ult}$, of the clay stratum is 471 mm.

Using Equation 12.18 to compute T_v and Figure 12.10 (or Equations 12.24 and 12.25) to find U, we obtain the following results:

Time		T_v	U (%)	Settlement (mm)		
(days)	(years)			Sand Strata	Clay Strata	Total
2		0.00017	1	14	5	19
5		0.00042	2	14	9	23
10		0.00084	3	14	14	30
20		0.0017	5	14	24	38
50		0.0042	7	14	33	47
100		0.0084	10	14	47	61
200		0.017	15	14	71	85
500	1.4	0.042	23	14	108	122
1000	2.7	0.084	33	14	155	169
1400	3.8	0.12	39	14	184	198
2000	5.5	0.17	47	14	221	235
3200	8.8	0.27	58	14	273	287
5000	13.7	0.42	71	14	334	348
7100	19.5	0.60	82	14	386	400
10000	27.4	0.84	90	14	424	438
14000	38.4	1.2	96	14	452	466
20000	54.8	1.7	99	14	466	480
32000	87.7	2.7	100	14	471	485

These results are plotted in Figure 12.9, along with the computer-generated results from Example 12.3. The greatest difference between these results is about 12 percent.

Correction for Construction Period

In reality, most loads applied to soils do not occur instantaneously. They usually are applied during some construction process that may last for weeks or months. For example, loads due to the weight of new fills are imparted only as fast as the fill is constructed.

A simple method of computing settlements during and after the construction period is to assume the load is applied at a uniform rate, then adjust the time t in the settlement computations as follows:

For $t \leq t_c$:

$$t_{adj} = \frac{t}{2}$$

(12.26)

For $t > t_c$:

$$t_{adj} = t - \frac{t_c}{2}$$

(12.27)

where:
  t = time since beginning of construction
  t_c = duration of construction period (i.e. time at end of construction)
  t_{adj} = adjusted time

Then, perform the settlement rate computation using t_{adj} and the value of H_{fill} present at time t. For example, if a 6 m thick fill is to be placed at a uniform rate over a period of 30 days, we would compute the settlement at t = 20 days using t_{adj} = 20/2 = 10 days and H_{fill} = 6 m (20/30) = 4 m (i.e., the amount of fill present at t = 20 days).

Example 12.5

A proposed fill is to be placed on the soil profile shown in Figure 12.11. The fill will be placed at a uniform rate over a period of six months. Develop a time–settlement curve considering this construction period.

Solution

H_{fill} increases linearly from 0 at t = 0 to 12 ft at t = 180 days.

SM/ML stratum:

Assume the consolidation settlement occurs as quickly as the fill is placed (i.e., no significant excess pore water pressures).

Divide into two layers: Layer 1 is the upper 4 ft and layer 2 is the lower 6 ft.

For Layer 1:

$$\sigma'_{z0} = \sum \gamma H - u = (118 \text{ lb/ft}^3)(2 \text{ ft}) - 0 = 236 \text{ lb/ft}^2$$

$$(\delta_c)_{ult} = \sum \frac{C_c}{1 + e_0} H \log\left(\frac{\sigma'_{zf}}{\sigma'_{z0}}\right) = (0.02)(4)\log\left(\frac{236 + \gamma_{fill}H_{fill}}{236}\right)$$

For Layer 2:

$$\sigma'_{z0} = \sum \gamma H - u = (118 \text{ lb/ft}^3)(4 \text{ ft}) + (119 \text{ lb/ft}^3)(3 \text{ ft}) - (62.4 \text{ lb/ft}^3)(3 \text{ ft}) = 642 \text{ lb/ft}^2$$

$$(\delta_c)_{ult} = \sum \frac{C_c}{1 + e_0} H \log\left(\frac{\sigma'_{zf}}{\sigma'_{z0}}\right) = (0.02)(4)\log\left(\frac{642 + \gamma_{fill}H_{fill}}{642}\right)$$

Compute δ_c for each time using the corresponding H_{fill}. The results of this computation are shown in the table on the next page.

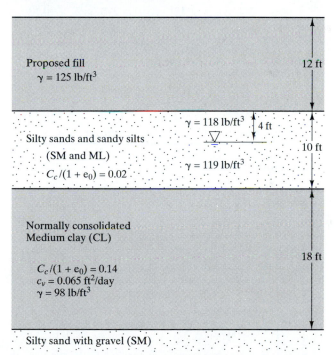

Figure 12.11 Soil profile for Example 12.5.

CL Stratum:

The consolidation in this stratum will continue well beyond the construction period, and needs to be computed using program **SETTRATE**. For $t = 0$ to 180 days, use Equation 12.26 to find t_{adj}, and use H_{fill} linearly increasing with time. For $t > 180$ days, use Equation 12.27 and $H_{fill} = 12$ ft.

Time Since Beginning of Construction t (days)	H_{fill} (ft)	Adjusted time for settlement computations t_{adj} (days)	Consolidation settlement, δ_c (ft)		
			SM/ML	CL	Total
0	0	0	0	0	0
30	2	15	0.04	0.14	0.18
60	4	30	0.07	0.20	0.27
90	6	45	0.09	0.24	0.33
120	8	60	0.11	0.28	0.39
180	12	90	0.13	0.34	0.47
240	12	150	0.13	0.44	0.57
300	12	210	0.13	0.51	0.64
360	12	270	0.13	0.57	0.70
480	12	390	0.13	0.66	0.79
600	12	510	0.13	0.73	0.86
900	12	810	0.13	0.83	0.96
1200	12	1110	0.13	0.88	1.01
1800	12	1710	0.13	0.92	1.05
2400	12	2310	0.13	0.93	1.06

These results are plotted in Figure 12.12.

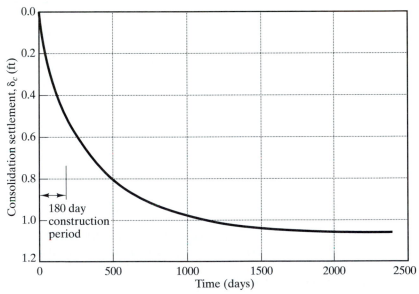

Figure 12.12 Results of settlement computations for Example 12.5.

QUESTIONS AND PRACTICE PROBLEMS

12.6 Consider the proposed fill and soil profile shown in Figure 12.5, except replace the sandy silt strata with an impervious bedrock. Using the simplified solution, compute the consolidation settlement at $t = 15$ years after placement of the fill. The ultimate consolidation settlement is 0.50 m. Do not apply any correction for the construction period.

12.7 Solve Problem 12.6 using program **SETTRATE**.

12.8 A fill is to be placed on the soil profile shown in Figure 12.13. The groundwater table is level with the original ground surface. Use program **SETTRATE**, develop a plot of consolidation settlement vs. time. Continue the plot until $U > 99\%$. Do not apply any correction for the construction period.

Note: As consolidation settlement occurs, some of the fill will become submerged beneath the groundwater table. The resulting buoyant force will reduce σ_{zf}' and thus reduce the consolidation settlement. However, this effect is small for this problem and may be ignored.

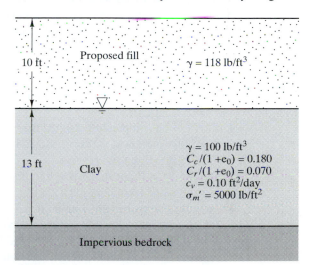

Figure 12.13 Soil profile for
Problems 12.8–12.10.

12.9 A shopping center is to be built on the fill described in Problem 12.8. The proposed buildings and other facilities can tolerate a settlement due to the weight of the fill of no more than 2 in. Therefore, once the fill has been placed, it will be necessary to wait until enough settlement has occurred that the remaining settlement will be less than 2 in. Only then may the building construction begin.

      Assuming the fill will be placed at a uniform rate from May 1 to June 1, determine the earliest start date for the building construction. Apply a correction for the construction period and use program **SETTRATE**.

      For this problem, consider only settlement due to the weight of the fill. Do not consider settlement due to the weight of the buildings.

12.10 Solve Problem 12.9 using the simplified solution.

12.3 THE COEFFICIENT OF CONSOLIDATION, c_v

Terzaghi's theory lumps all of the soil properties (other than the drainage distance) into one parameter, the coefficient of consolidation, c_v, as defined in Equation 12.1. Of the various parameters in this equation, the hydraulic conductivity, k, varies most widely, and thus is the most important factor. Therefore, c_v is very small in clays and very large in sands.

We need to have some means of measuring c_v before we can perform time–settlement analyses. One method of doing so might be to assess each of the parameters in Equation 12.1 and calculate c_v, but this is rarely done. Instead, engineers usually measure the rate of consolidation in a laboratory consolidation test and back-calculate c_v by performing a time–settlement analysis in reverse. Because H_{dr} in the lab is very small, the rate of consolidation is much faster than that in the field, but c_v should, in theory, be equal to the field value.

In principle, it should be a simple matter to obtain c_v from laboratory time–settlement data. The stress conditions in the laboratory sample are such that the U vs. T_v relationship is exactly as shown by the solid line in Figure 12.10 and Equations 12.24 and 12.25. Thus, we might expect to simply select an appropriate point on the laboratory time-settlement plot, identify the corresponding values of U, t, and T_v, and use Equation 12.18 to compute c_v.

In practice, this task is slightly more complicated because the time-settlement behavior in the lab is slightly different than that in the field. Therefore, it has been necessary to develop special curve-fitting methods to interpret the laboratory data. One of these is the *square root of time fitting method* developed by Taylor (1948), as follows:

1. Plot the soil compression against the square root of time, as shown in Figure 12.14.
2. The initial portion of the curve should be fairly straight. Extrapolate it back to $\sqrt{t} = 0$. This locates Point A.
3. Beginning at Point A, draw a line that has a slope of 1.15 times that of the initial portion of the laboratory curve.
4. Note the point where the line in Step 3 crosses the laboratory curve. This is Point B, which represents $U = 90\%$. Read the corresponding time, $\sqrt{t_{90}}$.
5. Using Equation 12.18, $T_v = 0.848$ (the theoretical value at $U = 90\%$), $t = t_{90}$, and $H_{dr} = $ one half the sample height (laboratory samples have double drainage), compute c_v.

Casagrande presented another commonly used method of finding c_v from laboratory test data (Holtz and Kovacs, 1981). It is called the *logarithm of time fitting method*. This method plots the data on a settlement vs. log time diagram and locates the point where $U = 100\%$. The procedure then locates $U = 50\%$ and the corresponding time, t_{50}, is placed into Equation 12.18 along with $T_v = 0.197$ (per Equation 12.24) to compute c_v.

Most geotechnical engineers prefer the square root of time method because it permits the next load to be placed as soon as t_{90} has been reached, whereas the logarithm of time method requires the load be left on long enough to identify t_{100} Since consolidation tests are very slow anyway, typically requiring days to complete, this difference can have a significant impact on the cost of performing the test.

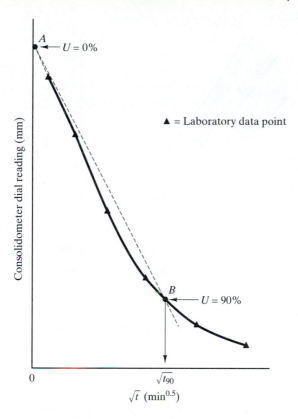

Figure 12.14 Taylor's square root of time method for computing c_v from laboratory consolidation test data.

Figure 12.15 presents an approximate correlation between c_v and the liquid limit. Although this diagram is not a substitute for performing laboratory tests, it may be used to check test results for reasonableness and for preliminary estimates. Table 12.1 lists some measured values. Notice there is some disagreement between these two references.

Terzaghi's one-dimensional consolidation equation (Equation 12.16) is based on c_v being a constant. However, in reality it varies with effective stress as described in Equation 12.1. This effect can be seen by computing c_v for each of the several load increments in the consolidation test. Figure 12.16 shows measured values at different effective stresses for several soils. Note the sudden change in c_v at the preconsolidation stress, which is explained by the C factor in Equation 12.1 (this factor is equal to C_r on one side of the preconsolidation stress, and C_c on the other side). Therefore, overconsolidated soils typically have c_v values five to ten times greater than the same soils in a normally consolidated condition.

It is possible to write a new version of Equation 12.16 where c_v is a variable, but this equation appears to require a numerical solution to compute u_e (i.e., we would not have a direct solution like Equation 12.17). Although computer software could be developed to do this, such analyses are rarely, if ever, performed in practice. Instead, we simply evaluate c_v at an effective stress equal to the σ_z' in the field, and consider it to be a constant.

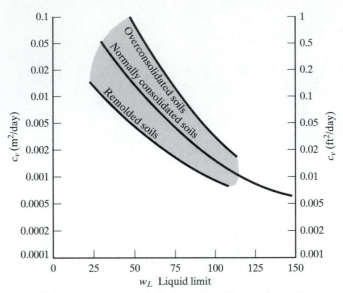

Figure 12.15 Approximate correlation between coefficient of consolidation, c_v, and liquid limit, w_L. Undisturbed normally consolidated soils typically plot near the center of the shaded zone, undisturbed overconsolidated soils in the upper portion, and remolded soils in the lower portion. (U.S. Navy, 1982)

TABLE 12.1 MEASURED VALUES OF c_v (Adapted from Holtz and Kovacs, 1981)

Soil	c_v	
	m²/day	ft²/day
Boston blue clay (CL) (Ladd and Luscher, 1965)	0.033 ± 0.016	0.33 ± 0.16
Organic silt (OH) (Lowe, Zaccheo, and Feldman (1964)	0.00016 - 0.00082	0.0016 - 0.0082
Glacial lake clays (CL) (Wallace and Otto, 1964)	0.00055 - 0.00074	0.0055 - 0.0074
Chicago silty clay (CL) (Terzaghi and Peck, 1967)	0.00074	0.0074
Swedish medium sensitive clays (CL-CH) (Holtz and Broms, 1972)		
Laboratory	0.0003 - 0.0006	0.003 - 0.006
Field	0.0006 - 0.003	0.006 - 0.03
San Francisco bay mud (CL)	0.0016 - 0.0033	0.016 - 0.033
Mexico City clay (MH) (Leonards and Girault, 1961)	0.0008 - 0.0014	0.008 - 0.014

12.4 ACCURACY OF SETTLEMENT RATE PREDICTIONS

As with all other geotechnical analyses, settlement rate predictions are subject to many errors. These include the sources of error for consolidation analyses in general, as described in Section 11.14, plus additional errors unique to rate computations.

An especially important source of error in settlement rate computations is our assessment of the length of the longest drainage path, H_{dr}. The time required to reach full consolidation is proportional to H_{dr}^2, as discussed earlier, so even small errors in this value can produce significant changes in the computed settlement rate. This problem can be quite insidious, because compressible clay strata that appear to be homogeneous often contain thin horizontal sandy seams. If these layers are continuous, or nearly so, the horizontal hydraulic conductivity can be much greater than the vertical hydraulic conductivity as discussed in Chapter 7 ($k_x \gg k_z$). Therefore, the excess pore water generated in the clay will move up or down to the nearest sand seam, then escape horizontally through the seam as shown in Figure 12.17.

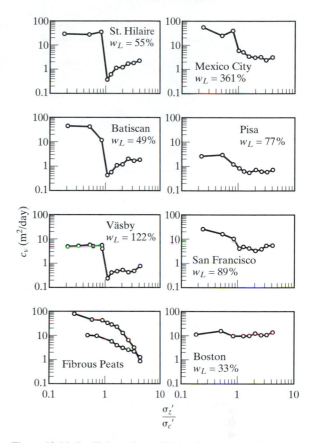

Figure 12.16 Coefficient of consolidation for various soils as a function of effective stress (Terzaghi, Peck, and Mesri, 1996).

For example, consider a 8 m thick clay strata with double drainage. If it contains thin continuous sand seams every 2 m, then H_{dr} is really 1/4 of the apparent value, and the time required for consolidation will be $(1/4)^2 = 1/16$ of the computed value. This illustrates the importance of identifying small details when logging exploratory borings.

The analysis method presented in this chapter was also based on drainage occurring only vertically. This assumption loses its validity when the loaded area is small, such as a structural foundation. In such cases, much of the drainage is horizontal, even if no sandy seams exist, and the consolidation settlement is correspondingly faster.

Another potential source of error occurs in soils that are not fully saturated. For example, some organic soils have $S < 100\%$, even though they may be located below the groundwater table. The air bubbles in such soils compress quickly, thus accelerating the consolidation process.

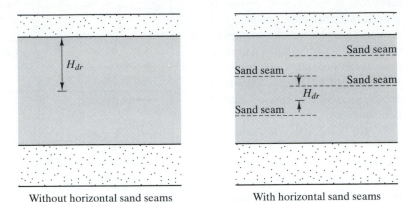

Without horizontal sand seams   With horizontal sand seams

Figure 12.17 Effect of thin horizontal sand seams on the rate of consolidation.

Fairly accurate predictions of settlement rates are sometimes possible when wide loads are placed at very uniform sites with extensive exploration and testing programs. However, non-uniformities in the subsurface conditions at most sites, combined with economic limitations on exploration and testing introduce substantially more error. Thus, it is not unusual for the actual rate of settlement to be half of the predicted rate, or twice the predicted rate.

12.5 CONSOLIDATION MONITORING

Predictions of both the rate and magnitude of settlement can be seriously in error, so engineers usually use conservative estimates when assessing the impact of these settlements on proposed construction. Often, such conservative estimates are acceptable because they do not have a serious impact on the final design. However, when the predicted settlements are large, such conservative designs can be very expensive. In such cases, geotechnical engineers sometimes install instruments in the ground to monitor the actual settlements, especially during the early stages of consolidation. We then use the data from these instruments to update our pre-construction settlement computations. These updated predictions are much more reliable because they have been calibrated based on the actual field performance, and thus may be used to justify less conservative designs.

There are two approaches to monitoring consolidation in the field. The first monitors only settlements, while the second monitors both settlements and pore water pressures.

Monitoring Settlement Only

A basic monitoring program consists of installing survey monuments at the ground surface and periodically measuring their elevation using a total station or other conventional land surveying equipment. These monuments need to be firmly fastened to the ground, and must not be subject to surficial soil movements, vandalism, or damage from construction equipment.

Another method of monitoring settlements is to install remote devices such as that shown in Figures 12.18 and 12.19. This one consists of a water reservoir installed at a fixed elevation with a tube leading to the sensing unit, which is buried in the ground. A pressure transducer at the sensing unit measures the pressure in the water, which permits computation of the difference in elevation between the reservoir and the sensor. As the sensor settles, the pressure changes and the new elevation can be computed. This device is installed before the fill is placed, thus allowing settlements to be monitored both during and after construction.

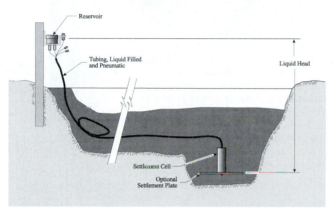

Figure 12.18 Installation of remote reading settlement plate. The reservoir is located on ground that is not settling (Slope Indicator Company).

Figure 12.19 Remote reading settlement plate (Slope Indicator Company).

Settlement monitoring devices that install in bore holes also are available. These permit monitoring settlements as a function of depth. The Sondex device, shown in Figures 12.20 and 12.21, is one such device.

Although settlement data is very useful, its interpretation is not always easy. For example, Figure 12.22 shows the pre-construction prediction of settlement vs. time from Example 12.3, along with the first 1000 days of observed settlements obtained from a settlement plate installed at Point C in Figure 12.7. The observed settlement is greater than the original prediction, but it is not clear if this is because the ultimate settlement will be greater than predicted, the rate of settlement is faster than predicted, or both. Thus, there are many reasonable interpretations of the remaining settlement. Two such interpretations are shown — one that assumes all of the error is in the C_c value, and one that assumes all is in c_v.

Monitoring Both Settlement and Pore Water Pressure

Fortunately, much of the ambiguity in interpreting settlement data can be overcome by also installing piezometers in the natural ground. It is best to install them before the fill is

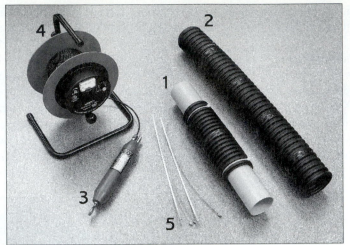

Figure 12.20 The Sondex device uses corrugated plastic casing with embedded steel wires (1) and (2). This casing is installed in a vertical boring and the surrounding annular void is backfilled. Then the Sondex probe (3) is lowered into the casing. It magnetically senses the location of each wire (5) and the corresponding depths are recorded, and the corresponding elevations are computed. This process is repeated at convenient time intervals to determine the time–settlement behavior at each ring. When not in use, the probe is wound onto the reel (4) and stored in a safe location (Slope Indicator Company).

Figure 12.21 Installation of a Sondex casing in a boring (Slope Indicator Company).

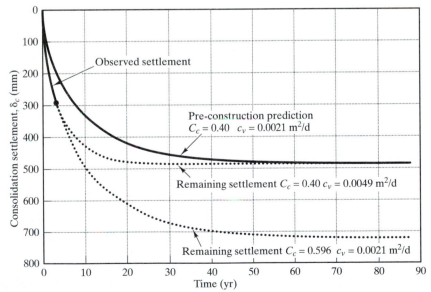

Figure 12.22 Observed settlement data compared to the pre-construction prediction. The dotted lines show two of many possible predictions of the remaining settlement.

placed, perhaps by using remote reading units such as the one shown in Figure 7.8. Initially these piezometers record the hydrostatic pore water pressure. Then, as the fill is placed, the pore water pressure will rise, reflecting the excess pore pressures. These excess pressures will eventually dissipate, and after a sufficiently long time the pore pressure will return to its hydrostatic value.

Using this pore pressure data, we can compute the field value of c_y. Then, by combining this value with the observed time–settlement curve, we obtain an unambiguous plot of the remaining time–settlement behavior. Although this analysis still contains some error, it is more precise than those made without piezometer data, and far superior to analyses based only on laboratory tests. Example 12.6 illustrates this technique.

Example 12.6

Figure 12.23 shows the first 1000 days of pore pressure readings from a piezometer installed at Point B in Figure 12.7. Use this data, along with the observed time–settlement plot in Figure 12.22 and the information in Example 12.3 to develop a revised prediction of the remaining time–settlement curve.

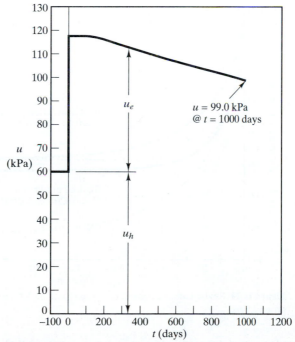

Figure 12.23 Piezometer data for Example 12.6.

Solution

Based on Figure 12.23, the excess pore water pressure at Point B, 1000 days after loading is:

$$u_e = u - u_h$$
$$= u - \gamma_w z_w$$
$$= 99.0 \text{ kPa} - (9.8 \text{ kN/m}^3)(6.1 \text{ m})$$
$$= 39.2 \text{ kPa}$$

By trial-and-error with Equation 12.17, c_v must be 0.0043 to obtain $u_e = 39.2$ kPa. This is about twice the laboratory value of 0.0021, and explains why the consolidation is occurring faster than anticipated.

Per Figure 12.22, the observed consolidation settlement at $t = 1000$ days (2.7 years) is 280 mm. Because 14 mm of this settlement occurred in the SP stratum (per Example 12.3), 280 - 14 = 266 mm must have occurred in the CL/ML stratum. Using trial-and-error with Program **SETTRATE**, the value of $C_c/(1+e_0)$ must be 0.203 to obtain $(\delta_c)_{ult} = 266$ mm. Thus, $C_c = (0.426)(1+1.10) = 0.426$. This is slightly higher than the 0.40 obtained from the laboratory tests.

Using these revised values of c_v and C_c, and program **SETTRATE**, we can develop the revised time–settlement curve shown in Figure 12.24.

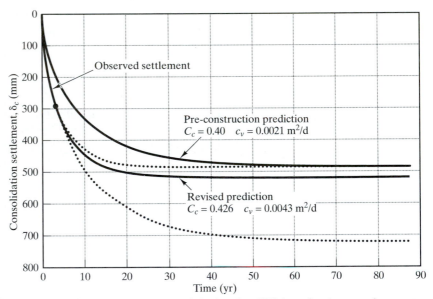

Figure 12.24 Revised time-settlement prediction based on 1000 days of settlement and pore water pressure data.

This methodology of making pre-construction predictions, then revising them based on observed performance is known as the *observational method* (Peck, 1969), a term coined by Terzaghi. This method is very useful in geotechnical engineering, and is one of the ways we can overcome many of the accuracy problems in our analyses without resorting to overly conservative designs.

QUESTIONS AND PRACTICE PROBLEMS

12.11 The data shown in the table to the right were obtained from a laboratory consolidation test on a normally consolidated undisturbed MH soil with a liquid limit of 65. The sample was 62 mm in diameter, 25 mm tall and was tested under a double drainage condition. Compute c_v using the square root of time fitting method. Then, compare your result with a typical value of c_v for this soil and determine if your value seems reasonable.

Time Since Loading (HH:MM:SS)	Dial Reading (mm)
00:01:01	7.21
00:03:16	7.74
00:08:35	8.40
00:16:39	9.01
00:30:15	9.60
00:59:17	10.11
01:54:29	10.35

12.12 A proposed fill is to be placed on the soil profile shown in Figure 12.25.

a. Using the laboratory test results shown in this figure and program **SETTRATE**, develop a time–settlement plot. Do not apply any correction for the construction period. The medium clay is normally consolidated.

b. A piezometer has been installed at Point A and a remote-reading settlement plate at Point B. Measurements from these instruments made 2580 days after placement of the fill indicated a pore water pressure of 1975 lb/ft^2 and a settlement of 1.20 ft. Using the technique described in Example 12.6, back-calculate the values of $C_c/(1+e_0)$ and c_v, and compare them to the laboratory values. Then use program **SETTRATE** to develop a revised time–settlement plot.

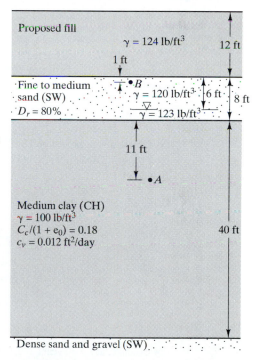

Figure 12.25 Soil profile for Problem 12.12.

12.6 OTHER SOURCES OF TIME DEPENDENCY

The discussions in this chapter have been based on the assumption that dissipation of excess pore water pressures is the sole cause of time dependancy in soil settlement. Although this mechanism is probably the most important one, others also exist.

Settlements due to secondary compression, as discussed in Section 11.10, are by definition independent of excess pore water pressures. When secondary compression is significant, it may be evaluated by developing a time-settlement plot that begins when the consolidation settlement is complete. Example 12.7 illustrates this technique.

This chapter also has assumed the soil is saturated ($S = 100\%$) and remains so. Unsaturated soils are subject to volume changes if they become wetted, and initially saturated soils may change in volume if they are dried. In the case of expansive soils, wetting causes swelling, while drying causes shrinkage. Another type of soil, called a collapsible soil, shrinks when it is wetted. Any of these processes can produce significant settlement or heaving at the ground surface, as discussed in Chapter 18, and the timing of these movements depends primarily on when the soil becomes wetted or dried.

This chapter also has considered only static loads. Vibratory loads, such as from heavy machinery or earthquakes, also can cause settlement, especially in loose sands. Static loads that cycle over longer periods, such as those due to the annual loading and unloading of grain silos, also can cause additional settlements (Coduto, 1999).

Example 12.7

The CL/ML soil in Example 12.3 has $C_\alpha = 0.015$. Compute the secondary compression settlement, then develop a new plot of settlement vs. time that considers both consolidation and secondary compression.

Solution

Per the discussion in Section 11.10, we will use t_p equals the time required to achieve $U = 95\%$. According to an analysis using Program **SETTRATE**, $U = 95\%$ occurs at $t = 12,100$ days (33.1 years).

Using Equation 11.28:

$$\delta_s = \frac{C_\alpha}{1 + e_p} H \log\left(\frac{t}{t_p}\right)$$

$$= \frac{0.015}{1 + 1.10}(10,000 \text{ mm}) \log\left(\frac{t}{33.1}\right)$$

$$= 71 \log\left(\frac{t}{33.1}\right)$$

For example, at $t = 50$ years, the consolidation settlement is $464 + 14 = 478$ mm (per Program **SETTRATE**), and the secondary compression settlement is 12 mm (per the above equation), for a total of 490 mm. Using this method, we obtain the plot shown in Figure 12.26. Comparing this plot with Figure 12.9 demonstrates that secondary compression is not a major issue at this site.

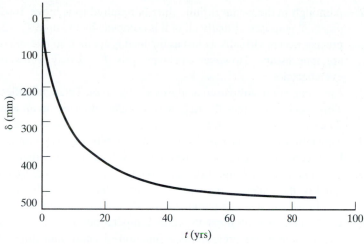

Figure 12.26 Time-settlement plot for Example 12.7.

12.7 METHODS OF ACCELERATING SETTLEMENTS

In theory, an infinite time is required to achieve $U = 100\%$. Fortunately, we do not need to wait that long in practice! Normally, construction of structures and other facilities can begin once U reaches a value such that the remaining settlement is less than some maximum allowable settlement. It is not necessary to wait until all of the consolidation settlement is completed. Nevertheless, even the time required to achieve this level of consolidation is sometimes excessive, perhaps requiring many years or even decades. This is especially likely when the compressible stratum is soft (high C_c) and thick (high H_{dr}).

When the time required to achieve the desired consolidation settlement is excessive, geotechnical engineers begin to consider methods of accelerating the consolidation process. Several methods are available, including the placement of surcharge fills and the installation of vertical drains. Although these methods are expensive, their cost often can be justified because they permit construction to begin years or even decades earlier than would otherwise have been possible. Chapter 19 discusses these methods.

SUMMARY

Major Points

1. As the consolidation process proceeds from $t = 0$ to $t = \infty$, the following changes occur in the soil:

    Excess pore water pressure, $u_e = \Delta\sigma_z \rightarrow 0$
    Vertical effective stress, $\sigma_z' = \sigma_{z0}' \rightarrow \sigma_{zf}'$
    Time factor, $T_v = 0 \rightarrow \infty$
    Degree of consolidation, $U = 0 \rightarrow 100\%$
    Consolidation settlement, $\delta_c = 0 \rightarrow (\delta_c)_{ult}$

2. Although in theory an infinite time is required to achieve 100% consolidation, for practical purposes virtually all of it is complete by the time $T_v \approx 2$. In sandy soils, this process occurs virtually as fast as the load is applied, so rate of consolidation analyses are unnecessary. However, in clayey soils, $T_v \approx 2$ may correspond to many years or even decades.

3. Most rate of consolidation analyses are based on Terzaghi's theory of consolidation. This theory computes the rate of consolidation based on an analysis of excess pore water pressure dissipation.

4. Terzaghi's theory is based on several simplifying assumptions. These include a homogeneous soil, one-dimensional consolidation, and others.

5. The length of the longest drainage path, H_{dr}, is one of the factors that controls how quickly the excess pore water pressures dissipate. All of the other factors are combined into a single parameter, c_v, the coefficient of consolidation.

6. Consolidation settlement vs. time computations may be based on assessments of excess pore water pressure as a function of depth and time. These pressures are computed using Terzaghi's theory. A rigorous solution is too tedious to do by hand, and requires a computer program for practical application.

7. An alternative method of developing time-settlement curves is to use the T_v-U function, which was based on the rigorous analysis and an additional simplifying assumption. This function makes hand computations more practical.

8. The coefficient of consolidation, c_v, is normally determined from laboratory consolidation tests.

9. Settlement rate predictions are not very accurate, so geotechnical engineers often install settlement and pore pressure monitoring equipment in the field. Data collected from this instrumentation may be used to update the laboratory values of C_c and c_v, which then may be used to produced revised time-settlement curves. This technique of updating calculations based on performance data is called the observational method.

10. Settlement predictions can be further refined by considering secondary compression settlement.

Vocabulary

Coefficient of consolidation

Degree of consolidation

Double drainage

Observational method

One-dimensional consolidation equation

Single drainage

Square root of time fitting method

Theory of consolidation

Time factor

COMPREHENSIVE QUESTIONS AND PRACTICE PROBLEMS

12.13 The clay stratum in Figure 12.25 has $C_c/(1+e_0) = 0.16$, $c_v = 0.022$, and $C_\alpha/(1+e_p) = 0.017$. Develop a time–settlement plot for $t = 0$ to 75 years considering both consolidation and secondary compression. Do not apply any corrections for the construction period.

12.14 The CL/ML stratum in Figure 12.7 contains thin horizontal sand seams spaced about 1 m apart. Using this new information, reevaluate the computation in Example 12.4 and develop a revised time–settlement plot. Compare this plot with the ones in Figure 12.9 and explain why they are diff nt.

12.15 Most of the international airport in San Francisco, California, is built on fill placed in San Francisco Bay. A cross-section through one portion of the airport is shown in Figure 12.27.

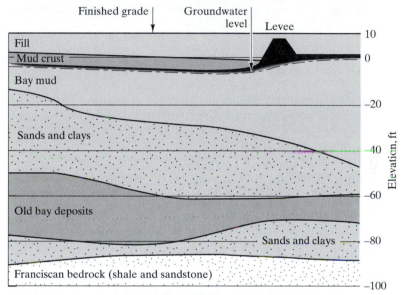

Figure 12.27 Cross-section at San Francisco Airport (Roberts and Darragh, 1962). The groundwater table is indicated by the dashed line.

The engineering properties of these soils are approximately as follows:

	Fill and Levee	Bay Mud Crust	Bay Mud	Sands and Clays	Old Bay Deposits
Dry unit weight, γ_d (lb/ft³)	108	49	40	80	61
Moisture content, w (%)	20	82	118	29	68
Compression index, C_c	0	1.0	1.3	0.5	1.2
Recompression index, C_r	0	0.09	0.17	0.09	0.14
Initial void ratio, e_0		2.40	3.25	1.10	1.70
Overconsolidation margin, σ_m' (lb/ft²)		3500	0	2800	2800
Coefficient of consolidation, c_v (ft²/yr)		130	7	300	5

Note: Some of these values are from Roberts and Darragh; others have been estimated by the author.

a. Compute the ultimate consolidation settlement at various points along the cross-section. Then develop a plot of ultimate consolidation settlement vs. horizontal position. When performing these computations, ignore the presence of the levee and any consolidation that may have already occurred due to its weight (in reality, these earlier settlements would increase the amount of differential settlement in this area, which could be worse than if the levee was never there).

b. Develop plots of settlement vs. time for the left and right ends of the cross-section. Assume all of the settlement in the crust and in the "sands and clays" strata will occur during construction, and assume both the bay mud and old bay deposits have double drainage.

   Hint: Perform separate time-settlement computations for the bay mud and old bay deposits strata, then add the ultimate settlements from the other strata.

12.16 The information presented in Figure 12.25 has the following uncertainties:

Depth to bottom of proposed fill	±1 ft
Depth to bottom of SW stratum	±1 ft
Depth to bottom of CH stratum	±2 ft
Unit weights	±10%
Relative density	±15% (i.e., $D_r = 68$–92%)
$C_c/(1+e_0)$	±20%
c_v	±35%

Considering these tolerances, compute the lower bound solution and upper bound solution for δ_c at $t = 10$ yr and $(\delta_c)_{ult}$. The lower bound solution is that which uses the worst possible combination of factors, while the upper bound uses the best possible combination.

12.17 The soil profile at a certain site includes an 8.5 m thick strata of saturated normally consolidated medium silty clay. This soil has a unit weight of 16.4 kN/m^3. A remote reading settlement plate has been installed a short distance below the natural ground surface and a remote reading piezometer has been installed at the midpoint of the silty clay. The initial readings from these instruments indicate a ground surface elevation of 7.32 m and a pore water pressure of 52 kPa. The initial vertical effective stress at the top of the silty clay stratum was 50 kPa.

   Then a 2.1 m deep fill with a unit weight of 18.7 kN/m^3 was placed on this site. A second set of readings made 220 days after placement of this fill indicate an elevation of 6.78 m and a pore water pressure of 77 kPa. Assuming all of the other soil strata are incompressible, and single drainage conditions exist in the silty clay, compute the values of $C_c/(1+e_0)$ and c_v, then develop a plot of consolidation settlement vs. time. This plot should extend from $U = 0\%$ to $U = 95\%$. Finally, mark the point on this plot that represents the conditions present when the second set of readings were made.

12.18 The analysis in Example 12.6 did not explicitly consider the possibility that the drainage distance H_{dr} used in the original analysis was not correct. Does the adjustment of c_v based on piezometer data, as described in this example, implicitly consider H_{dr}? Explain.

12.19 An engineer in your office is planning a drilling and sampling program at a site that has a thick stratum of soft to medium clay. The information gathered from this program, along with the associated laboratory test results, will be used in various geotechnical analyses, including evaluations of consolidation rates. This engineer has submitted the plan to you for your review and approval.

The engineer expects the clay stratum will be very uniform, and therefore is planning to obtain only a few samples. These samples will then be used to conduct laboratory consolidation tests. Although this plan will probably be sufficient to characterize the consolidation properties of the clay, you are concerned that thin sandy layers might be present in the clay, and that they might not be detected unless more samples are obtained. Write a 200–300 word memo to this engineer explaining the need to search for possible sandy layers, and the importance of these layers in consolidation rate analyses.

12.20 A piezometer has been installed near the center of a 20 m thick stratum of saturated clay. A fill was then placed over the clay and the measured pore water pressure in the piezometer increased accordingly. However, six months after the fill was placed, the piezometer reading has not changed. Does this behavior make sense? Use program **SETTRATE** to justify your answer.

13

Strength

Most of the properties of clays, as well as the physical causes of those few properties that have been investigated, are unknown. We know nothing about the elasticity of clays, or the conditions that determine their water capacity, or the relations between their water content and their viscosity, or the earth pressure that they exert and not even about the physical causes of the swelling of wetted clays. As a consequence, the civil engineer, dealing with this important material, is at the mercy of some unreliable empirical rules, and laboratory work carried out with clays leads only to a mass of incoherent facts.

Karl Terzaghi, 1920

Terzaghi's statement was an accurate assessment of soil mechanics as it existed in 1920. Engineers had very little understanding of soil behavior, and soil strength was one of the most mysterious aspects, especially in clays. Although some researchers had performed soil strength tests, even they did not fully understand what to do with the data once it had been obtained. Practicing engineers had to rely on empirical rules, intuition, and engineering judgment, which often were not adequate. As a result, unexplainable failures were far too common.

Fortunately, our knowledge and understanding of soil strength is now much better than it was in 1920. A large amount of research has been performed, and the results of this work have been successfully applied to practical engineering problems. Therefore, geotechnical designs that rely on soil strength assessments are now much more reliable.

13.1 STRENGTH ANALYSES IN GEOTECHNICAL ENGINEERING

The *strength* of a material is the greatest stress it can sustain. If the stress exceeds the strength, failure occurs. For example, structural engineers know the tensile yield strength of A36 structural steel is 36,000 lb/in² (248 MPa), so they must be sure the tensile stresses in such members are less than this value. In practice, the working stresses must be substantially less to provide a sufficient factor of safety against failure. Such strength analyses can be performed for tensile, compressive, and shear stresses.

Although tensile strength analyses are an important part of structural engineering, geotechnical engineers rarely perform them because soil has very little tensile strength. Even rock masses cannot sustain tension over significant volumes because of the presence of cracks and fissures. There are a few occasions where tensile failures occur, such as tensile cracks near the top of incipient landslides, and in heaves induced by upward seepage forces, as discussed in Section 10.10. However, the geometry of most geotechnical problems is such that nearly all of the ground is in compression.

Geotechnical engineering practice also differs from structural engineering in our assessment of compressive failures. Structural engineers define compressive strengths for various materials and design structural members accordingly, but soil and rock do not fail in compression per se, so we do not perform such analyses. Although the introduction of large compressive stresses may result in failure, the ground is actually failing in shear, not in compression. Therefore, nearly all geotechnical strength analyses evaluate shear only.

Many geotechnical engineering problems require an assessment of shear strength, including:

- **Earth slopes** — When the ground surface is inclined, gravity produces large geostatic shear stresses in the soil or rock. If these stresses exceed the shear strength, a landslide occurs.
- **Structural foundations** — Loads from a structure, such as a building, are transferred to the ground through structural foundations. This produces both compressive and shear stresses in the nearby soil. The latter could exceed the shear strength, thus producing a shear failure. This is known as a bearing capacity failure, and would probably cause the structure to collapse.
- **Retaining walls** — The weight of soil behind a retaining wall produces shear stresses in that soil. Its shear strength resists some of this stress, and the wall resists the rest. Thus, the load carried by the wall depends on the shear strength of the retained soil.
- **Tunnel linings** — Tunnels in soil or weak rock normally include linings of steel or concrete. Such linings must resist pressures exerted by the surrounding ground, thus keeping the tunnel from collapsing. The magnitude of these pressures depends on the strength of the surrounding soil or rock.
- **Highway pavements** — Wheel loads from vehicles pass through the pavement and into the ground below. These loads produce shear stresses that could cause a shear failure. Engineers often place layers of well-graded gravel (known as *aggregate base material*), high-quality soils, or other materials between the pavement and the natural ground. These materials are stronger and stiffer, and help transfer the loads into the ground with much less potential for failure.

Figure 13.1 shows these potential failure modes. We will discuss some of them in more detail later in this book.

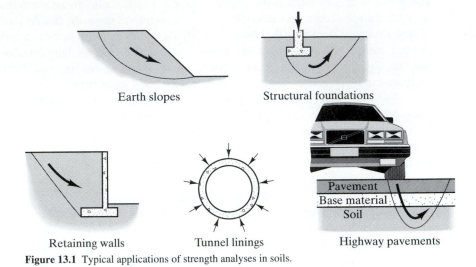

Earth slopes   Structural foundations

Retaining walls   Tunnel linings   Highway pavements

Figure 13.1 Typical applications of strength analyses in soils.

13.2 SHEAR FAILURE IN SOILS

The shear strength of common engineering materials, such as steel, is controlled by their molecular structure. Failure generally requires breaking the molecular bonds that hold the material together, and thus depends on the strength of these bonds. For example, steel has very strong molecular bonds and thus has a high shear strength, while plastic has much weaker bonds and a correspondingly lower shear strength.

However, the physical mechanisms that control shear strength in soil are much different. Soil is a particulate material, as discussed in Chapter 4, so shear failure occurs when the stresses between the particles are such that they slide or roll past each other as shown in Figure 13.2. Although some particle crushing may occur, the shear strength primarily depends on interactions between the particles, not on their internal strength. We divide these interactions into two broad categories: *frictional strength* and *cohesive strength*.

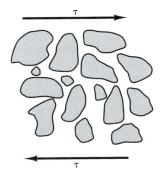

Figure 13.2 Shear failure occurs in a soil when the shear stresses are large enough to make the particles roll and slide past each other.

Frictional Strength

Frictional strength in soils is similar to classic sliding friction from basic physics. The force that resists sliding is equal to the normal force multiplied by the coefficient of friction, μ, as shown in Figure 13.3. However, instead of using the coefficient of friction, μ, geotechnical engineers prefer to describe frictional strength using the *effective friction angle* (or *effective angle of internal friction*), ϕ', where:

$$\phi' = \tan^{-1}\mu \qquad (13.1)$$

In addition, we find it more convenient to work in terms of stress instead of force, so the shear strength, s, due to friction is:

$$s = \sigma' \tan\phi' \qquad (13.2)$$

where:

    s = shear strength
    σ' = effective stress acting on the shear plane
    ϕ' = effective friction angle

Notice how Equation 13.2 uses the effective stress, not the total stress. This is an important distinction! We express strength in terms of effective stress because only the solid particles contribute to frictional strength (the pore water has no static shear strength), and the effective stress describes the normal stresses carried by the solid particles.

Figure 13.3 Comparison between friction on a sliding block and frictional strength in soil.

Although we sometimes use $\phi = 0$ as an analytical tool, as described later, in reality all soils have frictional strength. The value of ϕ' depends on both the frictional properties of the individual particles and the interlocking between particles. These are affected by many factors, including:

- **Mineralogy** — Soil includes many different minerals, and some slide more easily than others. For example, the friction angle in sands made of pure quartz is typically 30–36°. However, the presence of other minerals can change ϕ'. For example, sands containing significant quantities of mica, which is much smoother than quartz, have a smaller ϕ'. These are called *micaceous sands*.

Clay minerals are typically even weaker (ϕ' values as low as $4°$ have been measured in pure montmorillonite).

- **Shape** — The friction angle of angular particles is much higher than that of rounded ones.
- **Gradation** — Well-graded soils typically have more interlocking between the particles, and thus a higher friction angle, than those that are poorly graded. For example, GW soils typically have ϕ' values about $2°$ higher than comparable GP soils.
- **Void ratio** — Decreasing the void ratio, such as by compacting a soil with a sheepsfoot roller, also increases interlocking, which results in a higher ϕ'.
- **Organic material** — Organics introduce many problems, including a decrease in the friction angle.

The impact of water on frictional strength is especially important, and many shear failures are induced by changes in the groundwater conditions. Many people mistakenly believe water-induced changes in shear strength are primarily due to lubrication effects. Although the process of wetting some dry soils can induce lubrication, the resulting decrease in ϕ' is very small. Sometimes the introduction of water has an antilubricating effect (Mitchell, 1993) and causes a small increase in ϕ'. However, focusing on these small effects tends to obscure another far more important process, which is illustrated in Example 13.1.

Example 13.1

A geotechnical engineer is evaluating the stability of the slope in Figure 13.4. This evaluation is considering the potential for a shear failure along the shear surface shown. The soil has $\phi' = 30°$ and no cohesive strength. Compute the shear strength at Point A along this surface when the groundwater table is at level B, then compute the new shear strength if it rose to level C. The unit weight of the soil is 120 lb/ft^3 above the groundwater table and 123 lb/ft^3 below.

Solution

Groundwater table at B:

$$u = \gamma_w z_w = (62.4 \text{ lb/ft}^3)(20 \text{ ft}) = 1248 \text{ lb/ft}^2$$

$$\sigma'_z = \sum \gamma H - u = (120 \text{ lb/ft}^3)(26 \text{ ft}) + (123 \text{ lb/ft}^3)(20 \text{ ft}) - 1248 \text{ lb/ft}^2 = 4332 \text{ lb/ft}^2$$

The potential shear surface is horizontal, so $\sigma' = \sigma'_z$ and

$$s = \sigma' \tan \phi' = (4332 \text{ lb/ft}^3) \tan 30° = \textbf{2501 lb/ft}^2 \quad \Leftarrow \textit{Answer}$$

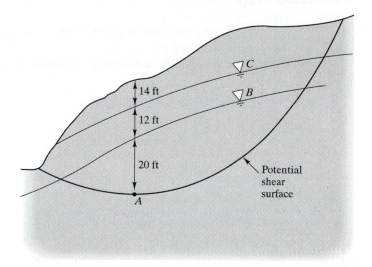

Figure 13.4 Cross-section of a potential landslide.

Groundwater table at C:

$$u = \gamma_w z_w = (62.4 \text{ lb/ft}^3)(32 \text{ ft}) = 1997 \text{ lb/ft}^2$$

$$\sigma'_z = \sum \gamma H - u = (120 \text{ lb/ft}^3)(14 \text{ ft}) + (123 \text{ lb/ft}^3)(32 \text{ ft}) - 1997 \text{ lb/ft}^2 = 3619 \text{ lb/ft}^2$$

$$s = \sigma' \tan \phi' = (3619 \text{ lb/ft}^3) \tan 30° = \textbf{2089 lb/ft}^2 \quad \Leftarrow \textit{Answer}$$

Commentary

The total stress at point A increases slightly when the groundwater table rises from B to C due to the greater unit weight of the zone of soil that becomes saturated. However, this is more than offset by the increase in pore water pressure, so the effective stress is reduced, along with a corresponding decrease in the shear strength. It is quite possible that this loss in shear strength would be sufficient to induce a shear failure (landslide) in this slope. Additional analyses, which we will discuss in Chapter 14, are necessary to determine if such a failure will occur.

Example 13.1 illustrates the important impact of pore water pressure on shear strength. This is the primary way water impacts the frictional strength, and is one of the reasons a thorough understanding of the principle of effective stress is so important in understanding soil behavior.

Cohesive Strength

Some soils have shear strength even when the effective stress, σ', is zero, or at least *appears* to be zero. This strength is called the *cohesive strength*, and we describe it using the variable c', the *effective cohesion*. If a soil has both frictional and cohesive strength, then Equation 13.2 becomes:

$$ s = c' + \sigma' \tan \phi' \tag{13.3} $$

where:

  s = shear strength
  c' = effective cohesion
  σ' = effective stress acting on the shear surface
  ϕ' = effective friction angle

There are two types of cohesive strength: true cohesion and apparent cohesion (Mitchell, 1993).

True cohesion is shear strength that is truly the result of bonding between the soil particles. These bonds include the following:

- *Cementation* is chemical bonding due to the presence of cementing agents, such as calcium carbonate ($CaCO_3$) or iron oxide (Fe_2O_3) (Clough, et al., 1981). Even small quantities of these agents can provide significant cohesive strengths. *Caliche* is an example of a heavily cemented soil that has a large cohesive strength. Cementation also can be introduced artificially using Portland cement or special chemicals, as discussed in Chapter 19.
- *Electrostatic and electromagnetic attractions* hold particles together. However, these forces are vary small and probably do not produce significant shear strength in soils.
- *Primary valence bonding* (*adhesion*) is a type of cold welding that occurs in clays when they become overconsolidated.

Apparent cohesion is shear strength that appears to be caused by bonding between the soil particles, but is really frictional strength in disguise. Sources of apparent cohesion include the following:

- *Negative pore water pressures* that have not been considered in the stress analysis. These negative pore water pressures are present in soils above the groundwater table, as shown in Figure 10.16. However, many geotechnical engineering analyses ignore these pressures (i.e., we assume $u = 0$, even though it is really < 0), so the effective stress is greater than we think it is (see Equation 10.32). The shear strength that corresponds to this additional effective stress thus appears to be cohesive, even though it is really frictional. This is the primary reason unsaturated clays appear to have "cohesive" strength and why moist unsaturated "cohesionless" sands can stand in vertical cuts.

- *Negative excess pore water pressures due to dilation.* Some soils tend to *dilate* or expand when they are sheared. In saturated soils, this dilation draws water into the voids. However, sometimes the rate of shearing is more rapid than the rate at which water can flow (i.e., the voids are trying to expand more rapidly than they can draw in the extra water). This is especially likely in saturated clays, because their hydraulic conductivity is so low. When this occurs, large negative excess pore water pressures can develop in the soil.

  The term *excess pore water pressure* was defined in Chapter 11. It is an additional pore water pressure, either positive or negative, that is superimposed on the hydrostatic pore water pressure. The excess pore water pressure is due to squeezing or expanding of the soil voids. In this case, the voids are expanding, so $u_e < 0$ and u becomes less than the hydrostatic value, u_h.

  If these excess pore water pressures are considered in our strength analysis, we could compute an accurate value of σ' and thus an accurate value of s using Equation 13.2. However, if we consider only the hydrostatic pore water pressure, our computed value of σ' will be too high and the soil will appear to have cohesive strength.

- *Apparent mechanical forces* are those due to particle interlocking, and can develop in soils where this interlocking is very difficult to overcome. The result is additional apparent cohesion.

Geotechnical engineers often use the term "cohesive soil" to describe clays. Although this term is convenient, it also is very misleading (Santamarina, 1997). Most of the so-called cohesive strength in clays is really apparent cohesion due to pore water pressures that are negative, or at least less than the hydrostatic pore water pressure. In such soils it is better to think of "cohesive strength" as a mathematical idealization rather than a physical reality.

In cemented soils, cohesive strength really does reflect bonding between the soil particles. In some cases, we may rely on this strength in our designs. However, in other cases, it is wise to ignore this source of strength. For example, if the cementing agent is water-soluble, it may disappear if the soil becomes wetted during the life of the project.

Definition of Failure

Another important difference between shear strength assessments of soils and those for more traditional engineering materials lies in our definition of failure. For example, with steel we usually define failure either as the point where the stress–strain curve becomes plastic and nonlinear (the yield strength), or when rupture occurs (the ultimate strength). However, in soils the stress–strain curve is nonlinear and plastic from the very beginning, and there is no rupture point in the classic sense. Therefore, we must use other means of defining shear strength.

Soils have two kinds of shear stress–strain curves, as shown in Figure 13.5. Those with *ductile* curves generally plateau at a well-defined peak shear stress as shown, and we can use this value as the design shear strength. However, those with *brittle* (or *strain-softening*) curves have two strengths, the *peak strength*, which is the high point on the curve, and the *residual strength* (or *ultimate strength*), which occurs at a much larger shear strain. Either value could be used for design, depending on the kind of problem being evaluated.

Sands and gravels have shear stress–strain curves that are either ductile or only mildly brittle, so the difference between peak and residual strength is small and can be neglected. However, some clays have very brittle curves, so the distinction between peak and residual strength becomes very important, as discussed later in this chapter.

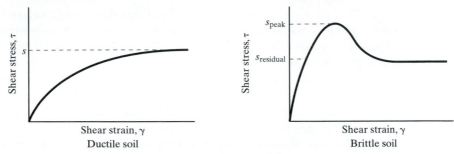

Figure 13.5 Shear stress-strain curves in soil, and definitions of failure.

The problem of defining failure is further complicated by the lack of a unique stress–strain curve. For example, a soil tested under certain conditions (i.e., intermediate principal stress, strain constraint, rate of strain, etc.) produces a certain stress–strain curve and thus a certain shear strength, yet the same soil tested under different conditions produces a different curve and a different strength. We attempt to overcome this problem by using test conditions that simulate the field conditions or by using standardized test conditions and calibrating the results with observed behavior in the field.

13.3 MOHR–COULOMB FAILURE CRITERION

Some soils have both frictional and cohesive strength, so we need to combine these two sources into a single all-purpose strength formula. Nearly all geotechnical analyses do this using the *Mohr–Coulomb failure criterion*, which then allows us to project the test data back into our analyses of existing or proposed field conditions. This may be done using either effective stress analyses or total stress analyses.

Effective Stress Analyses

The shear strength in a soil is developed only by the solid particles, because the water and air phases have no shear strength. Therefore, it seems reasonable to evaluate strength problems using the effective stress, σ', because it is the portion of the total stress, σ, carried by the solid particles. This is why Equations 13.2 and 13.3 were written in terms of effective stress.

If we perform a series of laboratory strength tests, each at a different value of σ', the results will be as shown in Figure 13.6. The vertical axis on this plot is shear stress, τ, and the curve represents the shear strength, s, which is the magnitude of τ at failure.

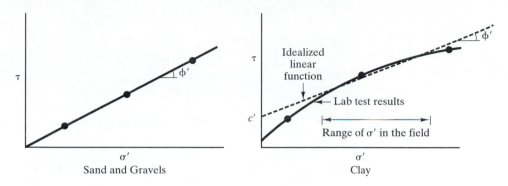

Figure 13.6 Shear strength as a function of effective stress. Each data point represents the results of a laboratory test.

In sands and gravels, these plots are nearly linear within the range of stresses normally encountered in the field. However, in clays it is slightly nonlinear, as shown. These nonlinear plots are inconvenient because they introduce more complexity into the analyses. Therefore, we nearly always use an idealized linear function by conducting tests at σ' values comparable to those expected in the field and connecting them with a straight line. We say the τ-intercept of this line is the effective cohesion, c', and the slope of the line is the effective friction angle, ϕ'. In reality, the relative contributions of cohesive and frictional strength are much more complex. For example, this c' value is really a combination of true cohesion and the mathematics of fitting a straight line to a curved function. In some cases it also may include some apparent cohesion. Nevertheless, this idealized representation is adequate for the vast majority of practical problems. The lines in Figure 13.6 are known as the *Mohr–Coulomb failure criterion*, and may be expressed mathematically using Equation 13.3.

Shear strength is defined as the shear stress at failure, so points in the soil that have (σ', τ) values that plot below the Mohr–Coulomb line theoretically will not fail in shear, while those that plot on or above the line will fail. We often call this line a *failure envelope* because it encloses the stresses that will not fail.

Once c' and ϕ' have been determined, we can evaluate the shear strength in the field using Equation 13.3. We compute the effective stresses in the field using the techniques described in Chapter 10, along with a groundwater elevation that represents the worst (i.e., highest) conditions anticipated during the life of the project.

Example 13.2

Samples have been obtained from both soil strata in Figure 13.7 and brought to a soil mechanics laboratory. A series of shear strength tests were then performed on both samples and plotted in diagrams similar to those in Figure 13.6. The c' and ϕ' values obtained from these diagrams are shown in Figure 13.7. Using this data, compute the shear strength on horizontal and vertical planes at Points A, B, and C.

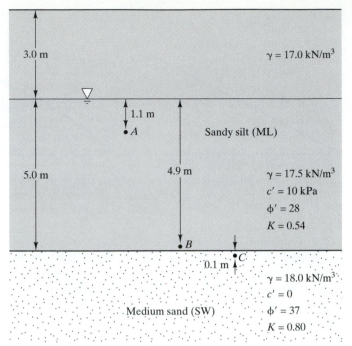

Figure 13.7 Soil profile for Example 13.2.

Solution

Point A—horizontal plane

$$\sigma'_z = \sum \gamma H - u$$
$$= (17.0 \text{ kN/m}^3)(3.0 \text{ m}) + (17.5 \text{ kN/m}^3)(1.1 \text{ m}) - (9.8 \text{ kN/m}^3)(1.1 \text{ m})$$
$$= 59.5 \text{ kPa}$$

$$s = c' + \sigma' \tan\phi'$$
$$= 10 \text{ kPa} + (59.5 \text{ kPa}) \tan 28°$$
$$= \mathbf{41.6 \text{ kPa}} \quad \Leftarrow Answer$$

Point A—vertical plane

$$\sigma'_x = K \sigma'_z = (0.54)(59.5 \text{ kPa}) = 32.1 \text{ kPa}$$

$$s = c' + \sigma' \tan\phi'$$
$$= 10 \text{ kPa} + (32.1 \text{ kPa}) \tan 28°$$
$$= \mathbf{27.1 \text{ kPa}} \quad \Leftarrow Answer$$

Using similar computations:

Point B—vertical plane	$s = 57.2$ kPa	← *Answer*
Point B—horizontal plane	$s = 35.5$ kPa	← *Answer*
Point C—vertical plane	$s = 68.1$ kPa	← *Answer*
Point C—horizontal plane	$s = 54.4$ kPa	← *Answer*

Commentary

At each point the shear strength on a vertical plane is less than that on a horizontal plane because $K < 1$. In addition, the shear strength at Point B is greater than that at Point A because the effective stress is greater. The strength at Point C is even higher than at Point B because it is in a new strata with different c', ϕ', and K values. Thus, the strength would increase gradually with depth within each stratum, but change suddenly at the boundary between the two strata.

Example 13.3

Draw the shear strength envelope for the ML stratum in Figure 13.7, then plot the upper half of the Mohr's circle for Point A on this diagram. Assume the principal stresses act vertically and horizontally.

Solution

Using the results from Example 13.2, we develop Figure 13.8:

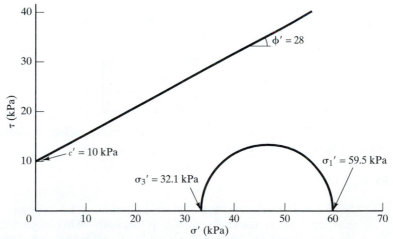

Figure 13.8 Failure envelope and Mohr's circle for Example 13.3.

Commentary

In this case, the entire Mohr's circle plots below the strength envelope. Therefore, the shear stress on all planes through Point A is less than the shear strength, and no shear failure will occur. However, if the Mohr's circle touches the envelope, such as the one in Figure 13.9, then a shear failure will occur on the plane represented by that point on the circle. This method of presenting stresses and strengths is named after the German engineer Otto Mohr (1835–1918) and the French scientist Charles Augustin de Coulomb (1736–1806). Neither of them drew diagrams like this to describe soil strength, but the underlying concepts are based on both men's work.

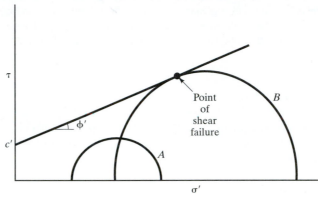

Figure 13.9 Shear failure occurs when the Mohr's circle is large enough to touch the failure envelope. Thus, no failure will occur at the point represented by Circle A, but failure will occur at the point represented by Circle B.

The ratio of the shear strength, s, on a specific plane to the shear stress, τ, on that plane is defined as the *factor of safety*, F:

$$F = \frac{s}{\tau}$$

(13.4)

Normally we define some minimum acceptable factor of safety, then we check proposed designs to verify that this criterion has been met.

Example 13.4

An 18-inch diameter storm-drain pipe is to be installed under a highway by jacking as shown in Figure 13.10. A mound of soil will be placed as shown to provide a reaction for the jacks. Then, the first section of pipe will be pushed into the ground below the highway. An auger will clean out the soil collected inside, then additional sections will be added, pushed in, and cleaned out one at a time until the pipe reaches the opposite side of the highway. This method allows the highway to remain in service while the pipe is being installed. The alternative would be to dig a trench, lay the pipe, and backfill, but this would require temporarily closing the highway.

At times the jack must apply a 300 k load to press the pipe into the ground. It will react against a steel plate placed on the soil mound. The load-carrying capacity of this plate is controlled by the shear strength of the adjacent soil. The soil has $c' = 400$ lb/ft^2, $\phi' = 29°$, and $\gamma = 120$ lb/ft^3. Will the soil beyond the plate be able to resist this load with a factor of safety of 1.5?

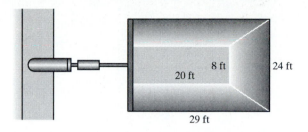

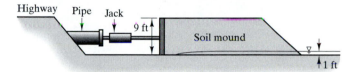

Figure 13.10 Proposed soil mound and jacking arrangement for Example 13.4.

Solution

Soil mound:

$$A_{top} = (20 \text{ ft}) (8 \text{ ft}) = 160 \text{ ft}^2$$

$$A_{bottom} = (29 \text{ ft}) (24 \text{ ft}) = 696 \text{ ft}^2$$

$$A_{average} = \frac{160 \text{ ft}^2 + 696 \text{ ft}^2}{2} = 428 \text{ ft}^2$$

$$W = (428 \text{ ft}^2)(9 \text{ ft})(120 \text{ lb/ft}^3) = 462,000 \text{ lb}$$

$$A_{shear} = A_{bottom} = 696 \text{ ft}^2$$

$$\sigma_z = \frac{W}{A} = \frac{462{,}000 \text{ lb}}{696 \text{ ft}^2} = 664 \text{ lb/ft}^2$$

$$u = \gamma_w z_w = (62.4 \text{ lb/ft}^3)(1 \text{ ft}) = 62 \text{ lb/ft}^2$$

$$\sigma_z' = \sigma_z - u = 664 \text{ lb/ft}^2 - 62 \text{ lb/ft}^2 = 602 \text{ lb/ft}^2$$

$$s = c' + \sigma' \tan\phi' = 400 \text{ lb/ft}^2 + (602 \text{ lb/ft}^2)\tan 29° = 734 \text{ lb/ft}^2$$

$$\tau = \frac{P}{A} = \frac{300{,}000 \text{ lb}}{696 \text{ ft}^2} = 431 \text{ lb/ft}^2$$

$$F = \frac{s}{\tau} = \frac{734 \text{ lb/ft}^2}{431 \text{ lb/ft}^2} = 1.7 > 1.5 \quad \therefore \text{OK} \qquad \Leftarrow \textit{Answer}$$

According to this analysis, the soil mound is sufficiently large to provide the necessary reaction force.

Total Stress Analyses

Analyses based on effective stresses, such as those in Examples 13.1 through 13.4, are possible only if we can predict the effective stresses in the field. This is a simple matter when only hydrostatic pore water pressures are present, but can become very complex when there are excess pore water pressures. For example, when a fill is placed over a saturated clay, excess pore water pressures develop in the clay as described in Chapters 11 and 12. In addition, some soils also develop additional excess pore water pressures as they are sheared, as discussed later in this chapter. Often these excess pore water pressures are difficult to predict, especially those due to shearing.

Because of these difficulties, geotechnical engineers sometimes evaluate problems based on total stresses instead of effective stresses. This approach involves reducing the lab data in terms of total stress and expressing it using the parameters c_T and ϕ_T. Equation 13.3 then needs to be rewritten as:

$$\boxed{s = c_T + \sigma \tan \phi_T}$$ (13.5)

where:

s = shear strength

c_T = total cohesion

σ = total stress acting on the shear surface

ϕ_T = total friction angle

The total stress analysis method assumes the excess pore water pressures developed in the lab are the same as those in the field, and thus are implicitly incorporated into c_T and ϕ_T. This assumption introduces some error in the analysis, but it becomes an unfortunate necessity when we cannot predict the magnitudes of excess pore water pressures in the field. It also demands the laboratory tests be conducted in a way that simulates the field conditions as closely as possible.

The shear strength of soils really depends on effective stress, so total stress analyses are less desirable than effective stress analyses, and the results need to be viewed with more skepticism. However, there are times when we must use total stress analyses because we have no other practical alternative.

QUESTIONS AND PRACTICE PROBLEMS

13.1   The effective stress at a certain plane in a soil is 120 kPa, the effective cohesion is 10 kPa, and the effective friction angle is 31°. A foundation to be built nearby will induce a shear stress of 50 kPa on this plane. Compute the factor of safety against a shear failure.

13.2   A soil has a unit weight of 118 lb/ft², c_T = 250 lb/ft², and ϕ_T = 29°. Compute the shear strength on a horizontal plane at a depth of 12 ft below the ground surface.

13.3   The soils in Figure 11.24 have the following strength parameters:

Silty sand	$c' = 0$	$\phi' = 31°$
Soft clay	$c_T = 20$ kPa	$\phi_T = 0°$
Medium clay	$c_T = 45$ kPa	$\phi_T = 0°$
Glacial till	$c' = 15$ kPa	$\phi' = 40°$

In addition, the glacial till has a unit weight of 22.0 kN/m³. Develop a plot of shear strength on a horizontal plane vs. depth from the ground surface to a depth of 20 m. Keep in mind the shear strength at a point depends on c and ϕ at that point, so it can suddenly change at strata interfaces.

13.4   Pile foundations consist of prefabricated poles, usually made of steel, wood, or concrete, that are driven into the ground with a pile hammer. The number of hammer blows per 0.1 m of pile penetration (known as the *blow count*) depends on the strength of the soil at the pile tip (along with other factors).

A series of piles is to be driven at the site described in Problem 13.3. The geotechnical engineer requires them to be driven until the tip is embedded 0.2 m into the glacial till. Could the field engineer use the blow count to determine when this penetration has been achieved? Explain.

13.4 SHEAR STRENGTH OF SATURATED SANDS AND GRAVELS

Little or no excess pore water pressure occurs in clean sands and gravels under static loading conditions because their hydraulic conductivities are so high. If changes in the normal or shear stresses cause the voids to expand or contract, water easily flows in or out as necessary. Therefore, the pore water pressure, u, is equal to the hydrostatic pore water pressure (Equation 7.7) and shear strength analyses may be based on effective stresses.

Determining c' and ϕ'

If no cementing agents or clay are present, saturated sands and gravels should have $c' = 0$. We determine the friction angle, ϕ', by conducting field or laboratory tests, as discussed later in this chapter. Figure 13.11 presents typical ϕ' values, which may be used for preliminary estimates or for checking test data.

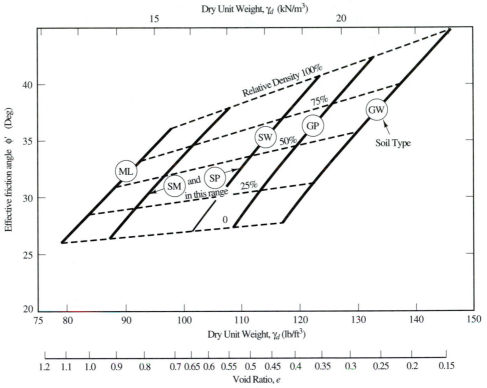

Figure 13.11 Typical ϕ' values for cohesionless soils without clay or cementing agents (Adapted from U.S. Navy, 1982).

Notice how the ϕ' values in Figure 13.11 increase as the unit weight increases. This is one of the reasons for compacting soils. This figure also illustrates that gravels are generally stronger than sands, and well-graded soils are generally stronger than poorly graded ones. The presence of large amounts of silt decreases ϕ', primarily because these particles are smoother and have lower coefficients of friction.

If cementing agents or overconsolidated clay (i.e., an SC soil) are present, then c' will be greater than zero. However, engineers are generally reluctant to rely on this additional strength, especially if it is from clay, if the cementing agents are water-soluble, or if there is concern that it may be an apparent cohesion. For design purposes, most engineers either ignore the cohesive strength of such soils, or use design values less than measured c'.

Selecting the Proper Value of σ'

We are performing an effective stress analysis, so the shear strength is defined by Equation 13.3. The value of ϕ' is determined by testing, and c' is usually 0, but what should we use for the effective stress, σ'? In most geotechnical design problems, the shear and normal stresses change simultaneously, so σ' at the beginning of loading is different from that at the end.

To understand which σ' to use for design, study the plots in Figure 13.12. These plots describe the conditions at a point in the soil below a spread footing foundation. As the external load, P, is applied to the foundation, perhaps over a construction period of a few weeks, the vertical total stress, σ_z, at point A increases accordingly. The pore water pressure, u, remains virtually constant because of the rapid drainage, so the vertical effective stress, σ_z' increases at the same rate as σ_z. Thus, the shear strength, s, also increases, and is greater at the end of construction than it was at the beginning. Concurrently, the external load also induces a shear stress, τ, in the soil. It occurs as quickly as the load is applied. Finally, the factor of safety changes during construction per Equation 13.4.

Based on this data, we can see that all of the changes in the soil occur during construction. As soon as the new shear stress occurs, the increased shear strength is present to resist it. Therefore, we may use the post-construction σ' for our strength analysis and take advantage of the corresponding increase in shear strength.

The opposite condition occurs when the soil is unloaded. In this case, σ' and s decrease during construction. To evaluate this condition, we again use the post-construction σ', but this time it produces a lower strength than the pre-construction value.

Example 13.5

The levee shown in Figure 13.13 is to be built along the side of a river to protect a nearby town from flooding. As a part of the design of this levee, the geotechnical engineer is considering the potential for a landslide along the failure surface shown in the figure. If the natural soils below the levee are clean sands with $\phi' = 34°$ and the shear stress at Point A is 400 lb/ft^2, compute the factor of safety against sliding at Point A.

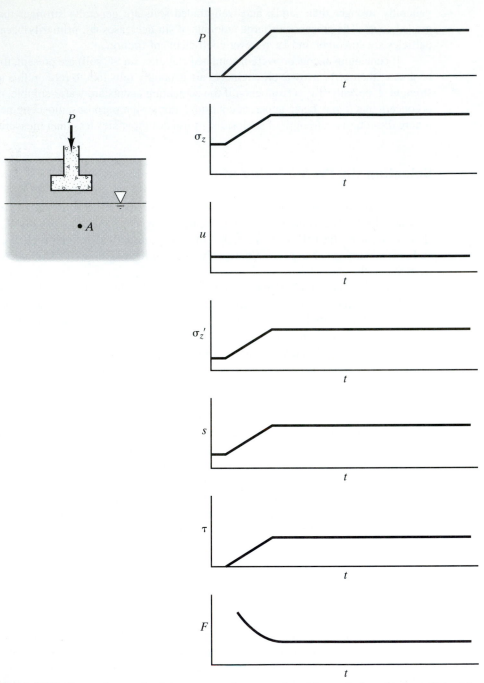

Figure 13.12 Changes in normal and shear stresses, shear strength, and factor of safety with time at Point A in a saturated sandy soil below a structural foundation.

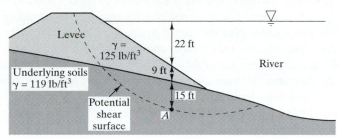

Figure 13.13 Proposed levee.

Solution

$$\sigma_z' = \sum \gamma H - u$$
$$= (62.4 \text{ lb/ft}^2)(22 \text{ ft}) + (125 \text{ lb/ft}^2)(9 \text{ ft})$$
$$+ (119 \text{ lb/ft}^2)(15 \text{ ft}) - (62.4 \text{ lb/ft}^2)(46 \text{ ft})$$
$$= 1412 \text{ lb/ft}^2$$

The shear surface at Point A is nearly horizontal, so assume $\sigma' = \sigma_z'$

$$s = c' + \sigma' \tan \phi' = 0 + (1412 \text{ lb/ft}^2) \tan 34° = 953 \text{ lb/ft}^2$$

$$F = \frac{s}{\tau} = \frac{953 \text{ lb/ft}^2}{400 \text{ lb/ft}^2} = \textbf{2.4} \quad \Leftarrow \textit{Answer}$$

Commentary

1. This analysis was based on effective stresses with the post-construction σ'.
2. The computed factor of safety of 2.4 would be satisfactory.
3. Real slope stability analyses need to consider the factor of safety along the entire shear surface, not just at one point. We will discuss methods of performing such analyses in Chapter 14.

Soil Liquefaction

Most large civil engineering projects require weeks or months to build. Some, such as large earth dams, are under construction for years. Therefore, the loads applied to soils from these projects also occur slowly, which is why cohesionless soils have plenty of time to draw water into or out of the voids as they expand or contract. Little or no excess pore water pressures develop in these situations because the potential rate of drainage is greater than the rate of loading.

However, the rate of loading is sometimes so rapid that even cohesionless soils cannot drain quickly enough. The most noteworthy example is the loading due to an earthquake, which is much faster than the rate of drainage. This is especially problematic in loose, saturated sands because they tend to compress when loaded (Lee, 1965), which normally would force some water out of the voids. However, because the loading occurs so quickly, the water cannot easily drain away and positive excess pore water pressures develop instead. As these pressures build up, both the effective stress and the strength decrease (see Equation 13.3). Sometimes the effective stress drops to zero, which means the soil loses all its shear strength and thus behaves as a dense liquid. We call this phenomenon *soil liquefaction*.

Soil liquefaction can cause extensive damage, so geotechnical engineers working in seismically active areas need to be aware of the soil conditions where this phenomena is likely to occur. Chapter 20 discusses liquefaction in more detail.

Quicksand

Section 10.10 discussed seepage forces and the unfortunate consequences that can occur when water flows upward through a soil and the seepage forces oppose the gravitational forces. According to Equation 10.46, upward seepage forces can become large enough that the vertical effective stress drops to zero. This can cause heaving, as illustrated in Example 10.10, especially if the seepage forces significantly exceed the gravitational forces. Another possibility is *quicksand*, which occurs in sandy soils when upward seepage produces a σ_z' close to zero. c' also is zero, these soils have no shear strength and behave as a heavy fluid.

Although true quicksand can occur in natural settings, most conditions identified as quicksand are actually just very loose saturated sand. However, quicksand is a very real danger in dewatered excavations and other constructed facilities where upward seepage forces have been artificially created. This strength loss can trigger the failure of shoring systems and other facilities, possibly resulting in property damage and loss of life.

QUESTIONS AND PRACTICE PROBLEMS

13.5 A certain well-graded sand deposit has an in-situ relative density of about 50%. A laboratory strength test on a sample of this soil produced an effective friction angle of 31°. Does this test result seem reasonable? Explain the basis for your answer.

13.6 The vertical effective stress at a certain point in a loose sand is 1000 lb/ft^2. If an earthquake were to occur, how much excess pore water pressure would need to develop at this point for liquefaction to occur? Show a numerical rationale for your answer.

13.7 A temporary excavation similar to the one shown in Figure 8.6 on page 251 is to be built. The soil is a clean sand with $\gamma = 118$ lb/ft^3, $c' = 0$, and $\phi' = 34°$. According to a flow net analysis, the groundwater flow in the soil immediately below the excavation will be upward and have a hydraulic gradient of 0.76. Compute the shear strength on a horizontal plane at a depth of 3 ft below the bottom of the excavation. Discuss the significance of your answer.

13.5 SHEAR STRENGTH OF SATURATED CLAYS AND SILTS

Shear strength assessments in clays and silts are more difficult than those in sands and gravels because:

- Clay particles undergo more significant changes during shear
- The low hydraulic conductivity impedes the flow of water into and out of the voids, so significant excess pore water pressures often develop in the soil

In addition, saturated clays and silts are generally weaker than sands and gravels, and thus are more often a cause of problems.

Volume Changes and Excess Pore Water Pressures

When loads such as structural foundations are applied to the ground, the total vertical stress, σ_z, and the shear stress, τ, increase as discussed in Chapter 10. These increases are shown on the plots in Figure 13.14. In sandy soils, the increase in σ_z immediately causes some of the pore water to flow out of the voids, which results in rapid consolidation and corresponding increases in the vertical effective stress, σ_z'. We call this the *drained condition* because the pore water can easily drain or move through the soil. The plots in Figure 13.14a describe various changes in the soil during loading under the drained condition. Our discussion of the shear strength of sands and gravels in the previous section assumed drained conditions prevailed.

However, if the same load is applied to a saturated clay, the flow of pore water is much slower because these soils have a much smaller hydraulic conductivity, k. Therefore, excess pore water pressures develop in the soil as discussed in Chapters 11 and 12. We call this the *undrained condition* because it is much more difficult for the pore water to drain or move through the soil. The plots in Figure 13.14b describe this condition. Notice the spike in the u plot, which illustrates the immediate build-up and gradual dissipation of excess pore water pressures. As a result, the increase in σ_z' is much slower than in sands.

Although we usually consider sands to be drained and clays to be undrained, either condition can occur in virtually any soil. The distinction between the drained and undrained conditions really depends on the rate of loading and the rate of drainage. If the rate of loading is slow compared to the rate of drainage, then drained conditions prevail. Conversely, if the rate of loading is rapid compared to the rate of drainage, then undrained conditions prevail. For normal rates of loading from construction (i.e., load applied over a period of weeks or months), sands usually are drained and clays usually are undrained. However, if the rate of loading is exceptionally fast, the undrained condition can exist in sands. For example, soil liquefaction, as described earlier, is the result of undrained conditions under rapid seismic loading. Similarly, if the rate of loading is very slow, the drained condition can occur in clays.

Some construction projects cause a decrease in the vertical stress and an increase in shear stress. For example, this occurs when we make a sloped excavation. In this case, the post-construction σ_z' and s are less than the pre-construction values, as shown in Figure 13.15. If undrained conditions prevail, negative excess pore water pressures are present.

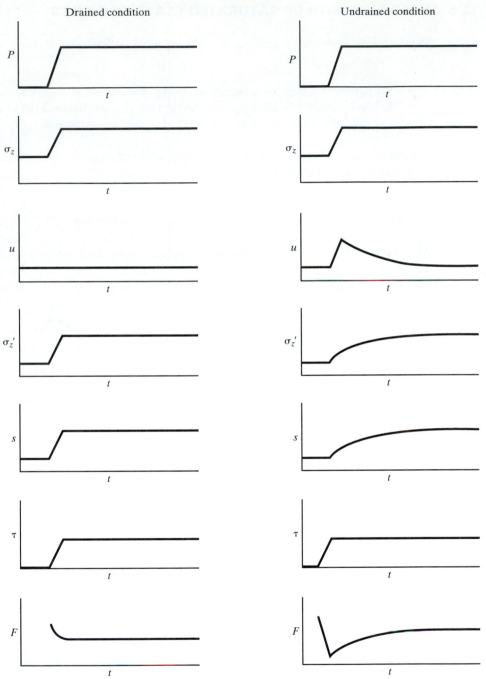

Figure 13.14 Changes in normal and shear stresses, shear strength, and factor of safety with time at a point in a saturated clay below a fill or a structural foundation. a) very slow loading (drained conditions); b) normal rate of loading (undrained condition).

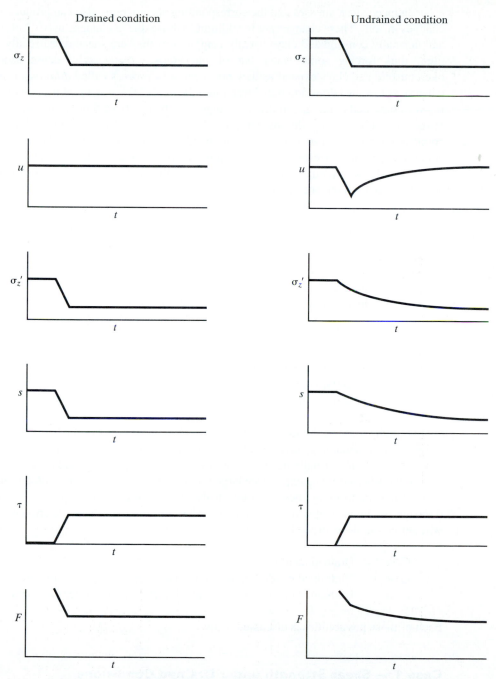

Figure 13.15 Changes in normal and shear stresses, shear strength, and factor of safety with time at a point in a saturated clay below an excavation. a) very slow loading (drained conditions); b) normal rate of loading (undrained condition).

Induced shear stresses and the corresponding shear strains also can produce volume changes in soil. These changes are in addition to those due to changes in σ'. Loose sands and normally consolidated clays usually compress as they are sheared, especially if σ' is high, thus forcing some water out of the voids. Conversely, dense sands and overconsolidated clays expand as they are sheared (a process called *dilation*), especially when σ' is low, and draw additional water into the voids. Once again, if the rate of loading is slow compared to the rate of drainage, there will be plenty of time for the water to flow in or out and drained conditions will prevail. Conversely, if the rate of loading is rapid compared to the rate of drainage, then undrained conditions will prevail. When undrained conditions prevail, corresponding excess pore water pressures develop in the soil.

Usually the induced normal stress and the induced shear stresses develop simultaneously, so the pore water pressure at a given time is:

$$u = u_h + (u_e)_{\text{normal}} + (u_e)_{\text{shear}} \tag{13.6}$$

where:
u = pore water pressure
u_h = hydrostatic pore water pressure (per Equation 7.7)
$(u_e)_{\text{normal}}$ = excess pore water pressure due to changes in normal stress
        > 0 if subjected to increases in σ
        < 0 if subjected to decreases in σ
$(u_e)_{\text{shear}}$ = excess pore water pressure due to shearing
        > 0 if soil tends to compress when sheared
        < 0 if soil tends to dilate when sheared

Usually, $(u_e)_{\text{normal}}$ dominates over $(u_e)_{\text{shear}}$. Thus, most soils that are being loaded, such as by fills or foundations, have a net positive u_e, while those that are being unloaded, such as by an excavation, usually have a net negative u_e. These processes are most pronounced in soft clays because they experience large volume reductions due to consolidation.

When performing shear strength analyses, it is important to properly assess the drainage conditions that will occur in the field because this assessment determines how we will define the shear strength. There are three possibilities:

Case 1 — Drained conditions
Case 2 — Undrained conditions with positive excess pore water pressures
Case 3 — Undrained conditions with negative excess pore water pressures

Each of these possibilities is discussed below.

Case 1 — Shear Strength under Drained Conditions

The undrained condition is the easiest to evaluate because there are no excess pore water pressures. We simply evaluate c' and ϕ' using an appropriate test, compute σ' based on the hydrostatic pore water pressures and post-construction conditions, and s using

Equation 13.3. We can use this method to evaluate shear strengths in the soil once the excess pore water pressures have dissipated. Thus, drained strength analyses may be used to evaluate long-term stability.

Figure 13.16 shows typical effective stress Mohr–Coulomb failure envelopes for both normally consolidated and overconsolidated clays and silts under drained conditions. Those that are normally consolidated and uncemented normally have $c' = 0$. However, overconsolidation produces a true cohesion, as discussed earlier. These two plots join when the effective stress equals the preconsolidation stress, σ_c'. Figure 13.17 presents typical ϕ' values for clays, which may be used for preliminary analyses or for checking test results.

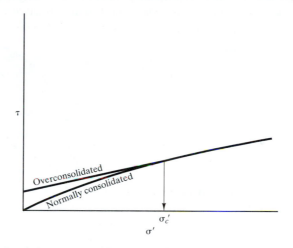

Figure 13.16 Mohr–Coulomb failure envelopes for saturated, uncemented clays and silts under drained conditions.

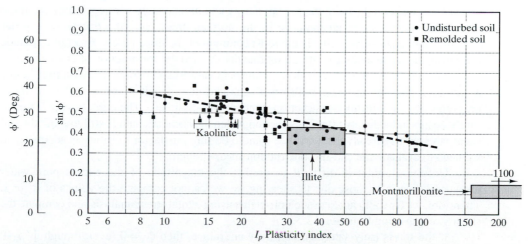

Figure 13.17 Typical effective friction angles for normally consolidated, saturated clays and silts (from *Fundamentals of Soil Behavior*, 2nd Ed. by J.K. Mitchell, Copyright ©1993; used by permission of John Wiley and Sons).

Example 13.6

The natural earth slope shown in Figure 13.18 has been in its present configuration for a very long time. A slope stability analysis is to be performed on the potential failure surface shown in the figure. What shear strength should be used in this analysis?

Solution

This slope has been in place for a long time, and no loads are being added or removed, so we can use a drained analysis based on c' and ϕ' with the hydrostatic pore water pressures. We could obtain c' and ϕ' by performing appropriate laboratory tests on undisturbed samples (as described later in this chapter), and compute σ' based on the cross-section.

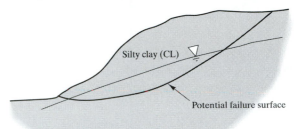

Silty clay (CL)

Potential failure surface

Figure 13.18 Cross-section of natural slope for Example 13.6.

Case 2 — Shear Strength under Undrained Conditions with Positive Excess Pore Water Pressures

The rate of construction for most projects is faster than the rate of drainage in saturated clays and silts, so undrained conditions prevail during and immediately after construction. If the new construction causes an increase in the normal stress, then the excess pore water pressures will be positive. This is the most common of the three cases. Examples include the construction and loading of structural foundations and the construction of embankments.

The plots in Figure 13.14b show changes in various soil parameters with time below a structural foundation being built on a saturated clay or silt. The excess pore water pressures build up during construction, then slowly dissipate. Thus, the lowest factor of safety occurs immediately after construction, when the shear strength is still low but the full shear stresses are already present. We want to be sure this factor of safety is adequate, so we need to evaluate the end-of-construction shear strength.

Unfortunately, it is difficult to predict the magnitude of the excess pore water pressures, especially those due to shearing, so we cannot compute the values of σ' or s. Therefore, we typically analyze such problems using a total stress analysis and compute the shear strength using Equation 13.5.

If the soil is truly saturated and truly undrained, then $\phi_T = 0$ (even though $\phi' > 0$) because newly applied loads are carried entirely by the pore water and do not change σ'.

This is very convenient, because the second term in Equation 13.5 drops out and we no longer need to compute σ. We call this a "$\phi = 0$ analysis." This shear strength is called the *undrained shear strength*, s_u, where $s_u = c_T$. Table 5.4 gives typical values of s_u.

Usually we assign an appropriate s_u value for each saturated undrained stratum based on laboratory or field test results. In reality, s_u is probably not constant throughout a particular soil stratum, even if it appears to be homogeneous. In general, s_u increases with depth because the lower portions of the strata have been consolidated to correspondingly greater loads, and thus have a higher shear strength. In normally consolidated clays, s_u is nearly proportional to σ_z'. The near-surface soils also have higher strengths if they had once dried out (desiccated) and formed a crust. Finally, the natural non-uniformities in a soil strata produce variations in s_u. We can accommodate these variations by simply taking an average value, or by dividing the strata into smaller layers, each with its own s_u.

Example 13.7

Revisit the proposed levee in Example 13.5, except the underlying soils are now saturated clays. The shear strength parameters at Point A are: $s_u = 700$ lb/ft^2, $c' = 300$ lb/ft^2, $\phi' = 24°$. Compute the factor of safety against sliding at Point A.

Solution

We solved Example 13.5 using an effective stress analysis because the underlying soils were sands, so drained conditions could be expected immediately after construction. However, if the underlying soils are clays, undrained conditions will prevail. The excess pore water pressures will be positive because the weight of the levee increases the normal stresses. Therefore, we need to evaluate this problem using the pre-construction shear strength, which is the undrained shear strength, s_u.

The factor of safety immediately after construction is then:

$$\text{Short term } F = \frac{s}{\tau} = \frac{700 \text{ lb/ft}^2}{400 \text{ lb/ft}^2} = 1.7 \quad \leftarrow \textbf{\textit{Answer}}$$

As the excess pore water pressures dissipate, the factor of safety will gradually increase. Once they have completely dissipated, which may take years, the new factor of safety may be computed using the hydrostatic pore water pressures and an effective stress analysis:

$$\sigma_z' = \sum \gamma H - u$$
$$= (62.4 \text{ lb/ft}^2)(22 \text{ ft}) + (125 \text{ lb/ft}^2)(9 \text{ ft})$$
$$+ (119 \text{ lb/ft}^2)(15 \text{ ft}) - (62.4 \text{ lb/ft}^2)(46 \text{ ft})$$
$$= 1412 \text{ lb/ft}^2$$

$$s = c' + \sigma' \tan \phi' = 300 \text{ lb/ft}^2 + (1412 \text{ lb/ft}^2) \tan 24° = 929 \text{ lb/ft}^2$$

$$\text{Long term } F = \frac{s}{\tau} = \frac{929 \text{ lb/ft}^2}{400 \text{ lb/ft}^2} = 2.3 \quad \leftarrow \textbf{\textit{Answer}}$$

The use of undrained strengths in Case 2 design problems should be conservative because this method assumes none of the excess pore water pressures in the field will dissipate until well after the load is placed. In reality, some dissipation usually occurs, with corresponding increases in shear strength. Sometimes these strength increases during construction are significant, so the use of undrained strength can be overly conservative. For example, significant strength increases might occur in soils beneath large fills that are placed slowly. In such cases, engineers sometimes perform special laboratory tests to quantify the excess pore water pressures, then use them to perform an effective stress analysis. This methodology also includes the installation of piezometers in the field to monitor the actual pore water pressures during construction, and thus is another example of the observational method, as discussed in Chapter 12.

Case 3 — Shear Strength under Undrained Conditions with Negative Excess Pore Water Pressures

When the construction causes the normal stress in a saturated clay or silt to decrease, negative excess pore water pressures develop. The most common example is an excavation. This negative u_e gradually dissipates, but now it causes a corresponding loss in shear strength with time as shown in Figure 13.15.

In Case 2, the factor of safety increased with time, which is a desirable characteristic. If a failure does occur, it probably will happen during or soon after construction. However, in Case 3, F decreases with time, which is potentially much more dangerous. The most likely time for a failure is long after construction, and probably after the site has been developed and occupied.

The lowest factor of safety now occurs after the excess pore water pressures have dissipated, so this condition needs to be evaluated using an effective stress analysis with the hydrostatic pore water pressures and the post-construction effective stresses. This approach addresses the long-term stability.

Example 13.8

A cut slope is to be made in a clayey soil to permit construction of a new highway, as shown in Figure 13.19. A slope stability analysis is to be performed along the potential shear surface shown in this figure. The soils are silty clays with $c' = 18$ kPa, $\phi' = 20°$, and $s_u = 100$ kPa. If the shear stress at point A is 60 kPa, compute the short-term and long-term factors of safety at this point.

Solution

Short-term stability

These soils are clayey, so negative excess pore water pressures will be present at point A immediately after construction. Therefore, the short-term stability should be assessed using an undrained total stress analysis.

$$\text{Short term } F = \frac{s}{\tau} = \frac{100 \text{ kPa}}{60 \text{ kPa}} = 1.7 \qquad \leftarrow \textit{Answer}$$

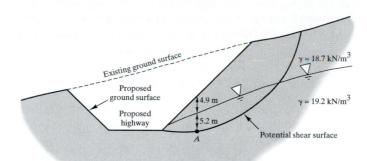

Figure 13.19 Proposed highway cut for Example 13.8.

Long-term stability

Eventually, the excess pore water pressures will dissipate, and the soils will attain their drained strength under the new stress conditions. The long-term stability analysis should be based on an effective stress analysis using the drained strengths.

$$\sigma_z' = \sum \gamma H - u$$
$$= (18.7 \text{ kN/m}^3)(4.9 \text{ m}) + (19.2 \text{ kN/m}^3)(5.2 \text{ m}) - (9.8 \text{ kN/m}^3)(5.2 \text{ m})$$
$$= 140 \text{ kPa}$$

$$\text{Long term } F = \frac{s}{\tau} = \frac{69 \text{ kPa}}{60 \text{ kPa}} = 1.1 \qquad \leftarrow \textit{Answer}$$

$$s = c' + \sigma' \tan \phi'$$
$$= 18 \text{ kPa} + (140 \text{ kPa}) \tan 20°$$
$$= 69 \text{ kPa}$$

Comments

Immediately after construction, the factor of safety at Point A is 1.7, which would usually be acceptable. However, once the negative excess pore water pressures have dissipated, F drops to only 1.1, which would generally not be acceptable. If the groundwater table rose, or if the actual c' and ϕ' values are slightly different than we think, failure could occur.

The factors of safety in this example only represent the conditions at Point A, and are intended only to illustrate the effects of negative pore water pressure dissipation. Actual slope stability analyses require assessment of the factor of safety along the entire shear surface, as discussed in Chapter 14.

Sensitivity

Some clays have a curious property called *sensitivity*, which means their strength in a remolded or highly disturbed condition is less than that in an undisturbed condition at the same moisture content. Sometimes this strength loss is very large, as shown in Figure 13.20. These highly sensitive clays, called *quick clays*, are found in certain areas of Eastern Canada, parts of Scandinavia, and elsewhere. This behavior occurs because these clays have a very delicate structure that is disturbed when they are remolded.

The degree of sensitivity is defined by the parameter S_t:

$$S_t = \frac{s_{\text{undisturbed}}}{s_{\text{remolded}}}$$

(13.7)

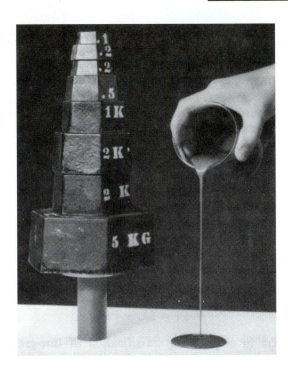

Figure 13.20 Undisturbed and remolded samples of Leda clay from Ottawa, Ontario. Both samples are at the same moisture content; the only difference is the remolding. This is an extreme example of a sensitive clay, with an S_t of about 1500 (National Research Council of Canada).

Table 13.1 presents two systems of classifying sensitivity based on S_t. Note the difference between the criteria commonly used in the United States, where highly sensitive clays are rare, with that used in Sweden, where they are common.

Shear failures in highly sensitive clays can be very dramatic because the strength loss makes the failure propagate over a wide area. This sometimes produces large flow slides, such as the one in Figure 2.18.

Sensitive clays also can recover from these strength losses through a process called *thixotropic hardening*.

TABLE 13.1 TYPICAL CLASSIFICATION OF
SENSITIVITY (Adapted from Holtz and Kovacs, 1981)

Classification	Sensitivity, S_t	
	United States	Sweden
Low sensitivity	2–4	< 10
Medium sensitivity	4–8	10–30
High sensitivity	8–16	30–50
Quick	> 16	50–100
Extra quick		> 100

Residual Strength

In Figure 13.5 we saw the difference in shear stress–strain curves between ductile and brittle soils and the resulting difference between peak strength and residual strength. Although this distinction is not important with sands or gravels because they have curves that are either ductile or very mildly brittle, it can be very important in clays.

Most normally consolidated clays are slightly ductile, and thus have residual strengths that are slightly less than the peak strength. This strain softening (loss of strength with increasing strain) is largely due to particle reorientation and a breakdown of the soil fabric. In sensitive clays, which have an especially delicate fabric, the residual strength can be much less than the peak strength.

Overconsolidated clays nearly always have a brittle stress–strain curve, with the residual strength significantly less than the peak strength. This is due to the factors just described, plus an increase in void ratio during shear and the resulting increase in moisture content.

Brittle soils have two "strengths," so the data points used to develop the Mohr–Coulomb strength envelope could be based on either the peak values or the residual values, as shown in Figure 13.21. Usually we use the peak strength, so it requires no special notation. However, if the data has been assessed using residual strengths, we use a subscript "r", and express the results as c_r' and ϕ_r'. Residual strength is purely frictional (i.e., there is no cohesive strength), but the envelope is typically nonlinear. Thus, any c_r' value is solely the product of fitting a straight line to a curved envelope, as discussed earlier.

Residual strength data is especially important when evaluating shear surfaces in the field produced by landslides. The landslide movement produces a very smooth surface, known as a *slickenside*, as shown in Figure 13.22. The shear strength along such surfaces has been reduced to the residual value, which is less than the strength of the surrounding undisturbed soils. A geotechnical engineer would need to know this strength when assessing the stability of existing landslides and when designing stabilization measures.

Some analyses use a third value, called the *fully softened strength*, which lies between the peak and residual values.

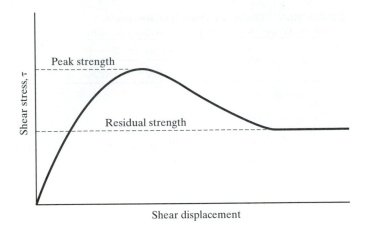

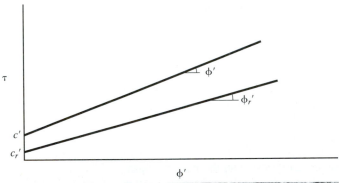

Figure 13.21 The peak strength envelope is obtained from the shear strength at the high points of the stress–strain curves, while the residual strength envelope is obtained from the strength at a large strain.

Figure 13.22 This slickenside was formed by a landslide in a clay. The shear strength along this surface has been reduced to the residual value.

Fissured Clays

Many stiff clays contain small cracks known as *fissures*. The strength along these fissures is less than that of the intact soil, so shear failures are more likely to occur along the fissures. However, the spacing of these fissures is often larger than the soil samples obtained from a boring, so these samples may not represent the larger soil mass. Laboratory tests on such samples probably represent the intact soil, and thus can be very misleading. We probably need the strength along the fissures for most analyses. This strength should be no smaller than the residual strength of the intact soil.

Creep

Sandy and gravelly soils can sustain shear stresses very close to their shear strength for long periods without failing. This is a very desirable trait, and is one of the reasons these soils are superior materials for many applications. Unfortunately, clayey soils are not so well behaved. When the shear stress in a clay exceeds about 70 percent of the shear strength, slow shear movements called *creep* begin to occur. These movements can be the source of many problems. For example, the upper soils on sloping ground sometimes exhibit creep, which causes them to slowly move downslope. Some clays exhibit significant creep at stresses as low as 50 percent of their shear strength. This creep behavior is one of the reasons we typically require higher factors of safety in clayey soils.

Alexandre Collin

After graduating from the École Polytechnique in Paris, Alexandre Collin (1808-1890) worked on several canal and dam projects in France. Some of these projects experienced landslides during and after construction, and Collin had the opportunity to study them. In the process of doing so, he became the first to develop analytical methods for evaluating landslides, and the first to recognize that soil has both frictional and cohesive strength. He also developed the first laboratory equipment to measure soil strength.

Collin attempted to publish his findings in 1840 by submitting a paper to a technical journal, but it was rejected due to the "specialty of the subject matter" (Skempton, 1949). He eventually self-published his findings in 1846, but his work was not widely circulated and was soon forgotten. Collin's contributions were finally recognized only after these principles were independently rediscovered in the early twentieth century.

Image courtesy of Thomas Telford, Ltd.

13.6 SHEAR STRENGTH OF SATURATED INTERMEDIATE SOILS

The discussions in Sections 13.4 and 13.5 divided soils into two distinct categories. The sands and gravels of Section 13.4 do not develop excess pore water pressures during static loading, and thus may be evaluated using effective stress analyses and hydrostatic pore water pressures. Conversely, the silts and clays of Section 13.5 do develop excess pore water pressures, and thus require more careful analysis. They also may have problems with sensitivity and creep. Although many "real-world" soils neatly fit into one of these two categories, others behave in ways that are intermediate between these two extremes. Their field behavior is typically somewhere between being drained and undrained (i.e., they develop some excess pore water pressures, but not as much as would occur in a clay).

Although there are no clear-cut boundaries, these intermediate soils typically include those with unified classification SC, GC, SC-SM, or GC-GM, as well as some SM, GM, and ML soils. Proper shear strength evaluations for engineering analyses require much more engineering judgment, which is guided by a thorough understanding of soil strength principles. When in doubt, it is usually conservative to evaluate these soils using the techniques described for silts and clays in Section 13.5.

13.7 SHEAR STRENGTH OF UNSATURATED SOILS

Thus far we have only considered soils that are saturated ($S = 100\%$). The strength of unsaturated soils ($S < 100\%$) is generally greater, but more difficult to evaluate. Nevertheless, many engineering projects encounter these soils, so geotechnical engineers need to have methods of evaluating them. This has been a topic of ongoing research (Fredlund and Rahardjo, 1993), and standards of practice are not yet as well established as those for saturated soils.

Some of the additional strength in unsaturated soils is due to negative pore water pressures, as shown in Figure 10.16. These negative pore water pressures increase the effective stress, and thus increase the shear strength. However, this additional strength is very tenuous and is easily lost if the soil becomes wetted.

Geotechnical engineers usually base designs on the assumption that unsaturated soils could become wetted in the future. This wetting could come from a rising groundwater table, irrigation, poor surface drainage, broken pipelines, or other causes. Therefore, we usually saturate (or at least "soak") soil samples in the laboratory before performing strength tests. This is intended to remove the apparent cohesion and thus simulate the "worst case" field conditions. We then determine the highest likely elevation for the groundwater table, which may be significantly higher than its present location, and compute positive pore water pressures accordingly. Finally, we assume $u = 0$ in soils above the groundwater table.

QUESTIONS AND PRACTICE PROBLEMS

13.8 Explain the difference between drained and undrained conditions.

13.9 A new building is to be built on a series of spread footing foundations that will be underlain by a saturated clay. Undisturbed soil samples have been obtained from this site and are ready to be tested. Should the laboratory test program focus on producing values of c' and ϕ', or s_u? Explain.

13.10 A steep excavation has been made in a saturated clay without the benefit of a slope stability analysis. It was completed one week ago, and thus far has not shown any signs of instability. Several people working on this project believe this is adequate demonstration of its stability, and feel it is safe. Do you agree? Why or why not?

13.11 A 5 m thick fill has recently been placed over a clayey wetlands to support a new highway. The groundwater table is at or near the natural ground surface. Soon after the fill was completed, but before the paving began, a small landslide occurred in the fill and the underlying soils. The slope failed, so its factor of safety was, by definition, equal to 1.0. Unfortunately, a sudden budget crisis stopped all work on the project and nothing was done for ten years. Then, a new source of funding permitted construction to resume. Is the factor of safety still equal to 1.0? Will remedial construction definitely be necessary to increase the factor of safety? What should be done to evaluate this situation? Explain.

13.12 Pile foundations consist of long poles driven into the ground. They transmit structural loads into the ground through end bearing (compression between the bottom of the pile and the soil below) and through skin friction (sliding friction along the sides of the pile). Both of these depend on the shear strength of the surrounding soil.

When piles are driven into saturated clays, they push the soil aside, causing it to compress and generating excess pore water pressures. After construction, these pressures eventually dissipate.
  a. Would you expect these excess pore water pressures to be positive or negative? Why?
  b. Would you expect the load capacity of the pile to increase, decrease, or remain constant with time? Why?

13.13 Soil can stand in vertical cuts only if it has cohesive strength. Even so, anyone can build a sand castle at the beach using clean fine-to-medium sand, and these castles can have vertical cuts. This appears to be a contradiction.
  a. Explain why sand castles can be built in this way.
  b. If no waves, thieves, rain, or wind disturb the castle, will the vertical cuts stand for a long time? Explain why or why not.

13.8 SHEAR STRENGTH MEASUREMENTS

Geotechnical engineers measure shear strength using both ex-situ and in-situ methods. As discussed in Chapter 3, *ex-situ methods* involve obtaining undisturbed samples from an exploratory boring, bringing them to a soil mechanics laboratory, and testing them there. *In-situ methods* use special equipment brought to the field and test the soils in place.

Ex-Situ Methods

Most shear strength measurements are performed in laboratories using ex-situ methods. These include direct shear tests, ring shear tests, unconfined compression tests, triaxial compression tests, and others.

Direct Shear Test

The earliest measurements of soil strength were probably those performed by the French engineer Alexandre Collin in 1846 (Head, 1982). His test equipment was similar to the modern direct shear machine. The *direct shear test* as we now know it [ASTM D3080] was perfected by several individuals during the first half of the twentieth century.

The test apparatus, shown in Figure 13.23, typically accepts a 60–75 mm (2.5–3.0 in) diameter cylindrical sample and subjects it to a vertical load, P. The vertical total stress, σ_z, is thus equal to P/A. The sample is contained inside a water bath to keep it saturated, but the hydrostatic pore water pressure is very small so we can assume σ_z' also equals P/A. It is important to select P values such that σ_z' is close to the field stresses. The sample is allowed to consolidate under this load.

(a)

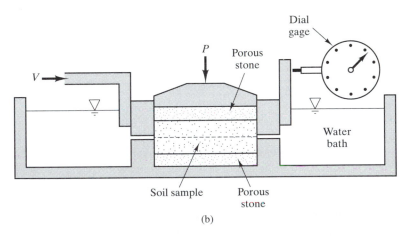

(b)

Figure 13.23 a) A direct shear machine. The sample is inside the sample holder, directly below the upper dial gage, b) cross-section through sample holder showing the sample and shearing action.

Once the soil has fully consolidated, a shear force, V, is gradually applied. This shear force induces a shear stress $\tau = V/A$. Usually V is applied slowly enough to maintain drained conditions. In sands, the required rate of loading is such that failure occurs in a couple of minutes. However, clays must be loaded much more slowly, possibly requiring a time to failure of several hours.

The shear stresses are then plotted against shear displacement, as shown in Figure 13.24. This procedure is then repeated two more times on "identical" new samples using different magnitudes of P.

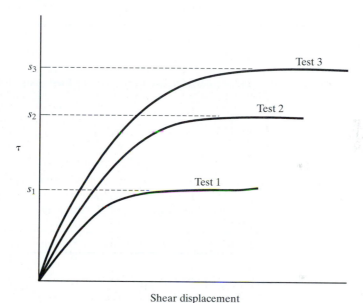

Figure 13.24 Shear stress vs. shear displacement curves from three direct shear tests.

The shear stress vs. shear displacement curves in Figure 13.24 continue until the direct shear machine reaches its displacement capacity. Unlike stress–strain curves in steel or other familiar materials, there is no rupture point. Some shear resistance always remains, no matter how much displacement occurs.

The peak shear strength, s, for each test is the highest shear stress obtained. These values are then plotted on a Mohr–Coulomb diagram as shown in Figure 13.25 and connected with a best-fit line. The τ-intercept is the effective cohesion, c', and the slope of the line is the effective friction angle, ϕ'. Sometimes direct shear tests are performed more quickly, thus simulating partially drained or undrained conditions. In this case, the test results are expressed in terms of the total stress parameters c_T and ϕ_T.

The direct shear test has the advantage of being simple and inexpensive. It is especially useful for obtaining the drained strength of sandy soils. It also can be used with clays, but produces less reliable results because it is difficult to fully saturate the sample and because we have no way of controlling the drainage conditions other than varying the speed of the test. The direct shear test also has the disadvantages of forcing the shear to occur along a specific plane instead of allowing the soil to fail through the weakest zone, and it produces non-uniform strains in the sample, which can produce erroneous results in strain-softening soils.

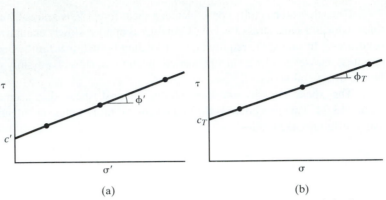

Figure 13.25 Results from a series of direct shear tests: a) Results from a drained test plotted using effective stresses; b) Results from an undrained test plotted using total stresses.

Example 13.9

A series of three direct shear tests has been conducted on a certain saturated soil. Each test was performed on a 2.375 inch diameter, 1.00 inch tall sample. The test has been performed slowly enough to produce drained conditions. The results of these tests are as follows:

Test Number	Normal Load (lb)	Shear Load at Failure (lb)
1	75	51
2	150	110
3	225	141

Determine c' and ϕ'.

Solution

$$A = \frac{\pi D^2}{4} = \frac{\pi (2.375 \text{ in})^2}{4} \left(\frac{1 \text{ ft}^2}{144 \text{ in}^2} \right) = 0.0308 \text{ ft}^2$$

Based on this area and the measured forces:

Test Number	σ' (lb/ft²)	τ (lb/ft²)
1	2438	1665
2	4876	3576
3	7314	4545

This data is plotted in Figure 13.26. It does not form a perfect line. This is due to experimental error, slight differences in the three samples, true nonlinearity, and other factors. We have drawn a best-fit line through these three points to obtain $c' = 400$ lb/ft² and $\phi' = 31°$.

Note: The shear area changes as the direct shear test progresses, and a more thorough analysis would account for this change. However, most engineers neglect this change because it has very little effect on the final test results and because it is slightly conservative to do so.

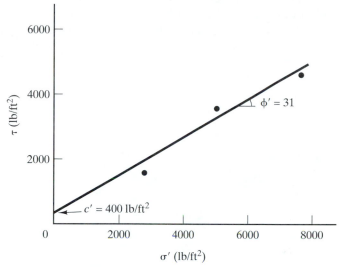

Figure 13.26 Direct shear test results.

Ring Shear Test

The direct shear test does not provide enough shear displacement to reach the residual strength in clays. Sometimes engineers attempt to overcome this mechanical deficiency by shearing the sample back and forth, or by other techniques. However, another device is available to measure the residual strength directly: the *ring shear test*, as shown in Figure 13.27. This test uses an annular-shaped soil sample that is subjected to a known normal load and rotated as shown. The shear stress in the sample may be computed from the torque required to rotate it. In theory, this test has an unlimited strain capacity, but in practice the residual strength is normally achieved after shear displacements of less than 1 m.

Figure 13.27 Ring shear machine. The soil sample is placed inside the annular space shown in the photo on the left. Then the load cap is installed and the normal load is applied. Once the soil has consolidated, a torque is applied from below, and this torque is resisted by the rods shown in the photo on the right.

Unconfined Compression Test

The *unconfined compression test* uses a cylindrical soil sample with no lateral confinement, as shown in Figures 13.28 and 13.29. An axial compressive load is gradually applied to the soil until it fails (i.e., the load reaches its peak value). The load is applied fairly rapidly (typically about 1 minute to failure), thus producing undrained conditions.

Figure 13.28 Conducting an unconfined compression test.

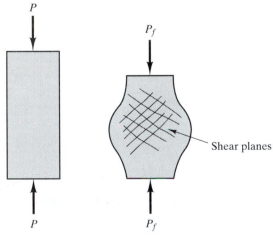

Figure 13.29 Force analysis of an unconfined compression test.

At beginning of test   At failure

The major principal stress acts vertically and is equal to the compressive load P divided by the cross-sectional area A. As the sample compresses, the center part bulges, so the area A, which is measured on a horizontal plane, increases. The cross-sectional area at failure is:

$$A_f = \frac{A_0}{1 - \epsilon_f}$$

(13.8)

where:

    A_f = cross-sectional area at failure
    A_0 = initial cross-sectional area = $\pi d^2/4$
    d = initial sample diameter
    ϵ_f = axial strain at failure

There is no lateral confinement, so $\sigma_3 = 0$ and the left side of the total stress Mohr's circle is always at the origin. Thus, the Mohr's circle at failure is as shown in Figure 13.30.

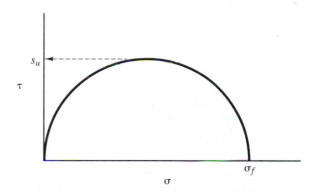

Figure 13.30 Mohr's circle at failure in an unconfined compression test.

This test appears to measure compressive strength, and the results are often expressed that way, as follows:

$$q_u = \frac{P_f}{A_f}$$

(13.9)

where:

q_u = unconfined compressive strength
P_f = normal load at failure
A_f = cross-sectional area at failure

However, the soil really fails in shear along diagonal planes as shown in Figure 13.29. If the soil is soft it fails along multiple diagonal planes and bulges in the middle as shown. If the soil is stiff, it is more likely to fail on a single distinct diagonal plane. If we assume this shear failure occurs along the plane or planes of maximum shear stress, which are inclined at 45° from the horizontal and are represented by the points on the top and bottom of the Mohr's circle, then the undrained shear strength, s_u is one-half of q_u:

$$s_u = \frac{P_f}{2\,A_f} \tag{13.10}$$

The unconfined compression test has the advantage of being simple and inexpensive, so we can use it to obtain a large number of s_u values at a low cost. However, σ_3 in the field is actually greater than zero, so the test tends to underestimate s_u.

Triaxial Compression Test

The *triaxial compression test* is similar to the unconfined compression test except the sample is surrounded by a waterproof membrane and installed in a pressure chamber known as a cell, as shown in Figure 13.31. The chamber is filled with water that is pressurized to produce a specified value of σ_3.

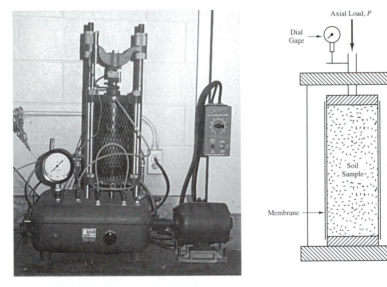

Figure 13.31 A triaxial test apparatus. The chamber in the photograph is covered with a protective wire mesh.

A vertical load, *P*, is slowly introduced through a rod extending through the top of the cell. This load, divided by the cross-sectional area (obtained using Equation 13.8) is the *deviator stress*, σ_d:

$$\sigma_d = \frac{P}{A} \tag{13.11}$$

The vertical stress in the sample, which also is the major principal stress, is the sum of the cell pressure and the deviator stress:

$$\sigma_1 = \sigma_3 + \sigma_d \tag{13.12}$$

Triaxial compression machines include devices for measuring the pore water pressure inside the soil sample while it is being tested. Thus, the effective principal stresses, σ_1' and σ_3', may be computed using Equation 10.32.

There are three common test procedures. The simplest is the *unconsolidated undrained* or UU test (also known as a *Q* or *quick* test). The sample is installed, the cell pressure applied, and the sample is loaded to failure as in the unconfined compression test. The undrained shear strength, s_u, is equal to the radius of the Mohr's circle at failure:

$$s_u = \left(\frac{\sigma_1 - \sigma_3}{2}\right)_f = \left(\frac{\sigma_d}{2}\right)_f \tag{13.13}$$

The measured values of s_u from a UU triaxial test are more reliable than those from the UC test because the presence of $\sigma_3 > 0$ more accurately simulates the field conditions and because the membrane assures the presence of undrained conditions. Only one sample is required to perform a UU test.

The second procedure is the *consolidated drained* or CD test (also known as a *S* or *slow* test). This time, the sample is allowed to consolidate under the applied σ_3. Then it is loaded very slowly, thus producing drained conditions. Usually three such tests are performed, each at a different σ_3. The Mohr's circles at failure are plotted as shown in Figure 13.32. The values of c' and ϕ' are obtained by drawing a line tangent to these circles.

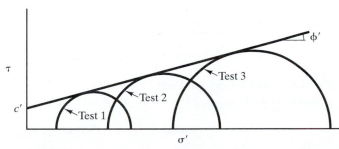

Figure 13.32 Mohr's circles at failure from a series of CD triaxial tests.

The third procedure, the *consolidated undrained* or CU test (also known as an *R* or *rapid* test), is an attempt to measure the drained strength without having to run the test so slowly. The sample is consolidated as with the CD test, but then it is loaded much more rapidly. Undrained conditions prevail, but the machine is able to measure the excess pore water pressures, thus allowing the engineer to compute effective stresses. The effective stress Mohr's circles are then plotted, and c' and ϕ' are determined as before. CU tests also produce values of c_T and ϕ_T.

The triaxial test also has provisions to saturate samples, and to verify that the saturation process was successful. This, along with the testing flexibility and other advantages makes it the standard by which other tests are compared.

Example 13.10

A series of three CU triaxial compression tests have been performed on a set of "identical" clay samples. Each sample had an initial diameter of 50 mm and an initial height of 120 mm. The conditions at failure are presented in the following table:

Test	Conditions at Failure			
No.	P_f (N)	ϵ_f (%)	$(\sigma_3)_f$ (kPa)	u_f (kPa)
1	89	5.0	75	42
2	180	6.1	150	69
3	220	5.8	225	109

Reduce the data and determine the effective stress parameters c' and ϕ'.

Solution

Using Equations 10.33, 13.8, 13.11, and 13.12:

Test No.	A_f (mm²)	$(\sigma_d)_f$ (kPa)	$(\sigma_3')_f$ (kPa)	$(\sigma_1')_f$ (kPa)
1	2060	43	33	76
2	2090	86	81	167
3	2080	106	116	222

The effective stress Mohr's circles are plotted in Figure 13.33.

The Mohr–Coulomb failure envelope is the best-fit line that is tangent to these Mohr's circles, as shown. Thus, the effective stress strength parameters are:

$$c' = 8 \text{ kPa} \qquad \leftarrow Answer$$
$$\phi' = 16°$$

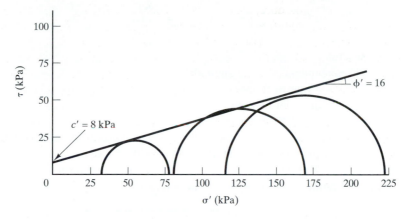

Figure 13.33 Effective stress Mohr's circles at failure for Example 13.10.

In-Situ Methods

Some in-situ test methods measure shear strength directly, while others simply develop some index, such as the Standard Penetration Test N-value, which may be combined with empirical correlations to estimate the shear strength.

Vane Shear Test

The *vane shear test* (VST) consists of a four-bladed vane that is inserted into the ground as shown in Figure 13.34. A steadily increasing torque is applied until the soil fails in shear, then the undrained shear strength, s_u, is computed from this torque. This test is only usable in very soft and soft clays and silts, and usually is performed at the bottom of a boring. In very soft soils, the vane can be pressed to large depths without a boring.

The shear surface has a cylindrical shape, and the data analysis neglects any shear resistance along the top and bottom of this cylinder. Usually the vane height-to-diameter ratio is 2, which, when combined with the applied torque, produces the following theoretical formula:

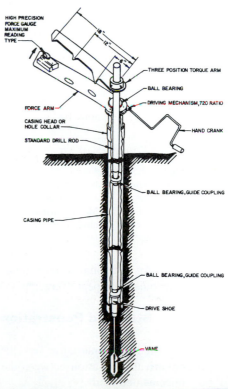

Figure 13.34 Vane shear test (U.S. Navy, 1982).

$$s_u = \frac{6T_f}{7\pi d^3} \qquad (13.14)$$

where:

s_u = undrained shear strength
T_f = torque at failure
d = diameter of vane

However, several researchers have analyzed failures of embankments, footings, and excavations using vane shear tests (knowing that the factor of safety was 1.0) and found that Equation 13.14 often overestimates s_u. Therefore, an empirical correction factor, λ, as shown in Figure 13.35, is applied to the test results:

$$s_u = \frac{6\lambda T_f}{7\pi d^3} \qquad (13.15)$$

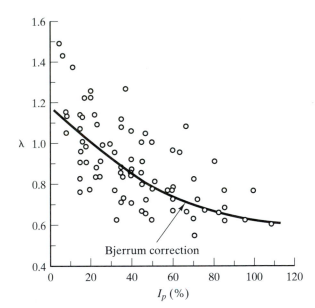

Figure 13.35 Vane shear correction factor, λ (from *Soil Mechanics in Engineering Practice*. 3rd Ed. by Terzaghi, Peck, and Mesri; copyright ©1996; used by permission of John Wiley and Sons).

An additional correction factor of 0.85 should be applied to test results from organic soils other than peat (Terzaghi, Peck, and Mesri, 1996).

Standard Penetration Test

The Standard Penetration Test (SPT) was described in Chapter 3. Figure 13.36 presents an empirical correlation between the SPT N_{60}-value and the effective friction angle, ϕ', in uncemented sands.

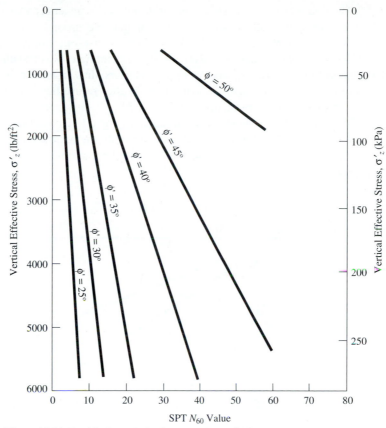

Figure 13.36 Empirical correlation between N_{60} and ϕ' for uncemented sands (Adapted from DeMello, 1971).

Cone Penetration Test

The cone penetration test (CPT), described in Chapter 3, also may be used to determine ϕ', as shown in Figure 13.37.

13.9   SHEAR STRENGTH AT INTERFACES BETWEEN SOIL AND OTHER MATERIALS

Many geotechnical construction projects use geosynthetic materials embedded into the soil for various purposes (Koerner, 1998). For example, geogrids can be used to reinforce the soil, geotextiles can be used for filtration, and geomembranes can be used to provide impervious barriers. Sometimes the shear strength along the interface between these materials and the adjacent soil becomes important and needs to be evaluated. This strength may be measured in the laboratory using a device similar to a direct shear machine.

The interface strength between geogrids and soil is generally very good because these materials are specifically designed to anchor into the ground. However, the interface strength for geomembranes is much lower and has been a source of problems. For example, in 1987 a series of high-density polyethylene geomembranes was installed at the Kettleman Hills landfill in California to serve as seepage barriers, thus protecting the groundwater from contamination. Unfortunately, about one year after refuse began to be placed over this geomembrane, a landslide occurred at the landfill. A subsequent investigation (Mitchell, et al., 1990) revealed the landslide occurred along the interface between the geomembrane and the adjacent soils. Because HDPE is so smooth, the interface friction angle is as low as 8°, or even less. Thus, the geomembrane solved one problem (seepage), but introduced new problem (shear failure).

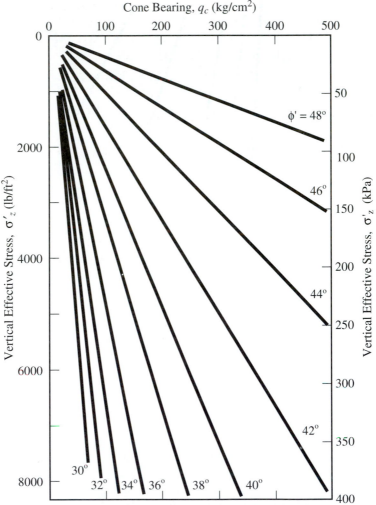

Figure 13.37 Empirical correlation between q_c and ϕ' for uncemented, normally consolidated quartz sands (Adapted from Robertson and Campanella, 1983).

13.10 UNCERTAINTIES IN SHEAR STRENGTH ASSESSMENTS

Shear strength assessments are subject to error from a wide range of sources, so it is important to understand how much (or how little!) faith to place in strength values obtained from laboratory or in-situ tests. The most common mistake, especially among inexperienced engineers, is to place far too much credibility in these numbers.

Even the most carefully performed laboratory tests are no better than the soil samples. Are they truly representative of the soil mass? How much sample disturbance has occurred, and what effect does it have on the test results? The test methods also introduce error, because they do not shear the soils in the same way as occurs in the field. In addition, tests on unsaturated samples may not properly account for future changes in the field moisture content.

Because of these and other factors, even carefully performed shear strength assessments can contain errors of 25 percent or more. Although we typically report test results to two or three significant figures, their true accuracy is much less. Geotechnical engineers attempt to compensate for these uncertainties by using conservative interpretations of test data and applying appropriate factors of safety.

QUESTIONS AND PRACTICE PROBLEMS

13.14 Which laboratory and in-situ tests would be appropriate for measuring ϕ' in a sand?

13.15 Which laboratory test would be most appropriate for measuring c' and ϕ' in a clay?

13.16 Which laboratory and in-situ tests would be appropriate for measuring s_u in a clay?

13.17 A series of direct shear tests has been performed on a dense well-graded sand. All tests were performed on 3.00 inch diameter, 1.25 inch tall cylindrical samples, and were run slowly enough to produce drained conditions. The results of these tests are summarized in the following table:

Test Number	Normal Load (lb)	Shear Load at Failure (lb)
1	100	84
2	200	159
3	400	319

Assuming the shear area remains constant during the test, determine the effective cohesion and effective friction angle from these test results. What values of these parameters would you expect? Are the test results consistent with your expectations? What values would you use for design?

13.18 An unconfined compression test has been performed on a 30 mm diameter, 75 mm long sample of clay. The axial load and axial strain at failure were 120 N and 8.1%, respectively. Compute the undrained shear strength.

13.19 A series of CU triaxial compression tests have been performed on "identical" 2.50 in diameter samples of a clay. All of the samples had an initial height of 6.00 in. The test results were as follows:

Test No.	P_f (lb)	ϵ_f (%)	$(\sigma_3)_f$ (lb/in^2)	u_f (lb/in^2)
		Conditions at Failure		
1	41.7	5.5	10.3	4.3
2	59.9	6.9	18.5	5.6
3	97.1	6.8	27.3	7.1

Plot the effective stress Mohr's circles at failure, draw the Mohr–Coulomb failure envelope, and determine c' and ϕ'. Express c' in lb/ft^2.

13.20 A series of vane shear tests has been performed in a stratum of inorganic clay that has a plasticity index of 50. The vane had a diameter of 50 mm and a height of 100 mm. The test results were as follows:

Depth (m)	Torque at Failure (N-m)
3.4	12.7
4.1	18.1
5.0	15.8
6.6	20.1

Compute the undrained shear strength, s_u, for each test, then combine this data to determine a single s_u value for this stratum.

Note: Geotechnical engineers frequently perform multiple tests on a single stratum, then combine these results into one value for design. The process of doing so is somewhat subjective, and requires the use of engineering judgement. Values significantly larger than the mean are typically discarded, then a design value is typically chosen somewhere between the mean and the minimum values.

SUMMARY

Major Points

1. Shear strength is one of the most important engineering properties of soils. We use it to design structural foundations, earth slopes, retaining walls, and many other engineering projects.
2. Soil strength has two physical sources: Frictional strength and cohesive strength. Frictional strength is due to the sliding and rolling of the particles past each other, while cohesive strength is due to interparticle bonds, such as cementation.

3. There are two kinds of cohesive strength: True cohesion and apparent cohesion. True cohesion is strength that is truly the result of bonding between soil particles, while apparent cohesion is really frictional strength in disguise. The most common source of apparent cohesion is negative pore water pressures.

4. Some soils have ductile stress–strain curves, where failure is defined as the peak shear stress. Other soils have brittle curves, which have two kinds of strength: peak strength and residual strength.

5. Nearly all soil strength analyses in engineering practice use the Mohr–Coulomb failure criterion, which defines shear strength using the parameters c and ϕ. These parameters may be expressed in terms of total stress or effective stress.

6. Effective stress analyses are more accurate descriptions of soil behavior, but become difficult to implement when unknown excess pore water pressures are present. In that case, it becomes necessary to resort to total stress analyses.

7. Saturated sands and gravels are almost always evaluated using effective stress analyses because the excess pore water pressures are minimal.

8. Saturated clays and silts can exhibit either drained or undrained conditions, depending on the rate of loading. For normal rates, undrained conditions prevail immediately after construction.

9. When the construction increases the normal stress in saturated silts or clays, strength analyses are usually based on the undrained strength and use total stress parameters. However, special drained strength analyses are possible when excess pore water pressures are measured in the lab or monitored in the field using piezometers.

10. When the construction decreases the normal stress in saturated clays and silts, strength analyses are usually based on the drained strength and use effective stress parameters.

11. Some clays lose a significant portion of their strength when remolded. This loss is defined by the sensitivity of the clay.

12. The strength at a large shear displacement is called the residual strength. It is useful when evaluating landslides.

13. Unsaturated soils have a higher shear strength due to the presence of apparent cohesion, but this additional strength may be lost if the soil becomes wetted in the future. Engineers usually assume a worst-case condition for such soils.

14. Several laboratory and in-situ methods are available to measure the shear strength. The appropriate method depends on the required parameters, the type of soil, and cost.

Vocabulary

aggregate base material
apparent cohesion
apparent mechanical forces
brittle
cementation
cohesive strength
consolidated drained test
consolidated undrained test
creep

deviator stress
dilation
direct shear test
drained condition
ductile
effective cohesion
effective friction angle
effective stress analyses

electrostatic and electro-
    magnetic attractions
excess pore water pressure
factor of safety
failure envelope
fissured clay
frictional strength
fully softened strength
micaceous sands

Mohr–Coulomb failure criterion	sensitivity	true cohesion
negative pore water pressure	slickenside	ultimate strength
peak strength	soil liquefaction strength	unconfined compression test
primary valence bonding	thixotropic hardening	unconsolidated undrained test
quick clays	total cohesion	undrained condition
quicksand	total friction angle	undrained shear strength
residual strength	total stress analyses	vane shear test
ring shear test	triaxial compression test	

COMPREHENSIVE QUESTIONS AND PRACTICE PROBLEMS

13.21 When subjected to typical rates of loading in the field, sands are usually considered to have drained conditions. Why?

13.22 A certain soil has a unit weight of 121 lb/ft^3 above the groundwater table and 128 lb/ft^3 below. It has an effective cohesion of 200 lb/ft^2, an effective friction angle of 31°, and extends from the ground surface down to a great depth. The groundwater table is at a depth of 18 ft below the ground surface, and $K = 0.78$. Compute the shear strength of this soil on both vertical and horizontal planes at depths of 15 and 30 ft below the ground surface.

13.23 A certain soil has $c' = 12$ kPa and $\phi' = 32°$. The major and minor total principal stresses at a point in this soil are 160 and 348 kPa, respectively, and the pore water pressure at this point is 96 kPa. Draw the failure envelope and the Mohr's circle and determine if a shear failure will occur. If so, determine the angle between the plane on which the major principal stress acts and the failure plane.

13.24 The rock outcrop shown in Figure 13.38 contains an inclined fracture. The fracture is inclined at an angle of 26° from the horizontal.

Figure 13.38 Rock outcrop for Problem 13.24.

a.   Assuming the effective cohesion along the fracture is zero, compute the lowest possible value of the effective friction angle along the fracture. Do this computation by assuming the factor of safety against sliding is equal to one.

Hint: Set the weight of the rock above the fracture equal to W and the area of the fracture equal to A. Then compute the vector component of W that acts parallel to the fracture, and determine what ϕ would be required to resist this force.

b.   If the effective cohesion and friction angle along the fracture are 0 and 38°, respectively, compute the factor of safety against sliding.

13.25 A grain silo, which is a very heavy structure, was recently built on a saturated clay. Because the harvest season was fairly short and intense, the silo was completely loaded with grain fairly quickly (i.e., within a couple of weeks). This is the first time the silo has been loaded. The grain weighs about twice as much as the empty silo. The weight of this grain, along with the weight of the silo, have induced both compressive and shear stresses in the soil below.

Suddenly, someone has become concerned that the soil may be about to fail in shear under the weight of the silo and the grain. This is a legitimate concern, because such failures have occurred before. Discuss the soil mechanics aspects of this situation and determine whether the risk of failure in the soil is increasing, decreasing, or remaining constant with time.

13.26 Hollywood movies sometimes show people "drowning" in quicksand and sinking to the bottom. Are such scenes accurate? What would happen to a person who accidently ventured into quicksand? Explain the reasoning for your answer.

Hint: Compare the unit weight of a human with the unit weight of the quicksand.

14

Stability of Earth Slopes

Wait a minute, I think the town is getting flooded!

The mayor of Armero, Colombia, speaking on a ham radio as the town was being demolished by a massive debris flow.
(Voight, 1990)

Many civil engineering projects are located on or near sloping ground, and thus are potentially subject to various kinds of slope instability such as slides, flows, and falls. Slope failures often produce extensive property damage, and occasionally result in loss of life. Therefore, geotechnical engineers and engineering geologists frequently need to evaluate existing and proposed slopes to assess their stability.

Schuster (1996) estimated the cost of slope-instability-related damage in the United States alone at $1.8 billion per year, or about $7 per capita per year. Some areas are especially prone to trouble. For example, Hamilton County, Ohio (which includes the City of Cincinnati), suffers from $12.4 million in slope-instability-related damage per year (about $14 per capita per year).

Individual slope failures also can be very disastrous and expensive. For example, the 1983 Thistle debris slide near Thistle, Utah, created a huge dam across a canyon, forming a new lake as shown in Figure 14.1 (Kaliser and Fleming, 1986; Shuirman and Slosson, 1992). It caused $200 million in direct damages, including the destruction of two major highways and the main line of the Denver and Rio Grande Western Railroad. This new lake had no outlet, so it would have eventually overtopped the slide and quickly eroded it, causing a massive flood. Therefore, the lake was drained, first with pumps and later with a permanent tunnel.

518

Figure 14.1 The 1983 Thistle debris slide near Thistle, Utah. The lake was drained after this photograph was taken (Utah Geological Survey).

Although most slope failures cause only property damage, some also are fatal. About 25 to 50 deaths per year occur as a result of slope failures in the United States, and about 5 per year in Canada (Schuster, 1996). The worst single event on record is a 1786 slide in China's Sichuan Province that dammed a river in a fashion similar to the Thistle slide. The river soon overtopped the new dam, which rapidly eroded and caused extensive flooding downstream. The flooding occurred very suddenly, and drowned about 100,000 people.

Sometimes civil engineering projects are built close to slopes that are naturally unstable, and eventually succumb to damage from failures that would have occurred whether or not the construction had taken place. More often, carelessly designed and constructed projects decrease the stability of nearby slopes and thus induce instability. For example, the construction of a road in hilly terrain might include creating a cut slope that undermines fractures or bedding planes in the rock, and thus induces a slide. Figure 14.2 is an example of such a project, and the disastrous results.

Figure 14.2 These houses were built near the top of a marginally stable slope that had been steepened as part of an earlier road construction project. The home builders and owners also contributed to the instability by adding to the groundwater through years of landscape irrigation and poor surface drainage. The slope finally failed during a very wet winter, and the failure extended beneath the houses.

Because of these potential problems, it is important to retain the services of engineering geologists and geotechnical engineers when building near or on slopes. With proper evaluation, analysis, design, and construction, slope stability problems can usually be avoided. It is even possible to stabilize ground that would otherwise be unacceptable.

14.1 TERMINOLOGY

Civil engineers use several special terms when describing earth slopes. These include the following (as shown in Figure 14.3):

- *Cut slopes* are those made by an excavation. They expose natural ground that was once buried.
- *Fill slopes* are those made by placing a fill.
- *Natural slopes* are, as the name implies, part of the natural topography.
- The *slope ratio* describes its steepness, and is always expressed as horizontal:vertical. For example, a "three to one" slope (3:1) is inclined at three horizontal to one vertical. Slopes steeper than 1:1 are described using fractions, such as ½:1. This notation can be confusing to engineers used to working with roofs, which are customarily described in the opposite fashion (vertical:horizontal).
- The *top of slope* and *toe of slope* are the points where it intersects flatter ground.
- The *slope face* is the ground surface between the top of slope and toe of slope.
- The *slope height*, *H*, is the difference in elevation between the top of slope and toe of slope (i.e., measured vertically, not diagonally along the face of the slope).
- A *terrace* is a narrow level area created in cut and fill slopes to accommodate surface drainage facilities.

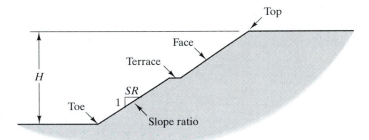

Figure 14.3 Terminology used to describe slopes.

14.2 MODES OF SLOPE INSTABILITY

Slopes can fail in many different ways, and several methods have been developed to classify these modes of failure. We will use the one proposed by Varnes (Varnes, 1958; Varnes, 1978; Cruden and Varnes, 1996).

The Varnes system divides slope failures into five types: *falls*, *topples*, *slides*, *spreads*, and *flows*. We will discuss each of them separately. In addition, these terms are preceded by *rock*, *debris*, or *earth* to indicate the type of material that has failed, where "rock" means bedrock, "debris" means predominantly coarse soils, and "earth" means predominantly fine soils. For example, rock slide, earth slide, rock fall, etc. Other descriptive terms also may be added to describe the rate of movement, history of movement, and other characteristics.

Falls

Falls are slope failures consisting of soil or rock fragments that drop rapidly down a slope, bouncing, rolling, and even becoming airborne along the way. Figure 14.4 shows the results of repeated falls in a rock slope and Figures 14.5 and 14.6 show large boulders that fell down steep slopes.. Falls most often occur in steep rock slopes, and are usually triggered when rock fragments are undermined by erosion, split apart by tree roots or ice, pushed out by water pressure, or shaken by an earthquake.

Falls usually occur very suddenly and rapidly, and thus have been responsible for many deaths. The "watch for falling rock" signs on many mountain roads warn motorists in potential rockfall areas.

Figure 14.4 Repeated rock falls from the outcrop on this steep slope have created a fan of debris called *talus*. The road at the bottom of this slope must be cleared of debris following heavy rains, earthquakes, or other events. This slope is near Forest Falls, California (a place named for waterfalls, not rockfalls!).

Topples

A *topple* is similar to a fall, except that it begins with a mass of rock or stiff clay rotating away from a vertical or near-vertical joint or fissure. This mode of failure occurs only in steep slopes, as shown in Figures 14.7 and 14.8. It is especially important in schist and slate (Goodman, 1993), but also can occur in other types of rocks.

Figure 14.5 The 1992 Landers and Big Bear Earthquakes in California (magnitudes 7.5 and 6.6) generated many rockfalls, including this one that landed on State Highway 38 (Photo by Jeff Knott).

Figure 14.6 This large boulder fell down a steep slope and struck the back of this house in Colorado (Colorado Geological Survey).

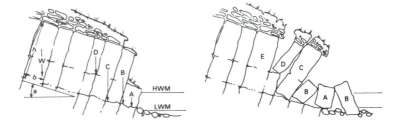

Figure 14.7 Topple instability, before and after failure (Varnes, 1978).

Figure 14.8 This rock slope in a basalt contains a series of near-vertical joints and has been subjected to a series of toppling failures. The debris in the foreground is the result of these topples. Devils Postpile National Monument, California.

Slides

Although many people use the term *slide* or *landslide* to describe any mode of slope instability, Varnes' system uses it only to describe those that involve one or more blocks of earth that move downslope by shearing along well-defined surfaces or thin shear zones. Slides may be described by their geometry, as shown in Figures 14.9–14.11. Common types include:

- *Rotational slides* move along curved shear surfaces that are concave up. These most often occur in homogeneous materials, such as fills.
- *Translational slides* move along more planar shear surfaces. These usually reflect weak zones or bedding planes, and their thickness-to-length ratios are usually less than 0.1. When the moving blocks remain relatively intact, translational slides are sometimes called *block-glide slides*.
- *Compound slides* have a shape between those of rotational and translational slides.
- *Complex* and *composite slides* have characteristics of slides and some other mode of slope instability, such as flows.

Figure 14.9 Rotational slide (Varnes, 1978).

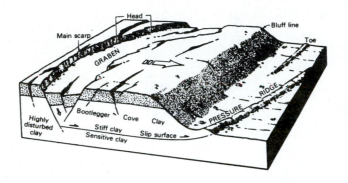

Figure 14.10 Block-glide translational slide (Varnes, 1978).

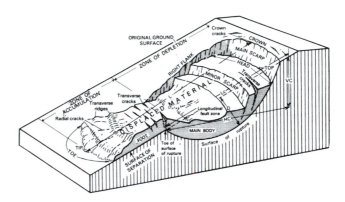

Figure 14.11 Complex slide that includes some flow characteristics (Varnes, 1978).

Special terms used to describe slides are illustrated in Figure 14.11 and defined below:

- *Crown* — The nearly undisturbed ground above the main scarp
- *Main scarp* — The steep natural ground formed above the slide when it moved downhill
- *Minor scarp* — A secondary scarp created within the main body of the slide as a result of secondary failures

- *Body* — The displaced soil or rock
- *Flank* — The borders along the left and right sides of the body where it meets the relatively undisturbed ground
- *Tension cracks* — Cracks that often appear in the crown. They are roughly parallel to the top of the slope and are caused by tensile stresses in the ground.

Spreads

Spreads (or *lateral spreads*) are similar to translational slides, except the blocks separate and move apart as they also move outward, as shown in Figure 14.12. This mode of failure reflects movement along a layer of very weak soil, and sometimes occurs during earthquakes when a zone of soil liquefies. Spreads also can occur along layers of sensitive clay.

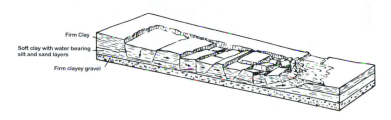

Figure 14.12 Lateral spread (Varnes, 1978).

Spreads usually occur on gentle to moderate slopes, and often terminate at a riverbank. They can be very destructive because they often affect large areas and move long distances. Spreads have been responsible for failures of bridges and other important structures, as shown in Figures 14.13 and 20.17.

Figure 14.13 Marine Research Facility, Moss Landing, California. Liquefaction during the 1989 Loma Prieta Earthquake caused a lateral spread beneath the left part of this building, which "stretched" it more than 1.5 m (5 ft). (Earthquake Engineering Research Center Library, University of California, Berkeley, Steinbrugge Collection)

Flows

Flows are downslope movements of earth that resemble viscous fluid. They differ from slides in that there are no well-defined blocks moving along shear surfaces. Instead, the mass flows downhill, with shear strains present everywhere. After the flow ceases, its products have a clearly fluidized appearance, as shown in Figures 2.18 and 14.14.

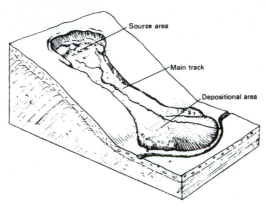

Figure 14.14 A flow failure (Varnes, 1978).

Flows often contain other objects, such as boulders and logs, that move with the fluidized earth. These are called *debris flows* and can be very destructive. Buildings, cars, and other objects that might survive the moving mud are often destroyed by the debris contained in this mud.

Because of their high speed and ability to travel long distances, flows are the most dangerous and destructive mode of slope instability. Two of the most dramatic examples occurred in the Peruvian Andes in 1962 and 1970 (Plafker and Ericksen, 1978). The 1962 event began as an avalanche on the steep Nevados Huascarán mountain. The falling snow and ice gathered boulders and mud as it continued down the mountain, producing a large mudflow. It attained an estimated velocity of 105 mi/hr (170 km/hr) and rapidly buried 9 towns, killing about 4,000 people. The 1970 event was even larger, and began with an avalanche on the same mountain, this time triggered by an earthquake. It also gathered mud and rocks, including one boulder with a mass of about 8.2×10^6 kg, and reached an estimated velocity of 170 mi/hr (270 km/hr). This event quickly buried the city of Yungay, killing its 18,000 inhabitants. The quotation at the beginning of this chapter illustrates the suddenness of such events in areas with steep terrain.

Less dramatic examples occur throughout the world, often threatening people and property. Because flows are usually triggered by rain or snowmelt, they often are accompanied by flooding. For example, a 1934 flood and debris flow in La Cañada, California, caused over $5 million in property damage, 40 deaths, and the loss of 400 houses (Troxell and Peterson, 1937).

QUESTIONS AND PRACTICE PROBLEMS

14.1 Explain the difference between a flow and a slide. Which one often travels farther, and thus can be a hazard to sites far from the slope?

14.2 Cut slopes in bedded sedimentary rocks can be very problematic. What mode of failure do you think would be most common, and how could we evaluate the potential for such a failure before construction?

14.3 A preliminary grading plan for a proposed highway shows a 50 ft tall, 2:1 cut slope ascending from each side of the highway with level land above both slopes. Following a geotechnical study, it became necessary to change these slope ratios to 3:1. Assuming the existing right-of-way barely accommodates the 2:1 slopes, how much additional right-of-way must now be purchased because of this change?

14.4 Some slides are large enough and move far enough to completely block a canyon or valley and thus form a new lake. The Thistle slide in Figure 14.1 is an example. Why are these slides especially dangerous, and what can be done to alleviate this danger once the slide has occurred?

14.5 A national park visitor's center has unwittingly been built on soils deposited by a series of earth flows. The building is located near the base of a canyon where it meets a larger valley. Ten years after construction, another earth flow occurred and deposited up to 3 ft of mud and debris around the visitor's center. Although the building was not seriously damaged, it was expensive and time-consuming to clean up the mess. Everyone now recognizes that this building should have been constructed somewhere else, but there is no funding available to move it or replace it. Suggest one or two ways of protecting the building from future earth flows.

14.3 ANALYSIS OF SLOPE STABILITY PROBLEMS

Geotechnical engineers and engineering geologists use both qualitative and quantitative methods to analyze slope instability problems. Some semi-quantitative methods, such as systematic plots of joint attitudes, also are very useful. These analyses often require the skills of both professions working in harmony, and need to consider both the present conditions and potential future conditions.

Analyses of potential falls and topples use qualitative and semi-quantitative methods almost exclusively. These include geologic mapping, evaluations of past performance, and so on, and are usually performed by engineering geologists. With topples, these methods might be supplemented with limited quantitative analyses.

Flows are most amenable to semi-quantitative analysis. Some engineers have attempted to use more quantitative analyses based on parameters such as shear strength and unit weight (Johnson and Rodine, 1984; Brunsden, 1984), but these methods do not appear to be widely used. Infinite slope analyses, discussed later in this section, also may give some insight on the potential for flows.

In contrast, slides are very amenable to quantitative analysis. Geotechnical engineers have developed methods of evaluating the potential for failure, and can express it as a factor of safety. These methods have proven to be reliable, and are used routinely in geotechnical engineering practice. Some methods also are applicable to evaluations of spreads.

The following discussions focus primarily on quantitative analyses of slides because geotechnical engineers use these methods extensively. However, this emphasis does not mean that slides are necessarily more important than other modes of failure, nor does it mean that qualitative analyses are not useful. The proper evaluation of slope stability problems requires a wide variety of techniques, so it is important not to become overly enamored with quantitative analyses at the expense of other methods.

14.4 QUANTITATIVE ANALYSIS OF SLIDES

The French engineer Alexandre Collin was probably the first to conduct quantitative analyses of landslides (Collin, 1846). Unfortunately, his work was not widely recognized at the time, and was largely forgotten. Therefore, the origin of modern slope stability analyses is traceable to another group of engineers working in Sweden during the 1920s. Apparently unaware of Collin's work, they had to begin anew and developed methods that soon became the basis for modern slope stability analyses.

Much of Scandinavia, especially Sweden and Norway, is underlain by sensitive marine clays with undrained shear strengths on the order of 300 lb/ft^2 (15 kPa). Because of this very low strength and high sensitivity, slopes in these soils are very prone to failure. These problems became especially troublesome during the late nineteenth and early twentieth centuries because of the cutting and filling associated with port construction (Petterson, 1955) and railroad construction. Both caused slides and flows. Following an especially costly failure in 1913, the Swedish State Railways formed a "Geotechnical Commission" (Statens Järnvägars, 1922) to study the problem and develop solutions (see discussion in Chapter 1). The commission's final report was issued in 1922, and is recognized as one of the early milestones in geotechnical engineering. The report presented a method of analysis we now call the *Swedish Slip Circle Method*, discussed later in this section. It later became the basis for other methods of analysis.

Limit Equilibrium Concept and Factor of Safety

Most quantitative analyses of slides or potential slides are *limit equilibrium analyses*, which means they evaluate the slope as if it were about to fail and determine the resulting shear stresses along the *failure surface*. Then, these stresses are compared to the shear strength to determine the *factor of safety*, F:

$$F = \frac{s}{\tau}$$

(14.1)

where:

F = factor of safety

s = shear strength

τ = shear stress

The factor of safety varies along the failure surface. Some sections may have "failed" (i.e., the shear stress equals the shear strength), while others may have a large reserve of excess shear strength (i.e., a large F).[1] However, limit equilibrium analyses do not attempt to define this distribution. They only give an overall value, which is more accurately defined as:

$$F = \frac{\int s\,dl}{\int \tau\,dl} \tag{14.2}$$

where l is the length along the shear surface, and both integrals are evaluated along its entire length.

In theory, a factor of safety of 1 indicates incipient failure, so any slope with $F > 1$ will supposedly be stable. However, there are many uncertainties in our analyses (soil profile, shear strength, groundwater conditions, etc.), so we need to account for them by requiring an even larger factor of safety. The most common design criterion requires a factor of safety of at least 1.5, although slightly lower values (perhaps about 1.3) may be acceptable on some highway projects where no structures are nearby and a failure would only require cleaning debris from the roadway.

Most slope stability analyses describe stability in terms of the factor of safety. This is known as a *deterministic analysis*. However, it also is possible to express stability as a *probability of failure*. For example, after considering the various uncertainties, we might determine that a certain slope has an annual probability of failure of 10^{-3} (i.e., one chance in 1000), which would then be compared to some acceptable level of risk (Wu, Tang, and Einstein, 1996; Wolff, 1996). This method is called a *probabilistic analysis*.

Effective Stress vs. Total Stress Analyses

In Chapter 10 we discussed the difference between effective stress, σ', and total stress, σ. Then, in Chapter 13 we saw that effective stress is a better indicator of soil strength, and wrote the Mohr–Coulomb strength equation accordingly as:

$$s = c' + \sigma' \tan \phi' \tag{14.3}$$

where:

s = shear strength
c' = effective cohesion
σ' = effective stress on the failure surface
ϕ' = effective friction angle

---

[1] See Deschamps and Leonards (1992) for further discussion of this point, and a proposal to replace the factor of safety with a new parameter, the *safety margin*.

Therefore, we would normally expect to perform slope stability analyses by evaluating c' and ϕ' using appropriate laboratory tests, computing σ' along the failure surface, then computing the shear strength using Equation 14.3. This approach is called an *effective stress analysis*, and works well so long as we are able to compute σ'.

However, when excess pore water pressures are present, it can be difficult to compute σ', as discussed in Chapter 13, so we sometimes resort to a *total stress analysis* that defines the shear strength as:

$$s = c_T + \sigma \tan \phi_T \qquad (14.4)$$

where:

   s = shear strength
   c_T = total cohesion
   σ = total stress on failure surface
   ϕ_T = total friction angle

When performing slope stability analyses in sands and gravels, we normally use an effective stress analysis because little or no excess pore water pressure is present. However, in saturated silts and clays, we need to classify the problem according to the guidelines on page 488, then use either a total stress analysis or an effective stress analysis, as discussed on pages 488–493.

The proper assessment of shear strength parameters can be difficult (see Duncan, 1996, and Abramson, 1996, for more information). Therefore, the examples and homework problems in this chapter simply give the proper values of c and ϕ or, in the case of undrained analyses, s_u.

Critical Failure Surface

Limit equilibrium analyses begin with the definition of a potential failure surface, which is where the shearing would occur if the slope were to fail. The analysis must be performed on the correct surface so that it reflects the proper failure geometry, but there are an infinite number of potential surfaces and it may be difficult to determine which is the *critical failure surface* (i.e., the one on which sliding is most likely).

Usually we locate the critical failure surface through an informed trial-and-error process. In other words, we probably have an approximate idea of its location, so we begin there and analyze many different potential surfaces (perhaps several hundred) and compute the factor of safety for each. The one with the lowest factor of safety is the critical failure surface, and that F is the factor of safety for the entire slope.

Although some slope stability analyses can be performed by hand, they are ideal candidates for computer programs. Therefore, geotechnical engineers usually use computers for all but the simplest slope stability analyses. Many such programs are available, both from the public domain and from private software developers.

Evaluation of Forces

Most limit equilibrium analyses divide the failure mass into N vertical *slices*, as shown in Figure 14.15. These slices are chosen such that the bottom of each one passes through only one type of material, and so that each slice is small enough that its bottom may be considered to be a straight line.

Both shear and normal forces act on the bottom of each slide and along the interface between slices, as shown in Figure 14.15. Combining all of these unknowns with the available equations of static equilibrium ($\Sigma F_x = 0$, $\Sigma F_z = 0$, $\Sigma M = 0$) reveals that this is a statically indeterminate problem whenever $N > 1$ (Duncan, 1996). Although some analyses can be performed with only one slice, and thus are statically determinate, most require 10 to 40 slices, and thus are indeterminate.

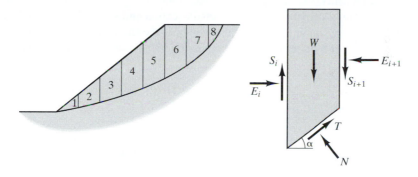

Figure 14.15 Division of the failure mass into vertical slices and summary of forces acting on each slice.

Indeterminate problems may be solved by increasing the number of equations and/or decreasing the number of unknowns. For slope stability problems, this is done by introducing simplifying assumptions as discussed below. Many different analysis methods have been developed, each based on a different set of simplifying assumptions (Fredlund, et al., 1981).

Analyses with One Slice

Failure surfaces that can be analyzed using only one slice do not require simplifying assumptions to make them statically determinate. Two types are commonly encountered: planar failure analyses and infinite slope analyses.

Planar Failure Analyses

Slopes with a single planar failure surface with constant values of c and ϕ, as shown in Figure 14.16, may be analyzed using a single slice. This geometry often occurs in nature, as in rocks where slippage will occur along fractures or bedding planes.

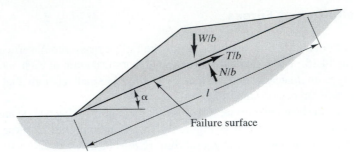

Figure 14.16 Slope with planar failure surface.

Most limit equilibrium analyses, including this one, are two-dimensional. This means they assume the slope extends for an infinite distance perpendicular to the cross-section. Mathematically, we will consider a slice of length "b" from this infinitely long slope, where b is measured perpendicular to the cross-section and is equal to 1 ft or 1 m, depending on the units of measurement. Thus, the weight of the sliding mass is expressed as W/b, perhaps using units of lb/ft, and the forces acting on its base are N/b and T/b, where:

$$N/b = (W/b) \cos \alpha \qquad (14.5)$$

$$T/b = (W/b) \sin \alpha \qquad (14.6)$$

The average pore water pressure, u, acting on the failure surface is:

$$u = \gamma_w z_w \qquad (14.7)$$

where:
γ_w = unit weight of water
z_w = average depth from groundwater table to failure surface

Using this information, we can derive a formula for F as follows:

$$
\begin{aligned}
\sigma' &= \sigma - u \\
&= \frac{N/b}{l} - u \\
&= \frac{(W/b) \cos \alpha}{l} - u \qquad (14.8)
\end{aligned}
$$

$$\tau = \frac{T/b}{l}$$

$$= \frac{(W/b)\sin\alpha}{l} \tag{14.9}$$

$$F = \frac{\int s\,dl}{\int \tau\,dl}$$

$$= \frac{sl}{\tau l}$$

$$\boxed{F = \frac{c'l + [(W/b)\cos\alpha - ul]\tan\phi'}{(W/b)\sin\alpha}} \tag{14.10}$$

In the special case of $c = 0$ and $u = 0$, Equation 14.10 reduces to:

$$\boxed{F = \frac{\tan\phi'}{\tan\alpha}} \tag{14.11}$$

For total stress analyses, Equation 14.10 becomes:

$$\boxed{F = \frac{c_T l + (W/b)\cos\alpha\tan\phi_T}{(W/b)\sin\alpha}} \tag{14.12}$$

Example 14.1

A 1.5:1 cut slope is to be made in a shale with an apparent dip of 16° as shown in Figure 14.17. Compute the factor of safety against failure along the lowermost daylighted bedding plane using a unit weight of 20.1 kN/m³ and along bedding strength parameters of $c' = 15$ kPa and $\phi' = 20°$.

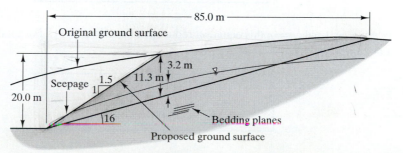

Figure 14.17 Cross-section for Example 14.1.

Solution

$$W/b = \frac{1}{2}(85.0\,\text{m})(11.3\,\text{m})(20.1\,\text{kN/m}^3) = 9650\,\text{kN/m}$$

$$l = \frac{85.0\,\text{m}}{\cos 16°} = 88.4\,\text{m}$$

Compute the pore water pressure based on a visual estimate of the average z_w (ranges from 0 to 3.2 m):

$$u = \gamma_w z_w = (9.8\,\text{kN/m}^3)(3.0\,\text{m}) = 29\,\text{kPa}$$

$$
\begin{aligned}
F &= \frac{c'l + [(W/b)\cos\alpha - ul]\tan\phi}{(W/b)\sin\alpha} \\
&= \frac{(15\,\text{kPa})(88.4\,\text{m}) + [(9650\,\text{kN/m}^3)\cos 16° - (29\,\text{kPa})(88.4\,\text{m})]\tan 20°}{(9650\,\text{kN/m}^3)\sin 16°}
\end{aligned}
$$

$$= 1.42 \qquad\qquad \Leftarrow \textit{Answer}$$

The computed factor of safety of 1.42 is slightly less than the minimum acceptable value of 1.50. Therefore, this design is probably not acceptable.

Infinite Slope Analyses

An infinite slope analysis is similar to a planar analysis, except the failure surface is parallel to the slope face, and the depth to that surface is small compared to the height of the slope, as shown in Figure 14.18. Equation 14.10 still applies, but it is more conveniently rewritten as follows:

$$F = \frac{c' + [\gamma D - \gamma_w z_w]\cos^2\alpha \tan\phi'}{\gamma D \sin\alpha \cos\alpha} \tag{14.13}$$

When $c' = 0$ and $z_w = 0$, this equation reduces to Equation 14.11. For total stress analyses, it becomes:

$$F = \frac{c_T + \gamma D \cos^2\alpha \tan\phi_T}{\gamma D \sin\alpha \cos\alpha} \tag{14.14}$$

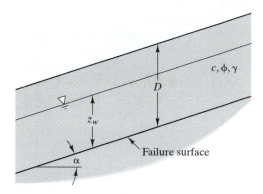

Figure 14.18 Geometry for infinite slope analyses.

Infinite slope analyses are useful when a thin layer of soil overlies a much harder strata or bedrock, and also may be used to evaluate the potential for shallow flowslides, which are sometimes called *surficial slumps*.

Circular Failure Surfaces

Slides in homogeneous or near-homogeneous soils often may be idealized as the arc of a circle. This *circular failure surface* geometry simplifies the mathematics because the normal force acting on the base of each slice passes through the center of the circle. It also simplifies the process of searching for the critical failure surface because each circle is defined by only three parameters, the radius, R, and the x and z coordinates of the center.

Swedish Slip Circle Analysis (c>0, ϕ=0)

The Geotechnical Commission appointed by the Swedish State Railways developed an analysis method based on circular failure surfaces in undrained soils (Statens Järnvägars, 1922). The shear strength of such soils is independent of σ and is defined solely by the parameter s_u (the "$\phi = 0$ condition"), so we do not need to know the forces N, E, or S, and the problem becomes statically determinate.

Using the variables defined in Figure 14.19, and a derivation similar to the one shown above, we can express the factor of safety as:

$$F = \frac{\pi R^2}{180} \frac{\sum s_u \theta}{\sum (W/b)d}$$

(14.15)

where:

F = factor of safety

R = radius of slip circle

s_u = undrained shear strength along slip surface

θ = arc of slip circle for a given slice

W/b = weight of slice per unit length of slope (i.e. kN/m)

d = moment arm of slice

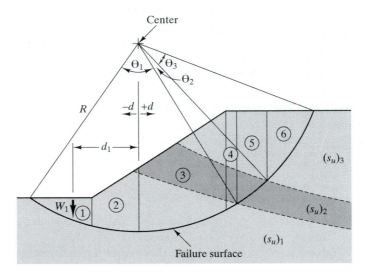

Figure 14.19 Cross-section for Swedish slip circle method.

Example 14.2

Using the Swedish slip circle method, compute the factor of safety along the trial circle shown in Figure 14.20.

Solution

Divide the slide mass into vertical slices as shown. One of the slice borders should be directly below the center of the circle (in this case, the border between slices 2 and 3). For convenience of computations, also draw a slice border wherever the slip surface intersects a new soil stratum and whenever the ground surface has a break in slope.

Then, compute the weight and moment arm for each slide using simplified computations as follows:

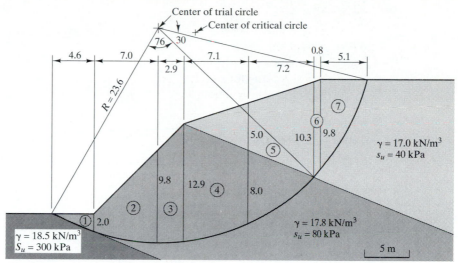

Figure 14.20 Cross-section for Example 14.2.

Weights:

$$W_1/b = 4.6 \left(\frac{2.0}{2} \right) 17.8 = 80 \text{ kN/m}$$

$$W_2/b = 7.0 \left(\frac{2.0 + 9.8}{2} \right) 17.8 = 130 \text{ kN/m}$$

$$W_3/b = 2.9 \left(\frac{9.8 + 12.9}{2} \right) 17.8 = 590 \text{ kN/m}$$

$$W_4/b = 7.1 \left(\frac{5.0}{2} \right) 17.0 + 7.1 \left(\frac{12.9 + 8.0}{2} \right) 17.8 = 1620 \text{ kN/m}$$

$$W_5/b = 7.2 \left(\frac{5.0 + 10.3}{2} \right) 17.0 + 7.2 \left(\frac{8.0}{2} \right) 17.8 = 1450 \text{ kN/m}$$

$$W_6/b = 0.8 \left(\frac{10.3 + 9.8}{2} \right) 17.0 = 140 \text{ kN/m}$$

$$W_7/b = 5.1 \left(\frac{9.8}{2} \right) 17.0 = 420 \text{ kN/m}$$

Moment arms:

$$d_1 = -7.0 - \frac{4.6}{3} = -8.5 \text{ m}$$

$$d_2 = -\frac{7.0}{2} = -3.5\,\text{m}$$

$$d_3 = \frac{2.9}{2} = 1.5\,\text{m}$$

$$d_4 = 2.9 + \frac{7.1}{2} = 6.5\,\text{m}$$

$$d_5 = 2.9 + 7.1 + \frac{7.1}{2} = 10.9\,\text{m}$$

$$d_6 = 2.9 + 7.1 + 7.2 + \frac{0.8}{2} = 17.6\,\text{m}$$

$$d_7 = 2.9 + 7.1 + 7.2 + 0.8 + \frac{5.1}{3} = 19.7\,\text{m}$$

Slice	s_u (kPa)	θ (Deg)	$s_u\theta$	W/b (kN/m)	d (m)	$(W/b)d$
1				80	-8.5	-680
2				130	-3.5	-450
3	80	76	6080	590	1.5	890
4				1620	6.5	10,530
5				1450	10.9	15,800
6				140	17.6	2,460
7	40	30	1200	420	19.7	8,280
		$\Sigma =$	7280		$\Sigma =$	36,830

$$F = \frac{\pi R^2}{180} \frac{\sum s_u\theta}{\sum (W/b)d} = \frac{\pi\,(23.6)^2}{180}\,\frac{7{,}280}{36{,}830} = \textbf{1.92} \quad \leftarrow \textit{Answer}$$

Comments

The computed factor of safety along this circle is 1.92. However, to find the factor of safety of the slope, we need to search for the critical circle. Through a process of controlled trial-and-error, we will find the critical circle is centered at the location shown in Figure 14.18, has a radius of 28.7 m, and a computed factor of safety of 1.66. Thus, the computed factor of safety against a slide is 1.66, which is greater than the usual standard of 1.5, and thus is probably acceptable.

Neither the trial circle nor the critical circle penetrates into the lowermost stratum, because this stratum has a shear strength significantly higher than the others.

Ordinary Method of Slices

When performing effective stress analyses, $\phi > 0$ so the Swedish slip circle method is not applicable. We now need to know the value of N for each slice to compute σ' for the equation $s = c + \sigma' \tan \phi$. Fellenius (1927, 1936) transformed this into a statically determinate problem by assuming the resultant of the normal and shear forces acting on the two sides of each slice are equal in magnitude and colinear. This means the side forces cancel each other, so we do not need to know their magnitudes or position. This method is known by different names, including the *ordinary method of slices* (OMS), the *Swedish method of slices* and the *Fellenius method*.

Using this assumption, and the dimensions shown in Figure 14.21, the factor of safety for effective stress analyses becomes:

$$F = \frac{\sum [c'l + ((W/b)\cos\alpha - ul)\tan\phi']}{\sum [(W/b)\sin\alpha]} \tag{14.16}$$

or, for total stress analyses:

$$F = \frac{\sum [c_T l + (W/b)\cos\alpha \tan\phi_T]}{\sum [(W/b)\sin\alpha]} \tag{14.17}$$

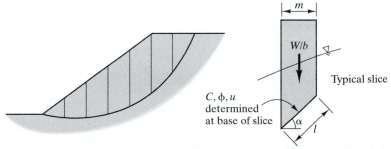

Figure 14.21 Cross-section and definition of variables for ordinary method of slices and modified Bishop's method.

Example 14.3

A 30 ft tall, 1.5:1 slope is to be built as shown in Figure 14.20. The soil is homogeneous, with $c' = 400$ lb/ft^2 and $\phi' = 29°$. The unit weight is 119 lb/ft^3 above the groundwater table, and 123 lb/ft^3 below. Using the ordinary method of slices and an effective stress analysis, compute the factor of safety along the trial circle shown in Figure 14.22.

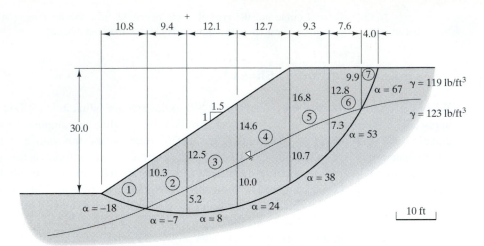

Figure 14.22 Cross-section of proposed slope for Example 14.3.

Solution

Weights:

$$W_1/b \; = \; 10.8\left(\frac{10.3}{2}\right)119 \; = \; 6{,}620\,\text{lb/ft}$$

$$W_2/b \; = \; 9.4\left(\frac{10.3+12.5}{2}\right)119 \; + \; 9.4\left(\frac{5.2}{2}\right)123 = \; 15{,}800\,\text{lb/ft}$$

$$W_3/b \; = \; 12.1\left(\frac{12.5+14.6}{2}\right)119 \; + \; 12.1\left(\frac{5.2+10.0}{2}\right)123 = \; 30{,}800\,\text{lb/ft}$$

$$W_4/b \; = \; 12.7\left(\frac{14.6+16.8}{2}\right)119 \; + \; 12.7\left(\frac{10.0+10.7}{2}\right)123 = \; 39{,}900\,\text{lb/ft}$$

$$W_5/b \; = \; 9.3\left(\frac{16.8+12.8}{2}\right)119 \; + \; 9.3\left(\frac{10.7+7.3}{2}\right)123 = \; 26{,}700\,\text{lb/ft}$$

$$W_6/b \; = \; 7.6\left(\frac{12.8+9.9}{2}\right)119 \; + \; 7.6\left(\frac{7.3}{2}\right)123 = \; 13{,}700\,\text{lb/ft}$$

$$W_7/b \; = \; 4.0\left(\frac{9.9}{2}\right)119 \; = \; 2{,}400\,\text{lb/ft}$$

Average pore water pressure at base of each slice:

$$u_1 \; = \; 0$$

$$u_2 \; = \; \left(\frac{5.2}{2}\right)62.4 \; = \; 160\ \text{lb/ft}^2$$

$$u_3 = \left(\frac{5.2+10.0}{2} \right) 62.4 = 470 \text{ lb/ft}^2$$

$$u_4 = \left(\frac{10.0+10.7}{2} \right) 62.4 = 650 \text{ lb/ft}^2$$

$$u_5 = \left(\frac{10.7+7.3}{2} \right) 62.4 = 560 \text{ lb/ft}^2$$

$$u_6 = \left(\frac{7.3}{2} \right) 62.4 = 230 \text{ lb/ft}^2$$

$$u_7 = 0$$

Slice	W/b (lb)	α (Deg)	c' (lb/ft^2)	ϕ' (Deg)	u (lb/ft^2)	l (ft)	$c'l +$ $((W/b) \cos \alpha -$ $ul) \tan \phi'$	(W/b) $\sin\alpha$
1	6,620	-18	400	29	0	11.4	8,000	-2,000
2	15,800	-7	400	29	160	9.5	11,700	-1,900
3	30,800	8	400	29	470	12.2	18,600	4,300
4	39,900	24	400	29	650	13.9	20,800	16,200
5	26,700	38	400	29	560	11.8	12,700	16,400
6	13,700	53	400	29	230	12.6	8,000	10,900
7	2,400	67	400	29	0	10.2	4,600	2,200
						$\Sigma =$	84,400	46,100

$$F = \frac{84,400}{46,100} = \textbf{1.83} \quad \Leftarrow \textit{Answer}$$

Note how slices 1 and 2 have a negative α because they are inclined backwards.

Comments

Further trials with other circles will demonstrate that this is the critical circle (i.e., it is the one with the lowest factor of safety). Therefore, according to the ordinary method of slices, the factor of safety of this slope is 1.83.

The Importance of Side Forces

The simplifying assumption in the ordinary method of slices reduces the problem to one that is both statically determinate and suitable for hand computations. Nevertheless, we need to ask "how valid is this assumption?"

If we consider a typical slice, as shown in Figure 14.15, the resultant of the shear and normal forces on the left side is really larger than that on the right, and thus contributes to the normal force, N. However, the OMS ignores this contribution and computes N based only on the weight of the slice. This produces an N value that is too low, an s value that is too low, and therefore an F that is too low. Thus, the OMS is conservative.

This conservatism is most pronounced when α is large. For shallow circles, the computed factor of safety is generally no more than 20 percent less than the "correct" value, but deep, small radius circles that extend well below the groundwater table have much more error, sometimes producing computed F values as much as 50 percent too low (Wright, 1985).

Several engineers have developed more refined methods of analyzing circular failure surfaces based on more reasonable assumptions. The most popular of these is the Modified Bishop's Method.

Modified Bishop's Method

Bishop (1955) addressed this problem by assuming the shear forces on sides of each slice are equal, and that the normal forces on the sides of each slice are colinear, but not necessarily equal. Although these assumptions are only approximations of the truth, they are much better than the assumptions used in the ordinary method of slices. This solution is called the *modified Bishop's method* or the *simplified Bishop's method*. Careful studies have shown that it produces computed F values within a few percent of the "correct" values (Wright, 1985), and thus is sufficiently precise for virtually all circular analyses. Therefore, this is the recommended method for circular failure surfaces.

Using the variables defined in Figure 14.21, the modified Bishop equation for effective stress analyses is:

$$F = \frac{\sum \left[\dfrac{m c' + ((W/b) - um)\tan \phi'}{\psi} \right]}{\sum \left[(W/b) \sin \alpha \right]} \tag{14.18}$$

$$\psi = \cos \alpha + \frac{\sin|\alpha| \tan \phi}{F} \tag{14.19}$$

For total stress analyses, Equation 14.18 becomes:

$$F = \frac{\sum \left[\dfrac{mc_T + (W/b)\tan\phi_T}{\psi} \right]}{\sum \left[(W/b)\sin\alpha \right]} \tag{14.20}$$

Although the modified Bishop method can be solved by hand, it is more tedious than the OMS because it is not a closed-form solution.   The factor of safety appears on both sides of the equation, so it is necessary to first estimate F, compute ψ using Equation 14.19, then compute F using Equation 14.18 or 14.20. This process must then be repeated with a new estimate (the computed value from the previous iteration is a good choice) until the estimated and computed values are essentially equal. Usually three iterations are sufficient. Of course, this is a trivial problem for computer-based solutions.

Example 14.4

Solve Example 14.3 using the modified Bishop's method.

Solution

First iteration — try $F = 1.90$
① = numerator in Equation 14.18
$(W/b) \sin \alpha$ = denominator in Equation 14.18

Slice	W/b (lb/ft)	α (Deg)	c' (lb/ft²)	ϕ' (Deg)	u (lb/ft²)	m (ft)	$(W/b)\sin\alpha$	Try $F = 1.90$ ψ	①
1	6,620	-18	400	29	0	10.8	-2,000	1.041	7,700
2	15,800	-7	400	29	160	9.4	-1,900	1.028	11,400
3	30,800	8	400	29	470	12.1	4,300	1.031	18,200
4	39,900	24	400	29	650	12.7	16,200	1.032	21,900
5	26,700	38	400	29	560	9.3	16,400	0.968	16,100
6	13,700	53	400	29	230	7.6	10,900	0.835	11,600
7	2,400	67	400	29	0	4.0	2,200	0.659	4,400
							46,100		91,300

$$F = \frac{91,300}{46,100} = 1.98$$

The computed F of 1.98 is greater than the assumed value of 1.90.

Second iteration — try $F = 1.95$

① = numerator in Equation 14.18

$(W/b) \sin \alpha$ = denominator in Equation 14.18

Slice	W/b (lb/ft)	α (Deg)	c' (lb/ft²)	ϕ' (Deg)	u (lb/ft²)	m (ft)	$(W/b) \sin \alpha$	Try $F = 1.95$ ψ	Try $F = 1.95$ ①
1	6,620	-18	400	29	0	10.8	-2,000	1.039	7,700
2	15,800	-7	400	29	160	9.4	-1,900	1.027	11,400
3	30,800	8	400	29	470	12.1	4,300	1.030	18,200
4	39,900	24	400	29	650	12.7	16,200	1.029	22,000
5	26,700	38	400	29	560	9.3	16,400	0.963	16,200
6	13,700	53	400	29	230	7.6	10,900	0.829	11,700
7	2,400	67	400	29	0	4.0	2,200	0.652	4,500
							46,100		91,700

$$F = \frac{91,700}{46,100} = 1.99$$

The computed F of 1.99 is greater than the assumed value of 1.95.

Further trials will produce **$F = 2.00$**   ⟵ *Answer*

The computed factor of safety for this circle is 2.00, which is slightly higher than the 1.83 computed using the ordinary method of slices.

Comments

Once again, further trials with other circles will demonstrate that this is the critical circle (i.e., the one with the lowest factor of safety). Therefore, according to the modified Bishop's method, the factor of safety of this slope is 2.00, which is slightly higher than the 1.83 obtained by the ordinary method of slices. The modified Bishop's method is generally considered to be more precise than the OMS.

Chart Solutions

When c, ϕ, and γ are constant throughout the slope, the analysis is substantially simplified and may be reduced to simple charts. Several such charts have been developed (see Abramson, et al., 1996), and they are useful for simple slopes that may not justify the time required to perform a computer-based analysis. One of these is the solution developed by Cousins (1978), part of which is reproduced in Figure 14.23. It is based on the geometry shown in Figure 14.24.

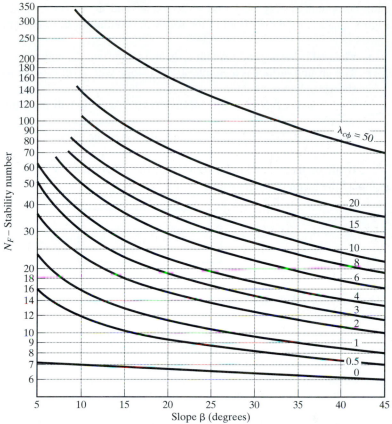

Figure 14.23 Stability chart for simple slopes that meet the criteria described in the text (Cousins, 1978). Used with permission of ASCE.

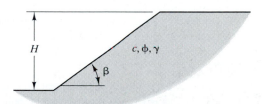

Figure 14.24 Slope geometry for Cousins' chart.

To use Cousins' chart, first compute $\lambda_{c\phi}$:

$$\lambda_{c\phi} = \frac{\gamma H \tan \phi}{c} \qquad (14.21)$$

This computation may be based either on c' and ϕ' or on c_T and ϕ_T. The chart in Figure 14.23 is valid only if all of the following conditions have been met:

- The soil is homogeneous with c, ϕ, and γ constant throughout
- The ground surface is a simple straight slope with level ground above and below, as shown in Figure 14.24
- $\lambda_{c\phi} \geq 2$ and/or $\beta > 53°$ (this means the critical circle will pass through the toe of the slope)
- The groundwater table is well below the toe of the slope

If all of these conditions have been met, determine N_F from Figure 14.23, then compute the factor of safety using:

$$F = N_F \frac{c}{\gamma H}$$

(14.22)

The chart is based on the critical slip surface, so only one iteration is required to obtain the factor of safety.

This chart is based on the friction circle method, and generally produces computed factors of safety comparable to those obtained from the ordinary method of slices. However, it is slightly less precise than a conventional solution. Cousins and others also have developed charts for more complex conditions, but such slopes are probably best evaluated using computer-based analyses as discussed earlier.

Example 14.5

Using the cross-section in Figure 14.22 with a very deep groundwater table, compute the factor of safety using Cousin's chart.

$$\lambda_{c\phi} = \frac{\gamma H \tan\phi}{c} = \frac{(119\,\text{lb/ft}^3)(30\,\text{ft})\tan 29°}{400\,\text{lb/ft}^2} = 4.9$$

$\lambda_{c\phi} = 4.9 > 2$, so the critical circle is a toe circle. The other criteria listed above also have been satisfied, so Cousin's chart is applicable to this problem.

$$\beta = \tan^{-1}\left(\frac{1}{1.5}\right) = 34°$$

From Figure 14.23, $N_F = 17.1$.

$$F = N_F \frac{c}{\gamma H} = 17.1\left(\frac{400\,\text{lb/ft}^2}{(119\,\text{lb/ft}^3)(30\,\text{ft})}\right) = \mathbf{1.92} \qquad \leftarrow \textit{Answer}$$

Note that this is the factor of safety for the critical circle. There is no need to search for this circle. The computed factor of safety of 1.92 is slightly higher than the 1.83 obtained by the ordinary method of slices, and slightly lower than the 2.00 obtained from the modified Bishop's method. Part of this difference is because this example used a different groundwater table than was used in the previous examples.

Irregular-Shaped Failure Surfaces

Many failure surfaces are neither planar nor circular, and thus cannot be analyzed using any of the methods described thus far. Even supposedly circular surfaces are often truncated at the top due to the formation of tension cracks. Therefore, geotechnical engineers have developed additional analysis methods to accommodate irregular-shaped failure surfaces. These are sometimes called *non-circular analyses*. These methods lose the mathematical simplifications of convenient geometry, and thus are more complex and difficult to implement. Most are practical only when solved by a computer.

Random-shaped failure surfaces also make searching routines more difficult, because the failure surface can no longer be defined by only three variables. Although some software includes searching capabilities, much more skill is required to locate the most critical failure surface.

Analysis methods have been proposed by Janbu (1957, 1973), Morgenstern and Price (1965), Spencer (1967), Sarma (1973), and others. Each method uses different simplifying assumptions to overcome the problem of static indeterminancy, and thus produces slightly different results. We will discuss only Spencer's method.

Spencer's Method

Spencer's method (Spencer, 1967, 1973; Sharma and Moudud, 1992) is popular among geotechnical engineers because it combines good precision with ease of use. Although the solution still requires a computer program, the required user input is simpler than that of some other methods.

Spencer assumed that the resultant of the normal and shear forces on the sides of each slice are colinear, and that all of them act at an angle θ from the horizontal. The solution of Spencer's method requires assuming an initial value for θ, then computing one value of F based on force equilibrium and another based on moment equilibrium. The θ value is then iterated until the two computed factors of safety are equal. This process is much too tedious to do by hand, but quite simple for a computer.

Partially Submerged Slopes

Many slopes are partially submerged, and thus are subjected to external hydrostatic pressures as shown in Figure 14.25. Examples included levees and earth dams. To analyze such slopes, simply treat the external water as if it were a "soil" with $c = 0$, $\phi = 0$, and $\gamma = \gamma_w$.

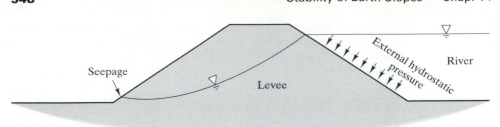

Figure 14.25 Partially submerged slope.

Partially submerged slopes are potentially subject to a special mode of failure called *rapid drawdown*. This occurs when the water outside the slope quickly drops to a lower elevation. This mode of failure is discussed in Chapter 15.

Back-calculated Strength

Most slope stability analyses begin with measured soil strength data and compute the factor of safety. However, when working with landslides that have already occurred, it is often useful to do the reverse analysis: Begin with $F = 1$ (because the slope failed) and back-calculate the soil strength (Duncan and Stark, 1992). Soil strengths obtained this way are generally very reliable, because they are based on the full shear surface, not on small samples. This strength can then be used for subsequent analyses of proposed remediation measures.

The soil strength is usually defined by two parameters, c and ϕ, so the solution to a back-calculated strength analysis cannot produce unique values of both parameters. Instead, we obtain a plot of the various combinations of c and ϕ that produce $F = 1$, and select one of these combinations for subsequent analyses.

Seismic Stability

A large number of slope failures have occurred during earthquakes, so geotechnical engineers working in seismically active regions routinely evaluate the seismic stability of earth slopes. This is a part of the broader discipline of geotechnical earthquake engineering, which we will explore in Chapter 20.

Some earthquake-induced failures are very large. For example, the 1959 Hebgen Lake Earthquake triggered a massive slide in Madison Canyon, Montana, as shown in Figure 14.26. This slide had a volume of about 25 million yd^3 (20 million m^3) and traveled at an estimated velocity of 110 mi/hr (180 km/hr), creating a 220 ft (67 m) tall dam across the canyon (Sowers, 1992). This dam formed a new lake similar to the one formed by the Thistle landslide described earlier. It also killed 28 campers who were in the canyon to enjoy its world-class fishing.

Figure 14.26 Madison canyon landslide in Montana as it appeared in 1997, thirty-eight years after the failure. The scarp from the slide, which is the light-colored area in the center of the photograph, is still clearly visible on the mountainside. This slide formed a dam that created Earthquake Lake, which is visible in the foreground.

Most seismic failures are much smaller than the one at Madison Canyon, and often consist of significant slope distortions without fully developing a true landslide. Nevertheless, these distortions can be large enough to cause significant property damage. For example, a large number of slope distortions occurred during the 1994 Northridge Earthquake in California (Stewart, et al., 1995). Most of them involved displacements of less than 8 cm, but they produced about $100 million in property damage.

The physical mechanisms of seismically induced slope movements are very complicated, and include all of the complexities of static slope stability plus those associated with the propagation of seismic waves and the dynamic strength of soil and rock (Rogers, 1992a). Thus, geotechnical analyses of seismic stability can sometimes be very difficult. However, there is a great deal of active research on this topic, and it is helping us better understand the physical mechanisms and develop methods of analyzing these problems (Marcuson, Hynes, and Franklin, 1992).

Liquefaction-Induced Failures

Many of the most dramatic and devastating earthquake-induced slides are the result of soil liquefaction. For example, the 1964 Turnagain Heights Landslide in Anchorage, Alaska was the result of liquefaction of buried sand strata (Seed and Wilson, 1964). This slide covered an area of about 130 acres and resulted in the destruction of 75 houses, as shown in Figure 14.27.

The key to avoiding such failures is to properly identify potentially liquefiable soils and understand their impact on slope stability. We will discuss soil liquefaction and various modes of liquefaction-related failure in Chapter 20.

Figure 14.27 The 1964 Turnagain Heights landslide in Anchorage, Alaska (Earthquake Engineering Research Center Library, University of California, Berkeley, Steinbrugge Collection).

Failures Caused Directly by Ground Shaking

Slope failures also can occur as a direct result of ground shaking, even without soil liquefaction. These most often occur in steep slopes, especially those covered with loose natural soils (usually colluvium) or poorly constructed fills.

Although this mode of failure can readily be identified after it occurs, it is sometimes difficult to recognize potentially hazardous slopes before the earthquake occurs. Often such assessments can be based on empirical comparisons of soil type and unit weight, slope ratio, groundwater conditions, and other factors. For example, if dry, natural colluvial soils of a certain thickness at a certain slope ratio were found to fail during an earthquake, similar soil conditions at another location would probably fail if subjected to a similar earthquake.

Geotechnical engineers also use quantitative analyses. However, these analyses are gross simplifications of the actual physical mechanisms, and thus may not be reliable. The most common of these are the pseudostatic method and Newmark's method, as discussed below.

Pseudostatic Method

The *pseudostatic method* is an enhancement of conventional limit equilibrium analyses that evaluates the seismic stability of an earth slope by applying a horizontal acceleration to each slice. This horizontal acceleration is assumed to continue indefinitely (or at least long enough for the slope to fail), and thus is idealized as a horizontal static force as shown in Figure 14.28. This static force is equal to $(a/g)W$, where a is the horizontal acceleration, g is the acceleration due to gravity, and W is the weight of the slice.

The computations then proceed like any other limit equilibrium analysis. However, this additional force produces a lower factor of safety intended to reflect the detrimental effects of the earthquake. Normally the minimum acceptable factor of safety also is lower, typically between 1.1 and 1.2.

Although this representation of the horizontal acceleration as a static force greatly simplifies the computations, it is not a very accurate representation of the seismic forces in a real earth slope. The differences between this analysis and reality include the following:

- The real seismic accelerations cycle back and forth in opposite directions, and continue for only a limited time
- The wavelength of the seismic waves is smaller than most slopes, so part of the slope may be accelerating uphill while another part is accelerating downhill. The entire slope will not be accelerating in the same direction, even for a moment.

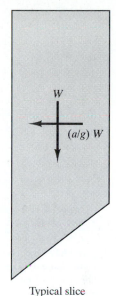

Typical slice

Figure 14.28 The pseudostatic method consists of applying a horizontal static force to each slice.

Because of these differences, pseudostatic analyses cannot be based on the anticipated peak ground accelerations, which are often in excess of $0.7g$. The use of such values would indicate failure in virtually all analyses. Instead, geotechnical engineers use values based on the observed behavior of slopes during earthquakes, typically 0.1 to 0.2 g (Hynes-Griffin and Franklin, 1984; Kramer, 1986). Thus, the "acceleration" value in the pseudostatic analysis is really more of an empirical index than a measure of the true ground accelerations.

Although the pseudostatic analysis has some value, it is only a rough approximation of the physical mechanisms acting in the field, and thus should be used only with considerable engineering judgement. In some cases, it can produce overly conservative results, and thus can dictate preventive measures that are not necessary.

When performing pseudostatic analyses, it is useful to note that slopes with static factors of safety greater than 1.70 and no liquefaction problems have never been known to fail during earthquakes (Rogers, 1992a; Hynes-Griffin and Franklin, 1984). Thus, it may be prudent to dispense with pseudostatic analyses when both of these criteria have been met.

Newmark's Method

Newmark (1965) developed an enhancement of the pseudostatic method that attempts to predict the slope displacement during an earthquake. It does so by first establishing the *yield acceleration*, a_y, that corresponds to $F = 1$ in a conventional pseudostatic analysis. Then, the engineer obtains an acceleration vs. time plot for the design earthquake, such as the one in Figure 14.29, then identifies the time intervals where $a > a_y$. By double integrating these intervals, we obtain the associated slope displacement which then is compared to some maximum allowable displacement.

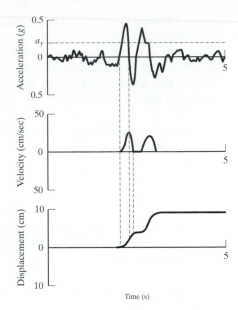

Figure 14.29 Use of Newmark analysis to predict slope displacements from an earthquake acceleration record (Wilson and Keefer, 1985).

Although this analysis is an improvement over the pseudostatic method, it still is a simplification of the true seismic response in a real slope. The results of this analysis are very sensitive to the selected a_y value and other factors (Kramer, 1996), and thus requires a great deal of care to implement.

QUESTIONS AND PRACTICE PROBLEMS

14.6 The better limit equilibrium analysis methods, such as the modified Bishop's and Spencer's methods, produce factors of safety that are within about 5 percent of the "true" values. How does this error compare to the uncertainty in the soil properties (c, ϕ, and γ) and the uncertainty in the design soil profile? In light of these other sources of uncertainty, is a ±5 percent error tolerable? Explain.

14.7 Most limit equilibrium analysis methods include one or more simplifying assumptions. Why are these assumptions necessary? Give an example of one of the methods and its assumptions.

14.8 A natural 2.25:1 slope is underlain by a residual soil derived from the underlying gneiss. Compute the factor of safety for a failure surface 4 ft below the ground surface using a total stress infinite slope analysis with $c_T = 200$ lb/ft^2, $\phi_T = 24°$, and $\gamma = 118$ lb/ft^3.

14.9 The proposed slope in Example 14.1 had an unacceptable factor of safety. We plan to remedy this situation by using a flatter slope. What slope ratio would be required to produce a factor of safety of 1.5?

14.10 Using the Swedish slip circle method, compute the factor of safety along the failure surface shown in Figure 14.30.

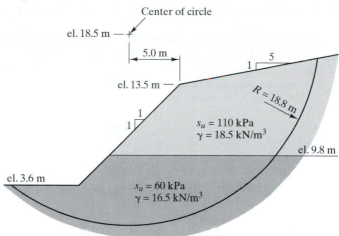

Center of circle

el. 18.5 m

5.0 m

el. 13.5 m

$R = 18.8$ m

$s_u = 110$ kPa
$\gamma = 18.5$ kN/m^3

el. 9.8 m

el. 3.6 m

$s_u = 60$ kPa
$\gamma = 16.5$ kN/m^3

Figure 14.30 Cross-section for Problem 14.10.

14.11 You are writing a computer program to perform slope stability computations. This program will consider only circular failure surfaces. What procedure might you use to locate the critical failure surface? Provide a detailed explanation.

14.12 Using the ordinary method of slices, compute the factor of safety along the failure surface shown in Figure 14.31.

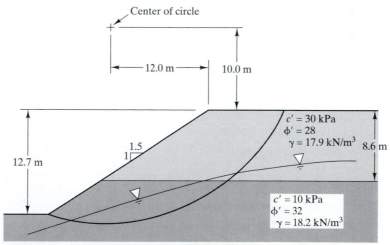

Center of circle

12.0 m   10.0 m

$c' = 30$ kPa
$\phi' = 28$
$\gamma = 17.9$ kN/m^3 8.6 m

1.5
1

12.7 m

$c' = 10$ kPa
$\phi' = 32$
$\gamma = 18.2$ kN/m^3

Figure 14.31 Cross-section for Problems 14.12 and 14.13.

14.13 Using the modified Bishop's method, compute the factor of safety along the failure surface shown in Figure 14.31.

14.14 An 8.5 m tall, 2:1 fill slope is to be made of soil with $c_T = 35$ kPa, $\phi_T = 23°$, and $\gamma = 19.5$ kN/m³. The groundwater table will be well below the toe of this slope. Using Cousin's chart, compute the factor of safety. Does this slope meet normal stability standards?

14.15 Using the ordinary method of slices, compute the factor of safety along the failure surface shown in Figure 14.32. Then, assume an earthquake occurs and the sand stratum liquefies, and compute a new factor of safety using $c' = 0$ and $\phi' = 0$ (in reality, it is σ' that goes to zero, but this is a mathematical trick that accomplishes the same thing). According to this analysis, will the slope survive the earthquake?

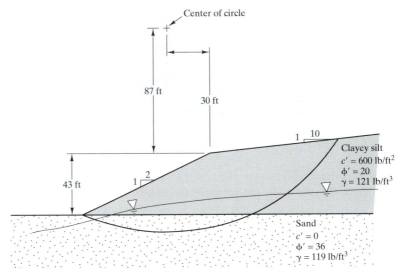

Figure 14.32 Cross-section for Problem 14.15.

14.16 The slope shown in Figure 14.33 has recently failed. A geotechnical investigation indicates the failure surface was as shown. Assuming the failure occurred while undrained conditions prevailed in the slope, back-calculate the value of s_u. Use the Swedish slip circle method with the cross-section that existed immediately before it failed. Use $\gamma = 119$ lb/ft³.

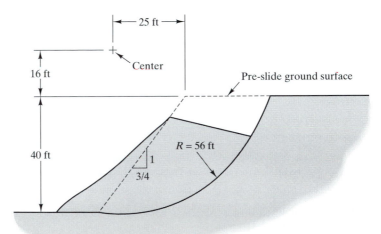

Figure 14.33 Cross-section for Problems 14.12 and 14.13.

14.5 STABILIZATION MEASURES

When proposed or existing slopes do not have sufficient stability, geotechnical engineers turn to various methods of slope stabilization (Hausmann, 1992; Rogers, 1992b; Abramson, et al., 1996; Turner and Schuster, 1996). Many, but not all, slopes can be economically stabilized, and many methods are available. The factor of safety (Equation 14.1) depends on both the shear stress and the shear strength, so stabilization measures must decrease the stress and/or increase the strength.

The selection of an appropriate stabilization plan depends on many factors, including:

- The subsurface conditions and potential modes of failure
- The present and required topography
- The presence of physical constraints, such as property lines or existing buildings
- The consequences of a failure (i.e., small for a rural low-traffic road, potentially catastrophic for a nuclear power plant), which determines the required reliability
- Availability of materials, equipment, and expertise (specialized methods may not be available in some areas)
- Performance history of various methods as implemented in the local area
- Aesthetics
- Time required for construction
- Cost

Unloading

The simplest way to decrease the shear stresses in the slope is to unload it, either by reducing the slope height or by increasing the slope ratio, as shown in Figure 14.34.

For example, if the slope is associated with a proposed highway, it may be possible to decrease its height by revising the vertical alignment of the highway. Unfortunately, this solution usually results in steeper grades, which may be unacceptable. Increasing the slope ratio generally does not require a new vertical alignment, but does need a wider right-of-way and involves larger earthwork quantities. This may be quite feasible in rural areas with rolling hills, but could be prohibitive in urban areas or where the natural terrain is steep.

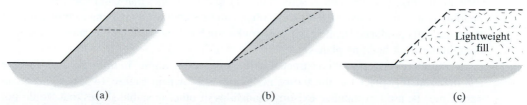

(a)   (b)   (c)

Figure 14.34 Slope stabilization by unloading. a) Reducing slope height; b) Increasing slope ratio: c) Using lightweight fill.

Another method of unloading involves construction of lightweight fills, as discussed in Chapter 6. These fills permit construction of slopes without inducing large shear stresses in the ground.

Buttressing

The short-term stability of cut slopes is generally greater than their long-term stability, so it usually is possible to make temporary construction slopes much steeper than would be permissible for the permanent slope. This is especially true when construction occurs during the dry season when the groundwater table is lower. We can use this behavior to build *buttress fills* that stabilize slopes.

The usual construction procedure is to overexcavate the proposed cut slope as shown in Figure 14.35a, then bring it back to the design grades using high-quality fill (i.e., one with higher *c* and φ values than the natural soils). The size of the buttress needs to be selected so that potential failure surfaces that pass through the buttress gain enough additional strength to raise the factor of safety to an acceptable value, and that potential failure surfaces that pass below the buttress also have an acceptable factor of safety. To meet these goals, buttresses often must include downward extensions called *shear keys*, as shown in Figure 14.35.

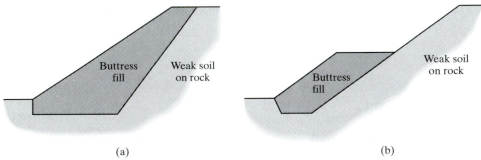

(a)   (b)

Figure 14.35 Buttress fills: a) This buttress requires overexcavating the slope, then rebuilding it with compacted fill. Thus, the finish grade is at its original location; b) This buttress was built by adding the fill to the front of the slope without any overexcavation. In this case the buttress is only about half the height of the slope, but any height can be built.

Sometimes the buttress fill is made of crushed gravel or other very high-quality soils that have very high strength. However, the concept also is valid for normal fills, so long as they are stronger than the natural soils. For example, cut slopes that will expose daylighted bedding planes in soft sedimentary rocks often can be stabilized by making a compacted fill from the soils produced by the excavated rock. Such fills are stronger because it does not contain the weak bedding planes.

Buttresses also can be constructed without overexcavation. In this case they become stabilization fills placed at the toe of the slope as shown in Figure 14.35b. These can sometimes be used to stabilize existing landslides or other unstable slopes that would not tolerate steep construction excavations, or when land is not at a premium. The top of such buttresses may be level, and thus can become usable for development.

Structural Stabilization

Another option is to stabilize slopes using structural elements. These include various kinds of retaining walls and tieback anchors. These methods are typically very expensive, but can be cost effective in certain situations, especially in urban areas.

Retaining Walls

Retaining walls are structural members that maintain adjacent ground surfaces at two different elevations, as shown in Figure 14.36. Sometimes a retaining wall is used in lieu of a slope, while other times they are used in conjunction with the slope to create a more stable condition. Chapter 16 discusses the various kinds of retaining walls.

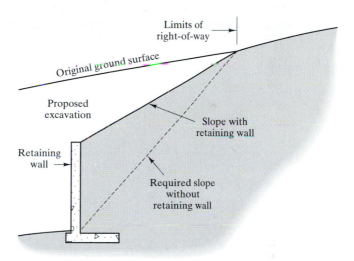

Figure 14.36 Use of retaining walls to stabilize slopes.

Tieback Anchors

Another structural measure is *tieback anchors*, which are tensile members that apply stabilizing forces onto the slope as shown in Figure 14.37. Tiebacks usually consist of steel rods inside grouted holes that extend well beyond the critical failure surface. This method is generally very expensive, but may be cost effective in urban areas where space is at a premium and land is expensive.

Drainage

Water is the "enemy" in slope stability problems, so stabilization measures often involve draining water, both surface and subsurface. The objective is to prevent excessive water from percolating into the ground, and to remove water that already is in the ground. These measures improve stability by decreasing the pore water pressures (thus increasing the strength), and by drying the soil (which increases its strength and decreases its weight).

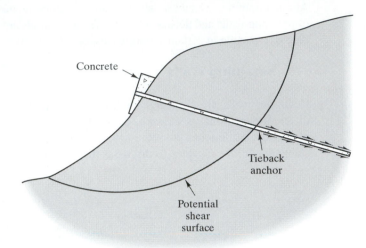

Figure 14.37 Use of tieback anchors to stabilize slopes.

Surface

Some surface drainage measures can be as simple as providing appropriate grades so surface water flows away from the slope instead of toward it. For example, if a building pad is to be located above a slope, it should be graded such that the surface water flows away from the slope. It is very poor practice to permit water to flow over the top of the slope.

Often, additional drainage measures also are needed to capture surface water and carry it away from the slope (Scullin, 1983). These often consist of ditches paved with concrete, as shown in Figure 14.38, and also can include buried culvert pipes.

Figure 14.38 Typical surface drainage facilities on a cut slope. This concrete terrace drain captures surface water and carries it to a safe discharge point.

The design of surface drainage facilities is often governed by building codes, but this does not relieve the engineer of the duty to provide additional facilities when required.

In emergencies, it may be helpful to use sandbags to divert surface water away from the slope and plastic sheets to cover the ground and reduce infiltration.

Subsurface

The objective of subsurface drains is to remove water that already is present in the ground. There are several methods of doing so, including the following:

- *Perforated pipe drains* consist of special pipes with holes, buried in the ground to collect the water and carry it to a safe location. These pipes are surrounded by gravel and a *filter fabric*, to assist the entry of water and prevent finer soil from washing into the pipe and clogging it. Sometimes these drains are placed in a trench as shown in Figure 14.39 to form a *French drain*. They also may be placed below fills, behind buttress fills, and in other key locations.

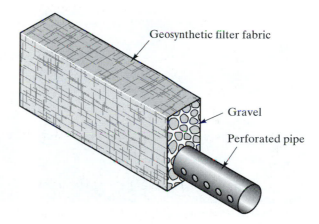

Geosynthetic filter fabric

Gravel

Perforated pipe

Figure 14.39 Perforated pipe drain.

- *Wells* are vertical holes drilled into the ground and equipped with pumps to remove the water as shown in Figure 14.40. Often they can double as exploratory borings. Unfortunately, the pumps are expensive to install and run, and require maintenance.
- *Horizontal drains*, shown in Figures 14.40 and 14.41, are drilled from the slope face and (in spite of their name) are inclined slightly upward. They are intended to intercept the groundwater and drain it by gravity. Horizontal drains do not require pumps, so they are less expensive to install and maintain.

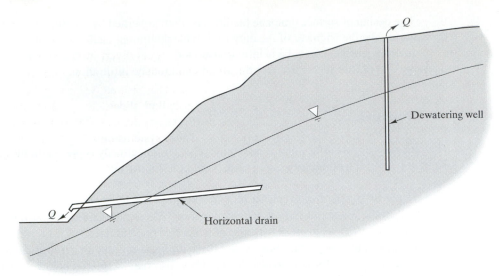

Figure 14.40 Use of wells and horizontal drains to remove subsurface water.

Figure 14.41 A series of horizontal drains in a cut slope along a highway. The close-up view shows one of the drains exiting the slope and discharging into a swale. This drain has a slow but continuous flow, as evidenced by an accumulation of algae at its discharge point.

Reinforcement

Structural engineers transform concrete into an efficient structural material by adding steel reinforcement at key locations. Soil also can be improved by installing synthetic reinforcement as discussed in Chapter 19. These systems increase strength, so they permit slopes to be built with much lower slope ratios (i.e., much steeper) than would otherwise be possible.

Vegetation

Appropriate vegetation is an important part of most slope stabilization plans. It provides erosion protection, draws water out of the ground, provides some reinforcement of the soil, and has important aesthetic value. Although vegetation has virtually no effect on deep-seated slides, it can be very helpful in preventing shallow slides, slumps, and flows.

In arid and semi-arid areas, it is often necessary to install irrigation systems to establish and maintain the desired vegetation. These systems must be closely monitored, because excessive irrigation can introduce large quantities of water into the ground and create serious stability problems.

14.6 INSTRUMENTATION

Geotechnical instrumentation is frequently employed in slope stability studies to help define the subsurface conditions and to monitor unstable ground. Although instrumentation is often expensive to install and monitor, it can provide valuable information that may not otherwise be available.

Inclinometers

An *inclinometer* is a instrument used to measure horizontal movements in the ground as a function of depth. These instruments are very helpful in slope stability studies, and can be used to locate shear surfaces and monitor the rate of shear displacement in slow-moving landslides.

To install an inclinometer, a vertical boring is drilled to a depth well below the potential zone of movement and a special plastic casing is inserted, as shown in Figure 14.42. The annular zone around the casing is then backfilled to hold it firmly in place. Thus, as the ground moves horizontally, the casing deforms with it.

Once the casing is installed, an initial set of readings is obtained by lowering the *inclinometer probe* inside. This probe, shown in Figure 14.43, precisely measures the inclination of the casing in two perpendicular directions. Thus, we know the horizontal position of the upper set of wheels with respect to the lower set. We begin with the probe at the bottom of the casing and progressively raise it by intervals equal to the wheelbase, taking measurements at each interval. By summing these

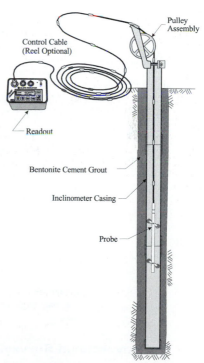

Figure 14.42 Cross-section of a typical inclinometer installation (Slope Indicator Co., Bothell, WA).

measurements, we know the initial horizontal position of the casing throughout its length.

Then, at some future date, we return with the probe and readout unit and obtain a second set of readings. By comparing the new horizontal configuration with the original configuration, we can determine the magnitude and direction of horizontal movements in the ground throughout the length of the casing. We continue to take additional readings at appropriate intervals as necessary. Figure 14.44 shows a typical plot of horizontal movement vs. depth.

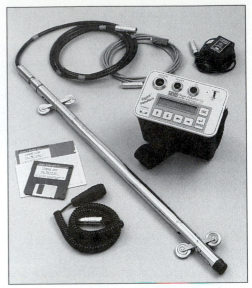

Figure 14.43 Inclinometer probe and readout unit (Slope Indicator Co., Bothell, WA).

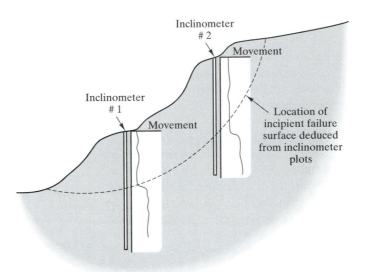

Figure 14.44 Plots of horizontal movement vs. depth for two inclinometers. This data might be used to detect early movements along an incipient failure surface.

Conventional Surveying

Slopes also may be monitored by installing monuments at various locations on the ground surface and measuring their position using conventional surveying equipment, such as total stations. Although this approach does not provide any information on subsurface movements, it can provide extensive information on surface movements, and thus is a useful way to monitor unstable ground. In addition, each monument is far less expensive than an inclinometer, and they can be installed in areas with difficult access (unlike inclinometer installations, which must be accessible with a drill rig).

Conventional surveying methods are becoming even more attractive with the increased availability and precision of global positioning system (GPS) receivers. These are devices that determine position based on signals from satellites, and can achieve accuracies on the order of ±1 cm.

Groundwater Monitoring

Groundwater has a significant impact on slope stability, so information on the groundwater table position and pore water pressures is very important. Therefore, geotechnical engineers often install observation wells and piezometers as a part of slope stability studies. These devices are described in more detail in Chapters 3 and 7.

QUESTIONS AND PRACTICE PROBLEMS

14.17 A certain slope has a factor of safety of 1.15 according to a Swedish slip circle analysis. In order to increase F to 1.50, you are considering the possibility of removing the upper portion of this slope, then rebuilding it to the original grades using a lightweight fill. Assuming the critical failure surface remains in the same location, how much must the weight of the potential slide body be reduced to produce the required factor of safety? Assume s_u along the failure surface remains unchanged. Express your answer as a percentage of the existing weight.

Note: In reality, the critical failure surface would probably shift to a new location, so this preliminary analysis would need to be followed by another search for the critical surface.

14.18 Buttress fills sometimes have a hydraulic conductivity smaller than that of the adjacent natural ground. This is especially common when the natural ground is stratified, and water seeps along the more pervious strata. Could this difference in hydraulic conductivity cause any problems? Explain. If so, what might be done to remedy these problems?

14.19 The soil beneath a slope consists of alternating layers of sand and clay. These layers are nearly horizontal, but vary in thickness such that no two boring logs found these layers at the same elevations. This slope is to be stabilized by installing a series of horizontal wells that are intended to lower the groundwater table. The wells will be drilled at 20 ft intervals near the toe of the slope, and each one will be drilled at the same angle and to the same length. Would you expect the same flow rate from each well? Why or why not?

14.20 In 1962, a developer purchased 100 acres of hilly land and subdivided it for use as a housing tract. The subsequent construction included extensive cuts and fills to create level building pads separated by 1.5:1 cut and fill slopes. Unfortunately, the building codes in that county were much more lax than they are today, so the quality of the earthwork was not as high as would now be required. As a result, some of the slopes in this tract have experienced slides, especially during years with heavier-than-normal rainfall.

One of the slopes is showing some signs of possible instability (i.e., tension cracks, some surface evidence of small movements, etc.), so the current owner wishes to stabilize it. You have designed a stabilization scheme that includes dewatering and construction of a buttress fill. You also need to install appropriate instrumentation to monitor the slope and thus determine if the stabilization is working. What type or types of instrumentation would be appropriate and where should it be installed?

SUMMARY

Major Points

1. When building near or on sloping ground, engineers need to determine whether the slope is stable. Many forms of instability can occur, and they are a source of extensive property damage and occasionally loss of life.

2. The various forms of instability do not necessarily occur independent of the others, so it is difficult to classify slope failures. Nevertheless, in spite of their limitations, classification schemes are useful. One system, developed by Varnes, divides failures into five types: falls, topples, slides, spreads, and flows.

3. Some types of slope instability may be analyzed only with qualitative methods, while other types are suitable for both qualitative and quantitative analyses. Geotechnical engineers most often deal with slides, which fortunately are well-suited for quantitative analysis.

4. Most quantitative analyses of slides use the limit equilibrium approach, which compares the shear strength along a failure surface to the shear stresses required for equilibrium to determine the factor of safety. It is necessary to find the most critical failure surface to compute the correct factor of safety.

5. For most problems, a complete limit equilibrium analysis would be statically indeterminate, and thus impossible. We overcome this problem by making simplifying assumptions that convert the problem into one that is statically determinate. Various assumptions have been proposed, thus producing a large number of limit equilibrium analysis methods.

6. Slides in homogeneous soils, such as compacted fills, usually can be idealized as occurring along the arc of a circle. However, slides in non-homogeneous soils and rocks typically occur on more irregular-shaped failure surfaces.

7. Simple slope stability problems may be solved using chart solutions, but more complex problems require more tedious hand solutions or computer-aided analyses.

8. Earthquakes often produce slope stability problems, and special analyses are required to address these problems.

9. Various methods are available to stabilize slopes, by increasing the shear strength, decreasing the shear stresses, or both.

10. Geotechnical instrumentation is often very helpful in assessing slope stability problems.

Vocabulary

back-calcuated strength
block-glide slide
body
buttress fill
circular failure surface
complex slide
composite slide
compound slide
critical failure surface
crown
cut slope
debris flow
deterministic analysis
effective stress analysis
factor of safety
failure surface
fall
fill slope
filter fabric
flank
flow

French drain
horizontal drain
inclinometer
infinite slope analysis
instrumentation
landslide
lateral spread
limit equilibrium analysis
main scarp
minor scarp
modified Bishop's method
natural slope
Newmark's method
non-circular failure surface
ordinary method of slices
perforated pipe drain
planar failure
probabilistic analysis
probability of failure
pseudostatic method
rapid drawdown

rotational slide
seismic stability
slide
slope face
slope height
slope ratio
Spencer's method
spread
stabilization
Swedish slip circle analysis
tension crack
terrace
tieback anchor
toe of slope
top of slope
topple
total stress analysis
translational slide
undrained shear strength

COMPREHENSIVE QUESTIONS AND PRACTICE PROBLEMS

14.21 A compacted fill slope is to be made of a soil with $c' = 200$ lb/ft^2, $\phi' = 30°$ and $\gamma = 122$ lb/ft^3. Using an infinite slope analysis with a shear surface 4.0 ft below the ground surface and the groundwater table 1.0 ft below the ground surface, determine the steepest allowable slope ratio that will maintain a factor of safety of at least 1.5.

Note: This analysis considers only surficial stability. A separate analysis would need to be conduced to evaluate the potential for a deep-seated slide in the fill.

14.22 A 4-inch perforated pipe drain has been installed as part of a subsurface drainage system. The pipe has been surrounded with a poorly-graded 1.5 inch gravel. The adjacent soils are sandy silts. What is missing from this design? What mode of failure is likely to occur? What should be done to improve this design?

15

Dams and Levees

If any one be too lazy to keep his dam in proper condition, and does not keep it so; if then the dam breaks and all the fields are flooded, then shall he in whose dam the break occurred be sold for money and the money shall replace the corn which he has caused to be ruined.

The Code of Hammurabi, Babylon, circa 2000 BC

Dams are earth or concrete barriers built across a drainage course to impound water. The lakes they create are called *reservoirs*. Dams are among the largest and most important projects in civil engineering. They provide flood control, water storage, hydroelectric power, and many other benefits.

The natural ground on either side of a dam is called an *abutment*. There are two of them: the *left abutment* and the *right abutment*, according to their orientation when facing downstream. The geologic conditions in the abutments are extremely important, as is the method of joining the dam to the abutments. The ground below the dam is called the *foundation,* and also is very important. The use of this term in dam engineering is different than that in buildings and other structures, where the "foundation" is a structural element as discussed in Chapter 17.

Levees (also called *dikes*) have many of the same design considerations as dams, but they are built parallel to a river instead of across it. They are intended to "train" the river along a certain alignment and to protect adjacent land from flooding. For example, parts of the City of New Orleans are well below the water level in the adjacent Mississippi River, and are protected from flooding by a system of levees. Although not as spectacular as dams, levees are important civil engineering works that require careful design, construction, and maintenance.

15.1 DAMS

Although people have been building dams for thousands of years, the twentieth century will probably be remembered as the greatest era of dam building. Most of the large dams in the world were built during the middle decades of the twentieth century. These include Hoover Dam, Grand Coulee Dam, and hundreds of others in the United States and Canada, along with similar facilities in other countries. Thousands of smaller dams also were built.

New dams, even some large ones, continue to be built, and this remains an important part of civil engineering. However, the pace of activity is not as high as it once was. The reasons for this slowdown include the following:

- For geologic, hydrologic, topographic, and economic reasons, there are a limited number of good sites for dams, especially large ones, and most of them have already been built.
- Because of environmental concerns, new dam building projects are subjected to intense scrutiny. Only those project with clear benefits reach the construction stage.

There is much more potential for dam construction in third-world nations, but the lack of available funds limits the number of projects.

Therefore, the future of dam engineering lies primarily in construction of smaller dams, and in the maintenance and upgrading of existing facilities. This latter role is no trivial matter, as some older dams have deteriorated or need to be upgraded for various reasons. For example, hydrologic and seismologic studies sometimes indicate existing dams will be unable to resist large floods or earthquakes, and thus need to be upgraded or replaced.

Hydrology and Hydraulics

Once a potential site for a dam has been selected, engineers perform detailed hydrologic analyses that define the nature of anticipated stream flows. It is especially important to define the flows that might occur during projected floods, because the dam must be designed to accommodate these flows.

Most dams are expected to provide at least some flood control, and this is the sole purpose of some dams. When heavy precipitation or snowmelt occurs upstream of the dam, producing a high flow rate, Q, in the river, we want the dam and reservoir to temporarily store some of the floodwaters, thus reducing the flow rate downstream of the dam and preventing floods. Hydrologists evaluate this process by conducting a *flood routing analysis*. This analysis begins with a design flood event, which is expressed as a *hydrograph* (a plot of Q vs. time), and is intended to produce a design that limits the Q downstream of the dam to some specified value. The variables in the dam design that control the flood routing include:

- **The reservoir elevation at the beginning of the flood.** Normally the reservoir is intentionally maintained at a low elevation at the beginning of the flood season to provide more storage space for flood waters.
- **The height of the dam.** This controls the volume of the reservoir, and is limited by topographic and economic considerations. The maximum reservoir level is somewhat lower in order to maintain an adequate *freeboard*, which is the difference in elevation between the water and the crest of the dam.
- **The hydraulics of the outlet works.** These *outlet works* consist of one or more pipes that draw water from the reservoir. The pipes may lead to a hydroelectric plant, municipal water supply systems, agricultural irrigation systems, or simply to the river downstream of the dam. The intake facilities are located at various elevations in the reservoir so discharges can be made regardless of the water elevation.
- **The crest elevation and hydraulics of the spillway.** The *spillway* is a much larger facility that discharges reservoir water directly into the river. The spillway crest is slightly below the crest of the dam, so it is used only when the reservoir level is high. In many dams, the spillway has never been used. However, the spillway is vital to the safety of the dam, especially earth dams, because it prevents water from running over the top of the dam. This must never be allowed to occur, because it would quickly erode the dam and cause an *overtopping failure*.

These variables are usually adjusted by a converging trial-and-error process until a suitable design has been achieved.

Types of Dams

Once the required height has been set by the hydrologic analysis, the design effort can focus on the dam itself. There are two broad categories of dams: *concrete dams* and *earth dams*, as well as a third hybrid category called *roller compacted concrete dams*.

Concrete Dams

Concrete dams consist of cast-in-place concrete that extends between the two abutments. When properly built, they are virtually watertight, and often are very beautiful structures. There are three principal types of concrete dams:

Concrete gravity dams are massive concrete structures that resist the hydrostatic loads from the reservoir by virtue of their weight. Grand Coulee Dam, shown in Figure 15.1, is one of the most well-known concrete gravity dams.

Concrete arch dams are curved in plan view and transmit much of the hydrostatic loads laterally to the abutments. This design uses the concrete more efficiently, so arch dams are much thinner than gravity dams. Hoover Dam, shown in Figure 15.2, is an example of a large concrete arch dam.

Concrete buttress dams have a sloping upstream face that rests on a series of walls or buttresses aligned parallel to the river. These buttresses transmit the hydrostatic loads to the foundation bedrock, and the sloping upstream face uses the downward component of the hydrostatic loads as a stabilizing force. This design uses less concrete than a gravity dam, but requires a large amount of complex formwork, and thus is rarely economical to build. Bartlett Dam, shown in Figure 15.3, is one of the few examples of this type.

Figure 15.1 Grand Coulee Dam on the Columbia River in Washington. This is one of the largest concrete gravity dams in the world (U.S. Bureau of Reclamation).

Figure 15.2 Hoover Dam on the Colorado River at the Nevada–Arizona border. This is a concrete arch dam (U.S. Bureau of Reclamation).

Figure 15.3 Bartlett Dam on the Verde River in Arizona is a concrete buttress dam (U.S. Bureau of Reclamation).

The construction of concrete dams is much more labor-intensive than that for earth dams. This may have been acceptable in the 1930s when labor wages were low and dam-building projects were partially driven by New Deal efforts to put people to work. However, wages today are much higher, making concrete dams relatively expensive. In addition, concrete dams can be built only on sites with sound rock foundations. Finally, developments in earthmoving equipment, as discussed in Chapter 6, now allow us to move large quantities of soil with a smaller labor force and at a lower cost. Therefore, nearly all new dams are now earth dams.

Earth Dams

Earth dams are built of compacted soil or rock fragments and are designed as gravity dams. They are much more massive than concrete gravity dams, and thus require larger amounts of material. Many earth dams contain more than 50,000,000 m^3 (64,000,000 yd^3) of compacted soil. However, unlike concrete dams, there is no need for formwork, which represents a significant cost savings. In addition, the unit cost of compacted fill is much less than that of concrete, so the total cost is usually less.

Before about 1900, there was no economical way to move large quantities of soil. Earth dams and other fills were built by hauling soil in animal-drawn wagons and other crude equipment. As a result, only modest earth dams were built. The development of hydraulic fill methods in the early twentieth century permitted more economical transport of soil. This method, which is described in Chapter 6, was used to build many earth dams between 1900 and 1940. However, the poor quality of the resulting fill, as exemplified by a large landslide during construction of Ft. Peck Dam in Montana, and the advent of modern earthmoving equipment has made hydraulic fill methods obsolete. No hydraulic fill dams have been built in North America since 1940.

Today, all earth dams are built using the earthmoving equipment described in Chapter 6. The fill quality and economics of these methods far exceeds that of earlier techniques. In addition, this equipment permits the precise placement of different soils in different parts of the dam. Therefore, all except the smallest dams are *zoned*, as shown in Figures 15.4 and 15.5. Each zone serves a specific purpose. The *core*, which is near the center of the cross-section, is the primary impervious zone and is used to block the flow of water through the dam. It is normally made of clayey soils. The *shells*, which are located on either side of the core, provide the strength and mass in the dam. *Drains*, which are located at various places in the dam, are used to draw off water that leaks through the core, and *filters* prevent the migration of fines into the drains.

Because such large quantities of soil are required, earth dams must be built from soils that are nearby. It is much too expensive to transport such quantities for long distances. In addition, the topography at each dam site is different. Therefore, no two dams use the same design. Each must be designed to accommodate the local conditions and requirements.

Rockfill dams are a special type of earth dam that is made of cobble to boulder-sized rock fragments with an internal concrete wall acting as the impervious barrier. *Earth-rock dams* are a hybrid that use rockfill shells and a compacted soil core.

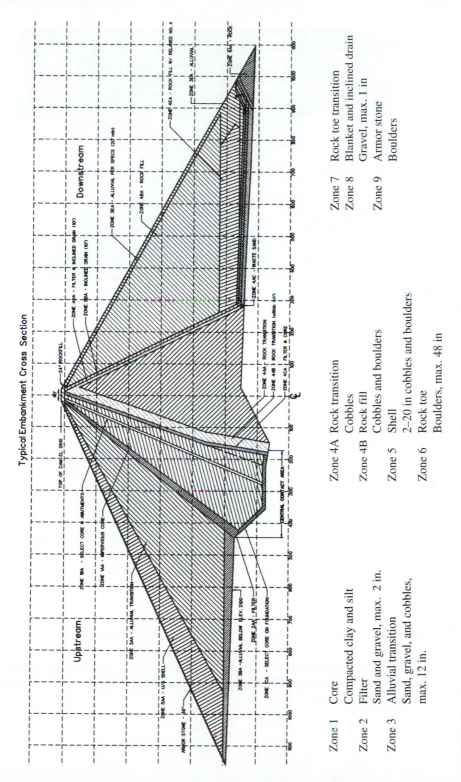

Figure 15.4 Cross-section and description of zones, Seven Oaks Dam, California (U.S. Army Corps of Engineers).

Zone 1	Core	Zone 7	Rock toe transition
	Compacted clay and silt	Zone 8	Blanket and inclined drain
Zone 2	Filter		Gravel, max. 1 in
	Sand and gravel, max. 2 in.	Zone 9	Armor stone
Zone 3	Alluvial transition		Boulders
	Sand, gravel, and cobbles, max. 12 in.		

Zone 4A	Rock transition
	Cobbles
Zone 4B	Rock fill
	Cobbles and boulders
Zone 5	Shell
	2–20 in cobbles and boulders
Zone 6	Rock toe
	Boulders, max. 48 in

Figure 15.5 Seven Oaks Dam under construction. The dark zone in the center of the dam is the zone 1 core material. It is surrounded by the other zones, per the cross-section in Figure 15.4. The downstream face of the dam is visible on the left side of the photograph.

Roller Compacted Concrete Dams

In the late twentieth century, engineers and contractors began experimenting with a new method of dam construction that combined features from both concrete and earth dams. This new hybrid method is called *roller compacted concrete* (RCC) dam construction.

Roller compacted concrete dams are made of coarse and fine aggregates similar to those used in conventional concrete, which are mixed with Portland cement and water. The cement content is very low (typically about 90 kg/m^3 or 150 lb/yd^3), so the freshly mixed material looks like a damp, well-graded gravel. It is then placed using scrapers and other standard earthmoving equipment, and compacted with heavy vibratory rollers. Finally, it is allowed to cure, thus forming an in-place material that is essentially a weak concrete.

Many small-to-medium-size RCC dams have been built, and seem to be performing well. RCC also has been used to build buttresses against existing dams that required additional support.

Geotechnical Analysis and Design of Earth Dams

There are many geotechnical considerations in the analysis and design of earth dams. Indeed, this is one of the few civil engineering projects where the geotechnical engineer is the lead design professional.

Seepage

Water from the reservoir flows through, beneath, and around earth dams, so the design process includes analyses of these flows. One of the reasons for doing so is to develop an estimate of the flow rate, Q, which represents losses from the reservoir. In some cases, such as dams intended primarily for flood control, even large seepage losses may be acceptable. However, other dams, such as those used to store water supplied by an aqueduct, may require a very low tolerance of seepage losses.

Another reason for evaluating seepage losses is to develop design groundwater conditions, which are then used in slope stability analyses. For example, the groundwater table inside the dam will be higher on the upstream side than on the downstream side, with corresponding impacts on the stability of the upstream slope.

Finally, perhaps the most important reason for performing seepage analyses is to assess the potential for *piping*, which is the formation of channels inside the dam due to internal erosion, as shown in Figure 15.6. Silty soils are especially prone to piping. We avoid this type of failure by including filter and drain zones that control the seepage and prevent soil migration. Section 8.5 presented design criteria to keep soil migration under control.

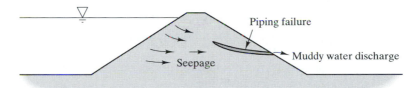

Figure 15.6 Piping failure in an earth dam.

Slope Stability

A landslide in either the upstream or downstream face of an earth dam would be disastrous, so these slopes must satisfy the stability requirements discussed in Chapter 14. In addition to the typical modes of failure, we also need to consider a special condition called *rapid drawdown failure* (Duncan, Wright, and Wong, 1990; Borja and Kishnani, 1992). This type of failure occurs when the exterior water level has been at a certain elevation for a long time, then quickly drops to a lower elevation as shown in Figure 15.7. If the soils have a low hydraulic conductivity, the groundwater table inside the slope cannot drop nearly as rapidly as the water outside, so high pore water pressures inside the slope remain unchanged, even though the stabilizing effect of the exterior hydrostatic pressures rapidly disappears. This unfortunate situation can occur in earth dams when the reservoir is lowered too rapidly, or in levees when river water levels go down rapidly. This combination of factors may produce a landslide.

Rapid drawdown failures have occurred on occasion. For example, the Walter Bouldin Dam in Alabama experienced rapid drawdown induced landslide in 1975 when the reservoir level was lowered 10 m in 5.5 hr (Duncan, Wright, and Wong, 1990).

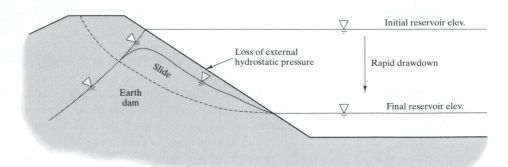

Figure 15.7 Rapid drawdown failure in an earth dam.

Lessons from Dam Failures

Although dams have a very good safety record, failures do occasionally occur. Such events always generate a "post-mortem" study to determine the cause of failure and to help us avoid similar events in the future. Studies of dam failures also have provided insight into other geotechnical problems.

A study by Biswas and Chatterjee (1971) examined more than 300 dam failures throughout the world and found the following causes:

- 35 percent were a direct result of floods that exceeded the spillway capacity, and thus were due to inaccurate hydrology.
- 25 percent were due to geotechnical problems, such as seepage, piping, high pore water pressures, inadequate seepage cutoff, fault movement, excessive settlement, or landslides
- The remaining 40 percent were from a variety of problems, including the use of poor construction materials or practices, wave action, acts of war, and poor maintenance.

The following case studies illustrate some of the causes of dam failures (Jansen, 1980).

South Fork Dam, Pennsylvania (1889)

One of the most memorable dam failures in the United States occurred near Johnstown, Pennsylvania in 1889, resulting in the famous Johnstown Flood. This earth dam had been completed in 1853 and was used to feed water to a shipping canal. The design was quite good for its time, and included both a spillway and an outlet works. However, the as-built spillway was smaller than that shown on the plans, which meant it had less capacity.

Unfortunately, soon after the dam was completed, the canal faced stiff competition from the recently constructed Pennsylvania Railroad. As a result, the canal was closed in 1857. The dam, which no longer had any useful purpose, was eventually sold to a sportsmen's club from Pittsburgh. They closed off the outlet works and lowered the crest of the dam (to provide sufficient width for a roadway), then began using it for a fishing resort. They also installed a series of iron bars across the spillway to prevent fish from leaving the lake. Unfortunately, these bars also trapped debris, and thus reduced the spillway capacity. All of these "improvements" diminished the dam's ability to pass a major flood.

An intense storm occurred in 1889, which produced 250 mm (10 in) of precipitation in 36 hours. The reservoir level rose quickly, but due to the inoperative outlet works, the reduced freeboard, the partially blocked spillway, and other factors, the dam was not able to pass the water and eventually failed by overtopping. The resulting flood killed 2,209 people.

St. Francis Dam, California (1928)

St. Francis Dam was a 62.5 m (205 ft) tall concrete arch dam built in 1926 as a part of the Owens River Aqueduct project. As was typical for the time, there was very little geologic or geotechnical input in the design process. As a result, the left abutment of the dam was built on an ancient landslide (Rogers, 1995). The dam also contained other design deficiencies, including inadequate attention to hydrostatic uplift forces. In addition, the height of the dam was arbitrarily raised during construction without a corresponding increase in the base width.

Shortly before midnight on March 12, 1928, the two-year-old dam burst, suddenly discharging 12 billion gallons of water, along with concrete fragments as large as 9 million kg (10,000 tons). The flood caused tremendous destruction along its 80 km (50 mile) journey to the Pacific Ocean, ultimately causing more than 500 deaths. The failure appears to have had multiple causes, but the primary one seems to be the ancient landslide, which was reactivated by the reservoir water. The moving slide shifted the dam, causing it to fail.

The St. Francis Dam failure was the worst American civil engineering disaster of the twentieth century. It was directly responsible for the creation of a state regulatory agency to oversee the safety of dams in California. Today, all states have such an agency.

Malpasset Dam, France (1959)

Malpasset Dam was a 61 m (200 ft) tall concrete arch dam in Southern France. The abutments consisted of gneiss bedrock, which appeared to be quite competent. However, this rock deformed excessively under the thrust forces from the dam, thus opening fractures within the abutments. These fractures appear to have filled with water, which eventually caused part of the rock to move outward, thus triggering the failure of the dam. A large flood ensued, causing extensive property damage and 421 deaths. The primary lesson from Malpasset was the importance of seemingly small geologic features.

Vaiont Dam, Italy (1963)

The greatest number of casualties associated with a dam in modern times occurred in 1963 at Vaiont Dam in Italy. This disaster killed 2,600 people, yet it was not due to a failure of the dam, but to a massive landslide in the reservoir.

The 265 m (869 ft) tall concrete arch dam was built in a very steep canyon. It was completed in 1960. Later that year, as the reservoir was being filled for the first time, a landslide occurred along the edge of the reservoir upstream of the left abutment. Although the effects of this slide were limited, the reservoir was maintained at a lower-than-normal elevation and the slope was monitored.

In 1963, the water level was permitted to rise an additional 20 m (66 ft), which also caused the groundwater in the adjacent rock to rise. Then, in September and October, heavy rainfall caused additional saturation in the slope and temporarily caused the reservoir to rise even higher, although still below the design water elevation. Then, on the evening of October 9, a huge landslide occurred in the same area as the 1960 slide. This mass extended for a distance of 2 km (1.2 mi) and had a volume of 240,000,000 m^3 (314,000,000 yd^3), which is about four times the volume of Oroville Dam (Figure 6.3). Because the canyon is so steep, the slide moved downward at an incredible speed, sending a massive wave up the opposite side of the reservoir and over the dam, quickly reaching an elevation of 240 m (780 ft) above the reservoir level! The wave that went over the dam reached a height of 100 m (330 ft) above the crest, and produced a 70 m (230 ft) tall flood wave in the river below.

The massive destruction and loss of life from this unfortunate event was unprecedented. Yet, the dam itself survived with only minor damage. A technical review board concluded that "bureaucratic inefficiency, muddling, withholding of alarming information, lack of judgement and evaluation, and lack of serious individual and collective consultation" were the real causes of the failure (Biswas and Chatterjee, 1971). These kinds of problems underlie many engineering failures throughout the world, not just the one at Vaiont.

Lower San Fernando Dam, California (1971)

The Lower San Fernando Dam was a hydraulic fill dam that was completed in 1918. As is typical of hydraulic fills, it contained loose soils, including loose sands. In addition, loose sandy soils were present in the natural ground below the dam. Later, a rolled fill was added to raise the crest of the dam, bringing it to a total height of 43 m (142 ft).

In 1971, a magnitude 6.2 earthquake occurred near the dam. This earthquake caused extensive liquefaction, both inside the dam and in the underlying natural soils, which resulted in a large landslide on the upstream slope. Figure 20.8 shows the dam as it appeared immediately after the landslide. Fortunately, a small freeboard remained, so the reservoir did not overtop the dam.

This failure focused attention on the susceptibility of hydraulic fills to seismically induced liquefaction, and prompted reevaluations of hydraulic fill dams in seismic regions. As a result, a number of these dams were replaced or modified to enhance their seismic stability.

Buffalo Creek Dam, West Virginia (1972)

Most mining activities generate large quantities of waste material, both solid and liquid, that need to be discarded. Often these materials are placed as *tailings dams*, which can retain significant reservoirs. Often these reservoirs are filled with liquid wastes from the mining operations. Unfortunately, the older tailings dams typically had little or no engineering design or construction control, and thus can be prone to failure.

The Buffalo Creek Dam was one such facility. It was made of waste materials from an adjacent coal mine, along with scrap timber, metal, and other debris. The mining company discharged liquid wastes behind the dam, then allowed them to percolate through in an effort to reduce stream pollution. The only outlet, other than seepage, was a 24-inch (610 mm) steel overflow pipe. There was no spillway.

In February 1972, a moderate rainstorm occurred that caused the "reservoir" to rise and eventually overtop the dam. The lack of sufficient outlet works and the poor construction of the dam allowed it to fail rapidly, causing a flood that killed 125 people and left 4,000 homeless.

Teton Dam, Idaho (1976)

Teton Dam was a 93 m (305 ft) tall earth dam on the Teton River in Southeastern Idaho. It failed on June 5, 1976, only eight months after the first filling of the reservoir began. The failure began as a cloudy seep on the face of the dam, and rapidly progressed to a piping failure through the embankment, as shown in Figure 15.8 and in Plate F of the color photos in Chapter 1. Its rapid discharge produced extensive flooding in the farmland and towns below the dam.

Figure 15.8 Teton Dam soon after failure. The reservoir was located to the left of the dam (U.S. Bureau of Reclamation).

This failure of a modern dam so soon after construction was a shock to the engineering community. It prompted one of the most intensive investigations of any dam failure. A panel of experts determined that it was caused by inadequate grouting of the highly fractured rock in the abutment, the use of extremely erodible silty soils in the core, and inadequate filters. The combination of these factors made the dam particularly susceptible to a piping failure.

General Comments on Dam Safety

We study failures, such as those listed above, because they help us better understand the behavior of dams and help us avoid similar problems in the future. However, studying such failures can leave the mistaken impression that dams are inherently unsafe facilities, just as studies of airliner accidents can leave the mistaken impression that commercial air travel is unsafe. In fact, the overall safety record for dams, especially those built after 1945, is very good. They have significantly reduced the risk of death, injury, and property loss from floods; provided large quantities of hydroelectric power; enabled the construction of extensive irrigation and municipal water supply facilities; and provided many other benefits to society. Thus, the value of dams to society far exceeds the small risk of failure.

15.2 LEVEES

During periods of heavy rainfall or rapid snowmelt, rivers often overflow their banks, flooding the adjacent land. This is a natural process, and the fertile soils deposited by these floods is often part of the reason these lands are highly desirable for farming. The farms, in turn, bring towns and cities. Commerce associated with the rivers and other reasons also promote development in these lands. However, this new usage of the land is not compatible with the natural flooding processes, so civil engineers are often asked to build flood control works. These works include dams, levees, and other facilities.

Levees are often built along the river to contain floodwaters and protect the adjacent land, as shown in Figure 15.9. They also can be built to encircle important areas, such as the town shown in Figure 15.10. Sometimes levees are built to contain aqueducts and irrigation canals. The crest elevation of these levees is governed by the anticipated water levels, and the cross-section by the soil conditions.

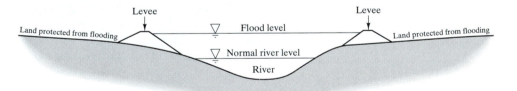

Figure 15.9 Levees protecting land from flooding due to rising river water.

Some levees, especially older ones, were "designed" and constructed with little or no engineering, and sometimes are the product of many years of informal construction. These levees are typically not very reliable, and occasionally fail. Others, such as those built by the U.S. Army Corps of Engineers and the U.S. Bureau of Reclamation, are carefully designed and built, and have much greater reliability. Unfortunately, systems of levees are only as strong as their weakest link, so land protected by engineered levees is sometimes flooded by the failure of nearby unengineered levees.

Figure 15.10 The town of Pembina, ND is located along the banks of the Red River. When the river reached flood stage in 1997, levees surrounding the town protected it from flooding. The normal river channel is barely visible on the right side of this aerial photograph. Interstate Highway 29, which is located outside the levees, is visible on the left. Part of the highway has flooded, but a bridge and ramps to the left of the town still extend above the water level. If the levees had not been built, most or all of the town would have been under water (U.S. Army Corps of Engineers - St. Paul District).

The most common causes of failure in levees include:

- overtopping by floodwaters, which leads to rapid erosion
- uncontrolled seepage through or beneath the levee, leading to a piping failure
- landslides in the levee slopes

All of these problems can be avoided through proper analysis, design, and construction. This is why modern engineered levees rarely fail. Unfortunately, many miles of unengineered levees still exist, and the cost of retrofitting or replacing them is enormous. Thus, the primary task for geotechnical engineers is to identify the most hazardous ones and reinforce them first.

SUMMARY

Major Points

1. Dams are earth or concrete barriers built across a drainage course to impound water. The design of these facilities must incorporate hydrologic, hydraulic, topographic, geotechnical, environmental, and economic factors, along with many other considerations.
2. There are two primary categories of dams: concrete and earth. Concrete dams may be of the gravity, arch, or buttress type, while earth dams are always of the concrete type. A third hybrid category, known as roller compacted concrete dams, combines features from both concrete and earth dams.

3. The geotechnical analysis and design of earth dams must consider seepage, slope stability, and other factors.
4. The safety record of dams is very good, especially for those built after 1945. However, when failures do occur, they are often spectacular and catastrophic. Therefore, they are extensively studied so we can learn how to avoid similar failures in the future.
5. Levees are similar to small earth dams, except they are built parallel to a drainage course and are intended to "train" or direct the flow in a river, canal, aqueduct, or other facility. Many older, undesigned and poorly constructed levees still exist, and they sometimes fail when subjected to flood waters. However, well engineered and constructed levees perform very well.

Vocabulary

concrete arch dam	foundation	reservoir
concrete buttress dam	freeboard	right abutment
concrete gravity dam	hydrograph	rockfill dam
core	left abutment	roller compacted concrete
dam	levee	dam
drain	outlet works	shell
earth dam	overtopping	spillway
filter	piping	
flood routing analysis	rapid drawdown	

COMPREHENSIVE QUESTIONS AND PRACTICE PROBLEMS

15.1   Concrete gravity dams resist the hydrostatic forces from the reservoir by virtue of their mass and sliding friction along their base. Assuming there is no other source of resistance, consider a proposed dam with a vertical upstream face and a mass x that must retain a reservoir with water depth y. How much must the mass be increased if the design water depth is increased to $1.5\,y$?

15.2   A moderate-size dam is to be built at a site in a third-world country. The site is underlain by high-quality hard rock. Discuss the factors that would influence the choice between a concrete dam or an earth dam. How might these factors be different at this site than in North America or Europe?

15.3   According to a recent hydrologic study, the spillway at an old earth dam is inadequate. A projected 200-year flood would be sufficient to overtop the dam, which would undoubtedly lead to failure and a devastating flood. Suggest three potential methods of dealing with this problem.

15.4   We do not have enough money to rebuild all of the inadequate levees, and therefore must focus our energies on the most hazardous ones. What factors might be considered in determining which levees to rebuild first?

16

Lateral Earth Pressures and Retaining Walls

Things should be made as simple as possible, but not one bit simpler.

Albert Einstein

Lateral earth pressures are those imparted by soils onto vertical or near-vertical structures. They may include both normal and shear pressures, as shown in Figure 16.1. These pressures are especially important in the design of *retaining walls*, which are civil engineering works that maintain adjacent ground surfaces at two different elevations. Figure 16.2 shows typical uses of retaining walls in civil engineering projects.

16.1 HORIZONTAL STRESSES IN SOIL

In this chapter, we will compute stresses using the same x, y, z coordinate system as used in Chapter 10. For convenience, we will align the axes such that the x axis is oriented perpendicular to the wall face.

Lateral earth pressures are the direct result of horizontal stresses in the soil. In Chapter 10 we defined the ratio of the horizontal effective stress to the vertical effective stress at any point in a soil as the *coefficient of lateral earth pressure, K*:

$$K = \frac{\sigma'_x}{\sigma'_z} \qquad (16.1)$$

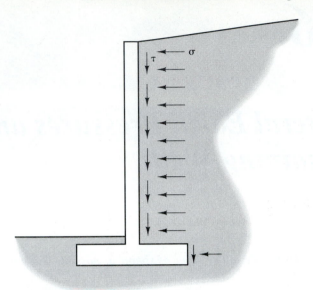

Figure 16.1 Lateral earth pressures imparted from a soil onto a vertical or near-vertical structure.

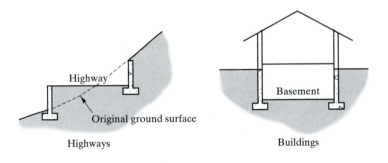

Highways   Buildings

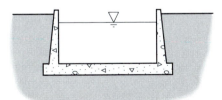

Flood control channels

Figure 16.2 Typical applications of retaining walls.

where:

K = coefficient of lateral earth pressure

σ_x' = horizontal effective stress

σ_z' = vertical effective stress

In the context of this chapter, K is important because it is an indicator of the lateral earth pressures acting on a retaining wall.

For purposes of describing lateral earth pressures, geotechnical engineers have defined three important soil conditions: the *at-rest condition*, the *active condition*, and the *passive condition*.

The At-Rest Condition

Let us assume a certain retaining wall is both *rigid* and *unyielding*. In this context, a rigid wall is one that does not experience any significant flexural movements. The opposite would be a *flexible* wall—one that has no resistance to flexure. The term *unyielding* means the wall does not translate or rotate, as compared to a *yielding* wall that can do either or both. Let us also assume this wall is built so that no lateral strains occur in the ground. Therefore, the lateral stresses in the ground are the same as they were in its natural undisturbed state.

The value of K in this situation is K_0, the *coefficient of lateral earth pressure at rest*. The most reliable method of assessing K_0 is to use in-situ tests such as the dilatometer test (DMT) or pressuremeter test (PMT) as discussed in Chapter 3. It also may be measured using special laboratory tests on undisturbed samples. However, because of cost constraints, engineers generally use these methods only on especially large or critical projects. For the vast majority of projects, we usually must rely on empirical correlations to develop design values of K_0. Several such correlations have been developed, including the following one from Mayne and Kulhawy (1982):

$$K_0 = (1 - \sin \phi') OCR^{\sin \phi'}$$

(16.2)

where:

ϕ' = effective friction angle of soil

OCR = overconsolidation ratio of soil

Equation 16.2 is based on laboratory tests performed on 170 soil samples that ranged from clay to gravel. It is applicable only when the ground surface is level. Usually K_0 is between 0.3 and 1.4.

If no groundwater table is present ($u = 0$), the lateral earth pressure, σ, acting on this wall is equal to the horizontal effective stress in the soil:

$$\sigma = \sigma_x' = \sigma_z' K_0$$

(16.3)

Lateral earth pressures below the groundwater table are discussed later in this chapter.

In the at-rest case, we assume the shear stress, τ, acting between the soil and the wall is zero.

In a homogeneous soil above the groundwater table, K_0 is a constant and σ_z' varies linearly with depth. Therefore, in theory, σ also varies linearly with depth, forming a triangular pressure distribution, as shown in Figure 16.3. Thus, if at-rest conditions are present, the horizontal force acting on a unit length of a vertical wall is the area of this triangle:

$$P_0/b \ = \ \frac{\gamma H^2 K_0}{2} \tag{16.4}$$

where:

P_0/b = normal force acting between soil and wall per unit length of wall
b = unit length of wall (usually 1 ft or 1 m)
γ = unit weight of soil
H = height of wall

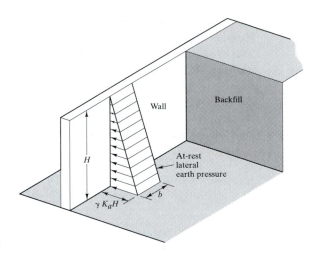

Figure 16.3 At-rest pressure acting on a retaining wall.

Example 16.1

An 8 ft tall basement wall retains a soil that has the following properties: $c' = 0$, $\phi' = 35°$, $\gamma = 127 \ \text{lb/ft}^3$, OCR = 2. The ground surface is horizontal and level with the top of the wall. The groundwater table is well below the bottom of the wall. Consider the soil to be in the at-rest condition and compute the force that acts between the wall and the soil.

Solution

$$K_0 = (1 - \sin\phi') \, OCR^{\sin\phi'}$$
$$= (1 - \sin 35°) \, 2^{\sin 35°}$$
$$= 0.635$$

$$P_0/b = \frac{\gamma H^2 K_0}{2}$$

$$= \frac{(127 \text{ lb/ft}^2)(8 \text{ ft})^2(0.635)}{2}$$

$$= \textbf{2580 lb/ft} \qquad \Leftarrow \textit{Answer}$$

Because the theoretical pressure distribution is triangular, this resultant force acts at the lower third-point on the wall.

The Active Condition

The at-rest condition is present only if the wall does not move. Although this may seem to be a criterion that all walls should meet, even very small movements alter the lateral earth pressure.

Suppose Mohr's circle A in Figure 16.4 represents the state of stress at a point in the soil behind the wall in Figure 16.5, and suppose this soil is in the at-rest condition. The inclined lines represent the Mohr-Coulomb failure envelope. Because the Mohr's circle does not touch the failure envelope, the shear stress, τ, is less than the shear strength, s.

Now, permit the wall to move outward a short distance. This movement may be either translational or rotational about the bottom of the wall. It relieves some of the horizontal stress, causing the Mohr's circle to expand to the left. Continue this process until the circle reaches the failure envelope and the soil fails in shear (circle B). This shear failure will occur along the planes shown in Figure 16.5, which are inclined at an angle of $45 + \phi/2$ degrees from the horizontal. A soil that has completed this process is said to be in the *active condition*. The value of K in a cohesionless soil in the active condition is known as K_a, the *coefficient of active earth pressure*.

Once the soil attains the active condition, the horizontal stress in the soil (and thus the pressure acting on the wall) will have reached its lower bound, as shown in Figure 16.6. The amount of movement required to reach the active condition depends on the soil type and the wall height, as shown in Table 16.1. For example, in a loose cohesionless soil, the active condition is achieved if the wall moves outward from the backfill a distance equal to only $0.004 H$ (about 12 mm for a 3 m tall wall). Although basement walls, being braced at the top, cannot move even that distance, a *cantilever wall* (one in which the top is not connected to a building or other structure) could very easily move 12 mm outward, and such a movement would usually be acceptable. Thus, a basement wall may need to be designed to resist the at-rest pressure, whereas the design of a free-standing cantilever wall could use the active pressure. Because the active pressure is smaller, the design of free-standing walls will be more economical.

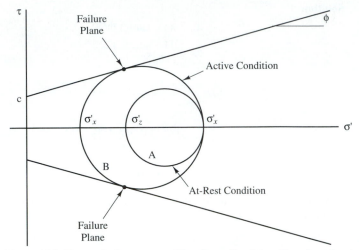

Figure 16.4 Changes in the stress conditions in a soil as it transitions from the at-rest condition to the active condition.

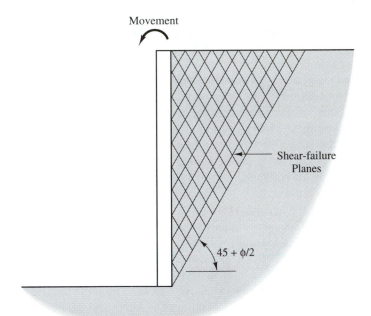

Figure 16.5 Development of shear failure planes in the soil behind a wall as it transitions from the at-rest condition to the active condition.

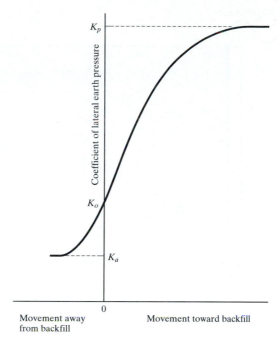

Figure 16.6 Effect of wall movement on lateral earth pressure.

TABLE 16.1 WALL MOVEMENT REQUIRED TO REACH THE ACTIVE
CONDITION (Adapted from CGS, 1992)

Soil Type	Horizontal Movement Required to Reach the Active Condition
Dense cohesionless	$0.001\,H$
Loose cohesionless	$0.004\,H$
Stiff cohesive	$0.010\,H$
Soft cohesive	$0.020\,H$

H = Wall height
Cohesionless soils include sands and gravels
Cohesive soils are those with a significant clay content

The Passive Condition

The *passive condition* is the opposite of the active condition. In this case, the wall moves *into* the backfill, as shown in Figure 16.7, and the Mohr's circle changes, as shown in Figure 16.8. Notice how the vertical stress remains constant whereas the horizontal stress changes in response to the induced horizontal strains.

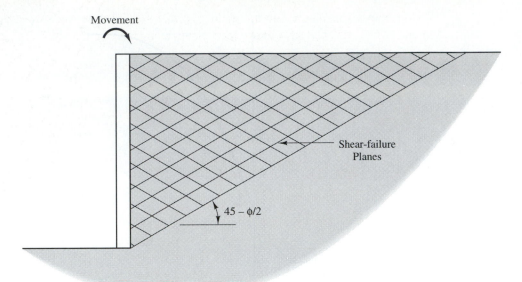

Figure 16.7 Development of shear failure planes in the soil behind a wall as it transitions from the at-rest condition to the passive condition.

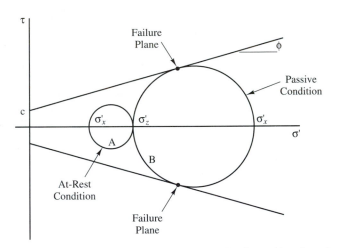

Figure 16.8 Changes in the stress condition in a soil as it transitions from the at-rest condition to the passive condition.

In a homogeneous soil, the shear failure planes in the passive case are inclined at an angle of 45 - ϕ/2 degrees from the horizontal. The value of K in a cohesionless soil in the passive condition is known as K_p, the *coefficient of passive earth pressure*. This is the upper bound of K and produces the upper bound of pressure that can act on the wall.

Engineers often use the passive pressure that develops along the toe of a retaining wall footing to help resist sliding, as shown in Figure 16.9. In this case, the "wall" is the side of the footing.

More movement must occur to attain the passive condition than for the active condition. Typical required movements for various soils are shown in Table 16.2.

TABLE 16.2 WALL MOVEMENT REQUIRED TO REACH THE PASSIVE CONDITION (Adapted from CGS, 1992)

Soil Type	Horizontal Movement Required to Reach the Passive Condition
Dense cohesionless	0.020 H
Loose cohesionless	0.060 H
Stiff cohesive	0.020 H
Soft cohesive	0.040 H

H = Wall height
Cohesionless soils include sands and gravels
Cohesive soils are those with a significant clay content

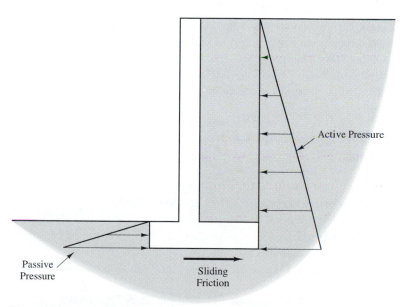

Figure 16.9 Active and passive pressures acting on a cantilever retaining wall.

Although movements on the order of those listed in Tables 16.1 and 16.2 are necessary to reach the full active and passive states, respectively, much smaller movements also cause significant changes in the lateral earth pressure. While conducting a series of full-scale tests on retaining walls, Terzaghi (1934b) observed:

> With compacted sand backfill, a movement of the wall over an insignificant distance (equal to one-tenthousanth of the depth of the backfill) decreases the [coefficient of lateral earth pressure] to 0.20 or increases it up to 1.00.

This effect is not as dramatic in other soils, but even with those soils only the most rigid and unyielding structures are truly subjected to at-rest pressures.

16.2 CLASSICAL LATERAL EARTH PRESSURE THEORIES

The solution of lateral earth pressure problems was among the first applications of the scientific method to the design of structures. Two of the pioneers in this effort were the Frenchman Charles Augustin Coulomb and the Scotsman W. J. M. Rankine (see the sidebar on Coulomb, later in this chapter). Although many others have since made significant contributions to our knowledge of earth pressures, the work of these two men was so fundamental that it still forms the basis for earth pressure calculations today. More than 50 earth pressure theories are now available; all of them have their roots in Coulomb and Rankine's theories.

Coulomb presented his theory in 1773 and published it 3 years later (Coulomb, 1776). Rankine developed his theory more than 80 years after Coulomb (Rankine, 1857). In spite of this chronology, it is conceptually easier for us to discuss Rankine's theory first.

In this book we will only consider lateral earth pressures in soils that are isotropic and homogeneous (ϕ, and γ have the same values everywhere, and they have the same values in all directions at every point) as well as cohesionless ($c = 0$). This is the simplest case. Soils that have a high clay content require special considerations that are beyond the scope of this discussion. *Foundation Design: Principles and Practices* (the companion volume to this book), discusses lateral earth pressures in layered soils, clayey soils, and in soils with $c > 0$.

Lateral earth pressure theories may be used with either effective stress analyses (c', ϕ') or total stress analyses (c_T, ϕ_T). However, effective stress analyses are usually more appropriate, and are the only type we will consider in this chapter.

Rankine's Theory for Cohesionless Soils

Assumptions

Rankine approached the lateral earth pressure problem with the following assumptions:

1. The soil is homogeneous and isotropic, as defined above.
2. The most critical shear surface is a plane. In reality, it is slightly concave up, but this is a reasonable assumption (especially for the active case) and it simplifies the analysis.
3. The ground surface is a plane (although it does not necessarily need to be level).
4. The wall is infinitely long so that the problem may be analyzed in only two dimensions. Geotechnical engineers refer to this as a *plane strain condition*.
5. The wall moves sufficiently to develop the active or passive condition.
6. The resultant of the normal and shear forces that act on the back of the wall is inclined at an angle parallel to the ground surface (Coulomb's theory provides a more accurate model of shear forces acting on the wall).

Active Condition

With these assumptions, we can treat the wedge of soil behind the wall as a free body and evaluate the problem using the principles of statics, as shown in Figure 16.10a. This is similar to the slope stability analysis methods we used in Chapter 14, and is known as a *limit equilibrium analysis*, which means that we consider the conditions that would exist if the soil along the base of the failure wedge was about to fail in shear.

Weak seams or other nonuniformities in the soil may control the inclination of the critical shear surface. However, if the soil is homogeneous, P_a/b is greatest when this surface is inclined at an angle of $45 + \phi/2$ degrees from the horizontal, as shown in the Mohr's circle in Figure 16.4. Thus, this is the most critical angle.

Solving this free body diagram for P_a/b and V_a/b gives:

$$P_a/b = \frac{\gamma H^2 K_a \cos\beta}{2} \tag{16.5}$$

$$V_a/b = \frac{\gamma H^2 K_a \sin\beta}{2} \tag{16.6}$$

$$K_a = \frac{\cos\beta - \sqrt{\cos^2\beta - \cos^2\phi}}{\cos\beta + \sqrt{\cos^2\beta - \cos^2\phi}} \qquad \beta \le \phi \tag{16.7}$$

The magnitude of K_a is usually between 0.2 and 0.9. Equation 16.7 is valid only when $\beta \le \phi$. If $\beta = 0$, it reduces to:

$$K_a = \tan^2(45° - \phi/2) \tag{16.8}$$

A solution of P_a/b as a function of H would show that the theoretical pressure distribution is triangular. Therefore, the theoretical pressure and shear stress acting against the wall, σ and τ, respectively, are:

$$\sigma = \sigma_z' K_a \cos\beta \tag{16.9}$$

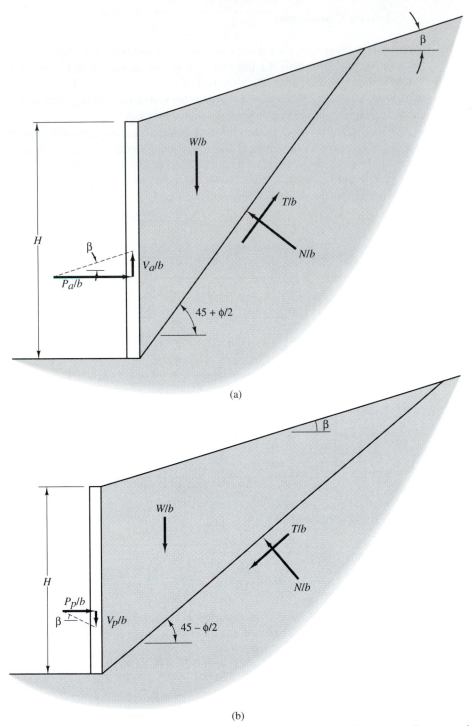

(a)

(b)

Figure 16.10 Free body diagram of soil behind a retaining wall using Rankine's solution: (a) active case; and (b) passive case.

$$\tau = \sigma_z' K_a \sin \beta \qquad\qquad (16.10)$$

where:

σ = soil pressure imparted on retaining wall from the soil

τ = shear stress imparted on retaining wall from the soil

P_a/b = normal force between soil and wall per unit length of wall

V_a/b = shear force between soil and wall per unit length of wall

b = unit length of wall (usually 1 ft or 1 m)

K_a = coefficient of active earth pressure

σ_z' = vertical effective stress

β = inclination of ground surface above the wall

H = wall height

However, observations and measurements from real retaining structures indicate that the true pressure distribution, as shown in Figure 16.11, is not triangular. This difference is because of wall deflections, arching, and other factors. The magnitudes of P_a/b and V_a/b are approximately correct, but the resultant acts at about $0.40H$ from the bottom, not $0.33H$ as predicted by theory (Duncan, et al., 1990).

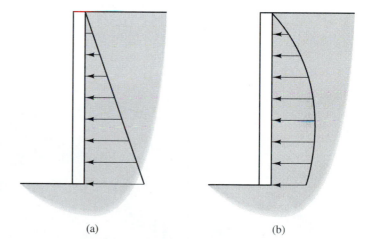

Figure 16.11 Comparison between (a) theoretical and (b) observed distributions of earth pressures acting behind retaining structures.

(a)   (b)

Example 16.2

A 6 m tall cantilever wall retains a soil that has the following properties: $c' = 0$, $\phi' = 30°$, and $\gamma = 19.2$ kN/m³. The ground surface behind the wall is inclined at a slope of 3 horizontal to 1 vertical, and the wall has moved sufficiently to develop the active condition. Determine the normal and shear forces acting on the back of this wall using Rankine's theory.

Solution

$$\beta = \tan^{-1}(1/3)$$
$$= 18°$$

$$K_a = \frac{\cos\beta - \sqrt{\cos^2\beta - \cos^2\phi}}{\cos\beta + \sqrt{\cos^2\beta - \cos^2\phi}}$$
$$= \frac{\cos 18° - \sqrt{\cos^2 18° - \cos^2 30°}}{\cos 18° + \sqrt{\cos^2 18° - \cos^2 30°}}$$
$$= 0.415$$

$$P_a/b = \frac{\gamma H^2 K_a \cos\beta}{2} = \frac{(19.2\,\text{kN/m}^2)(6\,\text{m}^2)(0.415)\cos 18°}{2} = \mathbf{136\ kN/m} \quad \Leftarrow Answer$$

$$V_a/b = \frac{\gamma H^2 K_a \sin\beta}{2} = \frac{(19.2\,\text{kN/m}^2)(6\,\text{m}^2)(0.415)\sin 18°}{2} = \mathbf{44\ kN/m} \quad \Leftarrow Answer$$

These results are shown in Figure 16.12.

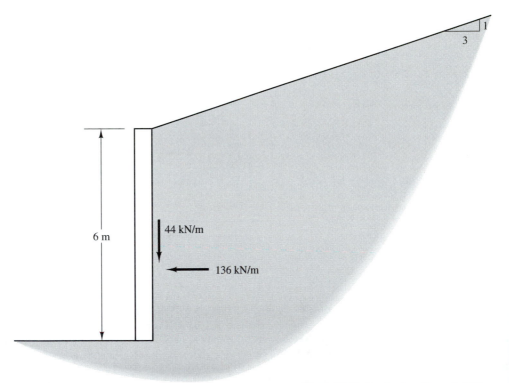

6 m

44 kN/m

136 kN/m

1

3

Figure 16.12 Results from Example 16.2.

Passive Condition

Rankine analyzed the passive condition in a fashion similar to the active condition except that the shear force acting along the base of the wedge now acts in the opposite direction (it always opposes the movement of the wedge) and the free body diagram becomes as shown in Figure 16.10b. Notice that the failure wedge is much flatter than it was in the active case and the critical angle is now 45 - ϕ/2 degrees from the horizontal.

The normal and shear forces, P_p/b and V_p/b, respectively, acting on the wall in the passive case are:

$$P_p/b = \frac{\gamma H^2 K_p \cos\beta}{2} \tag{16.11}$$

$$V_p/b = \frac{\gamma H^2 K_p \sin\beta}{2} \tag{16.12}$$

$$K_p = \frac{\cos\beta + \sqrt{\cos^2\beta - \cos^2\phi}}{\cos\beta - \sqrt{\cos^2\beta - \cos^2\phi}} \qquad \beta \le \phi \tag{16.13}$$

The magnitude of K_p is typically between 2 and 6. Equation 16.13 is valid only when $\beta \le \phi$. If $\beta = 0$, it reduces to:

$$K_p = \tan^2(45° + \phi/2) \tag{16.14}$$

The theoretical pressure and shear acting against the wall, σ and τ, respectively, are:

$$\sigma = \sigma'_z K_p \cos\beta \tag{16.15}$$

$$\tau = \sigma'_z K_p \sin\beta \tag{16.16}$$

where:

σ = soil pressure imparted on retaining wall from the soil
τ = shear stress imparted on retaining wall from the soil
P_p/b = normal force between soil and wall per unit length of wall
V_p/b = shear force between soil and wall per unit length of wall
b = unit length of wall (usually 1 ft or 1 m)
K_p = coefficient of passive earth pressure
σ'_z = vertical effective stress
β = inclination of ground surface above the wall

Example 16.3

A six-story building with plan dimensions of 150 ft × 150 ft has a 12 ft deep basement. This building is subjected to horizontal wind loads, and the structural engineer wishes to transfer these loads into the ground through the basement walls. The maximum horizontal force acting on the basement wall is limited by the passive pressure in the soil. Using Rankine's theory, compute the maximum force between one of the basement walls and the adjacent soil assuming full passive conditions develop, then convert it to an allowable force using a factor of safety of 3. The soil is a silty sand with $c' = 0$, $\phi' = 30°$, and $\gamma = 119$ lb/ft^3, and the ground surface surrounding the building is essentially level.

Solution

$$K_p = \tan^2(45° + \phi'/2) = \tan^2(45° + 30°/2) = 3.00$$

$$
\begin{aligned}
P_p/b &= \frac{\gamma H^2 K_p \cos\beta}{2} \\
&= \frac{(119 \text{ lb/ft}^3)(12 \text{ ft})^2 (3.00) \cos 0}{2} \\
&= 25{,}700 \text{ lb/ft}
\end{aligned}
$$

$$P_p = \frac{(25{,}700 \text{ lb/ft})(150 \text{ ft})}{1000 \text{ lb/k}} = 3860 \text{ k}$$

The allowable passive force, $(P_p)_a$, is:

$$(P_p)_a = \frac{P_p}{F} = \frac{3860 \text{ k}}{3} = \mathbf{1290 \text{ k}} \qquad \Leftarrow Answer$$

Note: The actual design computations for this problem would be more complex because they would need to consider the active pressure acting on the opposite wall, sliding friction along the basement floor, lateral resistance in the foundations, and other factors. In addition, the horizontal displacement required to develop the full passive resistance may be excessive, so the design value may need to be reduced accordingly. Finally, to take advantage of this resistance, the wall would need to be structurally designed to accommodate this large load, which is much greater than that due to the active or at-rest pressure.

Coulomb's Theory for Cohesionless Soils

Coulomb's theory differs from Rankine's in that the resultant of the normal and shear forces acting on the wall is inclined at an angle ϕ_w from a perpendicular to the wall, where $\tan \phi_w$

Charles Augustin Coulomb

Charles Augustin Coulomb (1736–1806) was a French physicist who is best remembered for his work in electricity and magnetism. However, he also made important contributions in other fields, including the computation of lateral earth pressures.

Coulomb graduated from the Mézières School of Military Engineers in France at the age of 26. Two years later, the young officer was sent to the Caribbean island of Martinique where he was placed in charge of building a fort to protect the harbor. In the process of finalizing the design of the fort, he became dissatisfied with the rules of thumb for sizing retaining walls because they dictated walls that were too large. Although some theoretical analyses had already been attempted, they were flawed. He later wrote (Kerisel, 1987):

"I have often come across situations in which all the theories based on hypotheses or on small-scale experiments in a physics laboratory have proved inadequate in practice."

Therefore, he began studying the problem, and eventually developed a new theory of lateral earth pressures. This work is generally recognized as the first important quantitative contribution to what would become geotechnical engineering. Coulomb was the first to define soil strength using both cohesion and friction, the first to consider wall friction, and the first to analytically search for the orientation of the most critical failure plane (which turned out to be at an angle of $45+\phi/2$, as shown in Figure 16.10). He also developed other important insights.

Coulomb published his results in 1776 as a paper titled *Essai sur une application des règles de maximis et minimis à quelques problèmes de statique relatifs à l'architecture* (Essay on an Application of the Rules of Maximum and Minimum to Some Statical Problems, Relevant to Architecture).[1] This paper also addressed other problems, including the stability of arches and the strength of beams.

He had a sense of both theory and practice. For example, his *Essai* also discussed the detrimental effects of groundwater and noted "Even though, to avoid this problem, vertical pipes are placed in practice behind retaining walls, and the drains at the feet of the same walls, so that the water can run off, these drains get blocked, either by soil carried along with the water, or by ice, and sometimes become useless."

Although Coulomb's work provided important insights into the earth pressure problem, it was difficult to apply to practical problems because nobody had the ability to measure c and ϕ of soil. The first significant soil strength tests would not be performed until about seventy years later by another Frenchman, Alexandre Collin, and his work was not widely recognized. As a practical matter, Coulomb's work did not reach its full potential until the twentieth century, when soil strength tests became common.

[1] Heyman (1972) provides an English translation and commentary.

is the coefficient of friction between the wall and the soil, as shown in Figure 16.13. This is a more realistic model, and thus produces more precise values of the active earth pressure.

Coulomb presented his earth pressure formula in a difficult form, so others have rewritten it in a more convenient fashion, as follows (Müller Breslau, 1906; Tschebotarioff, 1951):

$$P_a/b = \frac{\gamma H^2 K_a \cos\phi_w}{2} \tag{16.17}$$

$$V_a/b = \frac{\gamma H^2 K_a \sin\phi_w}{2} \tag{16.18}$$

$$K_a = \frac{\cos^2(\phi - \alpha)}{\cos^2\alpha \cos(\phi_w + \alpha)\left[1 + \sqrt{\dfrac{\sin(\phi + \phi_w)\sin(\phi - \beta)}{\cos(\phi_w + \alpha)\cos(\alpha - \beta)}}\right]^2} \tag{16.19}$$

where:
- σ = soil pressure imparted on retaining wall from the soil
- τ = shear stress imparted on retaining wall from the soil
- P_a/b = normal force between soil and wall per unit length of wall
- V_a/b = shear force between soil and wall per unit length of wall
- b = unit length of wall (usually 1 ft or 1 m)
- K_a = coefficient of active earth pressure
- σ_z' = vertical effective stress
- α = inclination of wall from vertical
- β = inclination of ground surface above the wall
- ϕ_w = wall–soil interface friction angle

Equation 16.19 is valid only for $\beta \le \phi$. When designing concrete or masonry walls it is common practice to use $\phi_w = 0.67\,\phi'$. Steel walls have less sliding friction, perhaps on the order of $\phi_w = 0.33\,\phi'$.

Coulomb did not develop a formula for passive earth pressure, although others have used his theory to do so. However, the addition of wall friction can substantially increase the computed passive pressure, possibly to values that are too high (Dunn, Anderson, and Kiefer, 1980). Therefore, engineers normally neglect wall friction in passive pressure computations ($\phi_w = 0$) and use Rankine's method to compute passive earth pressures (Equations 16.11–16.16).

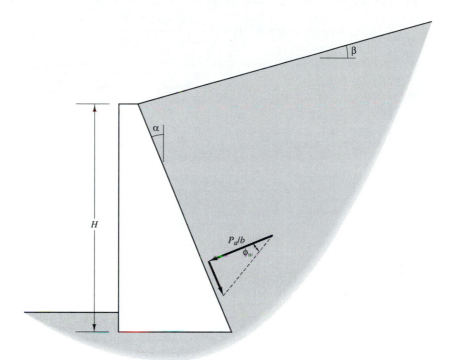

Figure 16.13 Parameters for Coulomb's lateral earth pressure equations. Walls inclined in the opposite direction have a negative α. V_a/b normally acts in the direction shown, thus producing a positive ϕ_w.

Example 16.4

Using Coulomb's method, compute the active pressure acting on the reinforced concrete retaining wall shown in Figure 16.14.

Solution

$$\beta = \tan^{-1}\left(\frac{1}{2}\right) = 27°$$

$$\phi_w = 0.67\,\phi' = 0.67\,(32°) = 21°$$

$$K_a = \frac{\cos^2(\phi - \alpha)}{\cos^2\alpha \, \cos(\phi + \phi_w)\left[1 + \sqrt{\dfrac{\sin(\phi + \phi_w)\sin(\phi - \beta)}{\cos(\phi_w + \alpha)\cos(\alpha - \beta)}}\,\right]^2}$$

$$= \frac{\cos^2(32° - 2°)}{\cos^2 2° \, \cos(21° + 2°)\left[1 + \sqrt{\dfrac{\sin(32° + 21°)\sin(32° - 27°)}{\cos(21° + 2°)\cos(2° - 27°)}}\,\right]^2}$$

$$= 0.491$$

$$P_a/b = \frac{\gamma H^2 K_a \cos\phi_w}{2}$$

$$= \frac{(19.8\,\text{kN/m}^2)(5.60\,\text{m})^2(0.491)\cos 21°}{2}$$

$$= \mathbf{142\,kN/m} \qquad \leftarrow Answer$$

$$V_a/b = \frac{\gamma H^2 K_a \sin\phi_w}{2}$$

$$= \frac{(19.8\,\text{kN/m}^2)(5.60\,\text{m})^2(0.491)\sin 21°}{2}$$

$$= \mathbf{55\,kN/m} \qquad \leftarrow Answer$$

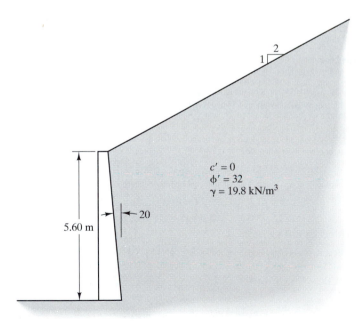

Figure 16.14 Retaining wall for Example 16.4.

QUESTIONS AND PRACTICE PROBLEMS

16.1 Explain the difference between the active, at-rest, and passive earth pressure conditions.

16.2 Which of the three earth pressure conditions should be used to design a rigid basement wall? Why?

16.3 A basement is to be built using 2.5 m tall masonry walls. These walls will be backfilled with a silty sand that has $c' = 0$, $\phi' = 35°$, and $\gamma = 19.7$ kN/m^3. Assuming the at-rest conditions will exist and using an overconsolidation ratio of 2, compute the normal force per meter acting on the back of this wall. Also, draw a pressure diagram and indicate the lateral earth pressure acting at the back of the wall.

16.4 A 10 ft tall concrete wall with a vertical back is to be backfilled with a silty sand that has a unit weight of 122 lb/ft^3, an effective cohesion of 0, and an effective friction angle of 32°. The ground behind the wall will be level. Using Rankine's method, compute the normal force per foot acting on the back of the wall. Assume the wall moves sufficiently to develop the active condition in the soil.

16.5 The wall described in Problem 16.4 has a foundation that extends from the ground surface to a depth of 2 ft. As the wall moves slightly away from the backfill soils to create the active condition, the footing moves into the soils below the wall, creating the passive condition as shown in Figure 16.9. Compute the ultimate passive pressure acting on the front of the foundation.

16.6 A 12 ft tall concrete wall with a vertical back is to be backfilled with a clean sand that has a unit weight of 126 lb/ft^3, an effective cohesion of 0, and an effective friction angle of 36°. The ground behind the wall will be inclined at a slope of 2 horizontal to 1 vertical. Using Rankine's method, compute the normal and shear forces per foot acting on the back of the wall. Assume the wall moves sufficiently to develop the active condition in the soil.

16.7 Repeat Problem 16.6 using Coulomb's method.

16.3 EQUIVALENT FLUID METHOD

As discussed earlier, the theoretical distribution of lateral earth pressure acting on a wall is triangular. This is the same shape as the pressure distribution that would be imposed if the wall was backfilled with a fluid instead of with soil. Further, if this fluid had the proper unit weight, the magnitude of the lateral earth pressure also would be equal to that from the soil.

Engineers often use this similarity when expressing lateral earth pressures for design purposes. Instead of quoting K values, we define the lateral earth pressure using the *equivalent fluid density*, G_h. This is the unit weight of a fictitious fluid that would impose the same horizontal pressures on the wall as the soil. We give this value to a civil engineer or structural engineer who wishes to design a wall, and they proceed using the principles of fluid statics.

This method is popular because it reduces the potential for confusion and mistakes. All engineers understand the principles of fluid statics (at least they should!), so G_h is easy to apply.

For homogeneous, isotropic soils with $c = 0$ the equivalent fluid density is:

$$G_h = K\gamma \qquad\qquad (16.20)$$

The normal force between the soil and the wall per unit length of the wall is:

$$P/b = \frac{G_h H^2}{2} \qquad\qquad (16.21)$$

where:

G_h = equivalent fluid density
$K = K_a$, K_0, or K_p, as appropriate
γ = unit weight of backfill soils
P/b = normal force between the soil and wall per unit length of the wall
H = height of the wall

Example 16.5

A 12 ft tall cantilever retaining wall supported on a 2 ft deep continuous footing will retain a sandy soil with $c' = 0$, $\phi' = 35°$, and $\gamma = 124$ lb/ft^3. The ground surface above the wall will be level ($\beta = 0$) and there will be no surcharge loads. Compute the active pressure and express it as the equivalent fluid density, then compute the total force imposed by the backfill soils onto the wall and the back of the footing.

Solution

$$K_a = \tan^2(45° - \phi/2) = \tan^2(45° - 35°/2) = 0.271$$

$$G_h = \gamma K_a = (124 \text{ lb/ft}^3)(0.271) = \textbf{34 lb/ft}^3 \qquad \leftarrow \textit{Answer}$$

Therefore, the retaining wall should be designed to retain a fluid that has a unit weight of 34 lb/ft^3. The active earth pressure acts on both the wall and its footing, so $H = 14$ ft.

$$
\begin{aligned}
P_a/b &= \frac{G_h H^2}{2} \\
&= \frac{(34 \text{ lb/ft}^3)(14 \text{ ft})^2}{2} \\
&= \textbf{3332 lb/ft} \qquad \leftarrow \textit{Answer}
\end{aligned}
$$

QUESTIONS AND PRACTICE PROBLEMS

16.8 A 4 m tall cantilever wall is to be backfilled with a dense silty sand. How far must this wall move to attain the active condition in the soil behind it? Is it appropriate to use the active pressure for design? Explain.

16.9 A 3 m tall cantilever retaining wall with a vertical back is to be backfilled with a soil that has an equivalent fluid density of 6.0 kN/m³. Compute the lateral force per meter acting on the back of this wall.

16.10 A proposed concrete retaining wall is to be built as shown in Figure 16.15. Using Rankine's method, compute the horizontal component of the active earth pressure acting on the 14.3-ft tall dashed line and the passive earth pressure acting on the front of the footing. Present your results as pressure diagrams. Then compute the resultant of the active earth pressure and the resultant of the passive earth pressure and show them as horizontal point loads.

Note: Another important force has not been considered in this analysis: The sliding friction force along the bottom of the footing. In a properly designed wall, the combination of this force and the resultant of the passive pressure is greater than the resultant active pressure with an appropriate factor of safety.

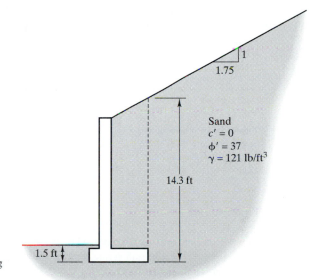

Figure 16.15 Proposed retaining wall for Problems 16.10 and 16.11.

16.11 Recompute the resultant of the active pressure in Problem 16.10 using Coulomb's method. Compare your answer to the value obtained in Problem 16.10.

16.4 GROUNDWATER EFFECTS

The discussions in this chapter have thus far assumed that the groundwater table is located below the base of the wall. If the groundwater table rises to a level above the base of the wall, the following three important changes occur:

1. The effective stress in the soil below the groundwater table will decrease, which decreases the active, passive, and at-rest pressures.
2. Horizontal hydrostatic pressures will develop against the wall and must be superimposed onto the lateral earth pressures.

3. The effective stress between the bottom of the footing and the soil becomes smaller, so there is less sliding friction.

The net effect of the first two changes is a large increase in the total horizontal pressure acting on the wall (i.e., the increased hydrostatic pressures more than offset the decreased effective stress). The resulting pressure diagram is shown in Figure 16.16.

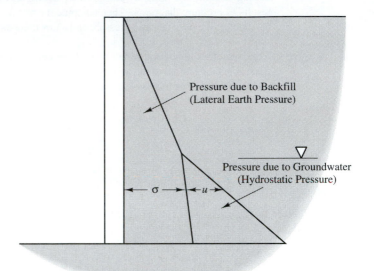

Pressure due to Backfill
(Lateral Earth Pressure)

Pressure due to Groundwater
(Hydrostatic Pressure)

σ u

Figure 16.16 Theoretical lateral pressure distribution with shallow groundwater table.

Example 16.5

This cantilever wall has moved sufficiently to create the active condition. Compute the lateral pressure distribution acting on this wall with the groundwater table at locations **a** and **b**, as shown in Figure 16.17.

The soil properties are: $c' = 0$, $\phi' = 30°$, $\gamma = 20.4$ kN/m³, and $\gamma_{sat} = 22.0$ kN/m³

Solution

Use Rankine's Method.

$$K_a = \tan^2(45° - \phi/2) = \tan^2(45° - 30°/2) = 0.333$$

With the groundwater table at **a**:

$$\sigma = \sigma_z' K_a \cos\beta$$
$$= \gamma z K_a \cos\beta$$
$$= 20.4 z (0.333) \cos 0$$
$$= 6.79 z$$

where z = depth below the top of the wall

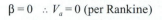

$$\beta = 0 \quad \therefore V_a = 0 \text{ (per Rankine)}$$

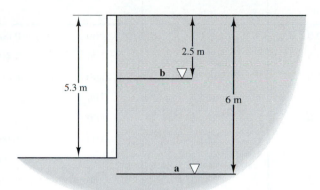

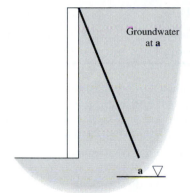

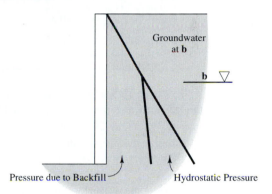

Figure 16.17 Retaining wall for Example 16.7.

With the groundwater table at **b**:

$$
\begin{aligned}
\sigma @ z \ge 2.5\,\text{m} &= \sigma'_z K_a \cos\beta \\
&= (\Sigma\gamma H - u) K_a \cos\beta \\
&= [(20.4\,\text{kN/m}^3)(2.5\,\text{m}) + (22.0\,\text{kN/m}^3)(z - 2.5\,\text{m}) - u](0.333)\cos 0 \\
&= 7.33 z - 0.33 u - 1.33
\end{aligned}
$$

$$u = 9.80\,\text{kN/m}^3 (z - 2.5\,\text{m}) \ge 0$$

Total horizontal pressure on wall $= \sigma + u$

z	Groundwater at **a**	Groundwater at **b**		
	σ	u	σ	Total Pressure
(m)	(kPa)	(kPa)	(kPa)	(kPa)
0.0	0.00	0.00	0.00	0.00
0.5	3.40	0.00	3.40	3.40
1.0	6.79	0.00	6.79	6.79
1.5	10.19	0.00	10.19	10.19
2.0	13.58	0.00	13.58	13.58
2.5	16.98	0.00	16.98	16.98
3.0	20.37	4.90	19.04	23.94
3.5	23.77	9.80	21.09	30.89
4.0	27.16	14.70	23.14	37.84
4.5	30.56	19.60	25.19	44.79
5.0	33.95	24.50	27.24	51.74
5.3	35.99	27.44	28.46	55.90

Example 16.5 demonstrates the profound impact of groundwater on retaining walls. If the groundwater table rises from **a** to **b**, the total horizontal force acting on the wall increases by about 30 percent. Therefore, the factor of safety against sliding and overturning could drop from 1.5 to about 1.0, and the flexural stresses in the stem would be about 30 percent larger than anticipated. There are two ways to avoid these problems:

1. Design the wall for the highest probable groundwater table. This can be very expensive, but it may be the only available option.
2. Install drains to prevent the groundwater from rising above a certain level. These could consist of *weep holes* drilled in the face of the wall or a *perforated pipe drain* installed behind the wall. Drains such as these are the most common method of designing for groundwater.

Further problems can occur if the groundwater becomes frozen and ice lenses form. The same processes that cause frost heave at the ground surface also produce large horizontal pressures on retaining walls. This is another good reason to provide good surface and subsurface drainage around retaining walls.

QUESTIONS AND PRACTICE PROBLEMS

16.12 Using a groundwater table at Level A and Rankine's method, compute the lateral earth pressure acting on the back of the wall in Figure 16.18. Present your results in the form of a pressure diagram, then compute the total force acting on the wall and the bending moment at the bottom of the stem.

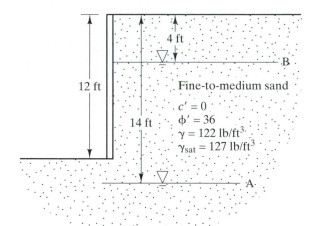

Figure 16.18 Proposed retaining wall for Problems 16.12 and 16.13.

16.13 Using the information from Problem 16.12 and a groundwater table at Level B, recompute the lateral and hydrostatic earth pressures acting on the back of the wall. Present your results in the form of a pressure diagram, then compute the total force acting on the wall and the bending moment at the bottom of the stem. Compare this moment with that computed in Problem 16.12.

16.5 RETAINING WALLS

Geotechnical engineers often participate in the design of *retaining walls* (also known as *earth retaining structures*). These are vertical or near-vertical walls that retain soil or rock, as shown in Figure 16.1. Many kinds of retaining structures are available, each best suited for particular applications. *Foundation Design: Principles and Practices* discusses the selection, analysis, and design of retaining walls in detail. This section is a brief introduction to the topic.

O'Rourke and Jones (1990) classified retaining walls into two broad categories: *externally stabilized systems* and *internally stabilized systems*, as shown in Figure 16.19. Some hybrid methods combine features from both systems.

Externally Stabilized Systems

Externally stabilized systems are those that resist the applied earth loads by virtue of their weight and stiffness. This was the only type of retaining structure available before 1960, and they are still very common. O'Rourke and Jones subdivided these structures into two categories: *gravity walls* and *in-situ walls*.

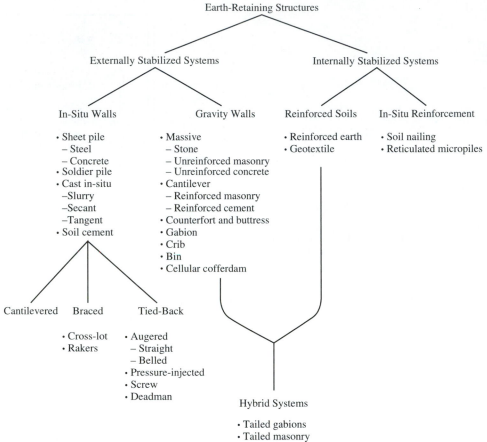

Figure 16.19 Classification of retaining walls (Adapted from O'Rourke and Jones, 1990; Used with permission of ASCE).

Gravity Walls

Massive Gravity Walls

The earliest retaining structures were *massive gravity walls*, as shown in Figure 16.20. They were often made of mortared stones, masonry, or unreinforced concrete and resisted the lateral forces from the backfill by virtue of their large mass. The construction of these walls is very labor-intensive and requires large quantities of materials, so they are rarely used today except for very short walls.

Cantilever Gravity Walls

The *cantilever gravity wall*, shown in Figure 16.21, is a refinement of the massive wall. These walls have a much smaller cross-section and thus require much less material. However, these walls have large flexural stresses, so they are typically made of reinforced concrete or reinforced masonry.

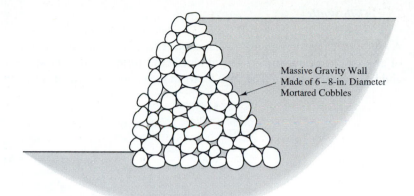

Figure 16.20 Massive gravity wall.

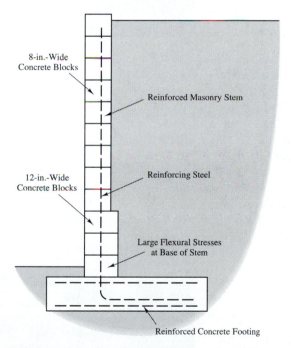

Figure 16.21 A cantilever gravity wall with a concrete block stem.

Crib Walls

A *crib wall*, shown in Figure 16.22, is another type of gravity retaining structure. It consists of precast concrete members linked together to form a crib. These members resemble a child's Lincoln Log toy. The zone between the members is filled with compacted soil.

Figure 16.22 A crib wall.

In-Situ Walls

In-situ walls differ from gravity walls in that they rely primarily on their flexural strength, not their mass.

Sheet Pile Walls

A *sheet pile* is a thin, wide pile driven into the ground using a pile hammer. A series of sheet piles in a row form a *sheet pile wall*, as shown in Figure 16.23. Most sheet piles are made of steel, but some are made of reinforced concrete.

Figure 16.23 A sheet pile wall.

It may be possible to simply cantilever short sheet piles out of the ground, as shown in Figure 16.24. However, taller sheet pile walls usually need lateral support at one or more levels above the ground. This may be accomplished in either of two ways: by *internal braces* or by *tieback anchors*.

Internal braces are horizontal or diagonal compression members that support the wall, as shown in Figure 16.24. Tieback anchors are tension members drilled into the ground behind the wall. The most common type is a grouted anchor with a steel tendon.

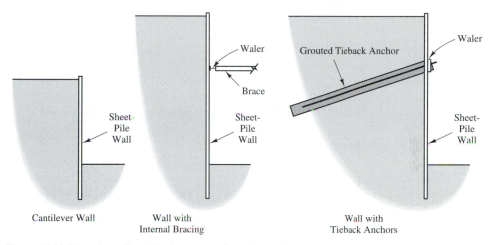

Figure 16.24 Short sheet pile walls often can cantilever, but taller walls usually require bracing or tieback anchors.

Soldier Pile Walls

Soldier pile walls consist of vertical wide flange steel members with horizontal timber lagging, as shown in Figure 16.25. They are often used as temporary retaining structures for construction excavations.

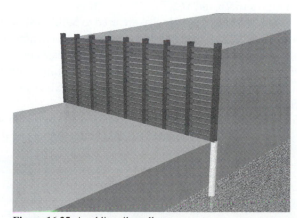

Figure 16.25 A soldier pile wall.

Slurry Walls

Slurry walls are cast-in-place concrete walls built using bentonite slurry. The contractor digs a trench along the proposed wall alignment and keeps it open using the slurry. Then, the reinforcing steel is inserted and the concrete is placed using tremie pipes or pumps. As the concrete fills the trench, the slurry exits at the ground surface.

Slurry walls have been used as basement walls in large urban construction, and often eliminate the need for temporary walls.

Internally Stabilized Systems

Internally stabilized systems reinforce the soil to provide the necessary stability. Various schemes are available, all of which have been developed since 1960. They can be subdivided into two categories: *reinforced soils* and *in-situ reinforcement*.

Reinforced Soils

Soil is strong in compression, but has virtually no tensile strength. Therefore, the inclusion of tensile reinforcing members in a soil can significantly increase is strength and load-bearing capacity, much the same way that placing rebars in concrete increases its strength. The resulting reinforcing soil is called *mechanically stabilized earth (MSE)*.

Often MSE is used so that slopes may be made steeper than would otherwise be possible. Thus, this method forms an intermediate alternative between earth slopes and retaining walls. This method is discussed in Section 19.7.

MSE also may be used with vertical or near-vertical faces, thus forming a type of retaining wall. In this case, it becomes necessary to place some type of facing panels on the vertical surface, even though the primary soil support comes from the reinforcement, not the panels. Such structures are called *MSE walls*. The earliest MSE walls were developed by Henri Vidal in the early 1960, using the trade name *reinforced earth*. This design uses strips of galvanized steel for the reinforcement and precast concrete panels for the facing, as shown in Figures 16.26 and 16.27.

Many other similar methods also are used to build MSE walls. The reinforcement can consist of steel strips, polymer geogrids, wire mesh, geosynthetic fabric, or other materials. The facing can consist of precast concrete panels, precast concrete blocks, rock filled cages called *gabions*, or other materials. Sometimes the reinforcement is simply curved around to form a type of facing. Figure 16.28 shows an MSE wall being built using wire mesh reinforcement and gabion facing.

MSE walls are becoming very popular for many applications, especially for highway projects. Their advantages include low cost and tolerance of differential settlements.

Figure 16.26 Reinforced earth walls consist of precast concrete facing panels and steel or polymer reinforcing strips that extend into the retained soil. This wall is under construction (The Reinforced Earth Company).

Figure 16.27 A completed reinforced earth wall (The Reinforced Earth Company).

Figure 16.28 An MSE wall under construction using galvanized wire mesh as the tensile reinforcement and rock-filled cages called gabions for the facing (Federal Highway Administration).

In-Situ Reinforcement

In-situ reinforcement methods differ from reinforced soils in that the tensile members are inserted into a soil mass rather than being embedded during placement of fill.

Soil Nailing

Soil nailing consists of drilling near-horizontal holes into the ground, inserting steel tendons, and grouting. The face of the wall is typically covered with shotcrete, as shown in Figure 16.29.

These walls do not require a construction excavation, and thus are useful when space is limited.

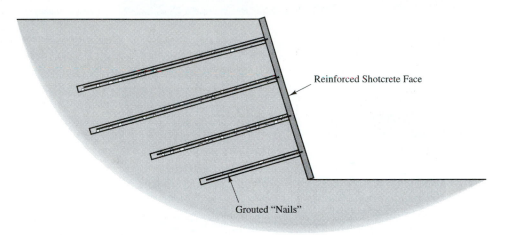

Figure 16.29 A soil nail wall.

SUMMARY

Major Points

1. Lateral earth pressures are those imparted by soil onto vertical or near-vertical structures. These pressures include both compression and shear.
2. The coefficient of lateral earth pressure, K, is the ratio of the horizontal to vertical effective stresses. In undisturbed soil, it has a value equal to the *coefficient of lateral earth pressure at-rest*, K_0. However, if the structure (usually a retaining wall) moves a sufficient distance out from the backfill, the soil reaches the *active condition*, and K drops to a lower value, K_a. Conversely, if the structure moves a sufficient distance toward the backfill, the soil reaches the *passive condition*. and K rises to a higher value, K_p.
3. The active and passive earth pressures may be computed using classical earth pressure theories. Coulomb's theory and Rankine's theory are the most commonly used.

4. If the groundwater table is located above the base of the wall, the resulting hydrostatic pressures will significantly increase the total force acting on the wall. In some cases, these hydrostatic pressures may be greater than the lateral earth pressures from the soil. Therefore, it is very important for walls to have good drainage.

5. Retaining walls are vertical or near-vertical structures designed to retain soil or rock. Many different types are available. Externally stabilized walls resist the applied loads by virtue of their weight and stiffness, whereas internally stabilized walls rely on reinforcement within the ground.

6. Externally stabilized walls include gravity walls and in-situ walls.

7. Internally stabilized walls include reinforced soils (also known as mechanically stabilized earth or MSE) and in-situ reinforced walls.

Vocabulary

active condition
at-rest condition
cantilever gravity walls
cantilever wall
coefficient of active earth
  pressure
coefficient of lateral earth
  pressure
coefficient of lateral earth
  pressure at rest
coefficient of passive earth
  pressure
Coulomb's theory
crib walls

earth retaining structure
equivalent fluid density
externally stabilized
  systems
flexible wall
gravity walls
in-situ walls
internal braces
internally stabilized
  systems
lateral earth pressure
limit equilibrium analysis
massive gravity walls
passive condition

plane strain condition
Rankine's theory
reinforced earth walls
retaining wall
reticulated micropiles
rigid wall
sheet pile walls
slurry walls
soil nailing
soldier pile walls
tieback anchors
unyielding wall
yielding wall

COMPREHENSIVE QUESTIONS AND PRACTICE PROBLEMS

16.14 A massive gravity wall is to be built on a hard bedrock, then backfilled with a very loose uncompacted cohesionless soil. Which should be used for the design earth pressure acting on the back of this wall, the at-rest pressure, the active pressure, or the passive pressure? Why?

16.15 Explain the difference between externally stabilized retaining walls and internally stabilized retaining walls.

16.16 What is mechanically stabilized earth, and how does it work?

17

Structural Foundations

*. . . and we can save 1000 Lira
by not conducting soil studies!*

Imaginary conversation between architects
and builders of the Tower of Pisa,
circa AD 1170

The structural elements that connect buildings, bridges, and other structures to the ground are called *foundations*. These elements are very important, because the safety and reliability of the structure can be no better than that of its foundations.

Geotechnical engineers are routinely involved in both the design and construction of structural foundations. The design phase is normally performed in conjunction with a structural engineer, with the geotechnical engineer being responsible for aspects of the design that relate to the soil or rock that supports the foundation, and the structural engineer taking care of structural integrity issues. During the construction phase, the geotechnical engineer works with the contractor and is responsible for comparing the soil conditions actually encountered with those anticipated in the design, providing various quality control services, and developing revised design recommendations as needed.

17.1 TYPES OF FOUNDATIONS

The purpose of foundations is to safely transmit structural loads into the ground. The primary loads usually act downward, but uplift, horizontal, and moment loads also may be present.

616

Many kinds of foundations are available, and the proper selection depends on the magnitude and direction of the structural loads, the subsurface conditions, and other factors. For convenience, we divide foundations into two broad categories: shallow foundations and deep foundations.

Shallow Foundations

Shallow foundations are those that transmit the structural loads to the near-surface soil or rock. There are two types: spread footings and mats, as shown in Figure 17.1.

A *spread footing foundation* is an enlargement at the bottom of a column or a bearing wall that spreads the structural load over a certain area of soil. They are nearly always made of reinforced concrete. The required footing size depends on the magnitude of the load, the engineering properties of the underlying soils, and other factors.

A *mat foundation* (also known as a *raft foundation*) is essentially one large spread footing that encompasses the entire structure. They spread the weight of the structure across a larger area, thus reducing the induced stresses in the underlying soils. Mata also have the advantage of structural continuity and thus reduce the potential for differential settlements. Mats are generally used on structures that are too heavy for spread footings, but not heavy enough to warrant a more expensive deep foundation system.

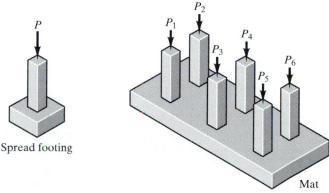

Figure 17.1 Shallow foundations.

Deep Foundations

Deep foundations are those that transmit some or all of the structural loads to deeper soil or rock, as shown in Figure 17.2. Such foundations are most often used with larger structures, or when the shallow soils are poor.

Many kinds of deep foundations are available. We can divide them into three broad categories:

- *Piles* are prefabricated poles made of steel, wood, or concrete, that are driven into the ground.

- **Drilled shafts** are constructed by drilling cylindrical holes into the ground, inserting reinforcing steel, and filling them with concrete.
- **Other types** include various hybrid methods, and other techniques.

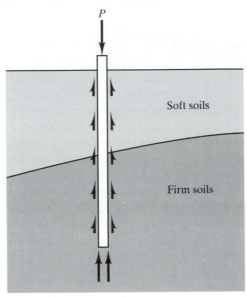

Figure 17.2 Deep foundations.

Focus of This Chapter

This chapter is only a brief introduction to structural foundations, so we will consider only spread footing foundations, which are the simplest and most common type. *Foundation Design: Principles and Practices* (Coduto, 1999), the companion volume to this book, discusses foundation engineering in much more detail, and covers the other types of foundations.

Spread footings may be built in a variety of shapes to suit individual needs, as shown in Figure 17.3. The most common shape is a *square footing*, which usually supports a single column. *Combined footings* are those that support more than one column. *Continuous footings* support bearing walls. Most continuous footings are linear, but some, such as those that support the exterior wall of a tank, are circular, thus forming a *ring footing*. Figure 17.3 also shows the dimensions B, L, D, and T, which we use to describe the size of spread footings.

17.2 SPREAD FOOTINGS — BEARING PRESSURE

During the late nineteenth and early twentieth centuries, engineers realized the design of spread footings could be based on the contact pressure between the footing and the ground that supports it. This pressure is called the *bearing pressure* (or *gross bearing pressure*), q:

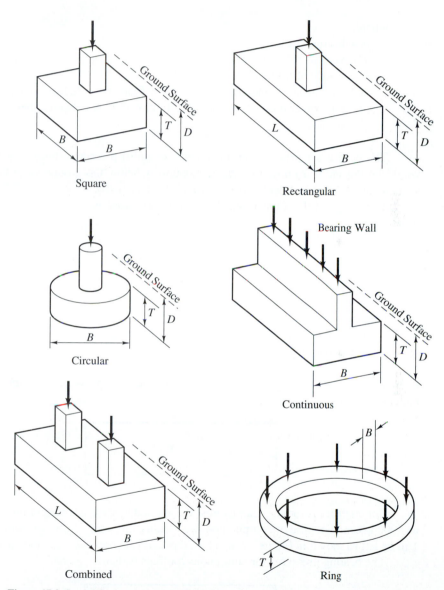

Figure 17.3 Spread footing shapes and dimensions.

$$q = \frac{P + W_f}{A} - u \qquad (17.1)$$

where:

q = bearing pressure
P = vertical column load
W_f = weight of foundation
A = base area of foundation (B^2 for square footings or BL for rectangular footings)
u = pore water pressure at bottom of footing (i.e., at a depth D below the ground surface)

The pore water pressure term accounts for uplift pressures (buoyancy forces) that would be present if a portion of the foundation is below the groundwater table. If the groundwater table is at a depth greater than D, then set $u = 0$.

The weight of the foundation, W_f, may be expressed as:

$$\frac{W_f}{A} = \gamma_c D \qquad (17.2)$$

where:

γ_c = unit weight of concrete = 150 lb/ft^3 = 23.6 kN/m^3
D = depth of footing

Combining Equations 17.1 and 17.2 gives the bearing pressure equation for square, rectangular, and circular footings:

$$q = \frac{P}{A} + \gamma_c D - u \qquad (17.3)$$

For continuous footings, we express the applied loads as a force per unit length, such as 2000 kN/m. For ease of computation, we identify this unit length as b, which is usually 1 m or 1 ft as shown in Figure 17.4. Thus, the load is expressed using the variable P/b.

The bearing pressures for continuous footings is then:

$$q = \frac{P/b}{B} + \gamma_c D - u \qquad (17.4)$$

If the column or wall load is vertical and acts through the centroid of the footing, and no moment loads are present, then we assume the bearing pressure is uniform across the bottom of the footing.

Sometimes engineers prefer to use the *net bearing pressure*, q', which is the difference between the gross bearing pressure, q, and the σ_{zo}' at depth D. In other words, q' is a measure of the increase in vertical effective stress at depth D. However, in this book we will use only the gross bearing pressure, as defined in Equations 17.3 and 17.4.

The two most important geotechnical design requirements for spread footings are bearing capacity and settlement. We analyze both of them in terms of the bearing pressure.

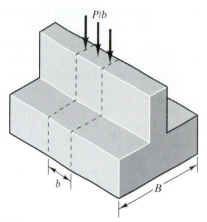

Figure 17.4 Definitions for loads on continuous footings.

17.3 SPREAD FOOTINGS — BEARING CAPACITY

A *bearing capacity failure* occurs when the shear stresses induced by the footing exceed the shear strength of the soil, as shown in Figure 17.5. Such failures are catastrophic, and thus must be avoided. This need to prevent a bearing capacity failure is called a *strength requirement* and is similar to structural engineers' requirements for strength of structural members.

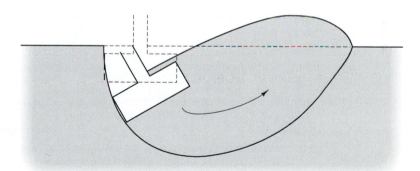

Figure 17.5 Bearing capacity failure.

Terzaghi's Bearing Capacity Formulas

The *ultimate bearing capacity*, q_{ult}, is the bearing pressure required to produce a bearing capacity failure. The value of q_{ult} depends on the size and depth of the footing and on the strength of the underlying soils. Once we know its value, we can design the footing so that the actual bearing pressure is sufficiently smaller than q_{ult} to provide an adequate factor of safety against a bearing capacity failure.

In 1943, Karl Terzaghi developed the first widely accepted formulas for computing ultimate bearing capacity. His analysis was based on a bearing capacity theory for continuous footings because this is a two dimensional problem, and thus is the simplest case. He evaluated the shear stress and shear strength along a failure surface with a certain geometry, then wrote an equation of equilibrium in terms of q_{ult}. He then extended this equation to square and circular footings by incorporating empirical coefficients. Terzaghi's formulas are as follows (Terzaghi, 1943):

For square footings:

$$q_{ult} = 1.3 c'N_c + \sigma_D' N_q + 0.4 \gamma' B N_\gamma \qquad (17.5)$$

For continuous footings:

$$q_{ult} = c'N_c + \sigma_D' N_q + 0.5 \gamma' B N_\gamma \qquad (17.6)$$

For circular footings:

$$q_{ult} = 1.3 c'N_c + \sigma_D' N_q + 0.3 \gamma' B N_\gamma \qquad (17.7)$$

where:
q_{ult} = ultimate bearing capacity
c' = effective soil cohesion
σ_D' = vertical effective stress at depth D below the ground surface
($\sigma_D' = \gamma D$ if depth to groundwater table is greater than D)
γ' = effective unit weight of the soil ($\gamma' = \gamma$ if the groundwater table is very deep; see discussion later in this section for shallow groundwater conditions)
D = depth of footing below ground surface
B = width (or diameter) of footing
N_c, N_q, N_γ = bearing capacity factors = $f(\phi')$ — see Table 17.1

Terzaghi's equations also may be used in a total stress analysis. In that case, substitute c_T, ϕ_T, and σ_D for c', ϕ', and σ_D'. If saturated undrained conditions exist, we may conduct a total stress analysis with the shear strength defined as $c_T = s_u$ and $\phi_T = 0$. In this case, $N_c = 5.7$, $N_q = 1.0$, and $N_c = 0.0$.

TABLE 17.1 BEARING CAPACITY FACTORS FOR TERZAGHI'S EQUATIONS

ϕ' (deg)	N_c	N_q	N_γ	ϕ' (deg)	N_c	N_q	N_γ
0	5.7	1.0	0.0	20	17.7	7.4	4.4
1	6.0	1.1	0.1	21	18.9	8.3	5.1
2	6.3	1.2	0.1	22	20.3	9.2	5.9
3	6.6	1.3	0.2	23	21.7	10.2	6.8
4	7.0	1.5	0.3	24	23.4	11.4	7.9
5	7.3	1.6	0.4	25	25.1	12.7	9.2
6	7.7	1.8	0.5	26	27.1	14.2	10.7
7	8.2	2.0	0.6	27	29.2	15.9	12.5
8	8.6	2.2	0.7	28	31.6	17.8	14.6
9	9.1	2.4	0.9	29	34.2	20.0	17.1
10	9.6	2.7	1.0	30	37.2	22.5	20.1
11	10.2	3.0	1.2	31	40.4	25.3	23.7
12	10.8	3.3	1.4	32	44.0	28.5	28.0
13	11.4	3.6	1.6	33	48.1	32.2	33.3
14	12.1	4.0	1.9	34	52.6	36.5	39.6
15	12.9	4.4	2.2	35	57.8	41.4	47.3
16	13.7	4.9	2.5	36	63.5	47.2	56.7
17	14.6	5.5	2.9	37	70.1	53.8	68.1
18	15.5	6.0	3.3	38	77.5	61.5	82.3
19	16.6	6.7	3.8	39	86.0	70.6	99.8

Although subsequent work has produced formulas that are more versatile and slightly more precise, Terzaghi's formulas are still widely used, and are adequate for many practical design problems.

Groundwater Effects

Apparent Cohesion

Sometimes soil samples obtained from the exploratory borings are not saturated, especially if the site is in an arid or semi-arid area. These soils have additional shear strength due to the presence of apparent cohesion, as discussed in Chapter 13. However, this additional strength will disappear if the moisture content increases. Water may come from landscape irrigation, rainwater infiltration, leaking pipes, rising groundwater, or other sources. Therefore, we do not rely on the strength due to apparent cohesion.

In order to remove the apparent cohesion effects and simulate the "worst case" condition, geotechnical engineers usually wet the samples in the lab prior to testing. This may be done by simply soaking the sample, or, in the case of the triaxial test, by backpressure saturation. However, even with these precautions, the cohesion measured in the laboratory test may still include some apparent cohesion. Therefore, we often perform bearing capacity computations using a cohesion value less than that measured in the laboratory.

Pore Water Pressure

If there is enough water in the soil to develop a groundwater table, and this groundwater table is within the potential shear zone, then pore water pressures will be present, the effective stress and shear strength along the failure surface will be smaller, and the ultimate bearing capacity will be reduced (Meyerhof, 1955). We must consider this effect when conducting bearing capacity computations.

When exploring the subsurface conditions, we determine the current location of the groundwater table and worst-case (highest) location that might reasonably be expected during the life of the proposed structure. We then determine which of the following three cases describes the worst-case field conditions:

- Case 1: $D_w \leq D$
- Case 2: $D < D_w < D + B$
- Case 3: $D + B \leq D_w$

All three cases are shown in Figure 17.6.

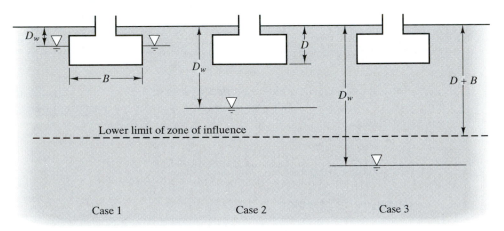

Figure 17.6 Three groundwater cases for bearing capacity analyses.

We account for the decreased effective stresses along the failure surface by adjusting the effective unit weight, γ', in the third term of Equations 17.5 - 17.7 (Vesić, 1973). The effective unit weight is the value that, when multiplied by the appropriate soil thickness, will give the vertical effective stress. It varies between the buoyant unit weight, γ_b, and the unit weight, γ, depending on the position of the groundwater table. We compute γ' as follows:

For case 1 ($D_w \leq D$):

$$\gamma' = \gamma_b = \gamma - \gamma_w$$

(17.8)

For case 2 ($D < D_w < D+B$):

$$\gamma' = \gamma - \gamma_w\left(1 - \left(\frac{D_w - D}{B}\right)\right)$$
(17.9)

For case 3 ($D+B \leq D_w$; no groundwater correction is necessary):

$$\gamma' = \gamma$$
(17.10)

In case 1, the second term in the bearing capacity formulas also is affected, but the appropriate correction is implicit in the computation of σ_D'.

If a total stress analysis is being performed, do not apply any groundwater correction because the groundwater effects are supposedly implicit within the values of c_T and ϕ_T. In this case, simply use $\gamma' = \gamma$ in the bearing capacity equations, regardless of the groundwater table position.

Allowable Bearing Capacity

We compute the *allowable bearing capacity, q_a,* using:

$$q_a = \frac{q_{ult}}{F}$$
(17.11)

The required factor of safety, F, depends on the type of structure, the type of soil, and other factors, and typically is between 2.0 and 3.5. Low factors of safety might be used for non-critical structures on sandy soils with extensive site characterization, while high factors of safety would more often be used for critical structures on clayey soils with minimal site characterization.

We then satisfy bearing capacity requirements by designing the footing such that $q \leq q_a$. Typically, P, c, ϕ, γ, and the groundwater conditions are fixed, so the only parameters we can vary are the footing dimensions B and D. If the soil is homogeneous, increasing D generally has very little impact, so we usually satisfy bearing capacity requirements by specifying a minimum required footing width, B. Normally both B and D are expressed as multiples of 100 mm or 3 in.

Example 17.1

Compute the factor of safety against a bearing capacity failure for the square spread footing shown in Figure 17.7 with the groundwater table at Position A.

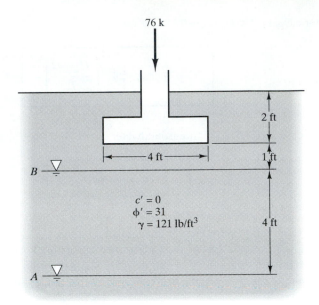

Figure 17.7 Proposed spread footing for Example 17.1.

Solution

$$\sigma'_D = \gamma D - u = (121\,\text{lb/ft}^3)(2\,\text{ft}) - 0 = 242\,\text{lb/ft}^2$$

$$D = 2\,\text{ft}$$
$$D_w = 7\,\text{ft}$$
$$D + B = 6\,\text{ft}$$

$D + B \leq D_w$, so groundwater case 3 applies — $\gamma' = \gamma$

Per Table 17.1 — $N_c = 40.4$, $N_q = 25.3$, $N_\gamma = 23.7$ when $\phi' = 31°$

$$
\begin{aligned}
q_{ult} &= 1.3c'N_c + \sigma'_D N_q + 0.4\gamma'BN_\gamma \\
&= 0 + (242\,\text{lb/ft}^2)(25.3) + 0.4(121\,\text{lb/ft}^2)(4\,\text{ft})(23.7) \\
&= 10{,}710\,\text{lb/ft}^2
\end{aligned}
$$

$$q = \frac{P}{A} + \gamma_c D - u = \frac{76{,}000\,\text{lb}}{(4\,\text{ft})^2} + (150\,\text{lb/ft}^3)(2\,\text{ft}) - 0 = 5050\,\text{lb/ft}^2$$

Because we are computing the factor of safety for a given bearing pressure, we rewrite Equation 17.11 in terms of q instead of q_a:

$$F = \frac{q_{ult}}{q} = \frac{10{,}710\,\text{lb/ft}^2}{5{,}050\,\text{lb/ft}^2} = \mathbf{2.1} \qquad \Leftarrow \textit{Answer}$$

Comments

For clean sands, a factor of safety of 2.1 would probably be marginally acceptable.

Example 17.2

Sometime after construction, the groundwater table in Example 17.1 rose to Position B. Compute the new factor of safety against a bearing capacity failure.

Solution

$$D = 2 \text{ ft}$$
$$D_w = 3 \text{ ft}$$
$$D + B = 6 \text{ ft}$$

$D < D_w < D + B$, so groundwater case 2 applies

$$\gamma' = \gamma - \gamma_w \left(1 - \left(\frac{D_w - D}{B} \right) \right)$$
$$= 121 \text{ lb/ft}^2 - 62.4 \text{ lb/ft}^2 \left(1 - \left(\frac{3 \text{ ft} - 2 \text{ ft}}{4 \text{ ft}} \right) \right)$$
$$= 74.2 \text{ lb/ft}^2$$

$$q_{ult} = 1.3 c' N_c + \sigma_D' N_q + 0.4 \gamma' B N_\gamma$$
$$= 0 + (242 \text{ lb/ft}^2)(25.3) + 0.4(74.2 \text{ lb/ft}^2)(4 \text{ ft})(23.7)$$
$$= 8{,}936 \text{ lb/ft}^2$$

$$F = \frac{q_{ult}}{q} = \frac{8{,}936 \text{ lb/ft}^2}{5{,}050 \text{ lb/ft}^2} = 1.8 \quad \Leftarrow Answer$$

Comments

The rising groundwater table has dropped the factor of safety to 1.8. Although this is still above the theoretical failure value of 1.0, it is less than the minimum acceptable F.

Example 17.3

A 1350 kN column load is to be supported on a square spread footing founded in a clay with $s_u = 150$ kPa. The depth of embedment, D, will be 500 mm, and the soil has a unit weight of 18.5 kN/m³. The groundwater table is at a depth below the bottom of the footing. Using a factor of safety of 3.0, determine the required footing width.

Solution

Per Table 17.1, $N_c = 5.7$, $N_q = 1.0$, $N_\gamma = 0.0$ for $\phi = 0$

$$\sigma_D' = \gamma D - u = (18.5 \text{ kN/m}^3)(0.5 \text{ m}) - 0 = 9 \text{ kPa}$$

$$
\begin{aligned}
q_{ult} &= 1.3 c N_c + \sigma_D N_q + 0.4 \gamma' B N_\gamma \\
&= 1.3(150 \text{ kPa})(5.7) + (9 \text{ kPa})(1.0) + 0 \\
&= 1121 \text{ kPa}
\end{aligned}
$$

$$q_a = \frac{q_{ult}}{F} = \frac{1121 \text{ kPa}}{3.0} = 374 \text{ kPa}$$

$$q_a = q = \frac{P}{B^2} + \gamma_c D - u$$

$$374 \text{ kPa} = \frac{1350 \text{ kN}}{B^2} + (23.6 \text{ kN/m}^3)(0.5 \text{ m}) - 0$$

$$B = 1.93 \text{ m}$$

Normally spread footing dimensions are expressed as multiples of 3 in or 100 mm. Therefore, round up to **B = 2.0 m.**   ← *Answer*

Example 17.4

A bearing wall for a proposed building is to be supported on a 24 in deep continuous footing founded in an unsaturated clayey sand (SC). The load from this wall will be 4.0 k/ft, and the soil has $c' = 100$ lb/ft^2, $\phi' = 28°$, and $\gamma = 119$ lb/ft^3. The groundwater table is at a very great depth. Compute the required footing width to maintain a factor of safety of 3.0 against a bearing capacity failure.

Solution

Per Table 17.1, $N_c = 31.6$, $N_q = 17.8$, $N_\gamma = 14.6$ for $\phi = 28°$

$$\sigma_D' = \gamma D - u = (119 \text{ lb/ft}^3)(2 \text{ ft}) - 0 = 238 \text{ lb/ft}^2$$

$$
\begin{aligned}
q_{ult} &= c' N_c + \sigma_D' N_q + 0.5 \gamma' B N_\gamma \\
&= (100 \text{ lb/ft}^3)(31.6) + (238 \text{ lb/ft}^3)(17.8) + 0.4(119 \text{ lb/ft}^3) B (14.6) \\
&= 7396 \text{ lb/ft}^2 + 695 B
\end{aligned}
$$

$$q_a = \frac{q_{ult}}{F} = \frac{7396 \text{ lb/ft}^2 + 695\,B}{3.0} = 2465 \text{ lb/ft}^2 + 232\,B$$

$$q = \frac{P/b}{B} + \gamma_c D - u = \frac{4000 \text{ lb/ft}}{B} + (150 \text{ lb/ft}^3)(2 \text{ ft}) - 0$$

Setting $q = q_a$ and solving for B gives $B = 1$ ft 9 in $\Leftarrow$ *Answer*

QUESTIONS AND PRACTICE PROBLEMS

17.1 A 5 ft square, 3 ft deep footing supports a column load of 110 k. The groundwater table is at a depth greater than 3 ft. Compute the bearing pressure.

17.2 An 800 mm wide, 400 mm deep continuous footing supports a wall load of 120 kN/m. The groundwater table is at a great depth. Compute the bearing pressure.

17.3 A proposed column is to be supported by a 1.5 m wide, 0.5 m deep square footing. The soil beneath this footing is a silty sand with $c' = 0$, $\phi' = 29°$, and $\gamma = 18.0 \text{ kN/m}^3$. The groundwater table is at a depth of 10 m below the ground surface. The factor of safety against a bearing capacity failure must be at least 2.75. Compute the maximum allowable column load.

17.4 A 39 inch wide, 24 inch deep continuous footing supports a wall load of 12 k/ft. This footing is underlain by a fine-to-medium sand with $c' = 0$, $\phi' = 31°$, and $\gamma = 122 \text{ lb/ft}^3$. The groundwater table is currently at a depth of 10 ft below the ground surface, but could rise to 4 ft below the ground surface during the life of the project. The factor of safety against a bearing capacity failure must be at least 3.0. Is the design acceptable? Provide computations to justify your answer. Comment on any special considerations.

17.5 A 949 kN column load is to be supported on a square spread footing that will be underlain by a clayey silt with $s_u = 125$ kPa and $\gamma = 18.0 \text{ kN/m}^3$. The bottom of this footing will be 1.0 m below the ground surface, and the groundwater table is more than 30 m below the ground surface. Using a factor of safety of 3.0, compute the required footing width.

17.6 A proposed cylindrical steel water tank is to be built on a medium clay that has an undrained shear strength of 31 kPa. The tank diameter will be 35.0 m, and it will contain 10.0 m of water. Its empty mass will be 253,000 kg. Assuming both the weight of the empty tank and that of the water are spread evenly along the bottom, compute the factor of safety against a bearing capacity failure. Is this factor of safety acceptable? If not, how could the design be modified to provide an acceptable F? Use $D = 0$.

Note: Although a ring footing would be present along the perimeter of this tank to support the weight of the walls, the live load (i.e., the weight of the water) is spread evenly across the bottom of the tank. This live load is a large fraction of the total load, so the bearing capacity analysis should be based on a circular load with a diameter equal to the diameter of the tank.

17.4 SPREAD FOOTINGS — SETTLEMENT

The structural loads applied to spread footings increase the vertical effective stress in the soils below, thus causing the footings to settle. Footings must be designed so this settlement does not exceed the tolerable settlement, thus protecting the structure from excessive movement. This criteria is called a *serviceability requirement* because it is controlled by the ability of the structure to perform properly, not by a threat of catastrophic failure.

There are two settlement requirements for structural foundations:

Total settlement, δ, is the change in footing elevation from the original unloaded position to the final loaded position. For buildings, the allowable total settlement, δ_a, depends on the need for maintenance of smooth pedestrian and vehicle access, avoidance of utility line shearing, maintenance of proper surface drainage, aesthetics, and other considerations.

Differential settlement, δ_D, is the difference in total settlement between two foundations or between two points on a single foundation. These differences are due to non-uniformities in the soil, differences in the structural loads, construction tolerances, and other factors. For buildings, the allowable differential settlement, δ_{Da}, depends on the ability of doors, windows, and elevators to operate if the building becomes distorted, the potential for cracks in the structure, aesthetics, and other similar concerns.

The structural engineer usually determines the maximum allowable total and differential settlements, then the geotechnical engineer performs settlement analyses to determine how to design the footings so the actual settlements do not exceed the allowable settlements ($\delta \leq \delta_a$ and $\delta_D \leq \delta_{Da}$). These analysis methods predict the total settlement, δ, based on the loads, the soil properties, and the footing geometry. The differential settlement expected to occur in the field is taken as some percentage of the total settlement or, in the case of erratic soil profiles, as the difference in results between two total settlement analyses.

If the settlement of a proposed footing is excessive, the design must be modified accordingly, usually by increasing B (thus decreasing q). Settlement requirements often dictate a larger B than needed to satisfy bearing capacity requirements, and the final design must use the larger of the B values obtained from these two analyses.

Footings on Clays and Silts

Settlement analyses for footings on clays and silts are similar to those described in Chapter 11. However, there are two additional issues that need to be considered: the computation of σ_{zf}' and the flexural rigidity of the footing.

Computation of σ_{zf}'

The final vertical effective stress, σ_{zf}', in the soil beneath the center of a spread footing is:

$$\sigma_{zf}' = \sigma_{z0}' + (\sigma_z)_{induced} \qquad (17.12)$$

where:

σ_{z0}' = initial vertical effective stress beneath center of footing

σ_{zf}' = final vertical effective stress beneath center of footing

$(\sigma_z)_{induced}$ = induced vertical stress beneath center of footing

When computing $(\sigma_z)_{induced}$, be sure to consider the column or wall load, the weight of the footing, and the weight of the excavated soil. In other words, we need to use the net increase in stress. For spread footings, this is most easily done by modifying Equations 10.25–10.28 as follows:

For circular footings:

$$(\sigma_z)_{induced} = (q - \sigma_D') \left[1 - \left(\frac{1}{1 + \left(\dfrac{B}{2z_f} \right)^2} \right)^{1.50} \right] \qquad (17.13)$$

For square footings:

$$(\sigma_z)_{induced} = (q - \sigma_D') \left[1 - \left(\frac{1}{1 + \left(\dfrac{B}{2z_f} \right)^2} \right)^{1.76} \right] \qquad (17.14)$$

For continuous footings:

$$(\sigma_z)_{induced} = (q - \sigma_D') \left[1 - \left(\frac{1}{1 + \left(\dfrac{B}{2z_f} \right)^{1.38}} \right)^{2.60} \right] \qquad (17.15)$$

For rectangular footings:

$$(\sigma_z)_{induced} = (q - \sigma_D') \left[1 - \left(\frac{1}{1 + \left(\dfrac{B}{2z_f} \right)^{1.38 + 0.62B/L}} \right)^{2.60 - 0.84 B/L} \right] \tag{17.16}$$

where:

$(\sigma_z)_{induced}$ = induced vertical stress beneath the center of footing

q = bearing pressure

σ_D' = vertical effective stress at depth D below the ground surface

B = width or diameter of footing

L = length of footing

z_f = depth from bottom of footing to point where stress is to be computed

The σ_D' term in Equations 17.13–17.16 accounts for the effective weight of the excavated soil, while the q term accounts for the column load and the weight of the footing.

Flexural Rigidity

The value of $(\sigma_z)_{induced}$ at a given depth z_f below the centerline of a loaded area is always greater than its value at the same depth below the edge of the loaded area. This difference is illustrated by the curves in Figures 10.9 and 10.10. Because the consolidation settlement depends on $(\sigma_z)_{induced}$, there will be more consolidation settlement below the center than below the edges. However, this conclusion is based on the assumption that the loaded area is perfectly flexible, and thus is able to settle more in center than at the edges. A steel tank such as the one described in Problem 17.6 is an example of such a load. However, a spread footing foundation is far from being flexible. Its structural rigidity is such that the settlement at the center will be essentially equal to that at the edge. Therefore, the actual settlement of a spread footing will be greater than that beneath the edge of a perfectly flexible load, but less than that beneath the center, as shown in Figure 17.8.

We account for this behavior by computing the settlement beneath the center of a perfectly flexible load, then applying a rigidity factor, r, of 0.85. In other words, the settlement of a footing is about 85 percent of the settlement at the center of a perfectly flexible loaded area that has the same dimensions and the same q.

Settlement Computations

Settlement computations for spread footings often consider only consolidation settlement, and assume the consolidation is one-dimensional. This is the method that will be presented here. *Foundation Design: Principles and Practices* presents an alternative method that considers both consolidation and distortion settlements, and three-dimensional effects. In some cases, secondary compression settlement also may be important.

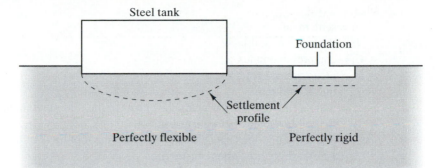

Figure 17.8 Influence of footing rigidity on settlement. The steel tank on the left is very flexible, so the center settles more than the edge. Conversely, the reinforced concrete footing on the right is very rigid, and thus settles uniformly.

The settlement equations are the same as Equations 11.23–11.25, except they now include the rigidity factor of 0.85:

For normally consolidated soils ($\sigma_{z0}' \approx \sigma_c'$):

$$\delta \ = \ 0.85 \sum H \ \frac{C_c}{1 + e_0} \ \log\left(\frac{\sigma_{zf}'}{\sigma_{z0}'} \right) \qquad (17.17)$$

For overconsolidated soils — case I ($\sigma_{z0}' < \sigma_{zf}' \leq \sigma_c'$):

$$\delta \ = \ 0.85 \sum H \ \frac{C_r}{1 + e_0} \ \log\left(\frac{\sigma_{zf}'}{\sigma_{z0}'} \right) \qquad (17.18)$$

For overconsolidated soils — case II ($\sigma_{z0}' < \sigma_c' < \sigma_{zf}'$):

$$\delta \ = \ 0.85 \sum H \left[\frac{C_r}{1 + e_0} \ \log\left(\frac{\sigma_c'}{\sigma_{z0}'} \right) + \frac{C_c}{1 + e_0} \ \log\left(\frac{\sigma_{zf}'}{\sigma_c'} \right) \right] \qquad (17.19)$$

where:

δ = settlement

H = thickness of soil layer

C_c = compression index

C_r = recompression index

e_0 = initial void ratio

σ_{z0}' = initial effective stress beneath the center of the footing at midheight of the soil layer (i.e., before construction)

σ_{zf}' = final effective stress beneath the center of the footing at midheight of the soil layer (i.e., after construction)

σ_c' = preconsolidation stress at midheight of the soil layer

TABLE 17.2 APPROXIMATE THICKNESSES OF SOIL LAYERS FOR MANUAL COMPUTATION OF SPREAD FOOTING SETTLEMENT

Layer Number	Approximate Layer Thickness	
	Square Footing	Continuous Footing
1	$B/2$	B
2	B	$2B$
3	$2B$	$4B$

1. Adjust the number and thickness of the layers to account for changes in soil properties. Locate each layer entirely within one soil stratum.
2. For rectangular footings, use layer thicknesses between those given for square and continuous footings.
3. Use somewhat thicker layers (perhaps up to 1.5 times the thicknesses shown) if the groundwater table is very shallow.
4. For quick, but less precise, analyses, use a single layer with a thickness of about $3B$ (square footings) or $6B$ (continuous footings).

To apply these equations, divide the soil beneath the footing into layers, compute the settlement of each layer, and sum. As the number of layers becomes smaller, the precision of the computed settlement becomes greater, with a corresponding increase in computational effort. For hand computations, three layers are usually sufficient. Table 17.2 presents guidelines for selecting the layer thicknesses. For computer-based analyses, such as the one described later in this chapter, many more layers may be used, with a resulting increase in precision.

Example 17.5

Compute the settlement of the spread footing in Figure 17.9. The allowable total settlement, δ_a, is 20 mm.

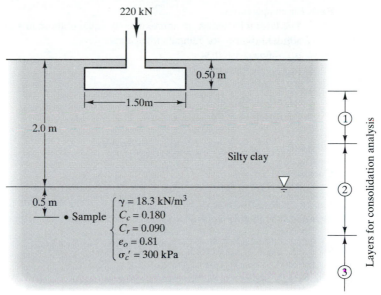

Figure 17.9 Proposed spread footing for Example 17.5.

Solution

$$(\sigma'_{z0})_{sample} = \sum \gamma H - u = (18.3 \text{ kN/m}^3)(2.5 \text{ m}) - (9.8 \text{ kN/m}^3)(0.5 \text{ m}) = 41 \text{ kPa}$$

$$\sigma'_m = \sigma'_c - \sigma'_{z0} = 300 \text{ kPa} - 41 \text{ kPa} = 259 \text{ kPa}$$

$$\sigma'_D = \gamma D - u = (18.3 \text{ kN/m}^3)(0.5 \text{ m}) - 0 = 9 \text{ kPa}$$

$$q = \frac{P}{A} + \gamma_c D - u = \frac{220 \text{ kN}}{(1.5 \text{ m})^2} + (23.6 \text{ kN/m}^3)(0.5 \text{ m}) - 0 = 110 \text{ kPa}$$

| Layer | H (m) | At midheight of layer | | | | | Case | $\dfrac{C_c}{1+e_0}$ | $\dfrac{C_r}{1+e_0}$ | δ (mm) |
		z_f (m)	σ'_{z0} (kPa)	Induced σ_z (kPa)	σ'_{zf} (kPa)	σ'_c (kPa)				
1	0.8	0.4	16	94	110	275	OC-I	0.10	0.05	28
2	1.6	1.6	37	30	67	296	OC-I	0.10	0.05	18
3	3.2	4.0	58	6	64	317	OC-I	0.10	0.05	6
									$\Sigma =$	52

Basis for computations:

> The layer thicknesses are approximately equal to those in Table 17.2. They have been adjusted slightly for computational convenience
> σ_{z0}' – Equation 10.34
> $(\sigma_z)_{induced}$ – Equation 17.14
> σ_{zf}' – Equation 17.12
> σ_c' – Equation 11.17
> δ – Equation 17.18

The computed settlement is **52 mm**   ← *Answer*

Comments

The computed settlement of 52 mm is greater than the allowable total settlement of 20 mm. Therefore, the footing width B needs to be increased until the settlement criteria is met.

Example 17.6

Compute the settlement of the spread footing in Figure 17.10. The allowable total settlement is 1.0 in.

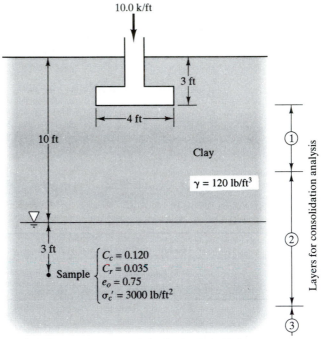

Figure 17.10 Proposed spread footing for Example 17.6.

Solution

$$(\sigma_{z0}')_{sample} = \sum \gamma H - u = (121 \text{ lb/ft}^3)(13 \text{ ft}) - (62.4 \text{ lb/ft}^3)(3 \text{ ft}) = 1390 \text{ lb/ft}^2$$

$$\sigma_m' = \sigma_c' - \sigma_{z0}' = 3000 \text{ lb/ft}^2 - 1390 \text{ lb/ft}^2 = 1610 \text{ lb/ft}^2$$

$$\sigma_D' = \gamma D - u = (121 \text{ lb/ft}^3)(3 \text{ ft}) - 0 = 363 \text{ lb/ft}^2$$

$$q = \frac{P/b}{B} + \gamma_c D - u = \frac{10,000 \text{ lb/ft}}{4 \text{ ft}} + (150 \text{ lb/ft}^3)(3 \text{ ft}) - 0 = 2950 \text{ lb/ft}^2$$

Note: In this case, the computed value of q is slightly conservative because the excavation has been partially filled with concrete ($\gamma = 150 \text{ lb/ft}^3$) and partly with soil ($\gamma = 121 \text{ lb/ft}^3$).

| Layer | H (ft) | At midheight of layer | | | | | Case | $\dfrac{C_c}{1+e_0}$ | $\dfrac{C_r}{1+e_0}$ | δ (in) |
		z_f (ft)	σ_{z0}' (lb/ft²)	Induced σ_z (lb/ft²)	σ_{zf}' (lb/ft²)	σ_c' (lb/ft²)				
1	4.0	2.0	600	2160	2760	2210	OC-II	0.069	0.020	0.73
2	8.0	8.0	1270	778	2048	2880	OC-I	0.069	0.020	0.34
3	16.0	20.0	1970	261	2231	3580	OC-I	0.069	0.020	0.18
									$\Sigma=$	1.25

The computed settlement is **1.2 in**   ⟵ *Answer*

Basis for computations:
σ_{z0}' – Equation 10.34
$(\sigma_z)_{induced}$ – Equation 17.15
σ_{zf}' – Equation 17.12
σ_c' – Equation 11.17
δ – Equation 17.18 and 17.19

Comments

The computed settlement of 1.2 in is greater than the allowable total settlement of 1.0 in, so the current design is not acceptable. This problem could be resolved by increasing B. However, another alternative would be to excavate and recompact the upper 7 ft of soil (i.e., from the ground surface to the bottom of Layer 1) before building the footing. If this treatment raises the overconsolidation margin to at least 2500 lb/ft², then Layer 1 will change to OC-I and the computed settlement will decrease to 1.0 in. This treatment also may decrease C_c, which would further decrease the settlement.

Footings on Sands and Gravels

In theory, the methods used to predict settlement of spread footings on clays and silts also could be used for sands and gravels. However, to use these methods, we would need to evaluate C_c and C_r in these soils, which would be very difficult or impossible because of the difficulties in obtaining undisturbed samples. In Chapter 11 we overcame this problem by using empirical correlations with the relative density (see Table 11.3). This method also could be used with spread footings, but we choose to take a different approach.

If standard penetration test (SPT) data is available from the field, we can predict the settlement of spread footings using direct empirical correlations. Several are available, including the modified Meyerhof formulas (Meyerhof, 1965):

For $B \leq 4$ ft (1.2 m):

$$\delta = \frac{0.0027 (q - \sigma_D')}{\bar{N}_{60} K_d}$$

(17.20-English)

$$\delta = \frac{1.3 (q - \sigma_D')}{\bar{N}_{60} K_d}$$

(17.20-SI)

For $B > 4$ ft (1.2 m):

$$\delta = \frac{0.0040 (q - \sigma_D')}{\bar{N}_{60} K_d} \left(\frac{B}{B+1} \right)^2$$

(17.21-English)

$$\delta = \frac{2.0 (q - \sigma_D')}{\bar{N}_{60} K_d} \left(\frac{B}{B+0.3} \right)^2$$

(17.21-SI)

where:

δ = settlement (in; mm)

q = bearing pressure (lb/ft^2; kPa)

σ_D' = vertical effective stress at depth D below the ground surface (lb/ft^2; kPa)

$\bar{N}_{60}$ = average SPT N_{60} value between the bottom of the footing and a depth $2B$ below the bottom

B = footing width (ft; m)

K_d = depth factor = $1 + 0.33 D/B \leq 1.33$

Note: When using Equations 17.20 and 17.21, you must express all variables using the units listed above. The unit conversion factors are built into the equations.

Do not correct the field N_{60} values for overburden, but do adjust them using Equation 17.22 when the soil is a dense silty sand below the groundwater table and $N_{60} > 15$.

$$N_{60\,adjusted} = 15 + 0.5\,(N_{60\,field} - 15) \qquad (17.22)$$

Meyerhof suggested the groundwater table effects would be implicitly incorporated into the SPT results. However, consider adjusting the measured N_{60} values if the sand was dry during testing but may become saturated later.

These formulas are simple to implement, but suffer from the inevitable errors associated with the standard penetration test. Nevertheless, their precision is suitable for many practical problems. For more important foundations, a more precise analysis may be performed using Schmertmann's method as discussed in *Foundation Design: Principles and Practices*. This method usually uses cone penetration test (CPT) data, but may be based on other tests.

Example 17.7

A 200 kN column load is to be supported on a 0.5 m deep footing in the sandy soil shown in Figure 17.11. The maximum allowable settlement is 20 mm. Determine the required footing width.

Solution

$$\sigma'_D = \gamma D - u$$
$$= (20.0 \text{ kN/m}^3)(0.5\,\text{m}) - 0$$
$$= 10 \text{ kPa}$$

Try $B = 1.0$ m

$$K_d = 1 + 0.33\,D/B = 1 + 0.33\,(0.5/1.0) = 1.17$$

$$\bar{N}_{60} = \frac{19 + 22}{2} = 21$$

$$\delta = \frac{1.3\,(q - \sigma'_D)}{\bar{N}_{60}\,K_d}$$
$$20 = \frac{1.3\,(q - 10\,\text{kPa})}{(21)(1.17)} \quad \rightarrow \quad q = 388 \text{ kPa}$$

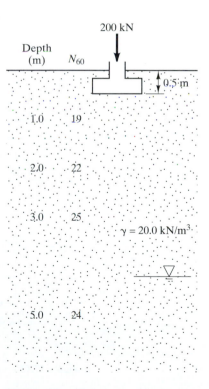

Figure 17.11 Proposed spread footing for Example 17.7.

$$q = \frac{P}{B^2} + \gamma_c D - u$$

$$388 \text{ kPa} = \frac{200 \text{ kN}}{B^2} + (23.6 \text{ kN/m}^3)(0.5 \text{ m}) - 0$$

$$B = 0.73 \text{ m}$$

The computed B of 0.73 m is close to the assumed value of 1.00 m. It is not necessary to run the computations again with a new assumed B, because doing so would produce only a small change in K_d and no change in $\overline{N}_{60}$.

Footing widths are normally expressed as multiples of 3 in or 100 mm, so the design width is **800 mm**.

QUESTIONS AND PRACTICE PROBLEMS

17.7   The footing described in Example 17.5 has been redesigned so B now equals 2.50 m. The column load and depth of embedment remain the same. Compute the new settlement, δ. Does this new design satisfy the allowable settlement criteria described in Example 17.5?

17.8   The proposed footing shown in Figure 17.12 has an allowable total settlement of 1.0 in. Compute the settlement and determine if it meets this criteria.

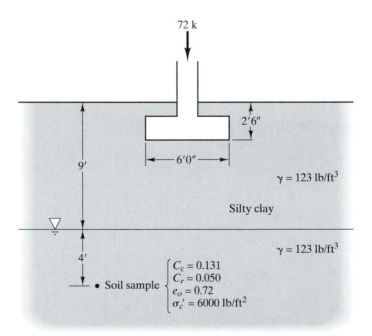

Figure 17.12 Proposed spread footing for Problem 17.8.

17.9 A proposed 3'6" wide continuous footing is to be built at the site described in **Problem 17.8**. The depth, D, the bearing pressure, q, and the allowable settlement, δ_a, are the same as before. Compute the predicted settlement and determine if it meets the settlement criteria.

17.10 Compute the settlement of the proposed footing shown in Figure 17.13.

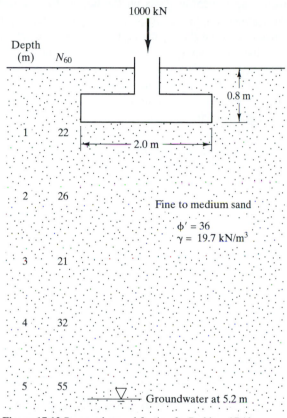

Figure 17.13 Proposed spread footing for Problem 17.10.

17.11 According to the structural engineer, the footing described in Problem 17.10 must not settle more than 25 mm. Determine the required B that produces the most economical design.

17.5 SPREAD FOOTINGS — SYNTHESIS AND DESIGN

Most structures include footings of various sizes, all of which need to satisfy both bearing capacity and settlement criteria. It is generally not practical to perform individual analyses for each footing, so engineers often assign a single *allowable bearing pressure*, q_A, to the entire site. This is the maximum bearing pressure that satisfies both bearing capacity and settlement requirements for all footings at the site. Notice the difference between this value and the allowable bearing capacity, q_a, which satisfies only bearing capacity requirements.

We develop q_A as follows:

1. Perform a bearing capacity analysis on the smallest footing to be built. This footing has the smallest B, so it will produce the lowest allowable bearing capacity, q_a, and thus will be conservative when applied to the larger footings.
2. Perform a settlement analysis on the largest footing to be built and determine the bearing pressure, q, that corresponds to the desired maximum settlement. This value will determine what B is necessary to satisfy settlement criteria for this footing. If the smaller footings also are designed for this q, their settlement will be less than the large footing, so again this is conservative.
3. Select the lowest value from Steps 1 and 2. This is the allowable bearing pressure, q_A. Usually it is expressed as a multiple of 25 kPa or 500 lb/ft². All footings may then be sized using this value such that:

For square, rectangular, and circular footings:

$$q = \frac{P}{A} + \gamma_c D - u \leq q_A \qquad (17.23)$$

Setting $q = q_A$ and rewriting gives:

$$A = \frac{P}{q_A - \gamma_c D + u} \qquad (17.24)$$

For continuous footings:

$$B = \frac{P/b}{q_A - \gamma_c D + u} \qquad (17.25)$$

Normally, the geotechnical engineer performs this analysis and presents the q_A value and other pertinent design criteria in a geotechnical investigation report (see discussion of these reports in Section 3.11). The structural engineer then uses this value and the computed column loads to size the footings.

Sometimes allowable bearing pressures are presented in terms of the net bearing pressure instead of the gross bearing pressure. Often it is not clear which method is being used. However, in this text we will use the gross bearing pressure exclusively.

Example 17.8

A proposed warehouse will have design column loads between 50 and 300 k. Each column is to be supported on a spread footing foundation underlain by sandy soils with the following engineering properties: $c' = 0$, $\phi' = 35°$, $\gamma = 118$ lb/ft³, and $N_{60} = 18$. The minimum acceptable factor of safety against a bearing capacity failure is 2.0 and the maximum allowable settlement is 1.0 in. The groundwater table is at a depth of 50 ft and all of the footings will be located 2 ft below the ground surface. What allowable bearing pressure should be used to design them? Using this allowable bearing pressure, what is the required footing width for a column that carries a 100 k load?

Solution

Bearing capacity analysis

Perform the bearing capacity analysis on the 50 k column load, because it will produce the smallest value of q_a.

Per Table 17.1: $N_q = 41.4, N_\gamma = 47.3$

$$\sigma_D' = \gamma D - u = (118 \text{ lb/ft}^3)(2 \text{ ft}) - 0 = 236 \text{ lb/ft}^3$$

$$
\begin{aligned}
q_{ult} &= 1.3c'N_c + \sigma_D'N_q + 0.4\gamma BN_\gamma \\
&= 0 + (236 \text{ lb/ft}^2)(41.4) + 0.4(118 \text{ lb/ft}^2)B(47.3) \\
&= 9770 + 2233B
\end{aligned}
$$

$$q_a = \frac{q_{ult}}{F} = \frac{9770 + 2233B}{2} = 4885 + 1120B$$

$$
\begin{aligned}
q &= \frac{P}{A} + \gamma_c D - u \\
&= \frac{50,000 \text{ lb}}{B^2} + (150 \text{ lb/ft}^3)(2 \text{ ft}) - 0 \\
&= \frac{50,000 \text{ lb}}{B^2} + 300 \text{ lb/ft}^2
\end{aligned}
$$

Setting $q_a = q$ and solving produces $q_a = 7780 \text{ lb/ft}^2$

Settlement analysis

Perform the settlement analysis will be performed on the 300 k column using the modifed Meyerhof's method.

For purposes of computing K_d, assume $B = 6$ ft.

$$K_d = 1 + 0.33 D/B = 1 + 0.33(2/6) = 1.11$$

$$\delta = \frac{0.0040\,(q - \sigma'_D)}{\bar{N}_{60}\,K_d}\left(\frac{B}{B+1}\right)^2$$

$$1 = \frac{0.0040\,[(300{,}000/B^2 + 300\ \text{lb/ft}^2) - 236\ \text{lb/ft}^2]}{(18)(1.11)}\left(\frac{B}{B+1}\right)^2$$

This equation solves to $B = 7$ ft 0 in, which corresponds to $q = 6422$ lb/ft^2. There is no need to go back and recompute K_d.

Synthesis and Conclusion

The settlement analysis produced a lower q, so it governs the analysis. Round off to a multiple of 500 lb/ft^2:

Design all footings using $q_A = 6500$ **lb/ft**^{2} ⇐ *Answer*

Sample Application

For a column carrying 100 k:

$$A = B^2 = \frac{P}{q_A - \gamma_c D + u} = \frac{100{,}000\ \text{lb}}{6500\ \text{lb/ft}^2 - (150\ \text{lb/ft}^3)(2\ \text{ft}) + 0} \rightarrow B = 4.02\ \text{ft}$$

Rounding to a multiple of 3 in gives *B = 4 ft 0 in* ⇐*Answer*

Program FOOTING

Program **FOOTING**, which is part of the software package associated with this book, performs the bearing capacity and settlement analyses described in this chapter. To use this program, you must first download and install the geotechnical analysis software package onto a computer. See Appendix C for computer system requirements and installation instructions.

To run the program, select **FOOTING** from the main menu, then select the system of units (SI or English) and whether the settlement is to be computed using consolidation test results or an SPT N-value. Then type in the required information and click on the **CALCULATE** button. The program computes the bearing pressure, the factor of safety against a bearing capacity failure, and the settlement. Compare these results with the allowable values, then modify the input data (typically by changing B) and click on **CALCULATE** again. Continue this process until the design criteria are satisfied.

To obtain a printout, click on the **PRINT** button. To exit the program and return to the main menu, click on the **RETURN TO MAIN MENU** button.

Example 17.9

Solve Example 17.8 using Program **FOOTING**.

Solution

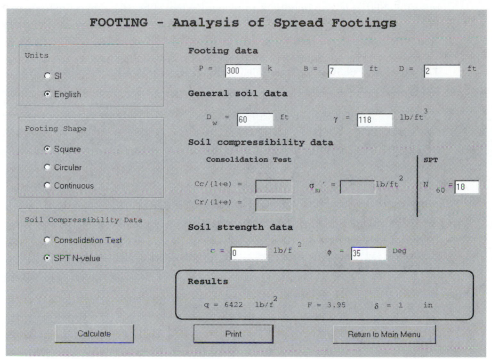

Figure 17.14 FOOTING screen for Example 17.9.

Presumptive Bearing Pressures

The most thorough and reliable method of developing design q_A values is to conduct bearing capacity and settlement computations as described in this chapter. These computations require certain soil properties, and thus can be performed only after drilling exploratory borings, conducting laboratory tests, etc. The cost of doing this work is justified on projects where the subsurface conditions are unknown and/or the proposed structure is heavy.

However, for lightweight structures, such as one or two-story wood frame buildings, to be constructed on sites that are known to be good, it may not be cost-effective to drill borings and conduct lab tests. In such cases, engineers often use *presumptive bearing pressures*, which are q_A values obtained directly from building codes or other similar sources. Table 17.3 presents typical presumptive bearing pressures. These values are generally more conservative than those obtained from soil investigations, and thus produce larger footings. However, for small structures on good sites, the additional construction cost probably will be less than the cost of performing a detailed soil investigation. Once the presumptive bearing pressure has been obtained, the footings may be sized directly using Equation 17.24 or 17.25.

Although presumptive bearing pressures are appropriate for some projects, it is best not to use them when the column loads are greater than about 200 kN (45 k), when the structure is sensitive to differential settlements, or when the subsurface conditions are questionable. It is also important to recognize that subsurface investigations reveal other important conditions, such as expansive soils, which may not be recognized in designs based solely on presumptive bearing pressures.

TABLE 17.3 PRESUMPTIVE BEARING PRESSURES FROM THE INTERNATIONAL BUILDING CODE, FIRST DRAFT (ICC, 1997)

Soil or Rock Classification	Allowable Bearing Pressure, q_A	
	(kPa)	(lb/ft²)
Crystalline bedrock	600	12,000
Sedimentary or foliated rock	300	6,000
Sandy gravel or gravel (GW and GP)	250	5,000
Sand, silty sand, clayey sand, silty gravel, or clayey gravel (SW, SP, SM, SC, GM and GC)	150	3,000
Clay, sandy clay, silty clay, or clayey silt (CL, ML, MH, and CH)	100	2,000
Mud, organic silt, organic clay, peat, or unprepared fill	0	0

Note: Reproduced from the First Draft of the International Building Code with permission of the International Code Council, Inc., Falls Church, VA. Copyright 1997. All rights reserved. The contents of this table are tentative and subject to change. Once the final code is published, consult it for the final version of this table.

Minimum Dimensions

Regardless of the results of bearing capacity and settlement analyses, there are certain practical minimum dimensions for spread footings. These minimums are governed by construction methods, the potential for eccentric loads, and other concerns. Figure 17.15 shows minimum dimensions. In some cases, building codes or customary practice may require even larger minimum dimensions.

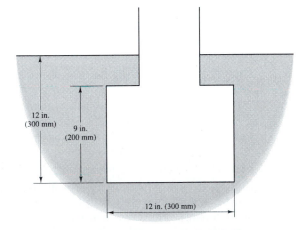

12 in. (300 mm)

9 in. (200 mm)

12 in. (300 mm)

Figure 17.15 Minimum dimensions for spread footings.

Example 17.10

A three-story wood-frame building will include a column with a design load of 21 k and is to be supported on a square footing. The soils at this site are silty sands, and the footing will have a depth of embedment of 24 inches. Compute the required footing width using the IBC presumptive bearing pressures.

Solution

According to Table 17.3, $q_A = 3,000$ lb/ft^2

$$A = B^2 = \frac{P}{q_A - \gamma_c D + u} = \frac{21,000 \text{ lb}}{3000 \text{ lb/ft}^2 + (150 \text{ lb/ft}^3)(2 \text{ ft}) - 0}$$

Solving gives $B = 2.79$ ft. This is greater than the minimum required width of 12 in (but need to check local building codes for possible additional requirements). Rounding to a multiple of 3 in gives **$B = 2$ ft 9 in** ← *Answer*

Other Geotechnical Concerns

This chapter is only an introduction to structural foundations, so our discussions have been limited to bearing capacity and settlement. However, other issues also may be important. These include:

- Frost heave — In areas with cold climates, the upper soils may heave due to the formation of ice lenses as discussed in Chapter 18. To avoid heave-induced damage, spread footings in such soils are typically founded below the frost depth.
- Expansive or collapsing soils — Some soils are subject to additional heave or collapse when wetted, and these motions can cause substantial damage to spread footings as described in Chapter 18. Special designs are often required to minimize the potential for such damage.
- Footings on or near slopes — If the ground surface is not level, special designs are necessary to maintain adequate stability.

QUESTIONS AND PRACTICE PROBLEMS

17.12 A proposed 1200 mm wide, 400 mm deep footing will be built on a sandy soil with $c' = 0$, $\phi' = 34°$, $\gamma = 20.1$ kN/m^3 and $N_{60} = 30$. The groundwater table is at a depth of 2 m below the ground surface. Using program **FOOTING**, determine the maximum allowable column load that may be placed on this footing while maintaining a factor of safety of at least 2.5 against a bearing capacity failure and a total settlement of no more than 15 mm. Which controls this design, bearing capacity or settlement?

17.13 A 103 k column load is to be supported on a square footing embedded 2.5 ft into the ground. The underlying soil is a silty clay with $C_c/(1+e_0) = 0.11$, $C_r/(1+e_0) = 0.03$, $\sigma_m' = 5000$ lb/ft^2, $\gamma = 119$ lb/ft^3, and $s_u = 2000$ lb/ft^2. The groundwater table is at a depth of 40 ft below the ground surface. The factor of safety against a bearing capacity failure must be at least 3, and the total settlement must not exceed 1 inch. Use program **FOOTING** to find the design B, then confirm the results using hand computations.

17.14 A two-story wood-frame apartment building is to be built at a site underlain by silty clay. One of the columns will impart a downward load of 62 kN on a square spread footing. Using $D = 400$ mm, determine the required footing width using the IBC presumptive bearing pressure.

17.15 A one-story masonry building is to be built at a site underlain by silty sand. The exterior walls will impart a downward load of 107 kN/m onto a 500 mm deep continuous footing. Determine the required footing width using the IBC presumptive bearing pressure.

17.16 A proposed building is to be built on the soil profile shown in Figure 17.13. The column loads will range from 250 to 1500 kN, and each will be supported on a 0.8 m deep square footing. The allowable total settlement is 25 mm, and the factor of safety against a bearing capacity failure must be at least 2.0. Using hand computations, compute a single allowable bearing pressure, q_A, that would be suitable for the design of all footings at this site, then confirm your answer using program **FOOTING**.

17.6 RECOGNIZING THE NEED FOR MORE EXTENSIVE FOUNDATIONS

Although spread footings provide suitable support for many structures, and are the most common type of foundation, there are times when they are not suitable. These include:

- Structures with loads so high and/or soil properties so poor that the footings would be very large. Usually some other type of foundation needs to be considered if the total footing area is more than 50 percent of the building footprint area.
- Locations where a footing might be undermined, such as from riverbed scour adjacent to bridge foundations or from future excavations near the foundation.
- A foundation that must penetrate through water, such as for a bridge pier.
- Requirements for a large uplift capacity (the uplift capacity of spread footings is limited to their dead weight).
- Requirements for a large lateral load capacity.

In these cases, we need to consider either a mat foundation or some type of deep foundation.

SUMMARY

Major Points

1. Structural foundations are used to transmit structural loads into the ground. There are two broad categories: Shallow foundations transmit the loads to the near-surface soils, while deep foundations transmit some or all of the loads to deeper soils.

2. Spread footings are one type of shallow foundation. It is the most common type, and the only one considered in this chapter.
3. The bearing pressure is the contact pressure between the bottom of the footing and the soil.
4. The primary geotechnical strength requirement for spread footings is called *bearing capacity*. It addresses the potential for shear failure in the soil. The allowable bearing capacity is the bearing pressure required to produce a bearing capacity failure divided by a factor of safety.
5. The primary serviceability requirement for spread footings is settlement. The allowable settlement depends on the tolerance of the structure to movements. The actual settlement depends on the loads, footing geometry, and soil conditions.
6. The allowable bearing pressure is the maximum bearing pressure that satisfies both bearing capacity and settlement requirements for all footings at a site.
7. For small structures on sites with soil conditions that are known to be good, footings may be sized using presumptive bearing pressures found in building codes.

Vocabulary

allowable bearing capacity	deep foundation	raft foundation
allowable bearing pressure	differential settlement	rectangular footing
allowable differential settlement	drilled shaft foundation	ring footing
	effective unit weight	shallow foundation
allowable total settlement	foundation	serviceability requirement
bearing capacity factor	gross bearing pressure	spread footing foundation
bearing capacity failure	mat foundation	square footing
bearing pressure	net bearing pressure	strength requirement
circular footing	pile foundation	total settlement
combined footing	presumptive bearing pressure	ultimate bearing capacity
continuous footing		

COMPREHENSIVE QUESTIONS AND PRACTICE PROBLEMS

17.17 A proposed bent for a bridge will impart a vertical load of 3100 k onto a spread footing foundation that will be embedded 6 ft into the ground. The underlying soils are dense well-graded sands with $c' = 0$, $\phi' = 37°$, $\gamma = 128$ lb/ft^3, and $N_{60} = 36$. The groundwater table is at a depth of 12 ft below the ground surface. This footing must have a factor of safety of at least 2.75 against a bearing capacity failure and a total settlement of no more than 1.5 inches. Using program **FOOTING**, determine the required footing width.

17.18 A proposed building is to be supported on a series of spread footing foundations resting on the underlying sandy clay. These foundations will be embedded to a depth of 500 mm below the ground surface. The column loads will range from 200 to 1200 kN. The allowable total settlement is 25 mm, and the factor of safety against a bearing capacity failure must be at least 3.0. The sandy clay has the following engineering properties: $C_c/(1+e_0) = 0.080$, $C_r/(1+e_0) = 0.010$, $\gamma = 19.5$ kN/m^3, $\sigma_m' = 300$ kPa, and $s_u = 200$ kPa. The groundwater table is at a depth of 5 m. Using program **FOOTING**, compute a single allowable bearing pressure, q_A, that would be suitable for the design of all footings at this site.

18

Difficult Soils

An approximate solution to the right problem is more desirable than a precise solution to the wrong problem.

U.S. Army, et al., 1971

Certain soil conditions are especially problematic and require extra effort from geotechnical engineers. These are sometimes called *difficult soils*. This chapter discusses some of the more common difficult soil conditions and examines methods of accommodating them in design and construction.

18.1 WEAK AND COMPRESSIBLE SOILS

Many construction sites are underlain by soils that are both weak and compressible. These include soft clays, highly organic soils, and others. Such soils are often found near the mouths of rivers, along the perimeter of bays, and beneath wetlands. They are prone to shear failure and excessive settlements. In addition, the areas underlain by such soils frequently are subject to flooding, so construction projects often require placing a fill to raise the ground surface to a suitable elevation. Unfortunately, the weight of such fills can cause large settlements. For example, Scheil (1979) described a building constructed on fill underlain by varved clay in the Hackensack Meadowlands of New Jersey. About 250 mm (10 in) of settlement occurred during placement of the fill, 12 mm (0.5 in) during

650

construction of the building, and an additional 100 mm (4 in) over the following ten years.

Sometimes land that is underwater is *reclaimed* by placing fills that raise it above the water level. Many waterfront cities, including Boston and San Francisco, have been extended into the water using this method. Such sites are usually underlain by soft soils that compress under the weight of such fills. In addition, many of these fills were placed many years ago using poor construction methods. For example, in the 1930s, the LaGuardia Airport in New York City was expanded into the adjacent bay by placing a fill made of incinerated refuse. This fill material is very compressible, and is underlain by a compressible organic clay deposit. As a result, parts of the airport have settled more than 2 m (7 ft)! This settlement is continuing, even more than half a century later, and poses significant problems in maintenance of runways and in construction of buildings.

Fills placed for bridge abutments can cause similar settlement problems. However, the bridge, which is probably supported on a deep foundation, generally settles much less than the approach fill, thus producing the "bump at the end of the bridge." Figure 11.2 shows the result of such a condition.

Fortunately, engineers and contractors have developed methods of coping with weak and compressible soils, and have successfully built large structures, highways, and other facilities on very poor sites. Most of these methods focus primarily on the settlement problem, because it often has the biggest impact on design. These methods include the following, either individually or in combination:

- Delay the construction of structures and other sensitive facilities until after most of the fill-induced settlement has occurred.
- Reduce the amount of settlement by using lightweight fill materials.
- Support structures on deep foundations that penetrate through the weak soils.
- Accommodate the settlement using specially designed structures or by accepting maintenance costs.
- Improve the engineering properties of the soils using special construction methods.

Delaying Construction

The consolidation process continues only until the soil reaches equilibrium under the new loading condition. Chapter 12 discussed methods of assessing the rate of consolidation and the time required to achieve a specified degree of consolidation. If the required time is not excessive, it may be economical to place the fill, then wait before constructing the buildings, roads, or other improvements.

This option does not necessarily require waiting until 100 percent consolidation has been achieved. For example, if the fill will produce 500 mm of settlement, but the proposed construction can accommodate only 100 mm, then construction can begin after 400/500 = 80% of the ultimate settlement is complete.

The primary advantage of this method is that it requires the lowest direct costs. However, in many cases the time required to achieve the required settlement is excessive, so this option often is not viable. However, there are methods of accelerating the settlement, as discussed in Chapter 19, and these methods often are cost effective.

Using Lightweight Fills

The majority of the settlement is usually due to the weight of fills placed on the site. Although these fills are usually made of soil, other materials are available that have a much lower unit weight and thus induce less settlement. These include geofoam (large blocks of styrofoam), special cementitious materials, and others, as discussed in Chapter 6. This option is generally cost effective only for small areas, such as backfills of bridge abutments.

Using Deep Foundations

If structures, such as buildings or bridges, are to be built on sites underlain by weak and compressible soils, they often must be supported on deep foundations that penetrate through these soils and into more competent underlying soils. This type of foundation isolates the structure from most of the settlement, and avoids overstressing the weak soils. Although this design may be reliable, it needs to account for the following two special problems:

1. If the site is in the process of settling, perhaps due to the weight of a fill, the ground surface will sink away from the structure, which will not experience significant settlement because of the deep foundation. This can cause access problems for pedestrians and vehicles, and thus can be a maintenance problem. The bridge in Figure 11.2 illustrates this problem.
2. If the upper soils are settling, they impart a downward load onto the foundations. This load, called *downdrag*, can be quite large, and needs to be added to the structural loads. This additional load capacity requirement increases the cost of the foundations.

Accommodating the Settlement

Sometimes it is possible to simply accommodate large settlements in the design, construction, and maintenance of the facility. For example, an airport terminal building at LaGuardia Airport (originally built to serve Eastern Airlines) is underlain by 24 m (77 ft) of soft organic clay which was covered by 6 to 12 m (19–38 ft) of incinerated refuse fill that had been placed in the 1930s. The organic clay has $C_c/(1+e_0) = 0.29$–0.33, which makes it "highly compressible" according to Table 11.1. When construction of this building began in 1979, the ground surface had already settled more than 2 m (7 ft), and was expected to settle an additional 450 mm (18 in) over the following 20 years.

Other buildings at the airport are supported on pile foundations, and require continual maintenance to preserve access for aircraft, motor vehicles, and people. To avoid these problems, this building was built on spread footing foundations, and included provisions for leveling jacks between the foundations and the building (York and Suros, 1989). As differential settlements occurred, the building could then be periodically releveled using the jacks. In addition, the building was designed to accommodate large differential settlements.

By 1988, the building had settled as much as 315 mm (12 in), with differential settlements of up to 56 mm (2 in). However, because of the settlement-tolerant design, the structure was performing well even though no releveling had yet been performed.

This design was at least $2 million less expensive than a pile foundation, and has performed better. This savings in construction cost was much greater than the cost of periodically releveling the building.

Improving the Soil

Another option is to improve the engineering properties of the soils before construction. Many special construction methods have been developed to do this, as discussed in Chapter 19. Once the soils have been improved, normal construction can proceed because the difficult soil conditions have been eliminated.

18.2 EXPANSIVE SOILS

Certain types of clayey soils expand when they are wetted and shrink when dried. These are called *expansive soils*, and are very troublesome. In the United States alone, they inflict about $9 billion per year in damages to buildings, roads, airports, pipelines, and other facilities—more than twice the combined damage from earthquakes, floods, tornados, and hurricanes (Jones and Holtz, 1973; Jones and Jones, 1987). The distribution of these damages is approximately as shown in Table 18.1.

Sometimes the damages from expansive soils are minor maintenance and aesthetic concerns, but often they are much worse, even causing major structural distress, as illustrated in Figure 18.1. According to Holtz and Hart (1978), 60 percent of the 250,000 new homes built on expansive soils each year in the United States experience minor damage and 10 percent experience significant damage, some beyond repair. Although the statistics for new houses built today are probably better, expansive soils continue to be a significant problem.

TABLE 18.1 ANNUAL DAMAGE IN THE UNITED STATES FROM EXPANSIVE SOILS (Jones and Holtz, 1993; Jones and Jones, 1987. Used with permission of ASCE.)

Category	Annual Damage
Highways and streets	$4,550,000,000
Commercial buildings	1,440,000,000
Single-family homes	1,200,000,000
Walks, drives and parking areas	440,000,000
Buried utilities and services	400,000,000
Multi-story buildings	320,000,000
Airport installations	160,000,000
Involved in urban landslides	100,000,000
Other	390,000,000
Total annual damages (1987)	$9,000,000,000

These soil movements and the damage they cause generally occur very slowly, and thus are not nearly as dramatic as hurricanes and earthquakes. In addition, they cause only property damage, not loss of life, and this damage is spread over wide areas rather than being concentrated in a small locality. Nevertheless, the economic loss is large and much of it could be avoided by proper recognition of the problem and incorporating appropriate preventive measures into the design, construction, and maintenance of new facilities.

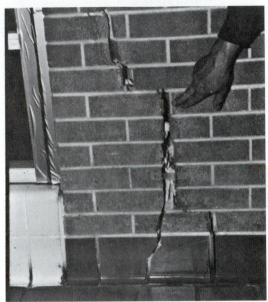

Figure 18.1 Heaving of an expansive soil caused this brick wall to crack. The $490,000 spent to repair this and other walls, ceilings, doors, and windows represented nearly one-third of the original cost of the six-year-old building (Colorado Geological Survey).

Physical Causes of Expansion and Shrinkage

There are many different clay minerals, as discussed in Chapter 4, and each has a different susceptibility to swelling, as shown in Table 18.2. Swelling occurs when water infiltrates between and within the clay particles, causing them to separate. Kaolinite is virtually nonexpansive because of the presence of strong hydrogen bonds that hold the individual clay particles together. Illite contains weaker potassium bonds that allow limited expansion, and montmorillonite particles are only weakly linked. Thus, water can easily flow into montmorillonite clays and separate the particles. Field observations have confirmed that the greatest problems occur in soils with a high montmorillonite content.

TABLE 18.2 SWELL POTENTIAL OF PURE CLAY MINERALS (Adapted from Budge, et al. (1964).

Surcharge Load		Swell Potential (%)		
(lb/ft^2)	(kPa)	Kaolinite	Illite	Montmorillonite
200	9.6	Negligible	350	1500
400	19.1	Negligible	150	350

Sources of Wetting and Drying

The shrinking and swelling potential in a soil becomes reality only when its moisture content changes. Such changes in moisture content can be due to natural processes, such as changes in the groundwater table and infiltration of rainwater. However, moisture changes due to human activities are often much larger and more extensive than those caused exclusively by natural causes, and thus are more often a source of problems.

Irrigation of landscaping is one of the most important causes of increased moisture content in soils, especially in arid and semi-arid areas. For example, irrigation of lawns and shrubs often is the equivalent of about 2000 mm (80 inches) of rain per year, a significant increase over the 200–500 mm/yr (8–20 in/yr) that naturally occurs in such places.

Other sources of excessive moisture and the associated soil expansion include:

- Changes in surface drainage patterns that prevent water from running off and allow water to percolate into the ground.
- Removal of vegetation that brings an end to transpiration.
- Placement of slab-on-grade floors, pavements, or other impervious materials on the ground, which stops both evaporation and the direct infiltration of rain water.

Some activities can have the opposite effect by removing moisture from the soil and causing shrinkage. Poorly placed trees with aggressive roots have caused such problems, especially in areas with moist climates.

Identifying, Testing, and Evaluating Expansive Clays

When working in an area where expansive soils can cause problems, geotechnical engineers must have a systematic method of identifying, testing, and evaluating the swelling potential of troublesome soils (Nelson and Miller, 1992). The ultimate goal is to determine which preventive design measures, if any, are needed to successfully complete a proposed project.

Experienced geotechnical engineers usually can identify potentially expansive soils based on a visual examination. To be expansive, a soil must have a significant clay content, probably falling within the unified symbols CL or CH (although some ML, MH, and SC soils also can be expansive). When dry, expansive soils often have distinct shrinkage cracks and other evidence of previous swelling and shrinking. However, any such visual identification is only a first step; we must obtain more information before we can develop specific design recommendations.

The next stage of the process—determining the degree of expansiveness—is more difficult. A wide variety of testing and evaluation methods have been proposed, but none of them are universally or even widely accepted. Some assessment techniques are as simple as performing Atterberg limits tests and classifying the expansiveness based on the results of these tests. Table 18.3 shows one such classification method. Alternatively, we could conduct a *swell test* by placing a sample in a device similar to a consolidometer and wetting it. The resulting swell can then be used as a semi-empirical assessment of expansion potential. Snethen (1984) suggested the following definition of potential swell:

Potential swell is the equilibrium vertical volume change or deformation from an oedometer-type[1] test (i.e., total lateral confinement), expressed as a percent of original height, of an undisturbed specimen from its natural water content and density to a state of saturation under an applied load equivalent to the in-situ overburden pressure.

Snethen also suggested that the applied load should consider any applied external loads, such as those from foundations. Using Snethen's test criteria, we could classify the expansiveness of the soil, as shown in Table 18.4.

TABLE 18.3 CORRELATIONS WITH COMMON SOIL TESTS (Adapted from Holtz, 1969, and Gibbs, 1969)

Percent Colloids	Plasticity Index	Shrinkage Limit	Liquid Limit	Swelling Potential
< 15	< 18	< 15	< 39	Low
13 - 23	15 - 28	10 - 16	39 - 50	Medium
20 - 31	25 - 41	7 - 12	50 - 63	High
> 28	> 35	> 11	> 63	Very high

TABLE 18.4 TYPICAL CLASSIFICATION OF SOIL EXPANSIVENESS BASED ON SWELL TEST RESULTS AT IN-SITU OVERBURDEN STRESS (Adapted from Snethen, 1984)

Swell Potential (%)	Swell Classification
< 0.5	Low
0.5 - 1.5	Marginal
> 1.5	High

The *expansion index test* [ASTM D4829] (ICBO, 1991b; Anderson and Lade, 1981) is a standardized loaded swell test. In this test a soil sample is remolded into a standard 4.01 in (102 mm) diameter, 1 in (25 mm) tall ring at a degree of saturation of about 50 percent. A surcharge load of 1 lb/in^2 (6.9 kPa) is applied, and then the sample is saturated and allowed to stand until the rate of swelling reaches a certain value or 24 hours, whichever is longer. The amount of swell is expressed in terms of the *expansion index*, or EI, which is defined as follows:

$$EI = 1000\, h\, F \qquad\qquad (18.1)$$

---

[1] The terms *oedometer* and *consolidometer* are synonymous.

where:

EI = expansion index

h = expansion of the soil (in)

F = percentage of the sample by weight that passes through a #4 sieve

Table 18.5 gives an interpretation of EI test results.

Because the expansion index test is conducted on a remolded sample, it may mask certain soil fabric effects that may be present in the field.

TABLE 18.5 INTERPRETATION OF EXPANSION INDEX TEST RESULTS (ICBO, 1997)

EI	Potential Expansion
0 - 20	Very Low
21 - 50	Low
51 - 90	Medium
91 - 130	High
> 130	Very High

Reproduced from the 1997 Edition of the *Uniform Building Code*, © 1997, with permission of the publisher, the International Conference of Building Officials.

Preventive Measures

Once the expansion potential has been evaluated, we develop preventive design, construction, and maintenance measures. These measures are intended to reduce the potential impact of expansive soils.

Building Foundations and Floors

Buildings, especially those that are lightweight, are prone to damage from expansive soils. The magnitude of heaving and shrinking generally varies across the building, thus causing problems similar to those associated with excessive differential settlements. These include cracks, inoperative doors and windows, etc. Common preventive measures include:

- Extending the foundations to greater depths, thus bypassing the zone of greatest moisture change and supporting the building on more stable soil.
- Adding extra reinforcing steel to foundations and slabs. In some cases, prestressed or post-tensioned slabs are used.
- Avoiding the use of slab-on-grade floors.
- Being especially careful to provide and maintain good surface drainage around the building.
- Avoiding the placement of irrigated landscaping close to the building
- Pre-moistening the soil prior to construction, thus causing it to expand before the building is erected.

Pavements

Highway pavements also are prone to damage from expansive soils. Common preventive measures include:

- Providing extensive surface and subsurface drainage to keep water away from the subgrade soils.
- Providing a non-expansive sub-base material.
- Treating the subgrade soils with lime or some other material to reduce their expansive properties.
- Providing more steel reinforcement.

Driveways, sidewalks, and other exterior *flatwork concrete* can have similar problems.

18.3 COLLAPSIBLE SOILS

Another moisture-driven phenomena, often seen in arid regions, is *collapse* (Clemence and Finbarr, 1981; Dudley, 1970; Houston and Houston, 1997). Soils prone to this behavior are called *collapsible soils*. In their natural state, these soils have a high void ratio and a low moisture content. They are usually alluvial or aeolian, and have a "honeycomb" or highly porous structure that is maintained by water-soluble interparticle bonds. These soils can cause problems when structures, highways, or other improvements are built on them, and the soil subsequently becomes wetted. The influx of water breaks down these bonds and causes the soil to compress.

Geotechnical engineers usually assess the collapse potential by placing an undisturbed sample in a consolidometer at its in-situ moisture content, loading it to a normal stress comparable to that in the field, then applying water. The amount of strain that occurs due to wetting is a measure of its collapse potential. Some soils experience strains in excess of 10 percent simply due to wetting.

Unlike expansive soils, which can heave or shrink as the moisture content changes, collapse is a one-way process. Therefore, preventive measures often attempt to pre-collapse the soil prior to construction. This may be accomplished by pre-wetting the soil, by excavating it and replacing it as a compacted fill, or by compacting the soil in place. Other techniques, such as grouting, deep foundations, and avoidance of wetting also have been used (Houston and Houston, 1989).

18.4 FROZEN SOILS

The temperature of soils near the ground surface reflects the recent air temperatures. Thus, when the air temperature falls below $0°C$ ($32°F$) for extended periods, the soil temperature drops to a comparable level and the pore water turns to ice. This transformation has significant impacts on civil engineering works built on such soils, and thus is an important aspect of geotechnical engineering in regions with cold climates.

Ground Freezing and Frost Heave

The depth of freezing in the ground depends on how far the air temperature falls below freezing, how long it remains there, and other factors. This depth is negligible in warm climates, such as Florida, but can extend to depths of 2 m (7 ft) or more when the winters are very cold, such as in Minnesota. In arctic and sub-arctic regions, the depth of freezing is even greater. In North America, problems with ground freezing are most common in the northern United States and in Canada. However, areas farther south also can be affected. For example, underground water pipes in Atlanta have frozen during exceptionally cold winters.

For geotechnical engineers, the most significant consequence of ground freezing is a phenomenon called *frost heave*, which is an upward movement in the ground due to the formation of underground ice. There are two causes of frost heave: The first occurs because the pore water expands about 9 percent in volume when it freezes. Thus, if the soil is saturated and has a typical porosity (say, 40 percent), it will expand about $9\% \times 40\% \approx 4\%$ in volume. In climates comparable to those in the northern United States, this could correspond to surface heaves of as much as 25–50 mm (1–2 in). Although such heaves are significant, they would probably be fairly uniform and cause relatively little damage.

The second cause of frost heave is more insidious and capable of producing much more damage to civil engineering works. If the groundwater table is relatively shallow, capillary action can draw water up to the frozen zone where it forms ice lenses as shown in Figure 18.2. In some situations, this mechanism can move large quantities of water, so it is not unusual for these lenses to produce ground surface heaves of 300 mm (12 in) or more. Such heaves are likely to be very irregular and create a hummocky ground surface that can cause extensive damage.

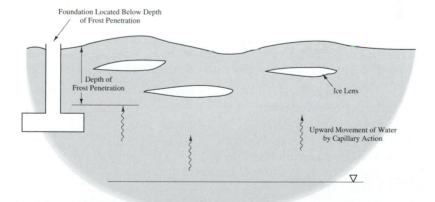

Figure 18.2 Formation of ice lenses. Water is drawn up by capillary action and freezes when it reaches the frozen soil, which is located within the depth of frost penetration. The frozen water forms ice lenses that cause heaving at the ground surface. Foundations placed below the depth for frost penetration are not subject to heaving.

Additional damage can occur when the frozen ground begins to thaw, especially if ice lenses are present. As the upper soils and ice lenses thaw, the resulting soil has a much greater moisture content than it originally had. However, the deeper soils have not yet thawed, so this excess water cannot drain away, resulting in a very soft and weak soil as shown in Figure 18.3. This condition is especially troublesome when it occurs beneath highways, and is often the cause of ruts and potholes. Once the soil completely thaws, the excess water drains down and the soil regains much of its original strength.

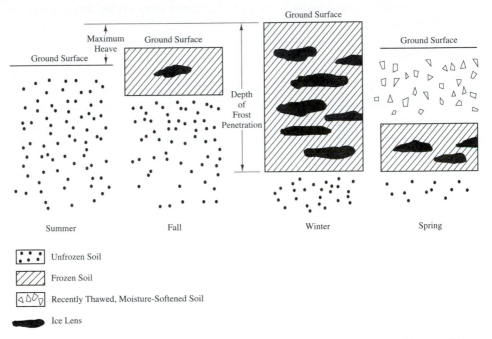

Figure 18.3 Idealized freeze–thaw cycle in temperate climates. During the summer, none of the ground is frozen. Then, during the fall and winter it progressively freezes from the ground surface downward. Finally, in the spring, it progressively thaws from the ground surface downward.

To evaluate the potential for frost heave at a given site, geotechnical engineers consider the following factors:

- The potential depth of freezing
- The frost susceptibility of the soil
- The proximity of potential sources of groundwater

The potential depth of freezing is often dictated by local building codes. For example, the Chicago Building Code specifies a design frost penetration depth of 42 in (1.1 m). Special thermodynamic analyses might be used on special projects, such as ice skating rinks or cold-storage warehouses, but they would rarely be performed on more ordinary projects.

To be considered *frost-susceptible*, a soil must be capable of drawing significant quantities of water up to the frozen zone through capillary action. Clean sands and gravels are not frost-susceptible because they are not capable of significant capillary rise. Conversely, clays are capable of raising water through capillary rise, but they have a low hydraulic conductivity, so they are unable to deliver large quantities of water. Therefore, clays are capable of only limited frost heave. However, intermediate soils, such as silts and fine sands, have both characteristics: They are capable of substantial capillary rise and have a high hydraulic conductivity. Large ice lenses are able to form in these soils, so they are considered to be very frost-susceptible.

The U.S. Army Corps of Engineers has classified frost-susceptible soils into four groups, as shown in Table 18.6. Higher group numbers correspond to greater frost susceptibility and more potential for formation of ice lenses. Clean sands and gravels (i.e., <3% finer than 0.02 mm) may be considered non-frost-susceptible and are not included in this table.

TABLE 18.6 FROST SUSCEPTIBILITY OF VARIOUS SOILS ACCORDING TO THE U.S. ARMY CORPS OF ENGINEERS (Adapted from Johnston, 1981)

Group	Soil Types	USCS Group Symbols
F1 (least susceptible)	Gravels with 3 - 10% finer than 0.02 mm	GW, GP, GW-GM, GP-GM
F2	a. Gravels with 10 - 20% finer than 0.02 mm	GM, GW-GM, GP-GM
	b. Sands with 3 - 15% finer than 0.02 mm	SW, SP, SM, SW-SM, SP-SM
F3	a. Gravels with more than 20% finer than 0.02 mm	GM, GC
	b. Sands, except very fine silty sands, with more than 15% finer than 0.02 mm	SM, SC
	c. Clays with $PI > 12$, except varved clays	CL, CH
F4 (most susceptible)	a. Silts and sandy silts	ML, MH
	b. Fine silty sands with more than 15% finer than 0.02 mm	SM
	c. Lean clays with $PI < 12$	CL, CL-ML
	d. Varved clays and other fine-grained, banded sediments	

Finally, there must be a source of groundwater. Usually the source is a shallow groundwater table, but it also could come from water infiltrating from the ground surface.

Preventive Measures

Once a potential frost heave problem has been identified, geotechnical engineers begin to consider preventive design measures. Many types of preventive measures are available, and the appropriate selection depends on the type of facility to be protected, the level of protection desired, cost, and other factors.

Highways and Other Pavements

Highways, parking lots, airports, and other paved areas are especially susceptible to damage from frost heave. Some of this damage occurs during the winter as a result of differential heaving associated with ice lenses, but more damage often occurs in the spring when the soils have partially thawed and contain trapped water. Heavy wheel loads from trucks or large aircraft are especially troublesome during the spring thaw because they produce bearing capacity failures in the weak soil, which then causes the overlying pavement to sink into the ground. Figure 18.4 shows such a failure.

Figure 18.4 The soils beneath this asphaltic concrete pavement in New York became wet and soft during the spring thaws. As a result, these soils failed under the weight of the heavy trucks that use this site. The pavement is now in very poor condition, with extensive alligator cracks and potholes.

Preventive design measures include:

- Excavating the upper soils and replacing them with non-frost-susceptible soils.
- Providing gradual transition sections between frost-susceptible and non-frost-susceptible subgrade soils.
- Restricting heavy traffic during the spring thaw.
- Installing thermal insulation between the pavement and the underlying soils (this method reduces the depth of frost penetration, but can enhance the formation of ice on the pavement surface, creating dangerous driving conditions).
- Increasing the thickness of aggregate base courses to spread out the wheel loads and to provide greater overburden pressure on the subgrade soil.
- Treating the subgrade soils with cement or lime.

Unfortunately, these preventive measures are often very expensive and may not be cost effective for all pavements. In addition, they are not always completely effective. Thus, maintenance crews are usually busy through the summer repairing these problems.

Buildings and Other Structures

Preventive measures for buildings and other structures are usually more extensive than those for pavements because these facilities have higher standards of performance, and because they cover smaller areas and are thus easier to remediate.

Engineers definitely want to protect building foundations from the effects of frost heave. The most common method is to place foundations at a depth below the depth of frost penetration, as shown in Figure 18.2. This is usually wise in all soils, whether or not they are frost-susceptible and whether or not the groundwater table is nearby. Even "frost-free" clean sands and gravels will often have silt lenses that are prone to heave, and groundwater conditions can change unexpectedly, thus introducing new sources of water. The small cost of building deeper foundations is a wise investment in such cases. However, foundations supported on bedrock or interior foundations in heated buildings normally do not need to be extended below the depth of frost penetration.

Another alternative is to remove the natural soils and replace them with a compacted fill made of soil known to be non-frost-susceptible. This may be an attractive method for unheated buildings with slab-on-grade floors to protect both the floor and the foundation from frost heave.

Builders in Canada and Scandinavia sometimes protect buildings with slab-on-grade floors using thermal insulation. This method traps heat stored in the ground during the summer and thus protects against frost heave, even though the foundations are shallower than the normal frost depth. Both heated and nonheated buildings can use this technique (NAHB, 1988 and 1990).

A peculiar hazard to keep in mind when foundations or walls extend through frost-susceptible soils is *adfreezing* (CGS, 1992). This is the bonding of soil to a wall or foundation as it freezes. If heaving occurs after the adfreezing, the rising soil will impose a large upward load on the structure, possibly separating structural members. Placing a 10 mm (0.5 in) thick sheet of rigid polystyrene between the foundation and the frozen soil reduces the adfreezing potential.

Ice Skating Rinks and Cold-Storage Warehouses

Although frost heave problems are usually due to freezing temperatures from natural causes, it is also possible to freeze the soil artificially. For example, refrigerated buildings such as cold-storage warehouses or indoor ice skating rinks can freeze the soils below and be damaged by frost heave, even in areas where natural frost heave is not a concern (Thorson and Braun, 1975; Duncan, 1992b). Heaves of up to 280 mm (11 in) have been observed in ice skating rinks in Minneapolis, which seriously impairs their usefulness. In some cases the deformation of the ice surface is so bad that hockey teams find it necessary to switch goals between periods and during the middle of the last period.

These facilities can freeze the soil to substantial depths because they usually operate year-round. For example, during nearly two years of continuous operation, an ice skating rink in Minnesota froze the soil to a depth of 6 m (20 ft), and might ultimately reach a depth of 12 m (40 ft) (Thorson and Braun, 1975).

Preventive design measures include excavating the upper soils and replacing them with non-frost-susceptible soils, placing thermal insulation or air passages between the building and the soil, and even placing heating tubes below the insulation.

Underground Pipelines

Underground pipelines, especially water lines, can freeze if they are located within frozen soil. This can cause them to burst (because of the expansion of water when it freezes), or at least it becomes a nuisance in that the water does not flow. Solutions to these problems include placing the pipelines below the frost depth or surrounding them with thermal insulation. Sometimes it also is possible to avoid freezing by keeping a water faucet running continuously, thus continually drawing warmer water through the pipe.

Permafrost

In areas where the mean annual temperature is less than 0°C, the penetration of freezing in the winter may exceed the penetration of thawing in the summer. This creates a zone of permanently frozen soil known as *permafrost* (Phukan, 1985; Andersland and Anderson, 1978). In the harshest of cold climates, such as Greenland, this permanently frozen ground is continuous, whereas in slightly "milder" climates, such as central Alaska, central Canada, and much of Siberia, the permafrost is discontinuous (i.e., the frozen zones are separated by seasonally frozen zones). Areas of seasonal and continuous permafrost in Canada are shown in Figure 18.5.

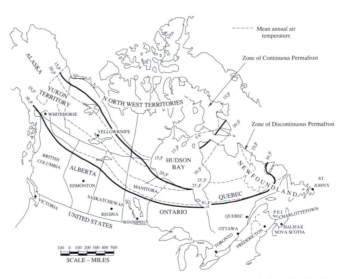

Figure 18.5 Zones of continuous and discontinuous permafrost in Canada (Adapted from Crawford and Johnson, 1971).

In areas where the summer thaws occur, the upper soils can be very wet and weak and probably not capable of supporting any significant loads, while the deeper soils remain permanently frozen. Foundations must penetrate through this seasonal zone and well into the permanently frozen ground below. In addition, it is very important that these foundations be designed so that they do not transmit heat to the permafrost. Figure 18.6 shows the results of permafrost thawing beneath a heated building. To avoid such problems, buildings are typically built with raised floors and a ducting system to maintain subfreezing air temperatures between the floor and the ground surface.

The Alaska Pipeline project is an excellent example of a major engineering work partially supported on permafrost (Luscher, et. al, 1975).

Figure 18.6 Heat from this lodge in Alaska thawed the permafrost below, causing it to settle. However, the permafrost did not thaw beneath the unheated porch (Photo courtesy of Professor Richard L. Handy).

18.5 CORROSIVE SOILS

Soil can be a very hostile environment in which to place engineering materials. Concrete, steel, and wood placed in contact with soil may become the target of chemical and/or biological attack that can adversely affect their integrity.

Steel and Iron

Corrosion is a nearly universal concern with steel and iron. In above-ground applications it generally can be kept under control by painting, galvanizing, and other measures, and visually monitored. Potentially hazardous conditions, such as heavily corroded bridges, usually can be detected by careful inspection. However, corrosion in underground facilities, such as tanks, pipelines, and pile foundations, is potentially much more troublesome and more difficult to monitor.

Sites where the elevation of the groundwater table fluctuates, such as tidal zones, are especially difficult because this scenario continually introduces both water and oxygen. Contaminated soils, such as sanitary landfills and shorelines near old sewer outfalls, are also more likely to have problems.

If the geotechnical engineer suspects that corrosion may be a problem, it is generally appropriate to retain the services of a corrosion engineer. Detailed assessments of corrosion and the development of preventive designs are beyond the expertise of most geotechnical engineers. Preventive measures might include:

- Applying protective coatings, such as coal tar enamel.
- Providing a *cathodic protection system*, which consists of applying a DC electrical potential between the item to be protected (the cathode) and a buried sacrificial metal (the anode). This system causes the corrosion to be concentrated at the anode and protects the cathode. These systems consume only nominal amounts of electricity, and in some cases can be self-energizing (i.e., generating their own electricity).
- Increasing the steel thickness by an amount equal to the anticipated deterioration.
- Using a different material. For example, underground tanks can be made of fiberglass.

Concrete

Concrete in contact with soil, such as buried pipelines, foundations, retaining walls, and slabs, is usually very resistant to corrosion and will remain intact for many years. However, serious degradation can occur in concrete subjected to soils or groundwater that contains high concentrations of sulfates (SO_4). These sulfates can react with the cement to form calcium sulfoaluminate (ettringite) crystals. As these crystals grow and expand, the concrete cracks and disintegrates. In some cases, serious degradation has occurred within 5 to 30 years of construction. Although we do not yet fully understand this process (Mehta, 1983), engineers have developed methods of avoiding these problems.

We can evaluate a soil's potential for sulfate attack by measuring the concentration of sulfates in the soil and/or in the groundwater and comparing them with those that have had problems with sulfate attack. If the laboratory tests indicate that the soil or groundwater has a high sulfate content, design the buried concrete to resist attack by using one or more of the following methods (Kosmatka and Panarese, 1988; PCA, 1991):

- **Reduce the water:cement ratio**—This reduces the hydraulic conductivity of the concrete, thus retarding the chemical reactions. This is one of the most effective methods of resisting sulfate attack.
- **Increase the cement content**—This also reduces the hydraulic conductivity. Therefore, concrete exposed to problematic soils should have a cement content of at least 6 sacks/yd^3 (564 lb/yd^3 or 335 kg/m^3).
- **Use sulfate-resisting cement**—Type II low-alkali and type V Portland cements are specially formulated for use in moderate and severe sulfate conditions, respectively. Pozzolan additives to a type V cement also help. Type II is easily obtained, but type V may not be readily available in some areas.
- **Coat the concrete with an asphalt emulsion**—This is an attractive alternative for retaining walls or buried concrete pipes, but not for foundations.

Unlike steel corrosion problems, which are generally passed on to corrosion engineers, sulfate attack problems are normally addressed by the geotechnical engineer.

Wood

It is generally best to avoid placing wood in contact with soil because it becomes subject to decay and insect attack. The worst condition occurs when the wood is subjected to repeated cycles of wetting and drying. Therefore, building codes usually require all wood in wood frame buildings to be at least 150 mm (6 in) above the ground. However, there are situations where wood is placed in contact with soil or buried underground. These include:

- Timber piles
- Telephone poles
- Wood retaining walls
- Fence posts
- Railroad ties

To reduce problems of decay and insect attack, wood may be treated before it is installed. The most common treatment consists of placing the wood in a pressurized tank filled with creosote or some other preserving chemical. This *pressure treatment* forces some of the chemicals into the wood and forms a coating on the outside. Creosote leaves a black tar-like substance which is often seen on telephone poles and railroad ties. Some chemical treatments produce a greenish color. Another option is to use wood species that are naturally resistant to decay, such as redwood or cypress.

SUMMARY

Major Points

1. Certain soil conditions are especially problematic and require special attention. These are called difficult soils.
2. Soils that are weak and compressible, such as soft clays and highly organic soils, are common near the mouths of rivers, along the perimeter of bays, and in wetlands. These soils generally require special measures to control or accommodate large settlements. They also are subject to shear failure.
3. Expansive soils are those that expand when wetted and shrink when dried. They have caused extensive damage to buildings, highways, and other projects. Engineers have developed methods of recognizing and testing these soils, along with various remediation measures.
4. Collapsible soils are found in arid and semi-arid areas. They initially are very dry and have a loose "honeycomb" structure. This structure is maintained by water-soluble bonds. If these soils subsequently become wetted, these bonds weaken and the soil collapses, sometimes causing excessive settlements. Engineers have developed

methods of evaluating such soils. Once they have been recognized, appropriate preventive measures may be implemented.

5. Frozen soils are those that have a temperature less than 0° C. They can cause a variety of problems, and are most often a concern in areas with cold winters. Frost heave is one of these problems. Once again, engineers have developed methods of assessing the potential for heave and implementing preventive measures.

6. Permafrost is permanently frozen ground. In this case, problems occur when engineering projects cause it to thaw.

7. Corrosive soils are those that are hostile to buried materials. Steel and iron can be subject to rusting, concrete can be subject to sulfate attack, and wood can be subject to rotting. The effects of corrosion can be reduced, but not always eliminated.

Vocabulary

collapsible soil	frost-susceptible soil
corrosive soil	frozen soil
expansion index test	permafrost
expansive soil	swell test
frost heave	

COMPREHENSIVE QUESTIONS AND PRACTICE PROBLEMS

18.1 A 2.5 m thick fill is to be placed over a thick stratum of soft clay, then a one-story office building is to be built on top of the fill. According to a settlement analysis, the weight of this fill will cause 600 mm of total settlement over a period of 30 years. The differential settlement will probably be about 100 mm. Although the building could be supported on spread footing foundations in the fill, the design engineer has decided to support it on a system of pile foundations that penetrate through the fill and into the underlying soils, thus insulating this building from the settlement problem.

Other than cost considerations, what problems will probably occur as a result of this design? What methods might be used to overcome these problems?

18.2 A one-story wood-frame house is to be built on a site in Texas that is underlain by a clay with a plasticity index of 40. Might this house be prone to distress due to expansive soils? Why or why not?

18.3 A project specification requires all imported soils be "non-frost-susceptible." Some import is required, and three sources are available. Source A has a fine silty sand (ML), Source B has a well-graded sand (SW), and Source C has a lean clay (CL). Which of these soils would be most likely to satisfy the project specifications? Explain the reason for your answer.

18.4 An underground steel pipeline is to be constructed in a soil that is mildly corrosive. What kinds of measures might be used to prevent failure due to excessive corrosion?

19

Soil Improvement

*Anyone who thinks he has all the answers
is not quite up-to-date on all the questions.*

Unknown Author

On most projects, geotechnical engineers focus on assessing the existing soil and rock conditions, then develop designs that are compatible with these conditions. For example, if the project involves designing a structural foundation and the soil conditions are good, it may be possible to use spread footings, whereas if soil conditions are bad it may be necessary to use deep foundations. However, on some projects the soil conditions are so poor that it becomes very expensive to accommodate them in the design. When this happens, we often consider various methods of *soil improvement*. These methods are intended to improve the quality of the soils. Although soil improvement is generally expensive, it is often cost effective because it reduces the cost of the remaining construction. In some cases, the proposed construction would not even be practical unless the soils are first improved.

Some soil improvement methods are proprietary (i.e., they are protected by patents, and may be performed only by certain contractors), and many require specialized equipment. Often these proprietary methods are implemented by design–build firms that do both the engineering design and the construction. However, other methods can be implemented by any qualified contractor with ordinary equipment.

This is a rapidly developing topic that is being driven by economic pressures to build on sites with marginal soils, the need to rebuild aging infrastructure in urban areas, increased recognition of seismic hazards, and various geoenvironmental problems. Many new techniques have been developed and refined during the last quarter of the twentieth century, and contractors equipped to implement them have become much more common (Schaefer, 1997). Methods that were only recently considered to be experimental are now proven and widely accepted. Thus, soil improvement has rapidly become a broad topic that easily could fill an entire course. This chapter is only a brief introduction.

19.1 REMOVAL AND REPLACEMENT

One of the oldest and simplest soil improvement methods is to simply excavate the unsuitable soils and replace them with compacted fill. This method is often used when the only problem with the soil is that it is too loose. In that case, the same soil is used to build the fill, except now it has a higher unit weight (because of the compaction) and thus has better engineering properties. This is a common way to remediate problems with collapsible soils.

Removal also may be a viable option when the excavated soils have other problems, such as contamination or excessive organics, and need to be hauled away. This method can be expensive because of the hauling costs and the need for imported soils to replace those that were excavated. It also can be difficult to find a suitable disposal site for the excavated soils.

Removal and replacement is generally practical only above the groundwater table. Earthwork operations become much more difficult when the soil is very wet, even when the free water is pumped out, and thus are generally avoided unless absolutely necessary.

19.2 PRECOMPRESSION

Another old and simple method of improving soils is to cover them with a temporary *surcharge fill*, as shown in Figure 19.1 (Stamatopoulos and Kotzias, 1985). This method is called *precompression*, *preloading*, or *surcharging*. It is especially useful in soft clayey and silty soils because the static weight of the fill causes them to consolidate, thus improving both their settlement and strength properties. Once the desired properties have been obtained, the surcharge is removed and construction proceeds on the improved site.

Surcharge fills are typically 3 to 8 m (10–25 ft) thick, and generally produce settlements of 0.3 to 1.0 m (1–3 ft). They have been used at sites intended for highways, runways, buildings, tanks, and other projects.

Precompression has many advantages, including:

- It requires only conventional earthmoving equipment, which is readily available. No special or proprietary construction equipment is needed.
- Any grading contractor can perform the work.
- The results can be effectively monitored by using appropriate instrumentation (especially piezometers) and ground level surveys.

- The method has a long track record of success.
- The cost is comparatively low, so long as soil for preloading is readily available.

However, there also are disadvantages, including:

- The surcharge fill generally must extend horizontally at least 10 m (33 ft) beyond the perimeter of the planned construction. This may not be possible at confined sites.
- The transport of large quantities of soil onto the site may not be practical, or may have unacceptable environmental impacts (i.e., dust, noise, traffic) on the adjacent areas.
- The surcharge must remain in place for months or years, thus delaying construction. However, the process can be accelerated as described below.

Figure 19.1 This surcharge fill will remain in place until the underlying soils have settled. The smokestack and crane in the background are located behind the fill.

Vertical Drains

The time required to achieve a certain level of consolidation is proportional to H_{dr}^2, where H_{dr} is the maximum drainage distance, as defined in Chapter 12. Thus, if the strata of compressible soil is very thick, the time required to achieve the desired consolidation may be excessive. In some cases, this time can easily be several years or even decades, even with a surcharge fill. Very few projects can accommodate such long delays. Therefore, when precompression is used on thick compressible soils, we generally need to employ some means of accelerating the consolidation process.

The most effective way of accelerating soil consolidation is to reduce H_{dr} by providing artificial paths for the excess pore water to escape. This can be done by installing vertical drains, as shown in Figure 19.2. The excess pore water within the compressible soil now drains horizontally to the nearest vertical drain, a much shorter distance than before. In addition, most soft clays contain thin horizontal sandy or silty seams, so the horizontal hydraulic conductivity, k_x, is typically much higher than the vertical value, k_z. This further increases the rate of consolidation. Thus, the time required to achieve the required degree of consolidation can typically be reduced from several years to only a couple of months. Vertical drains also may be used with only the permanent fill, thus eliminating the expense of a surcharge fill.

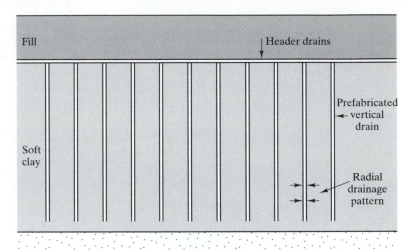

Figure 19.2 Use of vertical drains to accelerate consolidation.

The excess pore water pressures generated during consolidation provide the head to drive water through the vertical drains. Once consolidation is complete, the excess pore water pressures become zero and drainage ceases.

The earliest vertical drains consisted of a series of borings filled with sand. These *sand drains* were expensive to construct, so engineers developed another method: *prefabricated vertical drains* (also known as *wick drains* or *band drains*). They consist of corrugated or textured plastic ribbons surrounded by a geosynthetic filter cloth as shown in Figure 19.3. Most are about 100 mm (4 in) wide and about 5 mm (0.2 in) thick. These drains are supplied on spools, and are inserted into the ground using special equipment that resembles a giant sewing machine, as shown in Figure 19.4. Prefabricated vertical drains are considerably less expensive than sand drains, and thus have become the preferred method on nearly all projects.

The required spacing of vertical drains is determined by a radial drainage analysis, and represents a compromise between construction cost and rate of consolidation. Typically they are spaced about 3 m (10 ft) on center, which means hundreds of drains are usually required.

Although precompression can be very useful with soft silty and clayey soils, it is not very effective in sandy soils. Sands respond best to densification methods that use vibration.

a   b

Figure 19.3 a) A typical prefabricated vertical drain; b) Prefabricated vertical drains after installation. The drains are the small strips extending out of the ground. Wider header drains also have been installed on the ground surface to collect water from the vertical drains and carry it to a discharge location. The site is now ready to be covered with the fill (American Wick Drain Corporation).

Figure 19.4 Equipment used to install prefabricated vertical drains (American Wick Drain Corporation).

19.3 IN-SITU DENSIFICATION

Engineers and contractors have developed several methods of inducing strong vibrations in the ground to densify sandy soils in-situ. Many of these methods have proven to be cost effective, and are especially useful in remediating sandy soils that are prone to seismic liquefaction. Shallow soils often can be densified using heavy vibratory rollers such as the one in Figure 6.22, but they are effective only to depths of about 2 m (7 ft). Other methods, as discussed below, induce vibrations at greater depths and are used to densify deeper soils.

Vibro-Compaction

One method of densifying deeper soil deposits is to insert some type of vibratory probe into the ground. Two types are most commonly used: the terra probe and the vibroflot. A *terra probe* consists of a vibratory pile hammer attached to a steel pipe pile (Brown and Glenn, 1976). This device is vibrated into the ground, densifying the adjacent soils, and then retracted. A *vibroflot* is a specially constructed probe that contains vibrators and water jets. This probe is lowered into the ground using a crane, as shown in Figure 19.5. The presence of the vibrator near the tip probably induces greater vibrations in the ground, and the water jets assist in the insertion and extraction of the probe. This technique of soil improvement is called *vibroflotation*.

Figure 19.5 This crane is lowering a vibroflot into the ground (GKN Hayward Baker, Inc.).

Both of these techniques may be classified as *vibro-compaction methods* because they compact the soils in-situ using vibration. They are generally effective only when the silt content is less than 12–15 percent and the clay content is less than about 3 percent (Schaefer, 1997). The construction process typically uses a grid pattern, with spacings of 1.5 to 4 m (5–13 ft) and treatment depths of 3 to 15 m (10–50 ft).

Dynamic Compaction

Dynamic compaction (also called *dynamic consolidation* or *heavy tamping*) is another method of in-situ densification. It uses a special crane to lift 4 to 27 Mg (5–30 ton) weights, called *pounders*, to heights of 12 to 30 m (40–100 ft), then drop these weights onto the ground as shown in Figure 19.6. Typically the weight is dropped several times at each location. This process is repeated on a grid pattern across the site, leaving a series of 1 to 3 m (3–10 ft) deep craters. The ground surface is then releveled with conventional earthmoving equipment and the process is repeated at grid points midway between the primary drops. Finally, the upper soils are compacted and graded using conventional methods.

Figure 19.6 This crane is densifying the soil using dynamic compaction. It has just dropped a large weight onto the ground (GKN Hayward Baker, Inc.).

Although it appears crude, dynamic compaction can be a cost-effective method of densifying loose sandy and silty soils. It also has been used in soils that contain boulders and other large debris, and in sanitary landfills. The primary zone of influence typically extends to depths of 5 to 10 m (15–30 ft), with lesser improvements below these depths. It has been used to treat liquefaction-prone soils (Dise, Stevens, and Von Thun, 1994), collapsible soils (Rollins and Kim, 1994), and soils that are prone to excessive settlement. However, it is not an effective method for saturated clays because their low hydraulic conductivity does not permit rapid consolidation.

The effectiveness of a dynamic compaction program is typically evaluated by performing SPT or CPT tests both before and after construction. In favorable conditions, the post-construction $(N_1)_{60}$ values can be 10 to 20 blows higher than those measured before construction.

Because of the large impact forces, this method generates substantial shock waves, and therefore cannot be used close to existing structures.

Blast Densification

Blast densification is another method of in-situ densification, and is even more curious than dynamic compaction. This method consists of drilling a series of borings and using them to place explosives underground. These explosives are then detonated, and the resulting shock waves densify the surrounding soils. Blast densification has been used successfully on many projects, and is most effective in clean sands. However, because of vibration and safety issues, it is only suitable for remote sites and thus is not nearly as common as vibro-compaction or dynamic compaction.

19.4 IN-SITU REPLACEMENT

In-situ densification equipment also may be used to improve the ground by *in-situ replacement*. With this method, a vibroflot is used to create a shaft that is backfilled with gravel to form a *stone column* (Mitchell and Huber, 1985). This technique is called *vibro-replacement*. Alternatively, dynamic compaction equipment may be used to pound a gravel inclusion into the ground using a technique called *dynamic replacement*.

These methods may be used in nearly all types of soil, and are primarily intended to provide load bearing members that extend through the weak strata. The stone columns also act as vertical drains, thus helping to accelerate consolidation settlements and mitigate seismic liquefaction problems.

19.5 GROUTING

Grouting is the injection of special liquid or slurry materials, called *grout*, into the ground for the purpose of improving the soil or rock. It has been used extensively for the past several decades, and is a well-established method of soil improvement.

There are two primary kinds of grout. *Cementitious grouts* are made of Portland cement that hydrates after injection, forming a solid mass. *Chemical grouts* include a wide range of chemicals that solidify once they are injected into the ground. These include silicates, resins, and many others. Chemical grouts have a wider range of available properties, and thus can be used in some applications where cement grouts are ineffective. However, chemical grouts also are more expensive, and some are toxic or corrosive.

There are four principal grouting methods, as shown in Figure 19.7:

- *Intrusion grouting* (also known as *slurry grouting*) consists of filling joints or fractures in rock or soil by injecting grout through pipes. These pipes may be inserted from the ground surface, or from tunnels. The primary benefit from this work is a decrease in hydraulic conductivity. This method is often used to prepare the foundations and abutments for dams. It usually is done using cementitious grouts.

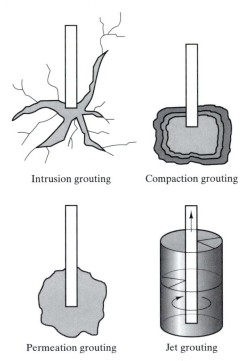

Intrusion grouting　　　　Compaction grouting

Permeation grouting　　　　Jet grouting

Figure 19.7 Types of grouting (Welsh, 1986).

- *Permeation grouting* is the injection of thin grouts into the soil such that they permeate into the voids (Littlejohn, 1993). Once the grout cures, the porous soil is transformed into a near solid mass. Although this can sometimes be done using cement grouts, the void space in most soils is much to small to permit passage of the Portland cement particles. Thus, most permeation grouting is performed using chemical grouts. Because of this, it is often called *chemical grouting*. The treated soil has a much lower hydraulic conductivity, and is stronger and less compressible than before. It is often used to form groundwater barriers and to stabilize soils in advance of making excavations or tunnels.

- *Compaction grouting* (also known as *displacement grouting*) uses a stiff (i.e., about 25 mm slump) grout that is injected into the ground under high pressure through a pipe to form a series of inclusions (Rubright and Welsh, 1993). This grout is too thick to penetrate significantly into the soil, but the grout inclusions compact the adjacent soil. Compaction grouting is often used to repair structures that have experienced excessive settlement, since it both improves the underlying soils and raises the structure back into position.

- *Jet grouting* (Bell, 1993) is the newest method. It was developed in Japan during the 1960s and 1970s, and uses a special pipe equipped with horizontal jets that inject grout into the soil at high pressure. The pipes are first inserted to the desired depth, then they are raised and rotated while the injection is in progress, thus forming a column of treated soil. Because of the high pressures, this method is usable on a wide range of soil types. This method has been used for groundwater control, under-pinning, stabilization in advance of tunneling, and many other applications.

19.6 STABILIZATION USING ADMIXTURES

Another method of improving soils is to treat them with an admixture (Ingles and Metcalf, 1972). The most common admixture is Portland cement. When mixed with the soil, it forms a material called *soil-cement*, which is comparable to a weak concrete. Other admixture materials include lime and asphalt. The objective of these admixtures is to provide artificial cementation, thus increasing strength and reducing both compressibility and hydraulic conductivity. It also reduces the expansion potential in clays.

Surface Mixing

Historically, most admixture stabilization has been performed by ripping the upper soils, applying the admixture (and possibly water), mixing with special equipment, and compacting. Once the mixture has cured, it forms a very hard and durable soil. These methods have most often been used for highways and airports, thus forming a layer often called a *subbase*. Typically, this layer is no more than 200 mm (8 in) thick. It is subsequently covered with a *base* of crushed gravel, then the pavement. Admixture-treated soils also have been used as erosion protection on the face of earth dams, levees, and channels.

These methods have been used successfully for many years. When properly designed and constructed, they can be effective and cost efficient. However, the construction process is very time-sensitive, because the mixture must be shaped to grade and compacted before curing progresses too far. In addition, specialized equipment is usually required to achieve sufficiently thorough mixing. If the mixing is inadequate, the resulting product will consist of alternating over-treated hard spots separated by untreated soft spots, a situation that may be worse than no treatment at all.

In-Situ Deep Mixing

During the 1970s and 1980s, a new method of stabilization was developed in Japan. It uses rotating mixer shafts, paddles, or jets that penetrate into the ground while injecting and mixing Portland cement or some other stabilizing agent (Toth, 1993; Yang, 1994, Schaefer, 1997). These techniques include deep cement mixing, soil mix walls, deep jet mixing, deep soil mixing, deep mixed method, and others. There are several kinds of mixing machines available, one of which is shown in Figure 19.8.

The treated soil has greater strength, reduced compressibility, and lower hydraulic conductivity than the original soils.

Figure 19.8 This rig uses three rotating augers to mix Portland cement with soil in-situ to form a soil mix wall (SMW Seiko, Inc.).

19.7 REINFORCEMENT

One of the similarities between concrete and soil is that both materials are strong in compression but weak in tension. In concrete we overcome this problem by placing steel reinforcing bars inside the concrete. This composite material, reinforced concrete, is far better than plain concrete.

The same principle can be applied to soils. The placement of tensile reinforcement members can significantly improve its stability and load-carrying capacity (Koerner, 1998). Various materials can be used, such as thin steel strips, special plastic grids, and geotextiles. The plastic grids, as shown in Figures 19.9 and 19.10, are the most common tensile reinforcement material because of their durability and low cost.

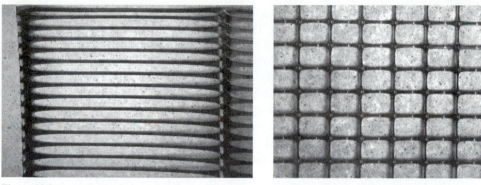

Figure 19.9 Tensar® geogrids are made in two different styles: A uniaxial geogrid, as shown on the left, is designed to resist tensile forces in one direction only, while a biaxial geogrid, as shown on the right, resists tensile forces in two perpendicular directions (Courtesy of Tensar Earth Technologies, Inc., Atlanta, Georgia, USA).

Figure 19.10 Geogrids being used to reinforce a fill for a highway. Multiple geogrids are used, each separated by a layer of soil. (Courtesy of Tensar Earth Technologies, Inc., Atlanta, Georgia, USA)

Tensile reinforcement is especially useful in the construction of compacted fill slopes and earth-retaining structures.

SUMMARY

Major Points

1. Most geotechnical engineering projects focus on assessing the engineering characteristics of the ground as it presently exists and designing the project to accommodate these conditions. However, sometimes it is cost effective to improve the soils, thus producing better engineering properties and placing fewer demands on the proposed construction.

2. Removal and replacement is one of the oldest and simplest methods of soil improvement. However, it is usually cost effective only when the required volumes are small and the excavation does not need to extend below the groundwater table.
3. Precompression consists of applying a surcharge load, thus accelerating consolidation settlements. It is generally effective only in silts and clays.
4. Vertical drains may be used to accelerate consolidation settlement, either with or without surcharge loads.
5. In-situ densification uses strong vibrations to densify the ground, and is effective in sandy soils. Several methods are available.
6. Grouting consists of the injection of special liquid or slurry materials to improve the soil.
7. Admixture stabilization consists of mixing soil with Portland cement or some other material.
8. Reinforcement methods consist of installing tensile reinforcement members in the soil, thus forming a composite material that has both compressive and tensile strength.

Vocabulary

blast densification	sand drain	terra-probe
dynamic compaction	soil cement	vertical drain
grouting	soil improvement	vibro-compaction
precompression	stone column	vibroflotation
removal and replacement	surcharge fill	wick drain

COMPREHENSIVE QUESTIONS AND PRACTICE PROBLEMS

19.1 A proposed medical office building is to be built on a vacant parcel of land adjacent to a hospital. This site is underlain by 30 ft of loose sand that is prone to seismic liquefaction. To rectify this liquefaction problem, the sand needs to be densified. Suggest an appropriate method of soil improvement for this site and indicate the reasons for your selection. The groundwater table is at a depth of 5 ft.

19.2 A zone of buried trash has been found at a proposed construction site. The total volume of this trash appears to be about 100 m³, and all of it appears to be within 3 m of the ground surface. This trash is weak and compressible, and thus would not provide adequate support for the proposed construction. The remainder of the site is underlain by ML and SM soils and the groundwater table is at a depth of 15 m. Recommend a method of solving this problem.

19.3 Explain how a time-settlement analysis could be used to estimate how long a surcharge fill must remain in place.

Geotechnical Earthquake Engineering

A big quake has to occur before the public will act to beef up building codes and formulate plans for disaster management.

Richard H. Jahns, Geology Professor from Stanford University speaking to a professional conference in California, two years before the disastrous 1971 Sylmar Earthquake

Many areas of the world are seismically active and subject to destructive earthquakes. Geologists, seismologists, geotechnical engineers, structural engineers, and others work together in these areas to protect the public from excessive earthquake-related damage and injury.

Geotechnical earthquake engineering is the branch of geotechnical engineering that deals with such matters. It is a very young discipline that began largely as a result of two earthquakes in 1964: the Alaska Earthquake, which was the largest recorded earthquake in North America, and the Niigata, Japan Earthquake, which was smaller but very significant. Subsequent earthquakes have further fueled interest, funding, and ultimately building code regulations related to geotechnical earthquake engineering.

20.1 EARTHQUAKES

Sources

Most earthquakes are the result of sudden massive shifting in bedrock due to forces within the earth. These are known as *tectonic earthquakes* and the movements occur along *faults* (see discussion of faults in Chapter 2). This shifting generates shock waves that propagate outward from the fault. Sometimes only a short section of the fault moves, and these movements are small, thus generating a mild earthquake. Other times a much longer section of the fault moves, sometimes hundreds of kilometers, and it shifts farther, perhaps several meters, creating powerful and destructive earthquakes.

The shearing action along a fault begins at a point called the *focus* or the *hypocenter*, then spreads over a certain area of the fault. The focus is typically 5 to 50 km (3–30 mi) below the ground surface, but may be as deep as 600 km (400 mi). The *epicenter* is the point on the ground surface immediately above the focus. However, faults are rarely vertical, so the epicenter is generally offset from the fault trace.

Earthquakes also develop from other sources, including *volcanic earthquakes* (associated with the eruption of volcanos), *explosion earthquakes* (such as those generated by underground nuclear tests), and *collapse earthquakes* (from underground collapses of mines, large landslides, and other sources). However, these are much less important than tectonic earthquakes and generally do not produce significant damage. Therefore, we focus our efforts almost completely on tectonic earthquakes.

Intensity and Magnitude

To provide for systematic study, we need to have some method of expressing the severity of earthquakes. Two approaches are commonly used: intensity and magnitude.

The *intensity* of an earthquake is an assessment of its effects at a particular location. Large earthquakes have greater intensity than small ones, and observers near the epicenter experience greater intensity than those farther away. Several intensity scales have been used, and nearly all of them are expressed as Roman numerals. In the United States, the most popular one is the *Modified Mercalli Intensity Scale*, shown in Table 20.1.

Figure 20.1 shows an intensity map compiled from damage surveys and personal interviews. Such maps provide a useful record of the earthquake, and can even be made for earthquakes that occurred long ago.

The development of *seismographs* (instruments that measure earthquakes) enabled seismologists to use more objective and quantitative methods to assess earthquakes. This lead to the development of *magnitude* scales. Charles Richter, a seismologist, developed the first magnitude scale, commonly called the *Richter magnitude*, based on data gathered from a certain seismograph (Richter, 1935).

The Richter scale has since been refined with the introduction of the *body wave magnitude*, M_b; the *local magnitude*, M_L; the *surface wave magnitude*, M_S; and the *moment magnitude*, M_w. All of them produce similar results for small earthquakes, but can be quite different for larger ones. For example, the 1964 Alaska Earthquake had $M_w = 9.2$ and $M_s = 8.6$, but M_b of only 6.5.

The magnitude of an earthquake is a measure of the amount of energy released. It is a logarithmic parameter, with an increase of one on a magnitude scale representing a thirty-fold increase in energy (not tenfold as is sometimes claimed). Thus, a magnitude 7 earthquake is 900 times more powerful than a magnitude 5. Table 20.2 lists the magnitudes of selected major earthquakes, with emphasis on those in North America and those that had an important influence on geotechnical earthquake engineering.

TABLE 20.1 ABRIDGED MODIFIED MERCALLI INTENSITY SCALE OF 1931 (Bolt, 1993).

Modified Mercalli Intensity (MMI)	Description
I	Not felt except by a very few under especially favorable circumstances.
II	Felt only by a few persons at rest, especially on upper floors of buildings. Delicately suspended objects may swing.
III	Felt quite noticeably indoors, especially on upper floors of buildings, but many people do not recognize it as an earthquake. Standing automobiles may rock slightly. Vibration like passing of truck. Duration estimated.
IV	During the day felt indoors by many, outdoors by few. At night some awakened. Dishes, windows, doors disturbed; walls make creaking sound. Sensation like heavy truck striking building. Standing automobiles rocked noticeably.
V	Felt by nearly everyone, many awakened. Some dishes, windows, and so on broken; cracked plaster in a few places; unstable objects overturned. Disturbances of trees, poles, and other tall objects sometimes noticed. Pendulum clocks may stop.
VI	Felt by all, many frightened and run outdoors. Some heavy furniture moved; a few instances of fallen plaster and damaged chimneys. Damage slight.
VII	Everybody runs outdoors. Damage negligible in buildings of good design and construction; slight to moderate in well-built ordinary structures; considerable in poorly built or badly designed structures; some chimneys broken. Noticed by persons driving cars.
VIII	Damage slight in specially designed structures; considerable in ordinary substantial buildings with partial collapse; great in poorly built structures. Panel walls thrown out of frame structures. Fall of chimneys, factory stacks, columns, monuments, walls. Heavy furniture overturned. Sand and mud ejected in small amounts. Changes in well water. Persons driving cars disturbed.
IX	Damage considerable in specially designed structures; well-designed frame structures thrown out of plumb; great in substantial buildings, with partial collapse. Buildings shifted off foundations. Ground cracked conspicuously. Underground pipes broken.
X	Some well-built wooden structures destroyed; most masonry and frame structures destroyed with foundations; ground badly cracked. Rails bent. Landslides considerable from river banks and steep slopes. Shifted sand and mud. Water splashed, slopped over banks.
XI	Few, if any, masonry structures remain standing. Bridges destroyed. Broad fissures in ground. Underground pipelines completely out of service. Earth slumps and land slips in soft ground. Rails bent slightly.
XII	Damage total. Waves seen on ground surface. Lines of sight and level distorted. Objects thrown into the air.

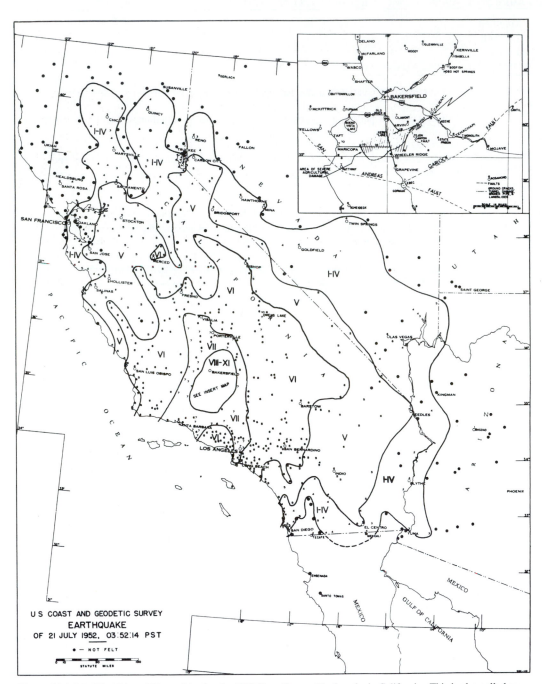

Figure 20.1 Modified Mercalli intensities for the 1952 Kern County Earthquake in California. This is also called an isoseismal map (U.S. Coast and Geodetic Survey).

TABLE 20.2 DESCRIPTION OF SELECTED EARTHQUAKES (Adapted and condensed from Kramer, 1996; Bolt, 1993; Algermissen, 1983, and other sources)

Date	Location	Magnitude and Intensity	Deaths	Comments
1811 and 1812	Missouri	M_s = 8.6, 8.4, 8.7; M_b = 7.2, 7.1, 7.4 XI, XI, XII (estimated)	Several	New Madrid Earthquakes; three large earthquakes in less than two months; seiches on Mississippi River; although about the same magnitude as the 1906 San Francisco Earthquake, these were felt across a much larger area; felt in Boston; awakened Thomas Jefferson in Virginia.
1857	California	M_w = 7.9 X (estimated)	1	Fort Tejon Earthquake; one of the largest earthquakes known to have been produced on the San Andreas Fault; the fault ruptured for 250 miles (400 km) with up to 30 ft (9 m) of offset.
1872	California	M_w = 7.8 X (estimated)	27	Owens Valley Earthquake.
1886	South Carolina	M_s = 7.7 M_b = 6.8 IX (estimated)	110	Strongest documented earthquake on the east coast; produced significant liquefaction; extensive damage in Charleston.
1906	California	M_s = 8.3 M_w = 7.9 XI	700	The great San Francisco Earthquake; first great earthquake to strike a densely populated area in the United States; produced up to 21 ft (7 m) of offset along a 270 mi (430 km) rupture of the San Andreas Fault; most damage was caused by subsequent fires.
1925	California	M_s = 6.5 IX	13	Santa Barbara Earthquake; caused liquefaction failure of Sheffield Dam; led to first explicit provisions for seismic design in U.S. building codes.
1933	California	M_s = 6.3 IX	120	Long Beach Earthquake; considerable building damage; schools particularly hard-hit, with many children killed and injured; led to greater seismic design requirements in building codes, particularly for school buildings.

TABLE 20.2 DESCRIPTION OF SELECTED EARTHQUAKES (continued)

Date	Location	Magnitude and Intensity	Deaths	Comments
1940	California	$M_s = 7.1$ X	9	Imperial Valley Earthquake; large ground displacements along Imperial Fault (see Figure 20.15); first important accelerogram for engineering purposes.
1959	Montana	$M_s = 7.1$ $M_w = 7.3$ X	28	Hebgen Lake Earthquake; caused large landslide that blocked a river and produced a lake; faulting within a reservoir produced a seiche that overtopped an earth dam. See Figure 20.16.
1960	Chile	$M_s = 8.3$ $M_w = 9.5$ XI	2,230	Probably the largest earthquake ever recorded.
1964	Alaska	$M_w = 9.2$ $M_S = 8.6$ $M_b = 6.5$ X	131	Good Friday Earthquake; largest recorded earthquake in North America. Caused severe damage due to liquefaction; earthquake-induced landslides (see Figure 20.18); tsunami.
1964	Japan	$M_S = 7.5$	26	Widespread liquefaction caused extensive damage, especially in Niigata (see Figures 20.9, 20.10, and 20.17); spurred research interest in liquefaction.
1971	California	$M_s = 6.2$ $M_L = 6.4$ X	65	Sylmar Earthquake; produced liquefaction in an earth dam (see Figure 20.8); many structural failures; prompted rehabilitation of many dams and revisions in building codes.
1985	Mexico	$M_S = 8.1$ IX	9,500	Epicenter was off the Pacific Coast, but greatest damage was in Mexico City, 220 mi (360 km) away because of its poor soil conditions.
1989	California	$M_S = 7.1$ $M_w = 6.9$ $M_L = 7.0$ IX	63	Loma Prieta Earthquake; liquefaction (see Figures 20.6–20.7); structural collapse; illustrated importance of local soil conditions.
1994	California	$M_s = 6.8$ $M_w = 6.7$ $M_L = 6.4$ IX	61	Northridge Earthquake; occurred on previously unknown fault; $20 million damage.
1995	Japan	$M_w = 6.9$ X	5,300	Hyogo-Ken Nabu Earthquake; caused extensive damage in Kobe; liquefaction; landslides; damage to retaining walls and subway stations; $100 million damage.

Although the study of past earthquakes is interesting and informative, we need to design for potential *future* earthquakes. In areas with hundreds or thousands of years of historic records, such as China, the potential for future activity can be based largely on what has happened in the past. Unfortunately, very little recorded data is available in North America, so we must supplement it with secondary evidence, such as geologic studies of faults.

A fault is said to be *active* if it is believed to be capable of generating new earthquakes. *Inactive* faults are those believed to be dormant. Geologists determine whether or not a fault is active by studying the age of strata displaced by it, examining records of earthquake epicenters, studying the surface topography, and other methods. For normal projects, a fault is typically deemed inactive if it has not moved within Holocene time (i.e., the last 11,000 years). For critical projects, such as nuclear power plants, the standard is much higher.

On active faults, geologists and seismologists assign a *maximum credible earthquake*, which is the largest that can be reasonably expected to occur, and a *maximum probable earthquake*, which usually corresponds to a 100-year recurrence interval. Routine engineering designs can then be based on one of these events. Designs of more critical projects are usually based on more detailed assessments of potential earthquake activity, and often use a statistical approach called a *seismic risk analysis*.

Earthquake-Related Hazards

Earthquakes can produce many different kinds of hazards. Those of interest to geotechnical engineers include the following:

- *Ground shaking* — accelerations produced at a specific location by a certain design earthquake
- *Liquefaction* — sudden loss of strength in certain soils
- *Surface rupture* — permanent ground deformation along a fault where it intersects the ground surface
- *Other permanent ground deformations* — those that occur away from faults
- *Tsunamis* and *seiches* — earthquake-generated waves in bodies of water

These hazards are discussed in the following sections.

20.2 GROUND SHAKING

Geotechnical engineers are very interested in the intensity, duration, and waveform of ground shaking during earthquakes. The intensity at a given location is often expressed in terms of the *peak acceleration*, usually in units of "g" where $1g$ = the acceleration of gravity = 9.8 m/s^2. For example, the 1994 Northridge Earthquake generated ground accelerations as large as $1.82\,g$ (CDMG, 1994b). There is a rough correlation between peak acceleration and modified Mercalli intensity, but other factors, such as duration, also need to be considered.

Ground Motion Propagation

As the energy released by an earthquake propagates outward from the rupture zone, the peak accelerations diminish, just as the amplitudes of sound waves diminish as they travel farther from their source. Seismologists call this *attenuation*.

The degree of attenuation depends on the distance from the earthquake source, the energy-absorbing nature of the bedrock, and other factors. For example, compare the intensity maps in Figure 20.2 for the 1906 San Francisco and 1811–1812 New Madrid Earthquakes. Although both had about the same magnitude, the bedrock in the eastern United States absorbs much less energy than that in the west, so the New Madrid Earthquakes were felt over a much wider region.

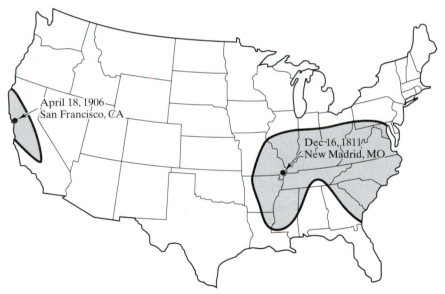

Figure 20.2 Comparison of attenuation in the 1906 San Francisco and 1811 New Madrid Earthquakes. The shaded areas encompass modified Mercalli intensities of V or greater (data from Stearns and Wilson, 1972, and Steinbrugge, 1970).

A number of attenuation functions have been developed. Because of the differences in bedrock, these functions usually apply only to a certain geographic area. Boore, et al. (1993) developed the following function for earthquakes in the western United States:

$$\log(a_{max/g}) = -0.038 + 0.216(M_w - 6) - 0.777 \log R + 0.158 G_B + 0.254 G_C \qquad (20.1)$$

$$R = \sqrt{d^2 + z_1^2} \qquad (20.2)$$

where:

$a_{max/g}$ = peak horizontal ground acceleration at the ground surface

M_w = moment magnitude

d = closest distance to fault trace (km)

z_1 = focal depth (km) (if unknown, 5 km is a conservative value)

G_B, G_C = empirical coefficients from Table 20.3

TABLE 20.3 COEFFICIENTS G_B AND G_C FOR EQUATION 20.1 (Boore, et al., 1993)

Site Class	Shear Wave Velocity in Upper 30 m	G_B	G_C
A	> 750 m/s (> 2500 ft/s)	0	0
B	360 - 750 m/s (1200 - 2500 ft/s)	1	0
C	180 - 360 m/s (600 - 1200 ft/s)	0	1

The shear wave velocity reflects the stiffness of the underlying soils, and may be determined by special in-situ tests.

Toro, et al. (1995) has developed a different function for earthquakes in the central and eastern United States:

$$\ln (a_{max/g})_{rock} = 2.20 + 0.81 (M_w - 6) - 1.27 \ln R + 0.11 \max\left(\ln \frac{R}{100}, 0 \right) - 0.0021 R \qquad (20.3)$$

where:

$(a_{max/g})_{rock}$ = peak horizontal acceleration in bedrock

M_w = moment magnitude

$R = \sqrt{d^2 + 9.3^2}$ (km)

d = closest distance to fault trace (km)

The fourth term in Equation 20.3 uses the greater of the two numbers in the parenthesis.

Both the Boore and Toro functions are plotted in Figure 20.3. If we know the maximum credible or maximum probable earthquakes for local faults and the distances from these faults to our project site, we can use such functions to predict the peak acceleration.

Earthquake waves travel differently through soil, so the ground shaking at sites underlain by soil is different from those underlain by rock. For example, during the 1989 Loma Prieta Earthquake, sites underlain by deep deposits of soft soils experienced peak ground accelerations two to three times greater than nearby sites on stiff soils or rock (Seed, et al., 1990). The portions of the Highway I-880 Cypress Viaduct that collapsed during that earthquake were founded on soft soils, while adjacent sections of the same design founded on stiffer soils did not collapse. Another example is the 1985 Michoacan Earthquake in Mexico, where the damage in Mexico City, 350 km from the epicenter, was much worse than in other cities that were much closer. This difference was due to the deep deposits of soft clay beneath Mexico City. Sophisticated analyses are available to address this effect; Figure 20.4 presents a simplified, although approximate, relationship. This plot may be used to adjust the bedrock accelerations computed from Equation 20.3.

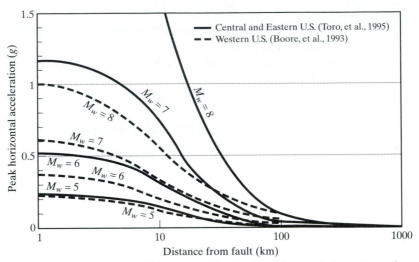

Figure 20.3 Attenuation of peak horizontal acceleration for earthquakes in the western and central United States (Boore, et al., 1993, and Toro, et al., 1994). The Boore curves are based on soil type B with a 5 km focal depth and reflect accelerations at the ground surface. The Toro curves represent acceleration in rock.

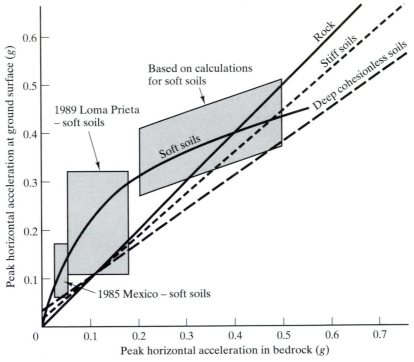

Figure 20.4 Approximate adjustment to convert peak rock acceleration to peak acceleration at the ground surface. The shaded boxes indicate observed relationships for soft soil sites during the 1989 Loma Prieta and 1985 Mexico earthquakes, along with a predicted relationship (Seed, et al., 1976, and Idriss, 1990).

Site Response

Structural engineers need information on earthquake motions to implement their seismic designs. For ordinary structures they use a simplified analysis method that requires only minimal information on the design earthquake. These analyses may be based entirely on data within building codes without any input from a geotechnical engineer. The method described in the *Uniform Building Code* also permits structural engineers to use lower design earthquake forces if the geotechnical engineer provides a *site coefficient, S,* which is based on a simple assessment of the soil profile (ICBO, 1997).

For large and important projects, structural engineers use more sophisticated seismic analyses that require more detailed input from the geotechnical engineer. This is often provided in the form of a *response spectrum* as shown in Figure 20.5. Such plots reflect both the accelerations and frequency content of the design ground motions, and help structural engineers determine how structures with different natural frequencies will respond to the earthquake. For example, structures with natural frequencies that resonate with the ground motions will be subjected to much higher accelerations than those without such frequencies.

Geotechnical earthquake engineers use seismicity data and dynamic assessments of the onsite soils to develop response spectra. See Kramer (1996) for more specific information.

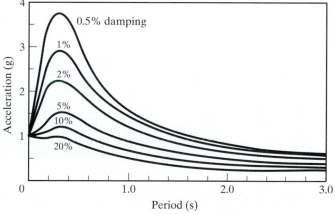

Figure 20.5 A typical response spectrum. This plot presents the acceleration in a structural member with a certain natural period and damping. In this case, the greatest accelerations occur in structures with natural periods of about 3 seconds (Housner, 1970).

20.3 LIQUEFACTION

Liquefaction is the rapid loss of shear strength in cohesionless soils subjected to dynamic loading, such as from an earthquake. Sometimes the shear strength falls to nearly zero, while other times it only drops to a lower-than-normal value. In either case, liquefaction can lead to many kinds of failures, so its evaluation is one of the most important aspects of geotechnical earthquake engineering.

There are two types of liquefaction:

- *Flow liquefaction* occurs when the static shear stresses in the soil exceed the shear strength of the liquefied soil. This usually leads to large and sudden shear movements in the soil.
- *Cyclic mobility* occurs when the static shear stresses are slightly less than the liquefied shear strength, but the static plus dynamic stresses are greater than the liquefied shear strength. This produces incremental shear movements that are generally not as dramatic as flow liquefaction, but still can be a source of significant damage.

Liquefaction is often accompanied by *sand boils*, which are made of liquefied sand ejected from the ground as shown in Figures 20.6 and 20.7. If sand boils are observed, we are sure liquefaction has occurred.

Figure 20.6 Sand boil at truck terminal in Port of Oakland, California, produced by the 1989 Loma Prieta Earthquake. (Earthquake Engineering Research Center Library, University of California, Berkeley, Steinbrugge Collection)

Several kinds of liquefaction-induced failures have been observed, including:

- Landslides — The static shear strength in sloping ground is often only slightly greater than the static shear stresses, so a loss in shear strength due to liquefaction can easily produce a landslide. Such failures have occurred in natural and artificial slopes, and in earth dams (see Figure 20.8).
- Lateral spreads — Liquefaction also can cause large horizontal movements in nearly level ground adjacent to river banks or other similar topographic features.
- Bearing capacity failure of foundations — Structural foundations impose shear stresses in the soil that supports them, and are subject to a bearing capacity failure if

the soil liquefies. Dramatic failures of this type occurred during the 1964 Niigata Earthquake, as shown in Figures 20.9 and 20.10.

- Flotation of buried structures — Underground structures, such as tanks, pipelines, and subway tubes, have a lower unit weight than the surrounding soil, and can float upward if the soil liquefies. For example, in the Niigata Earthquake, an underground reinforced concrete sewage treatment tank floated up when the surrounding soil liquefied, finally coming to rest 3 m (10 ft) above the ground surface (Seed, 1970).
- Sinking of buildings and other structures — Above-ground structures, especially those supported on spread footings, can sink into liquefied soil.

Figure 20.7 Sand boil in Marina Green, San Francisco, California, as a result of the 1989 Loma Prieta Earthquake. The green is underlain by a sandy hydraulic fill (Earthquake Engineering Research Center Library, University of California, Berkeley, Steinbrugge Collection)

Figure 20.8 Lower San Fernando Dam. The soils inside this earth dam liquefied during the 1971 Sylmar Earthquake in California. The dam was constructed by hydraulic filling (see Chapters 6 and 15), and thus contained loose, saturated, sandy soils prone to liquefaction. The left portion of the dam slid into the reservoir, leaving only 1.5 m of freeboard. The failure of this dam prompted the reconstruction or replacement of several other dams that also had been built using hydraulic filling.

Figure 20.9 The soils below these apartment buildings in Niigata, Japan, liquefied during the 1964 earthquake, which produced bearing capacity failures (as discussed in Chapter 17). The failure reportedly occurred very slowly, and the buildings were very strong and rigid, so they remained virtually intact as they tilted. There was very little damage to the interior, and the doors and windows still functioned after the failure. Afterwards, the occupants of the center building were able to evacuate by walking down the exterior wall. (Earthquake Engineering Research Center Library, University of California, Berkeley, Steinbrugge Collection)

Physical Processes

Geotechnical engineers began intensive studies of liquefaction following the 1964 earthquakes in Alaska and Japan. These studies have included field evaluations following major earthquakes and laboratory studies using special cyclic loading devices. As a result of these studies, we now have a much better understanding of the physical processes behind this phenomenon and are better able to identify the soil conditions that are prone to liquefaction in future earthquakes.

Liquefaction occurs only when all of the following criteria have been met:

- The soil is cohesionless
- The soil is loose
- The soil is saturated
- The earthquake produces ground shaking with sufficient intensity and duration
- The ground shaking produces undrained conditions in the soil

Figure 20.10 The soils below this building in Niigata, Japan, liquefied during the 1964 earthquake, causing settlement and bearing capacity failures. (Earthquake Engineering Research Center Library, University of California, Berkeley, Steinbrugge Collection)

A *cohesionless soil* is one that has a cohesion, *c*, of zero. In other words, these soils obtain all of their shear strength from friction and interlocking between the particles. This includes most sands and gravels, and some nonplastic silts. The other category of soils is called *cohesive soils* and have $c > 0$. Cohesive soils will not liquefy.

When cohesionless soils are loose (i.e., when they have a low relative density, D_R), they tend to compress when subjected to cyclic loading. This is why vibratory compaction equipment is so effective in these soils (see Chapter 6). If the soil is not saturated, this compression occurs easily. However, if it is saturated ($S = 100\%$), some of the pore water must escape before the solid particles can compress.

If the hydraulic conductivity, *k*, is very high, the necessary amount of water will escape and the soil will compress. This is what we called the *drained condition* in Chapter 13, and it would be present in clean gravels. However, if the hydraulic conductivity is lower, yet the rate of loading is still rapid (i.e., an earthquake), the water cannot drain quickly enough and the *undrained condition* occurs. This is the situation in most SW and SP sands, and can occur in some SM and ML soils (Vaid, 1994; Finn, Ledbetter and Wu, 1994; Singh, 1994), and in some gravels (Hynes, 1994; Evans and Zhou, 1994). Finer-grained soils do not compress as readily under cyclic loading and usually have some cohesive strength, so they are not being considered here.

When the undrained condition prevails in these soils, the squeezing action of the particles trying to compress produces excess pore water pressures. Each cycle of shaking adds to these excess pore water pressures. If the earthquake has sufficient intensity and duration, these pressures can become quite high, and can lead to liquefaction.

In Chapter 13, we wrote the Mohr-Coulomb strength equation:

$$s = c' + \sigma' \tan \phi' \qquad (20.4)$$

Combining it with Equations 10.32 and 13.6 and setting $c' = 0$ (because the soils under discussion are cohesionless) gives:

$$s = (\sigma - u_h - u_e) \tan \phi' \qquad (20.5)$$

where:
s = shear strength
σ = total stress
u_h = hydrostatic pore water pressure
u_e = excess pore water pressure
c' = effective cohesion
ϕ' = effective friction angle

Thus, as the excess pore water pressure builds up during the earthquake, the shear strength decreases, possibly falling to zero.

Assessment

Liquefaction research also has produced methods of assessing the susceptibility of soils to liquefaction. Most of these methods use the *cyclic stress approach*, which describes earthquake loading in terms of the *cyclic stress ratio*, τ_{cyc}/σ_{z0}', where τ_{cyc} is the cyclic shear stress and σ_{z0}' is the initial vertical effective stress. This method assesses the cyclic stress ratio anticipated at the site during a certain design earthquake and compares it to that required to produce liquefaction. Both of these values depend on many factors, and very detailed investigations and analyses can be employed to define them. However, for many projects, a simplified analysis (Seed, et al., 1985) is sufficient. We will confine our discussion to these simplified analyses.

The cyclic stress ratio induced in the soil by the design earthquake may be estimated by using the following simplified formula (Seed and Idriss, 1971):

$$\left(\frac{\tau_{cyc}}{\sigma_{z0}'} \right)_{eqk} = 0.65 \frac{a_{max}}{g} \frac{\sigma_{z0}}{\sigma_{z0}'} r_d \tag{20.6}$$

where:

$(\tau_{cyc}/\sigma_{z0}')_{eqk}$ = cyclic stress ratio produced by an earthquake
a_{max}/g = peak horizontal ground acceleration divided by acceleration of gravity
σ_{z0} = initial vertical total stress
σ_{z0}' = initial vertical effective stress
r_d = stress reduction factor (from Figure 20.11)

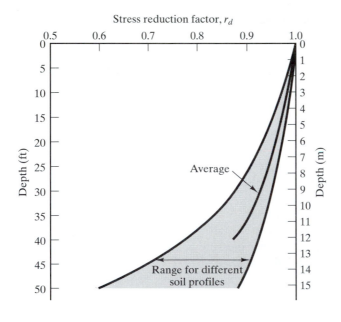

Figure 20.11 Stress reduction factor, r_d, for use in Equation 20.6. Usually we use the curve labeled "average." (Adapted from Seed and Idriss, 1971).

The design value of a_{max}/g is determined from an assessment of the magnitude of the design earthquake and its distance from the project site, as well as site response effects.

The cyclic stress ratio required to produce liquefaction depends on many factors, but the simplified analysis considers only the following:

- The standard penetration test (SPT) or cone penetration test (CPT) resistance, which reflects the relative density. Dense soils have much more resistance to liquefaction.
- The grain size distribution, expressed either as percentage of fines (i.e., percent passing a #200 sieve) or as D_{50} (the mean grain size). Soils with less than 5 percent fines are most susceptible to liquefaction. If more than 5 percent fines are present, the liquefaction resistance becomes much greater.
- The earthquake magnitude, which reflects its duration. Long duration earthquakes are more likely to cause liquefaction.

This method was originally developed using SPT data from sites that had experienced significant earthquakes and whose liquefaction history was known. Figure 20.12 shows cyclic stress ratios for these sites, with closed symbols representing those that had liquefied, and open symbols representing those that had not. The curves on this plot are thus empirical divisions between liquefiable and non-liquefiable soils. When using this figure, be sure to adjust the field N-value to $(N_1)_{60}$ using Equation 3.2.

Figure 20.12 is based on the SPT because nearly all of the available data is in this form. However, the SPT is subject to a variety of errors, and thus is not a very precise test (see discussion in Chapter 3). The cone penetration test (CPT) is more reliable and more precise, and should provide much better liquefaction predictions. Unfortunately, very little pre-earthquake CPT data is available from sites that have liquefied, so it is more difficult to develop the appropriate curves.

Figure 20.13 may be used to evaluate liquefaction from CPT results. It is based on laboratory tests and theoretical analyses, and seems to agree well with the SPT-based curves. To use this figure, it is necessary to apply an overburden correction to the field q_c values using Equation 3.4, thus converting them to q_{c1} values.

A soil's susceptibility to liquefaction also depends on the duration of the earthquake. Those that last longer generate more excess pore water pressures and thus are more likely to induce liquefaction. We adjust for this effect by noting that high-magnitude earthquakes also tend to have long durations, while those with low magnitudes have shorter durations. The τ_{cyc}/σ_{z0}' values in Figures 20.12 and 20.13 are calibrated for the duration of a typical magnitude 7.5 earthquake; for other magnitudes, adjust these values using Equation 20.7 and Figure 20.14.

$$\left(\frac{\tau_{cyc}}{\sigma_{z0}'} \right)_M = \psi \left(\frac{\tau_{cyc}}{\sigma_{z0}'} \right)_{M=7.5} \tag{20.7}$$

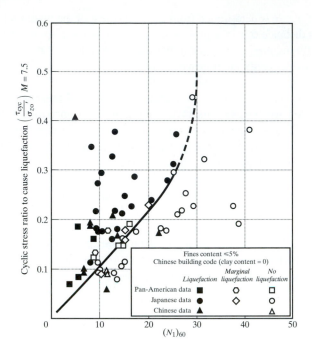

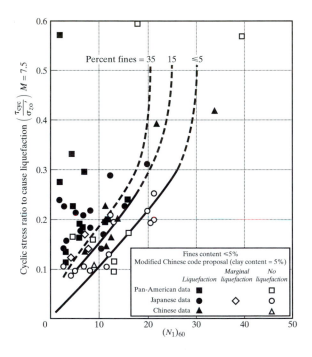

Figure 20.12 Cyclic stress ratio to cause in a magnitude 7.5 earthquake as a function of SPT $(N_1)_{60}$-value and percent fines. The upper diagram is for soils with $\leq 5\%$ fines. The lower diagram is for soils with $>5\%$ fines. The number beside each data point indicates the percent fines. (Seed, et al., 1985). Used with permission of ASCE.

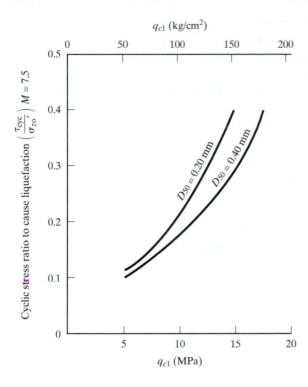

Figure 20.13 Cyclic stress ratio to cause liquefaction of clean sands in a magnitude 7.5 earthquake as a function of CPT and grain size (adapted from Mitchell and Tseng, "Assessment of Liquefaction Potential by Cone Penetration Resistance," Vol. 2, H. Bolton Seed Memorial Symposium Proceedings, May 1990, BiTech Publishers, Ltd., used with permission).

Most geotechnical engineers use the Seed and Idriss curve in Figure 20.14. However, later research suggests it may be overly conservative for low-magnitude earthquakes, and unconservative in high-magnitude earthquakes (Arango, 1996). The curve by Ambraseys may be more accurate.

It is not clear which magnitude is to be used. However, it seems reasonable to use M_L for magnitudes below about 6.5 and M_s for greater magnitudes.

The factor of safety against liquefaction is then:

$$F = \frac{\left(\dfrac{\tau_{cyc}}{\sigma'_{z0}} \right)_M}{\left(\dfrac{\tau_{cyc}}{\sigma'_{z0}} \right)_{eqk}} \qquad (20.8)$$

According to this analysis, liquefaction will occur whenever $F < 1$. However, significant excess pore water pressures can occur even at F values greater than 1. Generally, the minimum acceptable factor of safety is between 1.25 and 1.50 (Seed and Idriss, 1982).

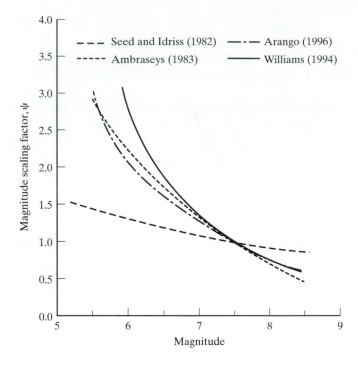

Figure 20.14 Magnitude scaling factor, ψ, for use in liquefaction analyses (Seed and Idriss, 1982; Williams, 1994; Ambraseys, 1988; and Arango, 1996).

Example 20.1

A series of exploratory borings have been drilled at a site near Memphis, Tennessee where liquefaction might be a problem. The results of standard penetration and sieve analysis tests were as follows:

Depth (ft)	$(N_1)_{60}$	Soil Classification	Percent Fines
5	14	Silty sand (SM)	30
8	12	Fine to medium sand (SW)	5
11	10	Fine to medium sand (SW)	3
13	12	Fine to medium sand (SW)	6
17	11	Fine to medium sand (SW)	3
22	19	Silty sand (SM)	20
26	24	Silty sand (SM)	22

The design earthquake will occur in the New Madrid Fault Zone (the same zone that generated the 1811–1812 earthquakes). The causative fault is 75 km from the site and the design earthquake has a moment magnitude of 8.0. The site is underlain by deep cohesionless soils.

Using a unit weight of 110 lb/ft^3 and a groundwater table 3 ft below the ground surface, evaluate the liquefaction potential at this site using the simplified method.

Solution

Per Equation 20.3, $a_{max} = 0.16g$ in bedrock
Per Figure 20.4, $a_{max} = 0.14g$ in deep cohesionless soil
Per Figure 20.14, $\psi = 0.65$ (Arango)

Depth (ft)	Cyclic Stress Ratio to Cause Liquefaction		σ_{z0} (lb/ft^2)	σ_{z0}' (lb/ft^2)	r_d	Cyclic Stress Ratio Produced by Earthquake	F
	$M = 7.5$	$M = 8.0$					
5	0.23	0.15	550	425	0.99	0.12	1.2
8	0.13	0.085	880	568	0.98	0.14	0.6
11	0.11	0.072	1210	711	0.97	0.15	0.5
13	0.14	0.091	1430	806	0.97	0.16	0.6
17	0.12	0.078	1870	996	0.96	0.16	0.5
22	0.29	0.189	2420	1234	0.95	0.17	1.1
26	-	-	-	-	-	-	High

Conclusion

The zone between depths of about 6 and 20 ft is clearly liquefiable. Below that depth, the increased density (as demonstrated by the higher N) and the presence of fines makes liquefaction less likely.

Remediation

It is generally best to simply avoid building on sites prone to liquefaction. However, when land values are high, it may be cost effective to remediate the liquefaction hazard prior to construction. Sometimes remediation also is practical at sites that have already been developed and later found to be prone to liquefaction.

In some cases, structures can be supported on deep foundations that extend through the liquefiable soil to deeper, stable strata. However, this solution is not as simple as it might first appear, because the same earthquake that causes liquefaction also produces large lateral loads in the structure and these loads need to be transmitted through the foundation and into the ground. Although deep foundations can easily transmit vertical loads through the liquefied soils, they may not be able to do so with the lateral loads. Therefore, most liquefaction remediation methods focus on improving the soil to prevent the liquefaction from occurring. Many such methods are available (Hryciw, 1995) and they have performed well during earthquakes (Mitchell, Baxter, and Munson, 1995).

Another option is to excavate the liquefaction-prone soils and replace them with a compacted fill. This is often very expensive, because it requires extensive dewatering systems to permit excavation below the groundwater table, and is normally justified only on very critical projects, such as earth dams.

Most liquefaction remediation methods use in-situ methods to increase the soil's relative density, thus improving its liquefaction resistance. Several methods are available, as discussed in Chapter 19.

Another approach is to solidify the soil by injecting chemical or cement grouts (Maher and Gucunski, 1995). This method is especially applicable to remediation of sites with existing structures.

Sometimes a liquefaction hazard can be virtually eliminated by aggressive groundwater pumping that lowers the groundwater table below the potentially liquefiable zone. If the soils are no longer saturated, the liquefaction hazard ceases to exist. However, this method depends on the continuous use of these pumps, and thus requires long-term funding and maintenance. If the groundwater quality is sufficient, the pumps can supply municipal or industrial needs, thus generating revenue.

Another remediation method consists of installing a series of vertical *stone columns*, which are borings filled with gravel (Seed and Booker, 1976). These columns have a high permeability, and thus provide for rapid drainage of the excess pore water pressures. This keeps the excess pore water pressures low enough to avoid liquefaction. The stone columns also provide additional support for the structure.

20.4 SURFACE RUPTURE

In small earthquakes, permanent shear deformations along the fault usually occur only underground and do not extend to the ground surface. However, those with magnitudes greater than about 6.0 usually are accompanied by fault rupture at the ground surface, as shown in Figure 20.15 (Bonilla, 1970; Youd, 1980; Bonilla, et al., 1984; CDMG, 1994a). This movement can be horizontal, vertical, or both, and the amount of movement increases as the magnitude increases. In addition, larger magnitudes also are associated with greater fault rupture lengths.

During very large earthquakes, the rupture distance can be several meters.

Figure 20.15 The Imperial Fault in California moved up to 5.9 m during a 1940 earthquake. Less movement occurred in the portion of the fault that passes through this orange grove, but still enough to be clearly discernible. The dark line, added to the photograph, indicates the fault location. (Earthquake Engineering Research Center Library, University of California, Berkeley, Steinbrugge Collection)

For example, the 1906 San Francisco Earthquake produced up to 6.1 m (20 ft) of displacement over a distance of 444 km (275 mi). However, some rupture can occur even during moderate earthquakes. For example, up to 30 cm (1 ft) of displacement occurred along 2 km (1.5 mi) of the Stephens Pass Fault in California during a 1978 earthquake that had a magnitude of only 4.3. This fault was not even known to exist prior to the earthquake!

Surface rupture presents a special problem for structures, transportation facilities, utilities, and other important facilities located directly over the fault. Buildings and bridges are especially susceptible to damage (see Figure 20.16), and must be set back a suitable distance from faults capable of surface rupture. In California, the 1972 Alquist-Priolo Earthquake Fault Zoning Act requires special fault studies for buildings near such faults, and the establishment of setback zones (CDMG, 1994a).

Sometimes engineers cannot avoid building across an active fault. For example, highways, pipelines, and other projects often must connect points on opposite sides of the fault. Fortunately, highways can withstand large fault movements and often can remain in service with little or no repairs. Some pipelines can be built with special flexible joints to accommodate small movements. Other facilities might be designed to accommodate rapid repair when necessary.

Designers of the California Aqueduct, a major water supply facility, chose to cross the San Andreas Fault using a small lake directly over it. The incoming aqueduct feeds the lake on one side of the fault, then the outgoing aqueduct drains it from the opposite side.

Earth dams and levees may be able to accommodate some fault movement, but are best built away from active faults.

Figure 20.16 Culligan's Blarneystone Ranch, Montana. Surface rupture during the 1959 Hebgen Lake Earthquake passed through this building, causing part of it to completely collapse (seen behind the scarp) and severe damage to the remaining section (on the left). This earthquake produced displacements up to 5.5 m (18 ft), which is far more than any building can accommodate. (Earthquake Engineering Research Center Library, University of California, Berkeley, Steinbrugge Collection)

20.5 OTHER PERMANENT GROUND DEFORMATIONS

Various kinds of earthquake-induced permanent ground deformations also can occur away from the fault rupture zone, and they have caused serious damage to building, bridges, and other improvements (Dobry, 1994). These include:

- Settlement
- Lateral spreading due to liquefaction
- Landslides and other slope stability failures
- Excessive movement or failure of nearby retaining structures
- Soil failure due to induced dynamic loads from a structure

Settlement

Loose sandy soils, either dry or saturated, often consolidate when subjected to cyclic loading from an earthquake. This behavior is similar to that beneath the vibratory compaction equipment discussed in Chapter 6, except on a much larger scale. Settlements as large as 5 percent of the sand strata thickness have been reported (Dobry, 1994). Such settlements are often very erratic, so the resulting differential settlements can be large. Semi-empirical methods have been developed to estimate the magnitude of these settlements (Tokimatsu and Seed, 1987).

Lateral Spreading Due to Liquefaction

Lateral spreads are massive horizontal movements of soil, as discussed in Chapter 14. They are most often caused by liquefaction, and this is the single most important cause of large ground deformation during earthquakes. Lateral spreads have been responsible for extensive damage to buildings, bridges, and other structures.

Lateral spreads frequently occur adjacent to river banks because liquefiable soils are often found there. This mode has been especially troublesome for bridges because it easily shears off pile foundations and spreads apart bents and abutments, separating them from the bridge deck (Bartlett and Youd, 1993). Pier movements as large as 2.1 m (6.9 ft) have been observed. One of the most dramatic failures of this type was the Showa Bridge, shown in Figure 20.17.

Semi-empirical methods have been developed to evaluate the potential for lateral spreads based on the liquefaction potential of the soil, the configuration of the ground surface, and other factors (Bartlett and Youd, 1992; Dobry and Baziar, 1992). Once this potential has been identified, lateral spread hazards can either be avoided by building elsewhere, or remediated using the liquefaction remediation measures described earlier, various kinds of walls, or other methods.

Figure 20.17 The Showa Bridge in Niigata, Japan, collapsed due to the formation of lateral spreads in the underlying soils during the 1964 earthquake. The lateral spreads moved the piers out of position, thus removing support from the simply supported deck. (Earthquake Engineering Research Center Library, University of California, Berkeley, Steinbrugge Collection)

Landslides and Other Slope Stability Failures

Earthquake-induced landslides also have been observed in areas where no liquefaction occurred. Most are small, but some large ones also have been observed. For example, the 1959 Hebgen Lake Earthquake in Montana triggered a massive slide in nearby Madison Canyon. It buried 28 people and created a dam across the canyon, creating a new lake — Earthquake Lake.

Buildings and other structures located near steep slopes are especially vulnerable to these slides. Figure 20.18 shows the destruction of such a building during the 1964 Alaska Earthquake.

Even when slopes do not fail, tension cracks sometimes form on the ground above. These cracks may be several centimeters wide, and thus can be hazards in themselves. These cracks also provide conduits for water to enter the ground, which later may trigger a landslide.

See Chapter 14 for discussions of rockfalls and seismic slope stability analysis methods.

Excessive Movement or Failure of Retaining Walls

Occasionally retaining walls fail during earthquakes, thus removing support for adjacent facilities. These have most often occurred in port facilities. Sometimes such failures have been due to liquefaction of the soils behind or below the wall.

Figure 20.18 The head scarp of this landslide, which was triggered by the 1964 Alaska Earthquake, extended beneath the Government Hill Elementary School in Anchorage. The building split apart and dropped about 2.8 m (9 ft). (Earthquake Engineering Research Center Library, University of California, Berkeley, Steinbrugge Collection)

Soil Failure Due to Induced Dynamic Loads from a Structure

Seismic loads acting on structures are eventually transmitted to the ground through their foundations, and thus increase the stresses in the ground. Sometimes these stresses exceed the soil strength and a failure occurs. In tall buildings, this is most likely to occur along the perimeter walls, since those foundations carry a large share of the seismic loads.

Seismic foundation failures are not very common, so long as the soil does not liquefy. Usually, the static factor of safety provides a sufficient margin to resist seismic loads. Those failures that have occurred have usually been in foundations that had deficient static designs.

20.6 TSUNAMIS AND SEICHES

A *tsunami* (from the Japanese words *tsu,* "port" and *nami* "wave") is a large ocean wave generated by an earthquake. They have sometimes been called "tidal waves," but this term is misleading and should not be used. They have nothing to do with tides. Although all earthquakes do not generate tsunamis, those that do often have disastrous results.

These waves travel very quickly, about 550–800 km/hr (350–500 mi/hr), but are virtually undetectable in the open sea. However, when the waves approach land, the water depth becomes shallower and the waves suddenly appear. Because of this high speed and "stealth"-like behavior, and their ability to travel long distances, tsunamis can be very destructive. For example, the Chile Earthquake of 1960 produced a tsunami that killed 61 people in Hawaii and 199 people in Japan. At Hilo, Hawaii, a 6 m (20 ft) wall of water struck the city, damaging or destroying buildings, sweeping away cars, and so on (Wiegel, 1970). A 1946 earthquake in Alaska also produced another tsunami at Hilo that destroyed a lighthouse located 10 m (30 ft) above sea level.

Early warning systems have been developed to quickly assess the potential for tsunamis after large earthquakes, but such systems sometimes produce the opposite of the intended result. Immediately following the 1964 Alaska Earthquake, a tsunami warning issued for San Francisco brought thousands of people to the beach to watch! In Crescent City, California, 2 m (7 ft) tall waves occurred first, and people rushed to the docks to inspect the damage. Then a 4 m (14 ft) tall wave arrived and killed ten people (Rahn, 1996).

A *seiche* is similar except it occurs in lakes or rivers. Sometimes these occur because the natural frequency of the lake matches that of the earthquake, creating a resonant condition. Seiches also can occur when surface fault rupture occurs beneath the water, which was the case in the 1959 Hebgen Lake Earthquake in Montana.

20.7 SEISMIC PROVISIONS IN BUILDING CODES

The first seismic provisions in a United States building code appeared in the 1927 Uniform Building Code. During the following decades, such code provisions became more common and more stringent, especially in the western states. Building codes often include seismic zoning maps, such as the one in Figure 20.19, and require different levels of seismic design for each zone.

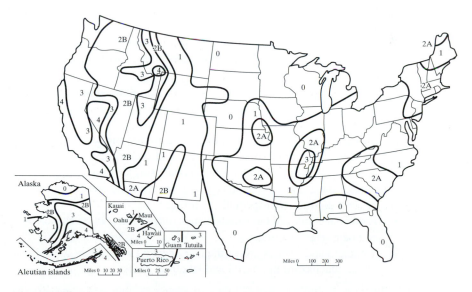

Figure 20.19 Seismic zoning map for the United States. Zone 0 has the least design requirements, and zone 4 has the most. Reproduced from the 1997 edition of the *Uniform Building Code* with permission of the publisher, The International Conference of Building Officials.

Unfortunately, both the public and government officials are sometimes slow to adopt the seismic provisions in building codes (see quotation at the beginning of this chapter). This reluctance is especially noteworthy in areas where large earthquakes occur at long intervals with only minimal activity between. For example, the city of Memphis, Tennessee, is located only 75 km (45 mi) from the epicenters of the 1811–1812 New Madrid Earthquakes (estimated magnitudes 7.3–7.8) and is in seismic zone 3, yet it had no seismic provisions in its building code before 1990. In that year, the city finally adopted a weakened version of the 1988 Southern Building Code seismic requirements (Olshansky, 1993).

SUMMARY

Major Points

1. Geotechnical earthquake engineering is the branch of geotechnical engineering that deals with earthquakes and their effects on civil engineering projects.

2. Although the ground motions produced by earthquakes are very complex, it often is convenient to describe them using the *magnitude* and the *intensity* of an earthquake. The magnitude is a measure of the amount of energy released, and is independent of the distance from the epicenter to the observer. The intensity describes the earthquake effects at a particular location, and depends on the magnitude, distance from the epicenter, soil conditions, and other factors.

3. Ground shaking is the most fundamental hazard from earthquakes. Like the intensity, it depends on the magnitude, distance from the epicenter, soil conditions, and other factors. Geotechnical engineers estimate the potential ground motions at the project site; these estimates are used in the design of structures and other civil engineering works.

4. Liquefaction is another important hazard related to earthquakes. It can cause serious distress to buildings and other structures, flotation of buried facilities, landslides, and other problems.

5. Methods have been developed to assess the liquefaction potential at a project site. When a hazard exists, special soil improvement techniques are available to remediate the problem.

6. Surface rupture describes shearing at the ground surface along a fault trace. Buildings and other structures must not be located on top of active faults, because it is not practical to design them to accommodate surface rupture.

7. Lateral spreads are a special type of ground failure caused by liquefaction. They can be very destructive, especially to bridges.

8. Earthquake-induced landslides, which may or may not be caused by liquefaction, also can be a source of damage.

9. Tsumanis and seiches are earthquake-induced waves in bodies of water. They can be a potential hazard in coastal and lakefront areas.

Vocabulary

active fault	inactive fault	peak acceleration
attenuation	intensity	Richter magnitude
cohesionless soil	lateral spreads	sand boil
cohesive soil	liquefaction	seiche
cyclic mobility	magnitude	seismic risk analysis
cyclic stress ratio	maximum credible	seismograph
epicenter	earthquake	site response spectrum
flow liquefaction	maximum probable	surface rupture
focus	earthquake	tectonic earthquakes
ground shaking	Modified Mercalli Intensity	tsunami
hypocenter	Scale	

COMPREHENSIVE QUESTIONS AND PRACTICE PROBLEMS

Note: Perform all liquefaction analyses using the Seed and Idriss curve in Figure 20.11.

20.1　A site in California is located 10 mi from Fault A and 23 mi from Fault B. These two faults have maximum probable earthquake magnitudes, M_w of 6.5 and 7.8, respectively. The site is underlain by a deep deposit of soil that has a shear wave velocity of 2000 ft/s. Using the Boore et al (1993) attenuation relationship, compute the peak horizontal acceleration at the ground surface for both earthquakes, then select the value to be used for design.

20.2　Using the fault definitions in Figure 2.9, determine whether the fault in Figure 20.15 has experienced right-lateral or left-lateral movement. Explain.

20.3　The soil at a certain site consists of a fine sand with 4% passing the #200 sieve. The $(N_1)_{60}$ value at a depth of 5.0 m is 12, the unit weight is 16.4 kN/m³, and the groundwater table is at a depth of 0.6 m. The design earthquake has a magnitude of 7.0 and would produce a peak horizontal acceleration of 0.45g at this site. Compute the factor of safety against a liquefaction failure.

20.4　A series of cone penetration tests and exploratory borings have been performed at a site that might be prone to liquefaction. Based on this data, the following representative soil profile has been developed:

Depth (m)	Soil Classification	q_c (kg/cm²)	γ (kN/m³)
0–3.0	Silty sand $D_{50} = 0.10$ mm	60	17.5
3.0–5.5	Fine to medium sand $D_{50} = 0.22$ mm	55	18.0
5.5–8.0	Clayey sand $D_{50} = 0.08$ mm	120	18.2
> 8.0	Silty clay $D_{50} = 0.004$ mm	130	18.7

The groundwater table is at a depth of 1.0 m. The design earthquake has a magnitude of 6.8 and would produce a peak horizontal acceleration of 0.50g at this site. Assess the potential for liquefaction and discuss your findings.

20.5 The soil profile at a certain site in the western United States consists of 10 ft of sandy silt (γ = 115 lb/ft^3 above the groundwater table, 118 lb/ft^3 below) underlain by 12 ft of fine to medium sand (γ = 109 lb/ft^3). The groundwater table is at a depth of 8 ft. The sand strata has an average $(N_1)_{60}$ value of 10, 6% passing the #200 sieve. A deep deposit of soft-to-medium clay (Boore's soil type C) is present below the sand. The design earthquake has a moment magnitude of 6.3 and would occur on a fault located 8 km from the site.

   a. Compute the factor of safety against a liquefaction failure, and indicate where the liquefaction is most likely to occur.
   b. Using Equation 4.21 with D_{50} = 0.4 mm, t = 100 yr, and OCR = 1, determine the present relative density of the sand strata.
   c. The chief engineer is considering the use of vibroflotation to densify the sand in-situ, thus eliminating the liquefaction problem. What minimum relative density must be achieved to obtain a factor of safety of 1.25.

20.6 According to the seismic zone map in Figure 20.19, New England is in zone 2A. Conduct a literature search to determine the history of significant earthquakes in this region.

20.7 The 1964 Niigata, Japan, earthquake had a magnitude of 7.5 and was centered about 35 miles from the city. The peak horizontal ground accelerations in Niigata were about 0.16g. Much of the city is underlain by a deep deposit of loose sand, as shown in Figure 20.20. It has about 10% fines. Using this data and an assumed unit weight of 100 lb/ft^3, develop a plot of factor of safety against liquefaction vs. depth in the sand strata. Are the results of your analysis consistent with the damage shown in Figures 20.9 and 20.10? If such an analysis had been performed before the earthquake (which would have been impossible, since this analysis technique had not yet been developed), would it have predicted this damage?

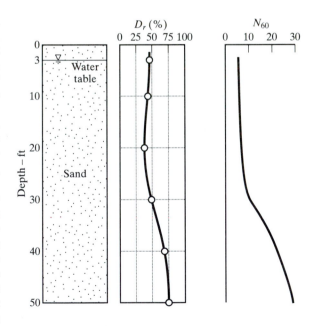

Figure 20.20 Typical soil profile in parts of Niigata, Japan (Seed and Idriss, 1966; Seed and Idriss, 1967).

20.8 In 1961, soil dredged from San Diego Bay was used to create Harbor Island. This fill was placed hydraulically (see discussion of hydraulic fills in Chapter 6) and consists primarily of SP, SP-SM, and SM soils, as shown in the simplified soil profile in Figure 20.21 (Forrest and Noorany, 1989). Using a peak horizontal ground acceleration of 0.20g from a magnitude 5.4 earthquake, assess the potential for liquefaction at this site. Assume the SP soils have 4% fines, SP-SM have 8% fines, and SM have 15% fines.

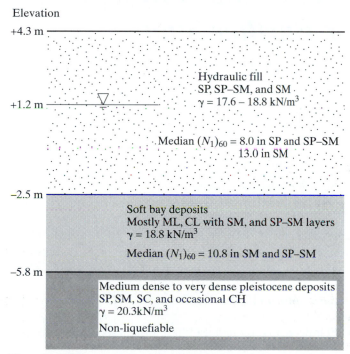

Elevation

+4.3 m

Hydraulic fill
SP, SP–SM, and SM
$\gamma = 17.6 - 18.8$ kN/m³

+1.2 m

Median $(N_1)_{60} = 8.0$ in SP and SP–SM
13.0 in SM

–2.5 m

Soft bay deposits
Mostly ML, CL with SM, and SP–SM layers
$\gamma = 18.8$ kN/m³

Median $(N_1)_{60} = 10.8$ in SM and SP–SM

–5.8 m

Medium dense to very dense pleistocene deposits
SP, SM, SC, and occasional CH
$\gamma = 20.3$ kN/m³

Non-liquefiable

Figure 20.21 Simplified soil profile and SPT results at Harbor Island, San Diego, California (Forrest and Noorany, 1989).

Appendix A

Recommended Resources for Further Study

Chapter 1 — Introduction to Geotechnical Engineering

Casagrande, Arthur (1960), "Karl Terzaghi — His Life and Achievements," *From Theory to Practice in Soil Mechanics*, L. Bjerrum, et al., p. 3–21, John Wiley, New York.
> The most comprehensive description of Terzaghi's professional life.

Casagrande, Arthur (1964), "Karl Terzaghi, 1883-1963," *Géotechnique*, Vol. 14, No. 1, p. 1–9.
> Follow-up to Casagrande (1960), written soon after Terzaghi's death.

De Boer, R., Schiffman, R.L., and Gibson, R.E. (1996), "The Origins of the Terzaghi-Fillunger Dispute," *Géotechnique*, Vol. 46, No. 2, p. 175–186.
> An interesting story of a dispute between Terzaghi and one of his colleagues.

Flodin, Nils, and Broms, Bengt (1981), "Historical Development of Civil Engineering in Soft Clay," Chapter 1 in *Soft Clay Engineering*, E.W Brand and R.P. Brenner, Eds., Elsevier, Amsterdam.
> Very thorough discussion of the history of geotechnical engineering in soft clays.

Kerisel, Jean (1987), *Down to Earth; Foundations Past and Present: The Invisible Art of the Builder*, A.A. Balkema, Rotterdam.
> A "coffee table" style book with interesting text and illustrations of foundation engineering from ancient times to the present.

Publications Committee of the XI International Conference on Soil Mechanics and Foundation Engineering (1985), *Golden Jubilee Volume,* A.A. Balkema, Rotterdam.
This volume includes three papers on the history of geotechnical engineering:
- "The History of Geotechnical Engineering Up Until 1700," Jean Kerisel
- "A History of Soil Properties," A.W. Skempton
- "The Last Sixty Years," Ralph Peck

Skempton, A.W. (1979), "Landmarks in Early Soil Mechanics," *Seventh European Conference on Soil Mechanics and Foundation Engineering*, Vol. 5, p. 1–26.
Excellent discussion of the early history of geotechnical engineering.

Bjerrum, Laurits, and Flodin, Nils (1960), "The Development of Soil Mechanics in Sweden: 1900-1925," *Géotechnique*, Vol. 10, No. 1, p. 1–18.
Traces the early Swedish advances in geotechnical engineering.

Chapter 2 — Engineering Geology

Goodman, Richard E. (1993), *Engineering Geology: Rock in Engineering Construction*, John Wiley, New York.
As the title indicates, this book focuses on the application of geology to engineering construction. It includes many illustrative case studies and examples.

West, Terry R. (1995), *Geology Applied to Engineering*, Prentice Hall, Upper Saddle River, NJ.
A college textbook on engineering geology. Discusses most of the topics in this chapter in more detail, along with many other topics not mentioned here.

Chapter 3 — Site Exploration and Characterization

Acker, W.L. (1974), *Basic Procedures for Soil Sampling and Core Drilling*, Acker Drill Company, Scranton, PA.
Discusses the "nuts and bolts" of drilling exploratory borings and recovering soil and rock samples.

ASCE (1972), "Subsurface Investigation for Design and Construction of Foundations and Buildings," *Journal of the Soil Mechanics and Foundation Division*, Vol. 98, No. SM5–SM8.
Practical recommendations from a task committee.

Hunt, Roy E. (1984), *Geotechnical Engineering Investigation Manual*, McGraw Hill, New York.
A very thorough discussion of various investigation tools and methods.

Kulhawy, Fred H. (1996), "Estimation of In-Situ Test Uncertainty"; *Uncertainty in the Geologic Environment*, Geotechnical Special Publication 58, p. 269-286, ASCE.
Discusses sources an magnitudes of uncertainty in site characterization.

Chapter 4 — Soil Composition

Mitchell, James K. (1993), *Fundamentals of Soil Behavior*, 2nd ed., John Wiley, New York.
  The standard geotechnical engineering reference on soil structure and composition, and their impacts on engineering behavior. Includes extensive discussions of clay mineralogy.

Chapter 5 — Soil Classification

ASTM Standard D2487, "Classification of Soils for Engineering Purposes (Unified Soil Classification System)," *ASTM Annual Book of Standards*, Volume 04.08, American Society for Testing and Materials, Philadelphia.
  This is the authoritative source for information on the Unified Soil Classification System, and provides more detail than presented here.

Chapter 6 — Excavation, Grading, and Compacted Fill

Church, Horace K (1981), *Excavation Handbook*, McGraw Hill, New York.
  An extensive guide to earthmoving equipment, with data on productivity, selection, and other practical concerns.

Hilf, Jack W. (1991), "Compacted Fill," Chapter 8 in *Foundation Engineering Handbook*, 2nd ed., Hsai-Yang Fang, Ed., Van Nostrand Reinhold, New York.
  Discussion of advanced concepts.

Nunnally, S.W. (1993), *Construction Methods and Management*, 3rd ed., Prentice Hall, Upper Saddle River, NJ.
  Discusses selection and performance of equipment for earthmoving and other purposes.

Chapters 7 and 8 — Groundwater

Bathe, Klaus-Jürgen (1996), *Finite Element Procedures*, Prentice Hall, Upper Saddle River, NJ.
  Discusses applications of the finite element method to solve many engineering problems, including seepage.

Cedergren, Harry R. (1989), *Seepage, Drainage, and Flow Nets*. 3rd ed., John Wiley, New York.
  Discusses principles of groundwater and applications to soil drainage around structures, pavements, dams, and other projects. Includes discussions of complex flow nets.

Desai, C.S. (1979), *Elementary Finite Element Method,* Prentice Hall, Upper Saddle River, NJ.
> Discusses applications of the finite element method, including its use in groundwater problems.

Driscoll, Fletcher G. (1986), *Groundwater and Wells*, 2nd ed., Johnson Filtration Systems, Inc., St. Paul, MN.
> Practical guide to the analysis, design, and installation of wells for water supply.

Kashef, Abdel-Aziz (1986), *Groundwater Engineering*, McGraw Hill, New York.
> Discusses various aspects of groundwater analysis, utilization, and management.

Powers, J. Patrick (1992), *Construction Dewatering: New Methods and Applications*, 2nd ed., John Wiley, New York.
> A practical guide to the analysis, design, and operation of construction dewatering systems.

Chapter 9 — Geoenvironmental Engineering

Bedient, Philip B., Rifai, Hanadi S., and Newell, Charles J. (1994), *Ground Water Contamination: Transport and Remediation*, Prentice Hall, Upper Saddle River, NJ.
> Detailed discussions of the analysis and remediation of underground contamination problems.

LaGrega, Michael D., Buckingham, Phillip, and Evans, Jeffrey C. (1994), *Hazardous Waste Management*, McGraw Hill, New York.
> A comprehensive textbook on the treatment, disposal, and remediation of hazardous wastes.

McBean, Edward A., Rovers, Frank A., and Farquhar, Grahame J. (1995), *Solid Waste Landfill Engineering and Design*, Prentice Hall, Upper Saddle River, NJ.
> Covers the design of sanitary landfills.

Chapter 10 — Stress

Poulos, H.G., and Davis, E.H. (1974), *Elastic Solutions for Soil and Rock Mechanics*, John Wiley, New York.
> An extensive collection of formulas and charts to solve stress problems in soil and rock.

Chapters 11 and 12 — Settlement

Duncan, J. Michael (1993), "Limitations of Conventional Analysis of Consolidation Settlement," ASCE *Journal of Geotechnical Engineering*, Vol. 119, No. 9, p. 1331-1359; Discussions in Vol. 121, No. 6, p. 513–518.

Discusses the application of consolidation analyses to practical problems, with emphasis on understanding their limitations.

Chapter 13 — Strength

Mitchell, James K. (1993), *Fundamentals of Soil Behavior*, John Wiley, New York.
Extensive discussions of the physical basis for soil behavior, including shear strength.

Chapter 14 — Stability of Earth Slopes

Abramson, Lee W., Lee, Thomas S., Sharma, Sunil, and Boyce, Glenn M. (1996), *Slope Stability and Stabilization Methods*, John Wiley, New York.
A comprehensive book on the investigation, analysis, and stabilization of earth slopes.

Bromhead, E.N. (1986), *The Stability of Slopes*, Surrey University Press, Glasgow.
A comprehensive book on earth slope engineering written from a United Kingdom perspective.

Brunsden, Denys, and Prior, David B., Editors (1984), *Slope Instability*, John Wiley, New York.
Contains chapters written by engineers, geologists, and geographers and discussed many aspects of slope instability.

Duncan, J. Michael (1992a), "State-of-the-Art: Static Stability and Deformation Analysis," *Stability and Performance of Slopes and Embankments-II*, Geotechnical Special Publication No. 31, Vol. I, p. 222–266, Raymond B. Seed and Ross W. Boulanger, Eds., ASCE.
A state-of-the art review.

Seed, Raymond B. and Boulanger, Ross W. (1992), *Stability and Performance of Slopes and Embankments - II*, Geotechnical Special Publication No. 31, ASCE.
Two-volume conference proceedings with professional papers on a wide variety of topics related to slope stability.

Turner, A. Keith and Schuster, Robert L., Editors (1996), *Landslides — Investigation and Mitigation*, Special Report 247, National Research Council, Transportation Research Board, Washington, DC.
Excellent coverage of the investigation, analysis, and remediation of slope stability problems.

Wright, Stephen G. (1985), "Limit Equilibrium Slope Analysis Procedures," *Design of Non-Impounding Waste Dumps*, p. 63-77, American Institute of Mining Engineers.
Discusses the various limit equilibrium methods, along with their strengths and limitations.

Chapter 15 — Dams and Levees

Bureau of Reclamation (1987), *Design of Small Dams*, 3rd ed., United States Department of the Interior, Bureau of Reclamation, Denver, CO.
>   Discusses all major aspects of small dam design.

Jansen, Robert B., Ed. (1988), *Advanced Dam Engineering for Design, Construction, and Rehabilitation*, Von Nostrand Reinhold, New York.
>   Comprehensive book on most aspects of earth and concrete dam design.

Sherard, James L., Woodward, Richard J, Gizienski, Stanley F., and Clevenger, William A. (1963), *Earth and Earth-Rock Dams*, John Wiley, New York.
>   Although somewhat dated, this book is still a classic in earth dam engineering.

Chapters 16 and 17 — Earth Pressures, Retaining Walls, and Foundations

Coduto, Donald P. (1999), *Foundation Design: Principles and Practices*, 2nd Ed., Prentice Hall, Upper Saddle River, NJ.
>   The companion volume to this book. Includes much more information on the geotechnical, structural, and construction aspects of retaining walls and structural foundations.

Chapter 18 — Difficult Soils

Andersland, Orlando B. and Anderson, Duwayne M., Eds. (1978), *Geotechnical Engineering For Cold Regions*, McGraw Hill, New York.
>   A textbook covering many aspects of geotechnical engineering in cold regions.

Greenfield, Steven J. and Shen, C.K. (1992), *Foundations in Problem Soils*, Prentice Hall, Upper Saddle River, NJ.
>   Discusses state-of-practice and practical guidelines for construction on difficult soils, including expansive soils.

Nelson, John D. and Miller, Deborah J. (1992), *Expansive Soils: Problems and Practice in Foundation and Pavement Engineering*, John Wiley, New York.
>   Comprehensive discussion of expansive soil behavior and preventive design measures.

Zeavaert, Leonardo (1983), *Foundation Engineering for Difficult Subsoil Conditions*, 2nd ed., Van Nostrand Reinhold, New York.
>   Methods of foundation construction on difficult soils, with emphasis on techniques used in Mexico City.

Chapter 19 — Soil Improvement

Ingles, O.G. and Metcalf, J.B. (1972), *Soil Stabilization*, Butterworths, Sydney.
> Detailed discussions of various methods of stabilization through the use of admixtures.

Koerner, Robert M. (1998), *Designing with Geosynthetics*, 4th ed., Prentice Hall, Upper Saddle River, NJ.
> In-depth discussions of all kinds of geosynthetics, including those used for soil reinforcement.

Moseley, M.P. (1993), *Ground Improvement*, Chapman and Hall, London.
> Discusses several major soil improvement techniques and their applicability to practical problems

Schaefer, Vernon R., (1997), *Ground Improvement, Ground Reinforcement, Ground Treatment Developments 1987-1997*, Geotechnical Special Publication No. 69, ASCE.
> Includes an extensive committee report that presents the 1997 state-of-the-art in various kinds of ground improvement methods.

Stamatopoulos, Aris C. and Kotzias, Panaghiotis C. (1985), *Soil Improvement by Preloading*, John Wiley, New York.
> Thorough discussion of preloading, both with and without vertical drains.

Chapter 20 — Geotechnical Earthquake Engineering

Bolt, Bruce A. (1993), *Earthquakes*, W.H. Freeman, New York.
> An easy-to-understand book on the causes and consequences of earthquakes. Written by a noted seismologist.

Kramer, Steven L. (1996), *Geotechnical Earthquake Engineering*, Prentice Hall, Upper Saddle River, NJ.
> A comprehensive up-to-date textbook on this topic.

Seed, R.B.; Dickenson, S.E.; Reimer, M.F.; Bray, J.D.; Sitar, N.; Mitchell, J.K.; Idriss, I.M.; Kayen, R.E.; Kropp, A.; Harder, L.F.; and Power, M.S. (1990), *Preliminary Report on the Principal Geotechnical Aspects of the October 17, 1989 Loma Prieta Earthquake*, Report No. UCB/EERC-90/05, Earthquake Engineering Research Center, University of California at Berkeley.
> An extensive report on the geotechnical aspects of a major earthquake.

Wiegel, Robert L., Ed. (1970), *Earthquake Engineering*, Prentice Hall, Upper Saddle River, NJ.

> Although somewhat dated, this book still contains a great deal of useful information on seismological, geotechnical, and structural aspects of earthquake engineering.

Soil Testing

The various discussions of laboratory and in-situ tests in this book are not intended to be complete test procedures. Virtually all of these tests are more complex than presented here, so it is necessary to consult a laboratory manual before actually performing them. The following references contain much more detailed information:

ASTM, *Annual Book of ASTM Standards*, Volume 04.08 — Soil and Rock, American Society for Testing and Materials, Philadelphia.

> The authoritative source for most soil testing standards in North America.

Bardet, Jean-Pierre (1997), *Experimental Soil Mechanics*, Prentice Hall, Upper Saddle River, NJ.

> A textbook suitable for instructional use in undergraduate soil mechanics laboratory courses. Much easier to follow than ASTM.

Head, K.H. (1982), *Manual of Soil Laboratory Testing*, Pentech Press, London.

> A three-volume set that includes detailed discussions of nearly all common laboratory tests.

Appendix B

Unit Conversion Factors

ENGLISH UNITS

The following English units are commonly used in geotechnical engineering:

TABLE B1 COMMON ENGLISH UNITS

Unit	Measurement	Symbol
foot	distance	ft
inch	distance	in
pound	force or mass	lb
kip (kilopound)	force	k
ton	force or mass	t
second	time	s
pound per square foot	stress or pressure	lb/ft^2 or psf
pound per square inch	stress or pressure	lb/in^2 or psi
pound per cubic foot	unit weight	lb/ft^3 or pcf

SI AND METRIC UNITS

The following SI (Système International) units are commonly used in geotechnical engineering:

TABLE B2 COMMON SI UNITS

Unit	Measurement	Symbol
meter	distance	m
gram	mass	g
Newton	force	N
Pascal	stress or pressure	Pa
kilonewton per cubic meter	unit weight	kN/m^3
second	time	s

These units are often accompanied by the following prefixes:

TABLE B3 COMMON SI PREFIXES

Prefix	Symbol	Multiplier
milli	m	10^{-3}
centi	c	10^{-2}
kilo	k	10^{3}
mega	M	10^{6}

Some non-SI metric units also are used, especially in Europe. The most common example is the use of kg/cm^2 as a unit of stress.

CONVERSION FACTORS

The conversion factors in Tables B4 - B8 are useful for converting measurements between English, metric and SI units. Most of these factors are rounded to four significant figures. Those in **bold type** are absolute conversion factors (for example 12 inches = 1 ft). When units of force are equated to units of mass, the acceleration (F = ma) is presumed to be 9.807 m/s^2 (32.17 ft/s^2), which is the acceleration due to gravity on the earth's surface.

There are at least three definitions for the word "ton": the 2000 lb short ton (commonly used in the United States and Canada), the 2240 lb long ton (used in Great Britain), and the 1000 kg (2205 lb) metric ton (also known as a tonne).

A useful approximate conversion factor: 1 short ton/ft^2 ≈ 1 kg/cm^2 ≈ 100 kPa ≈ 1 atmosphere. These are true to within 2 to 4%.

TABLE B4 UNITS OF DISTANCE

To Convert	To	Multiply by
ft	in	**12**
ft	m	0.3048
in	ft	**1/12**
in	mm	25.40
m	ft	3.281
mm	in	0.03937

TABLE B5 UNITS OF FORCE

To Convert	To	Multiply by
k	kN	4.448
k	lb	**1000**
kg_f	lb	2.205
kg_f	N	9.807
kg_f	ton (metric)	**0.001**
kN	k	0.2248
lb	k	**0.001**
lb	kg_f	0.4536
lb	N	4.448
lb	ton (short)	**1/2000**
lb	ton (long)	**1/2240**
N	kg_f	0.1020
N	lb	0.2248
ton (short)	lb	**2000**
ton (long)	lb	**2240**
ton (metric)	kg_f	**1000**

TABLE B6 UNITS OF VOLUME

To Convert	To	Multiply by
ft^3	gal	7.481
gal	ft^3	0.1337

TABLE B7 UNITS OF STRESS AND PRESSURE

To Convert	To	Multiply by
atmosphere	lb/ft^2	2117
atmosphere	kPa	101.3
bar	kPa	**100**
kg_f/cm^2	kPa	98.07
kg_f/cm^2	lb/ft^2	2048
kPa	atmosphere	0.009869
kPa	bar	**0.01**
kPa	kg_f/cm^2	0.01020
kPa	lb/ft^2	20.89
kPa	lb/in^2	0.1450
kPa	metric ton/m^2	0.1020
lb/ft^2	atmosphere	4.725×10^{-4}
lb/ft^2	kPa	0.04787
lb/ft^2	lb/in^2	**1/144**
lb/in^2	kPa	6.895
lb/in^2	lb/ft^2	**144**
lb/in^2	MPa	6.895×10^{-3}
metric ton/m^2	kPa	9.807
MPa	lb/in^2	145.0

TABLE B8 UNITS OF UNIT WEIGHT

To Convert	To	Multiply by
kN/m^3	lb/ft^3	6.366
kN/m^3	metric ton/m^3	0.1020
kN/m^3	Mg_f/m^3	0.1020
lb/ft^3	kN/m^3	0.1571
metric ton/m^3	kN/m^3	9.807
$Mg_f/m3$	kN/m^3	9.807

Appendix C

Computer Software

The author has developed seven computer program specifically for this book. These programs are collectively referred to as the *Geotechnical Analysis Software Package*.

SUMMARY OF PROGRAMS

The geotechnical analysis software package includes the following programs:

Stress analysis:

STRESSP	Geostatic and induced stresses beneath a point load
STRESSL	Geostatic and induced stresses beneath a line load
STRESSR	Geostatic and induced stresses beneath a rectangular area load
STRESSC	Geostatic and induced stresses beneath a circular area load

Instructions for using these four programs are located on pages 345–347.

Settlement:

FILLSETT	Ultimate consolidation settlement due to weight of fill
SETTRATE	Rate of consolidation settlement due to weight of a fill

Instructions for using these two programs are located on pages 407–408 and 431–441, respectively.

Spread footing foundations:

FOOTING	Bearing capacity and settlement of spread footings

Instructions for using this program are located on pages 644–645.

MINIMUM SYSTEM REQUIREMENTS

These programs are designed to be used on IBM compatible personal computers. To use them, your computer must meet or exceed the following requirements:

Operating system:	32-bit Microsoft Windows, such as Windows 95, Windows 98, or Windows NT (version 3.51 or later). For Windows NT version 4.0, it is best to have installed Service Pack 2.
Processor:	80486 or higher
Video display:	VGA (640×480) resolution or higher
RAM:	Minimum 8 MB
Pointing device:	Any Windows compatible mouse or other pointing device
Hard disk space:	Less than 4 Mb
Printer:	Any windows compatible printer (necessary only if output is required)

DOWNLOADING AND INSTALLATION INSTRUCTIONS

To download the software, use any web browser and the internet to log onto the following address:

http://www.prenhall.com/coduto

Follow the links to the *Geotechnical Engineering: Principles and Practices* software download page, then follow the on-screen instructions.

Once the software has been downloaded, install it on the hard disk of a computer. The web page gives more information on how to install the software.

USING THE SOFTWARE

The software works best when the monitor is set for 800×600 or 1024×768 resolution. It will work at higher resolutions, but the fonts will probably be too small for most users, unless the monitor is very large. The software also will display properly at 640×480 resolution (standard VGA), but it fills the entire screen. At this setting, the user may wish to hide the Windows taskbar, which is usually located at the bottom of the screen. This may be done by moving the mouse to the top of the bar until a double-headed arrow appears, then dragging it off the screen. The Windows taskbar may be restored by performing the same action in reverse.

To use the software, begin by clicking on the Windows START button, going to programs, and selecting "Geotechnical Analysis Software". An introductory screen will appear, followed by the main menu. Select the desired program from the main menu, then follow the instructions for that program.

References

AASHTO (1993), "Recommended Practice for the Classification of Soils and Soil–Aggregate Mixtures for Highway Construction Purposes," AASHTO designation M 145–91, *Standard Specifications for Transportation Materials and Methods of Sampling and Testing*, American Association of State Highway and Transportation Officials, Washington, DC.

ABRAHAMSON, N. A. — *see* Toro, G. R.

ABRAMSON, LEE W., LEE, THOMAS S., SHARMA, SUNIL, AND BOYCE, GLENN M. (1996), *Slope Stability and Stabilization Methods*, John Wiley, New York.

ACHARYA, PRASANNA KUMAR (1980), *Architecture of Manasara*, Translated from the original Sanskrit, 2nd ed., Oriental Books, New Delhi.

ACKER, W. L. (1974), *Basic Procedures for Soil Sampling and Core Drilling*, Acker Drill Company, Scranton, PA.

ADAMS, JEFFREY A. — *see* Reddy, Krishna R.

ALBIN, PEDRO — *see* Thornley, J. H.

ALGERMISSEN, S. T. (1983), *An Introduction to the Seismicity of the United States*, Earthquake Engineering Research Institute.

AMBRAYSEYS, N. N. (1988), "Engineering Seismology," *Earthquake Engineering and Structural Dynamics*, Vol. 17, No. 1, pp. 1–105.

AMERICAN GEOLOGICAL INSTITUTE (1976), *Dictionary of Geological Terms*, Revised Ed., Anchor Books.

ANDERSLAND, ORLANDO B., AND ANDERSON, DUWAYNE M., EDS. (1978), *Geotechnical Engineering For Cold Regions*, McGraw Hill, New York.

ANDERSON, L. R. — *see* Dunn, I. S.

ANDERSON, DUWAYNE M. — *see* Andersland, Orlando B.

ARANGO, IGNACIO (1996), "Magnitude Scaling Factors for Soil Liquefaction Evaluations," *Journal of Geotechnical Engineering*, Vol. 122, No. 11, pp. 929–936, ASCE.

ASCE (1972), "Subsurface Investigation for Design and Construction of Foundations and Buildings," *Journal of the Soil Mechanics and Foundation Division*, Vol. 98, No. SM5–SM8.

ASTM, *Annual Book of ASTM Standards*, Volume 04.08 — Soil and Rock, American Society for Testing and Materials, Philadelphia.

ATTERBERG, A. (1911), "Über die Physicalische Bodenuntersuchung und über die Plastizität der Tone," *Internationale Mitteilungen fur Bodenkunde*, Vol. 1 (in Swedish).

BACHUS, ROBERT C. — *see* Clough, G. Wayne

BAGUELIN, F., JÉZÉQUEL, J. F., AND SHIELDS, D. H. (1978), *The Pressuremeter and Foundation Engineering*, Trans Tech, Clausthal, Germany.

BARDET, JEAN–PIERRE (1997), *Experimental Soil Mechanics*, Prentice Hall, Upper Saddle River, NJ.

BARTLETT, S. F., AND YOUD, T. L. (1993), "Prediction of Liquefaction–Induced Ground Displacement Near Bridges," *1993 National Earthquake Conference*, Vol. II, pp. 575–584, Central United States Earthquake Consortium, Memphis, TN.

BARTLETT, S. F., AND YOUD, T. L. (1992), *Empirical Analysis of Horizontal Ground Displacement Generated by Liquefaction–Induced Lateral Spreads*, Technical Report No. NCEER–92–0021, National Center for Earthquake Engineering Research, Buffalo, NY.

BATHE, KLAUS–JÜRGEN (1996), *Finite Element Procedures*, Prentice Hall, Upper Saddle River, NJ.

BAXTER, CHRISTOPHER D. P. — *see* Mitchell, James K.

BAZIAR, MOHAMMAD H. — *see* Dobry, Ricardo

BEDIENT, PHILIP B., RIFAI, HANADI S., AND NEWELL, CHARLES J. (1994), *Ground Water Contamination: Transport and Remediation*, Prentice Hall, Upper Saddle River, NJ.

BEECH, J. F. — *see* Kulhawy, F. H.

BELL, A. L. (1993), "Jet Grouting," Chapter 7 in *Ground Improvement*, M. P. Moseley, Ed., Chapman and Hall, London.

BERKEY, CHARLES P. (1939), "Geology in Engineering," *Frontiers in Geology*, pp. 31–34, Geological Society of America.

BERTRAM, G. E. (1940), *An Experimental Investigation of Protective Filters*, Publications of the Graduate School of Engineering, No. 267, Harvard University.

BISHOP, ALAN W. (1955), "The Use of the Slip Circle in the Stability Analysis of Slopes," *Géotechnique*, Vol. 5, pp. 7–17.

BISWAS, ASIT K., AND CHATTERJEE, SAMAR (1971), "Dam Disasters — An Assessment," *Engineering Journal (Canada)*, Vol. 54, No. 3, pp. 3–8.

BISWAS, ASIT K. (1970), *History of Hydrology*, North Holland Publishing Company, Amsterdam.

BJERRUM, LAURITS, AND FLODIN, NILS (1960), "The Development of Soil Mechanics in Sweden: 1900–1925," *Géotechnique*, Vol. 10, No. 1, pp. 1–18.

BLACK, WILLIAM T. — *see* Luscher, Ulrich

BLACKALL, T. E. (1952), "A. M. Atterberg 1846–1916," *Géotechnique*, Vol. 3, pp. 17–19.

BOLT, BRUCE A. (1993), *Earthquakes*, W. H. Freeman, New York.

BONALA, MOHAN V. S. – *see* Reddi, Lakshmi N.

BONILLA, M. G., MARK, R. K., AND LIENKAEMPER, J. J. (1984), "Statistical Relations Among Earthquake Magnitude, Surface Rupture Length, and Surface Fault Displacement," *Bulletin of the Seismological Society of America*, Vol. 74, No. 6, pp. 2379–2411.

BONILLA, M. G. (1970), "Surface Faulting and Related Effects," Chapter 3 in *Earthquake Engineering*, Robert L. Wiegel, Ed., Prentice Hall, Upper Saddle River, NJ.

BOOKER, JOHN R. — *see* Seed, H. B.

BOORE, D. M., JOYNER, W. B., AND FUMAL, T. E. (1993), *Estimation of Response Spectra and Peak Accelerations from Western North America Earthquakes: An Interim Report*, Open File Report 93–509, U.S. Geological Survey, Reston, VA.

BORJA, RONALDO I., AND KISHNANI, SUNIL (1992), "Movement of Slopes During Rapid and Slow Drawdown," *Stability and Performance of Slopes and Embankments–II*, Geotechnical Special Publication No. 31, Vol. I, pp. 404–413, Raymond B. Seed and Ross W. Boulanger, Eds., ASCE.

BOULANGER, ROSS W. — *see* Seed, Raymond B.

BOUSSINESQ, J. (1885), *Application des Potentiels à L'Étude de L'Équilibre et du Mouvement des Solides Élastiques*, Gauthier–Villars, Paris (in French).

BOWLES, JOSEPH (1996), *Foundation Analysis and Design*, 5th ed., McGraw Hill, New York.

BOYCE, GLENN M. — *see* Abramson, Lee W.

BRAUN, J. S. — *see* Thorson, Bruce M.

BRAY, JONATHAN D. — *see* Stewart, Jonathan P.; Seed, R. B.

BRIAUD, JEAN–LOUIS (1992), *The Pressuremeter*, A. A. Balkema, Rotterdam.

BRIAUD, JEAN–LOUIS, AND MIRAN, JEROME (1991), *The Cone Penetration Test*, Report No. FHWA–TA–91–004, Federal Highway Administration, McLean, VA.

BROMHEAD, E. N. (1986), *The Stability of Slopes*, Surrey University Press, Glasgow.

BROMS, BENGT — *see* Flodin, Nils; Holtz, R. D.

BROWN, RALPH E., AND GLENN, ANDREW J. (1976), "Vibroflotation and Terra–Probe Comparison," *Journal of the Geotechnical Engineering Division*, Vol. 102, No. GT10, pp. 1059–1072, ASCE.

BRUMUND, WILLIAM F., JONAS, ERNEST, AND LADD, CHARLES C. (1976), "Estimating In–Situ Maximum Past (Preconsolidation) Pressure of Saturated Clays From Results of Laboratory Consolidometer Tests," *Estimation of Consolidation Settlement*, Special Report 163, pp. 4–12, Transportation Research Board, Washington, DC.

BRUMUND, WILLIAM F. (1995), "Environmental Geotechnology — Examples of Where We Are and How We Got Here," *Geoenvironment 2000*, pp. 1622–1629, Yalcin B. Acar and David E. Daniel, Eds., ASCE.

BRUNSDEN, DENYS, AND PRIOR, DAVID B., Editors (1984), *Slope Instability*, John Wiley, New York.

BRUNSDEN, D. (1984), "Mudslides" Chapter 9 in *Slope Instability*, D. Brunsden and D. B. Prior, Eds., John Wiley, New York.

BUCKINGHAM, PHILLIP — *see* LaGrega, Michael D.

BUREAU OF RECLAMATION (1987), *Design of Small Dams*, 3rd ed., United States Department of the Interior, Bureau of Reclamation.

BURMISTER, D. M. (1962), "Physical, Stress–Strain, and Strength Responses of Granular Soils," *Symposium on Field Testing of Soils*, STP 322, pp. 67–97, ASTM.

CAMPANELLA, R. G. — *see* Robertson, P. K.

CAPANO, C. — *see* Kulhawy, F. H.

CARROLL, R. G., JR. (1983), "Geotextile Filter Criteria," *Engineering Fabrics in Transportation Construction*, Transportation Research Record 916, pp. 46–53.

CASAGRANDE, A. (1936), "The Determination of the Pre–Consolidation Load and Its Practical Significance," Discussion D–34, *Proceedings of the First International Conference on Soil Mechanics and Foundation Engineering*, Vol. III, pp. 60–64.

CASAGRANDE, ARTHUR M. (1948), "Classification and Identification of Soils," ASCE *Transactions*, Vol. 113, pp. 901–991.

CASAGRANDE, ARTHUR (1960), "Karl Terzaghi — His Life and Achievements," *From Theory to Practice in Soil Mechanics*, L. Bjerrum et al., pp. 3–21, John Wiley, New York.

CASAGRANDE, ARTHUR (1964), "Karl Terzaghi, 1883–1963," *Géotechnique*, Vol. 14, No. 1, pp. 1–9.

CATERPILLAR (1993), *Caterpillar Performance Handbook*, 24th ed., Caterpillar, Inc., Peoria, IL.

CDE, INCORPORATED (1996), World Wide Web site www.cde-inc.com.

CDMG (1994b), *CSMIP Strong Motion Records From the Northridge, California Earthquake of January 17, 1994*, California Division of Mines and Geology, Sacramento.

CDMG (1994a), *Fault–Rupture Hazard Zones in California*, Special Publication 42, California Division of Mines and Geology, Sacramento.

CEDERGREN, HARRY R. (1989), *Seepage, Drainage, and Flow Nets*. 3rd ed., John Wiley, New York.

CGS (1992), *Canadian Foundation Engineering Manual*, 3rd ed., Canadian Geotechnical Society, BiTech, Vancouver, BC.

CHATTERJEE, SAMAR — *see* Biswas, Asit

CHUNG, C. K., AND FINNO, R. J. (1992), "Influence of Depositional Processes on the Geotechnical Parameters of Chicago Glacial Clays," *Engineering Geology*, Vol. 32, pp. 225–242.

CHUNG, RILEY M. — *see* Seed, H. Bolton

CHURCH, HORACE K (1981), *Excavation Handbook*, McGraw Hill, New York.

CIVIL ENGINEERING (1981), "Rocky Mountain Arsenal: Landmark Case of Groundwater Polluted by Organic Chemicals, and Being Cleaned Up," *Civil Engineering*, Vol. 51, No. 9, pp. 68–71, ASCE.

CLAYTON, C. R. I. (1990), "SPT Energy Transmission: Theory, Measurement, and Significance," *Ground Engineering*, Vol. 23, No. 10, pp. 35–43.

CLAYTON, C. R. I., MÜLLER STEINHAGEN, H., AND POWRIE, W. (1995), "Terzaghi's Theory of Consolidation and the Discovery of Effective Stress," *Proceedings of the Institution of Civil Engineers, Geotechnical Engineering*, Vol. 113, pp. 191–205. Includes an English translation of a portion of Terzaghi (1925a).

CLEMENCE, SAMUEL P., AND FINBARR, ALBERT O. (1981), "Design Considerations for Collapsible Soils," *Journal of the Geotechnical Engineering Division*, Vol. 107, No. GT3, pp. 305–317, ASCE.

CLEVENGER, WILLIAM A. — *see* Sherard, James L.

CLOUGH, G. WAYNE, SITAR, NICHOLAS, BACHUS, ROBERT C., AND RAD, NADER SHAFII (1981), "Cemented Sands Under Static Loading," *Journal of the Geotechnical Engineering Division*, Vol. 107, No. GT6, pp. 799–817, ASCE

CLOUGH, G. WAYNE — *see* Duncan, J. Michael

CODUTO, DONALD P., AND HUITRIC, RAYMOND (1990), "Monitoring Landfill Movements Using Precise Instruments," *Geotechnics of Waste Fills: Theory and Practice*, STP 1070, pp. 358–370, ASTM.

CODUTO, DONALD P. (1999), *Foundation Design: Principles and Practices*, 2nd ed., Prentice Hall, Upper Saddle River, NJ.

COHEN, ROBERT M. — *see* Mercer, James W.

COLLIN, A. (1846), *Recherches Expérimentales sur les Glissements Spontanés des Terrains Argileux, accompagnées de Considerations sur Quelques Principes de la Méchanique Terrestre*, Carilian–Goery and Dalmont, Paris (in French). Translated to English by W. R. Schriever as *Landslides in Clays by Alexandre Collin, 1846*. University of Toronto Press, 1956 .

COULOMB, C. A. (1776), "Essai sur une application des règles de maximis et minimis à quelques problèmes de statique relatifs à l'architecture," *Mémoires de mathématique et de physique présentés à l'Académie Royale des Sciences*, Paris, Vol. 7, pp. 343–382 (in French; Translated to English by Heyman, 1972).

COUSINS, B. F. (1978), "Stability Charts for Simple Earth Slopes," *Journal of the Geotechnical Engineering Division*, Vol. 104, No. GT2, pp. 267–282, ASCE.

CRAWFORD, C. B., AND JOHNSON, G. H. (1971), "Construction on Permafrost," *Canadian Geotechnical Journal*, Vol. 8, No. 2, pp. 236–251.

CRUDEN, DAVID M., AND VARNES, DAVID J. (1996), "Landslide Types and Processes," Chapter 3 in *Landslides: Investigation and Mitigation*, A. Keith Turner and Robert L. Schuster, Eds.

DARCY, H. (1856), *Les Fontaines Publiques de la Ville de Dijon* (The Water Supply of the City of Dijon), Dalmont, Paris (in French).

DARRAGH, ROBERT D. — *see* Roberts, Don V.

DAVIS, E. H. — *see* Poulos, H. G.

DE RUITER, J. (1981), "Current Penetrometer Practice," *Cone Penetration Testing and Experience*, pp. 1–48, ASCE.

DE MELLO, V. (1971), "The Standard Penetration Test — A State-of-the-Art Report," *Fourth Pan–American Conference on Soil Mechanics and Foundation Engineering*, Vol. 1, pp. 1–86.

DE BOER, R., SCHIFFMAN, R. L., AND GIBSON, R. E. (1996), "The Origins of the Terzaghi–Fillunger Dispute," *Géotechnique*, Vol. 46, No. 2, pp. 175–186.

DEEGAN, JOHN (1987), "Looking Back at Love Canal," *Environmental Science and Technology*, Vol. 21, No. 4, pp. 328–331 and 421–426.

DESCHAMPS, R. J., AND LEONARDS, G. A. (1992), "A Study of Slope Stability Analysis," *Stability and Performance of Slopes and Embankments–II*, Geotechnical Special Publication No. 31, Vol. I, pp. 267–291, Raymond B. Seed and Ross W. Boulanger, Eds., ASCE.

DICKENSON, S. E. — *see* Seed, R. B.

DISE, KARL, STEVENS, MICHAEL G., AND VON THUN, J. LAWRENCE (1994), "Dynamic Compaction to Remediate Liquefiable Embankment Foundation Soils," *In-Situ Deep Soil Improvement*, Special Geotechnical Publication 45, pp. 1–25, Kyle M. Rollins, Ed., ASCE.

DOBRIN, MILTON B. (1988), *Introduction to Geophysical Prospecting*, McGraw Hill, New York.

DOBRY, RICARDO (1994), "Foundation Deformation Due to Earthquakes," *Vertical and Horizontal Deformations of Foundations and Embankments*, Geotechnical Special Publication No. 40, A. T. Yeung and G. Y. Felio, Eds., Vol. 2, pp. 1846–1863, ASCE.

DOBRY, RICARDO, AND BAZIAR, MOHAMMAD H. (1992), "Modeling of Lateral Spreads in Silty Sands by Sliding Soil Blocks," *Stability and Performance of Slopes and Embankments–II*, Vol. I, pp. 625–652, Geotechnical Special Publication No. 31, R. B. Seed and R. W. Boulanger, Eds., ASCE.

DOWDING, CHARLES H. (1979), *Site Characterization & Exploration*, ASCE.

DOWDING, CHARLES H. (1996), *Construction Vibrations*, Prentice Hall, Upper Saddle River, NJ.

DRISCOLL, FLETCHER G. (1986), *Groundwater and Wells*, 2nd ed., Johnson Filtration Systems, Inc., St. Paul, MN.

DUDLEY, JOHN H. (1970), "Review of Collapsing Soils," *Journal of the Soil Mechanics and Foundations Division*, Vol. 96, No. SM3, pp. 925–947, ASCE.

DUNCAN, JAMES M., WRIGHT, STEPHEN G, AND WONG, KAI S. (1990), "Slope Stability During Rapid Drawdown," Chapter 12 in *H. Bolton Seed Memorial Symposium Proceedings*, J. Michael Duncan, Ed., Vol. 2, Bi Tech, Vancouver, BC.

DUNCAN, J. MICHAEL, CLOUGH, G. WAYNE, AND EBELING, ROBERT M. (1990), "Behavior and Design of Gravity Earth Retaining Structres," *Design and Performance of Earth Retaining Structures*, Philip C. Lambe and Lawrence A. Hansen, Eds., pp. 251–277, ASCE.

DUNCAN, J. MICHAEL (1992a), "State-of-the-Art: Static Stability and Deformation Analysis," *Stability and Performance of Slopes and Embankments–II*, Geotechnical Special Publication No. 31, Vol. I, pp. 222–266, Raymond B. Seed and Ross W. Boulanger, Eds., ASCE.

DUNCAN, J. M. (1992b), "Thirteenth Bjerrum Memorial Lecture: A Case History of Mysterious Settlements in a Building," *Canadian Geotechnical Journal*, Vol. 29, No. 1–10.

DUNCAN, J. MICHAEL, AND STARK, TIMOTHY D. (1992), "Soil Strengths from Back Analysis of Slope Failures," *Stability and Performance of Slopes and Embankments–II*, Geotechnical Special Publication No. 31, Vol I, pp. 890–904, R. B. Seed and R. W. Boulanger, Eds., ASCE.

DUNCAN, J. MICHAEL (1993), "Limitations of Conventional Analysis of Consolidation Settlement," ASCE *Journal of Geotechnical Engineering*, Vol. 119, No. 9, pp. 1331–1359; Discussions in Vol. 121, No. 6, pp. 513–518.

DUNCAN, J. MICHAEL (1996), "Soil Slope Stability Analysis," Chapter 13 in *Landslides: Investigation and Mitigation*, Special Report 247, A. Keith Turner and Robert L. Schuster, Eds., Transportation Research Board, Washington, DC.

DUNCAN, J. MICHAEL — *see* Stark, Timothy D.

DUNN, I. S., ANDERSON, L. R., AND KIEFER, F. W. (1980), *Fundamentals of Geotechnical Analysis*, John Wiley, New York.

DUNNIGAN, L. P. — *see* Sherard, J. L.

EBELING, ROBERT M. — *see* Duncan, J. Michael

EDIL, TUNCER B. – *see* Kim, Jae Y.

EINSTEIN, HERBERT H. — *see* Wu, Tien H.

ERICKSEN, G. E. — *see* Plafker, G.

EVANS, MARK D., AND ZHOU, SHENGPING (1994), "Cyclic Behavior of Gravelly Soil," *Ground Failures Under Seismic Conditions*, Geotechnical Special Publication No. 44, pp. 158–176, S. Prakash and P. Dakoulas, Eds., ASCE.

EVANS, JEFFREY C. — *see* LaGrega, Michael D.

FARQUHAR, GRAHAME J. — *see* McBean, Edward A.

FAUST, CHARLES R. — *see* Mercer, James W.

FELDMAN, H. S. — *see* Lowe, J.

FELLENIUS, WOLMAR (1927), *Erdstatische Berechnungen mit Reibung and Kohaesion*, Ernst, Berlin.

FELLENIUS, WOLMAR (1936), "Calculation of Stability of Earth Dams," *Transactions, Second Congress on Large Dams*, Washington, Vol. 4, pp. 445–462.

FENG, TAO–WEI — *see* Mesri, Gholamreza

FETTER, C. W. (1993), *Contaminant Hydrogeology*, MacMillan, New York.

FINBARR, ALBERT O. — *see* Clemence, Samuel P.

FINN, W. D. LIAM, LEDBETTER, R. H., AND WU, GUOXI (1994), Liquefaction of Silty Soils: Design and Analysis," *Ground Failures Under Seismic Conditions*, Geotechnical Special Publication No. 44, pp. 51–76, S. Prakash and P. Dakoulas, Eds., ASCE.

FINNO, R. J.— *see* Chung, C. K.

FLEMING, ROBERT W. — *see* Kaliser, Bruce N.

FLINT, RICHARD F., AND SKINNER, BRIAN J. (1974), *Physical Geology*, John Wiley, New York.

FLODIN, NILS, AND BROMS, BENGT (1981), "Historical Development of Civil Engineering in Soft Clay," Chapter 1 in *Soft Clay Engineering*, E. W Brand and R. P. Brenner, Eds., Elsevier, Amsterdam.

FLODIN, NILS — *see* Bjerrum, Laurits

FORCHHEIMER, PHILIP (1914), *Hydraulik*, pp. 26, 494–495, Leipzig (in German).

FORCHHEIMER, P. (1917), "Zur Grundwasserbewegung nach isothermischen Kurvenscharen" (Concerning groundwater movement in accordance with isothermal families of curves), *Sitzber. kais Akad. d. Wiss., Wein, Abt. IIa,* Vol. 126, pp. 409–440 (in German).

FORREST, CAROL L., AND NOORANY, IRAJ (1989), "Liquefaction Risk Analysis for a Harbor Fill," *Marine Geotechnology*, Vol. 8, pp. 33–49.

FOX, PATRICK J. (1995), "Consolidation and Settlement Analysis," Chapter 18 in *The Civil Engineering Handbook*, W. F. Chen, Ed., CRC Press.

FRANKLIN, ARLEY G. — *see* Hynes–Griffin, Mary Ellen, and Marcuson, W. F.

FRANKS, ALVIN L. (1993), "Hybid Disposal Systems and Nitrogen Removal in Individual Sewage Disposal Systems," *Bulletin of the Association of Engineering Geologists*, Vol. 30, No. 2, pp. 181–209.

FREDLUND, D. G., KRAHN, J., AND PUFAHL, D. E. (1981), "The Relationship Between Limit Equilibrium Slope Stability Methods," *Proceedings, International Conference on Soil Mechanics and Foundation Engineering*, pp. 409–416.

FREDLUND, D. G., AND RAHARDJO, H. (1993), *Soil Mechanics for Unsaturated Soils*, John Wiley, New York.

FRONTARD, J. (1914), "Notice sur l'accident de la digue de Charmes," *Ann. Ponts et Chaussées*, 9th Series, Vol. 23, pp. 173–280 (in French).

FUMAL, T. E. — *see* Boore, D. M.

GIBBS, HAROLD J. (1969), Discussion of Holtz, W. G. "The Engineering Problems of Expansive Clay Subsoils," *Proceedings of the Second International Research and Engineering Conference on Expansive Clay Soils*, pp. 478–479, Texas A&M Press.

GIBSON, R. E. — *see* De Boer, R.

GIRAULT, P. — *see* Leonards, G. A.

GIZIENSKI, STANLEY F. — *see* Sherard, James L.

GLENN, ANDREW J. — *see* Brown, Ralph E.

GOODMAN, RICHARD E. (1990), "Soils Versus Rocks as Engineering Materials," *H. Bolton Seed Memorial Symposium Proceedings*, J. Michael Duncan, Ed., Vol. 2, pp. 111–133, BiTech, Vancouver, BC.

GOODMAN, RICHARD E. (1993), *Engineering Geology: Rock in Engineering Construction*, John Wiley, New York.

GREENFIELD, STEVEN J., AND SHEN, C. K. (1992), *Foundations in Problem Soils*, Prentice Hall, Upper Saddle River, NJ.

GRIGORIU, MIRCEA D. — *see* Kulhawy, Fred H.

GUCUNSKI, N. — *see* Maher, M. H.

HAMBLIN, W. KENNETH, AND HOWARD, JAMES D. (1975), *Exercises in Physical Geology*, 4th ed., Burgess Publishing Co., Minneapolis.

HAMILTON, DOUGLAS H., AND MEEHAN, RICHARD L. (1992), "Cause of the 1985 Ross Store Explosion and Other Gas Ventings, Fairfax District, Los Angeles," *Engineering Geology Practice in Southern California*, Special Publication No. 4, Association of Engineering Geologists.

HANDY, RICHARD L. (1980), "Realism in Site Exploration: Past, Present, Future, and Then Some — All Inclusive," *Site Exploration on Soft Ground Using In-Situ Techniques*, pp. 239–248, Report No. FHWA–TS–80–202, Federal Highway Administration, Washington, DC.

HANDY, RICHARD L. (1995), *The Day the House Fell*, ASCE.

HARDER, LESLIE F., AND SEED, H. BOLTON (1986), *Determination of Penetration Resistance for Coarse-Grained Soils Using the Becker Hammer Drill*, Report No. UCB/EERC–86/06, Earthquake Engineering Research Center, Richmond, CA.

HARDER, L. F. — *see* Seed, H. B., and Seed, R. B.; Seed, H. Bolton

HARR, MILTON E. (1962), *Groundwater and Seepage*, McGraw Hill, New York.

HART, STEPHEN S. — *see* Holtz, Wesley G.

HARTMAN, STEPHEN W. (1983), "Love Canal: A Case Study," *IEEE Engineering Management Review*, Vol. 10, No. 5, pp. 8–42.

HAUSMANN, MANFRED R. (1992), "Slope Remediation," *Stability and Performance of Slopes and Embankments – II*, Geotechnical Special Publication No. 31, Vol II, pp. 1274–1317, R. B. Seed and R. W. Boulanger, Eds., ASCE.

HAZEN, A. (1911), "Discussion of 'Dams on Sand Foundations' by A. C. Koenig," *ASCE Transactions*, Vol. 73, p. 199.

HEAD, K. H. (1982), *Manual of Soil Laboratory Testing*, Vol. 2, Pentech, London.

HEYMAN, JACQUES (1972), *Coulomb's Memoir on Statics*, Cambridge University Press.

HICKS, RANDALL T., AND RIZVI, RAIS (1996), "Do-Nothing Cleanups," *Civil Engineering*, Vol. 66, No. 9, pp. 54–57, ASCE.

HILF, JACK W. (1991), "Compacted Fill," Chapter 8 in *Foundation Engineering Handbook*, 2nd ed., Hsai–Yang Fang, Ed., Van Nostrand Reinhold, New York.

HIRIART, FERNANDO, AND MARSAL, RAUL J. (1969), "The Subsidence of Mexico City," *Nabor Carrillo; El Hundimiento de la Ciudad de Mexico y Proyecto Texcoco* (Nabor Carrillo; The Subsidence of Mexico City and Texcoco Project), pp. 109–147, In English and Spanish.

HOLTZ, R. D., AND BROMS, B. B. (1972), "Long-Term Loading Tests at Skå-Edeby, Sweden," *Proceedings of the ASCE Specialty Conference on Performance of Earth and Earth-Supported Structures*, Vol. I, Part 1, pp. 435–464, ASCE.

HOLTZ, ROBERT D., AND KOVACS, WILLIAM D. (1981), *An Introduction to Geotechnical Engineering*, Prentice Hall, Upper Saddle River, NJ.

HOLTZ, W. G. (1969), "Volume Changes in Expansive Clay Soils and Control by Lime Treatment," *Proceedings of the Second International Research and Engineering Conference on Expansive Clay Soils*, pp. 157–173, Texas A&M Press.

HOLTZ, WESLEY G., AND HART, STEPHEN S. (1978), *Home Construction on Shrinking and Swelling Soils*, Colorado Geological Survey Publication SP–11, Denver.

HOLTZ, WESLEY G. — *see* Jones, D. Earl

HORVATH, JOHN S. (1995), *Geofoam Geosynthetic*, Horvath Engineering, Scarsdale, NY.

HOUGH, B. K. (1969), *Basic Soils Engineering*, 2nd ed., The Ronald Press, New York.

HOUSNER, G. W. (1970), "Design Spectrum," Chapter 5 in *Earthquake Engineering*, Robert L. Wiegel, Ed.

HOUSTON, SANDRA L. — *see* Houston, William N.

HOUSTON, SANDRA L., AND HOUSTON, WILLIAM N. (1997), "Collapsible Soil Engineering," *Unsaturated Soil Engineering Practice*, Geotechnical Special Publication 68, pp. 199–232, ASCE.

HOUSTON, WILLIAM N., AND HOUSTON, SANDRA L. (1989), "State-of-the-Practice Mitigation Measures for Collapsible Soil Sites,," *Foundation Engineering: Current Principles and Practices*, Vol. 1, pp. 161–175, F. H. Kulhawy, Ed., ASCE.

HOUSTON, WILLIAM N. — *see* Houston, Sandra L.

HOWARD, JAMES D. — *see* Hamblin, W. Kenneth

HRYCIW, ROMAN D. (1995), *Soil Improvement for Earthquake Hazard Mitigation*, Geotechnical Special Publication No. 49, ASCE.

HUBER, TIMOTHY R. — *see* Mitchell, James K.

HUITRIC, RAYMOND — *see* Coduto, Donald P.

HUMPHREY, DANA – *see* Whetton, Nathan

HUNT, ROY E. (1984), *Geotechnical Engineering Investigation Manual*, McGraw Hill, New York.

HVORSLEV, M. JUUL (1949), *Subsurface Exploration and Sampling of Soils for Civil Engineering Purposes*, ASCE.

HYNES, MARY ELLEN (1994), "Behavior of Gravelly Soils During Seismic Conditions — An Overview," *Ground Failures Under Seismic Conditions*, Geotechnical Special Publication No. 44, pp. 117–120, S. Prakash and P. Dakoulas, Eds., ASCE.

HYNES, M. E. — *see* Marcuson, W. F.

HYNES–GRIFFIN, MARY ELLEN, AND FRANKLIN, ARLEY G. (1984), *Rationalizing the Seismic Coefficient Method*, Miscellaneous paper GL–84–13, Waterways Experiment Station, U.S. Army Corps of Engineers.

ICBO (1997), *Uniform Building Code*, International Conference of Building Officials, Whittier, CA.

ICC (1997), *International Building Code, First Draft*, International Code Council.

IDRISS, I. M. (1990), "Response of Soft Soil Sites During Earthquakes," *Proceedings, H. Bolton Seed Memorial Symposium*, J. M. Duncan, Ed., Vol. 2, pp. 273–289, BiTech, Vancouver, BC.

IDRISS, I. M. — *see* Seed, H. B. and Seed, R. B.

IFAI (1997), *Geotechnical Fabrics Report Specifier's Guide*, Industrial Fabrics Association, International, St. Paul, MN (published annually).

INGLES, O. G., AND METCALF, J. B. (1972), *Soil Stabilization*, Butterworths, Sydney.

JANBU, N. (1957), "Earth Pressure and Bearing Capacity Calculations by Generalized Procedure of Slices," *Proceedings, Fourth International Conference on Soil Mechanics and Foundation Engineering*, London, Vol. 2, pp. 17–26.

JANBU, N. (1973), "Slope Stability Computations," *Embankment Dam Engineering — Casagrande Volume*, pp. 47–86, John Wiley, New York.

JANSEN, ROBERT B. (1980), *Dams and Public Safety*, Water and Power Resources Service (Bureau of Reclamation), U.S. Department of the Interior.

JANSEN, ROBERT B., Ed. (1988), *Advanced Dam Engineering for Design, Construction, and Rehabilitation*, Von Nostrand Reinhold, New York.

JÉZÉQUEL, J. F. — *see* Baguelin, F.

JOHNSON, A. M., AND RODINE, J. R. (1984), "Debris Flow," Chapter 8 in *Slope Instability*, D. Brunsden and D. B. Prior, Eds., John Wiley, New York.

JOHNSON, A. W., AND SALLBERG, J. R. (1960), *Factors that Influence Field Compaction of Soils*, Bulletin 272, Highway Research Board, Washington, DC.

JOHNSTON, G. H., (1981), *Permafrost Engineering, Design and Construction*, John Wiley, New York.

JOHNSON, G. H. — *see* Crawford, C. B.

JONAS, ERNEST — *see* Brumund, William F.

JONES, C. J. F. P. — *see* O'Rourke, T. D.

JONES, D. EARL, AND HOLTZ, WESLEY G. (1973), "Expansive Soils — The Hidden Disaster," *Civil Engineering*, Vol. 43, No. 8, ASCE.

JONES, D. EARL, AND JONES, KAREN A. (1987), "Treating Expansive Soils," *Civil Engineering*, Vol. 57, No. 8, August 1987, ASCE.

JONES, KAREN A. — *see* Jones, D. Earl

JOYNER, W. B. — *see* Boore, D. M.

KALISER, BRUCE N., AND FLEMING, ROBERT W. (1986), "The 1983 Landslide Dam at Thistle, Utah," *Landslide Dams: Processes, Risk, and Mitigation*, Geotechnical Special Publication No. 3, pp. 59–83, Robert L. Schuster, Ed., ASCE.

KASHEF, ABDEL–AZIZ (1986), *Groundwater Engineering*, McGraw Hill, New York.

KAYEN, R. E. — *see* Seed, R. B.

KEEFER, D. K. — *see* Wilson, R. C.

KERISEL, JEAN (1987), *Down to Earth; Foundations Past and Present: The Invisible Art of the Builder*, A. A. Balkema, Rotterdam.

KIEFER, F. W. — *see* Dunn, I. S.

KIM, JAE Y., EDIL, TUNCER B., AND PARK, JAE K. (1997), "Effective Porosity and Seepage Velocity in Column Tests on Compacted Clay," *Journal of Geotechnical and Geoenvironmental Engineering*, Vol. 123, No. 12, pp. 1135–1142.

KIM, JI–HYOUNG — *see* Rollins, Kyle M.

KISHNANI, SUNIL S. — *see* Borja, Ronaldo I.

KOERNER, ROBERT M. (1998), *Designing with Geosynthetics*, 4th ed., Prentice Hall, Upper Saddle River, NJ.

KOSMATKA, STEVEN H., AND PANARESE, WILLIAM C. (1988), *Design and Control of Concrete Mixtures*, 13th Ed., Portland Cement Association, Skokie, IL.

KOTZIAS, PANAGHIOTIS C. — *see* Stamatopoulos, Aris C.

KOVACS, W. D., SALOMONE, L. A., AND YODEL, F. Y. (1981), *Energy Measurements in the Standard Penetration Test*, Building Science Series 135, National Bureau of Standards, Washington, DC.

KOVACS, WILLIAM D. — *see* Holtz, Robert D.

KRAHN, J. — *see* Fredlund, D. G.

KRAMER, STEVEN L. (1996), *Geotechnical Earthquake Engineering*, Prentice Hall, Upper Saddle River, NJ.

KROPP, ALAN L. — *see* Seed, R. B.; Stewart, Jonathan P.

KULHAWY, FRED H., ROTH, MARY JOEL S., AND GRIGORIU, MIRCEA D. (1991), "Some Statistical Evaluations of Geotechnical Properties," *Proceedings of ICASP6, Sixth International Conference on Applications of Statistics and Probability in Civil Engineering*, Vol. 2, pp. 707–712, L. Esteva and S. E. Ruiz, Eds.

KULHAWY, F. H., TRAUTMANN, C. H., BEECH, J. F., O'ROURKE, T. D., MCGUIRE, W., WOOD, W. A., AND CAPANO, C. (1983), *Transmission Line Structure Foundations for Uplift-Compression Loading*, Report No. EL–2870, Electric Power Research Institute, Palo Alto, CA.

KULHAWY, F. H., AND MAYNE, P. W. (1990), *Manual on Estimating Soil Properties for Foundation Design*, Report EL–6800, Electric Power Research Institute, Palo Alto, CA.

KULHAWY, FRED H. (1996), "Estimation of In-Situ Test Uncertainty"; *Uncertainty in the Geologic Environment*, Geotechnical Special Publication 58, pp. 269–286, ASCE.

KULHAWY, FRED H. — *see* Mayne, Paul W.

KYRIELEIS, W. — *see* Sichart, W.

LADD, C. C., AND LUSCHER, U. (1965), "Engineering Properties of Soils Underlying the MIT Campus," *Research Report R65–68, Soils Publication 185*, Department of Civil Engineering, Massachusetts Institute of Technology.

LADD, CHARLES C. — *see* Brumund, William F.

LAGREGA, MICHAEL D., BUCKINGHAM, PHILLIP, AND EVANS, JEFFEREY C. (1994), *Hazardous Waste Management*, McGraw Hill, New York.

LAMBE, T. W. (1958), "The Engineering Behavior of Compacted Clay," *Journal of the Soil Mechanics and Foundations Division*, Vol. 84, No. SM2, pp. 1655–1 to 1655–35, ASCE.

LAMBE, T. WILLIAM, AND WHITMAN, ROBERT V. (1969), *Soil Mechanics*, John Wiley, New York.

LEDBETTER, R. H. — *see* Finn, W. D. Liam

LEDESMA, JOSÉ (1936), "The National Theater Building and Efforts Made to Prevent Its Further Sinking," *Proceedings, International Conference on Soil Mechanics and Foundation Engineering*, Vol. I, pp. 119–123.

LEE, KENNETH L. (1965), *Triaxial Compressive Strength of Saturated Sands Under Seismic Loading Conditions*, PhD Dissertation, University of California, Berkeley.

LEE, THOMAS S. — *see* Abramson, Lee W.

LEGGET, ROBERT F., AND HATHEWAY, ALLEN W. (1988), *Geology and Engineering*, 3rd ed., McGraw Hill, New York.

LEONARDS, G. A., AND GIRAULT, P. (1961), "A Study of the One-Dimensional Consolidation Test," *Proceedings of the Fifth International Conference on Soil Mechanics and Foundation Engineering*, Vol. I, pp. 116–130.

LEONARDS, G. A. — *see* Deschamps, R. J.

LIAO, S. S. C., AND WHITMAN, R. V. (1986), "Overburden Correction Factors for SPT in Sand," *Journal of Geotechnical Engineering*, Vol. 112, No. 3, pp. 373–377, ASCE.

LIENKAEMPER, J. J. — *see* Bonilla, M. G.

LITTLEJOHN, G. S. (1993), "Chemical Grouting," Chapter 5 in *Ground Improvement*, M. P. Moseley, Ed., Chapman and Hall, London.

LO, DOMINIC O. KWAN — *see* Mesri, Gholamreza

LOWE, J., ZACCHEO, P. F., AND FELDMAN, H. S. (1964), "Consolidation Testing with Back Pressure," *Journal of the Soil Mechanics and Foundations Division*, Vol. 90, No. SM5, pp. 69–86, ASCE.

LUSCHER, ULRICH, BLACK, WILLIAM T., AND NAIR, KESHAVAN (1975), "Geotechnical Aspects of the Trans-Alaska Pipeline," *Transportation Engineering Journal*, Vol. 101, No. TE4, pp. 669–680, ASCE.

LUSCHER, U. — *see* Ladd, C. C.

MAHER, M. H., AND GUCUNSKI, N. (1995), "Liquefaction, and Dynamic Properties of Grouted Sand," *Soil Improvement for Earthquake Hazard Mitigation*, R. D. Hryciw, Ed., pp. 37–50, Geotechnical Special Publication No. 49, ASCE.

MARCHETTI, SILVANO (1980), "In-Situ Tests by Flat Dilatometer," *Journal of the Geotechnical Engineering Division*, Vol. 106, No. GT3, pp. 299–321 (also see discussions, Vol. 107, No. GT8, pp. 831–837), ASCE.

MARCUSON, W. F., HYNES, M. E., AND FRANKLIN, A. G. (1992), "Seismic Stability and Permanent Deformation Analyses: The Last Twenty Five Years," *Stability and Performance of Slopes and Embankments–II*, Vol I, pp. 552–592, Geotechnical Special Publication No. 31, R. B. Seed and R. W. Boulanger, Eds., ASCE.

MARK, R. K. — *see* Bonilla, M. G.

MARSAL, RAUL J. — *see* Hiriart, Fernando

MAYNE, PAUL W., AND KULHAWY, FRED H. (1982), "K_0-OCR Relationships in Soil," *Journal of the Geotechnical Engineering Division*, Vol. 108, No. SM5, pp. 63–91, ASCE.

MAYNE, P. W. — *see* Kulhawy, F. H.

MCBEAN, EDWARD A., ROVERS, FRANK A., AND FARQUHAR, GRAHAME J. (1995), *Solid Waste Landfill Engineering and Design*, Prentice Hall, Upper Saddle River, NJ.

MCGUIRE, W. — *see* Kulhawy, F. H.

MCMAHON, DAVID J. — *see* Stewart, Jonathan P.

MEANS, R. E., AND PARCHER, J. V. (1963), *Physical Properties of Soils*, Charles E. Merrill Books, Inc.

MEEHAN, RICHARD L. — *see* Hamilton, Douglas H.

MEHTA, P. K. (1983), "Mechanism of Sulfate Attack on Portland Cement Concrete — Another Look," *Cement and Concrete Research*, Vol. 13, pp. 401-406.

MEIGH, A. C. (1987), *Cone Penetration Testing: Methods and Interpretation*, Butterworths, London.

MERCER, JAMES W., FAUST, CHARLES R., TRUSCHEL, ANTHONY D., AND COHEN, ROBERT M. (1987), "Control of Groundwater Contamination: Case Studies," *Detection, Control, and Renovation of Contaminated Ground Water*, pp. 121–133, ASCE.

MESRI, GHOLAMREZA, LO, DOMINIC O. KWAN, AND FENG, TAO–WEI (1994), "Settlement of Embankments on Soft Clays," *Vertical and Horizontal Deformations of Foundations and Embankments*, Vol. 1, pp. 8–56, ASCE.

MESRI, GHOLAMREZA — *see* Terzaghi, Karl

METCALF, J. B. — *see* Ingles, O. G.

MEYERHOF, G. G. (1955), "Influence of Roughness of Base and Ground-Water Conditions on the Ultimate Bearing Capacity of Foundations," *Géotechnique*, Vol. 5, pp. 227–242 (Reprinted in Meyerhof, 1982).

MEYERHOF, G. G. (1965), "Shallow Foundations," *Journal of the Soil Mechanics and Foundations Division*, Vol. 91, No. SM2, pp. 21–31, ASCE (Reprinted in Meyerhof, 1982).

MEYERHOF, G. G. (1982), *The Bearing Capacity and Settlement of Foundations*, Tech–Press, Technical University of Nova Scotia, Halifax.

MILLER, DEBORAH J. — *see* Nelson, John D.

MIRAN, JEROME — *see* Briaud, Jean–Louis

MITCHELL, JAMES K. (1978), "In-Situ Techniques for Site Characterization," *Site Characterization and Exploration*, pp. 107–129, C. H. Dowding, Ed., ASCE.

MITCHELL, JAMES K., AND HUBER, TIMOTHY R. (1985), "Performance of a Stone Column Foundation," *Journal of Geotechnical Engineering*, Vol. 111, No. 2, pp. 205–223, ASCE.

MITCHELL, JAMES K., SEED, RAYMOND B., AND SEED, H. BOLTON (1990), "Stability Considerations in the Design and Construction of Lined Waste Depositories," *Geotechnics of Waste Fills: Theory and Practice*, pp. 209–224, ASTM.

MITCHELL, JAMES K., AND TSENG, DAR–JEN (1990), "Assessment of Liquefaction Potential by Cone Penetration Resistance," *H. Bolton Seed Memorial Symposium Proceedings*, Vol. 2, pp. 335–350, BiTech, Vancouver, BC.

MITCHELL, JAMES K. (1993), *Fundamentals of Soil Behavior*, 2nd ed., John Wiley, New York.

MITCHELL, JAMES K., BAXTER, CHRISTOPHER D. P., AND MUNSON, TRAVIS C. (1995), "Performance of Improved Ground During Earthquakes," *Soil Improvement for Earthquake Hazard Mitigation*, R. D. Hryciw, Ed., pp. 1–36, Geotechnical Special Publication No. 49, ASCE.

MITCHELL, J. K. — *see* Seed, R. B.

MORGENSTERN, N. R., AND PRICE, V. E. (1965), "The Analysis of the Stability of General Slip Surfaces," *Géotechnique*, Vol. 15, No. 1, pp. 79–93.

MOSELEY, M. P. (1993), *Ground Improvement*, Chapman and Hall, London.

MOUDUD, ABDUL — *see* Sharma, Sunil

MÜLLER STEINHAGEN, H. — *see* Clayton, C. R. I.

MÜLLER–BRESLAU, H. (1906) "Erddruk auf Stützmauern," Alfred Kröner Verlag, Stuttgart (in German).

MUNSON, TRAVIS C. — *see* Mitchell, James K.

NAHB (1988), *Frost-Protected Shallow Foundations for Houses and Other Heated Structures*, National Association of Home Builders, Upper Marlboro, MD.

NAHB (1990, draft), *Frost-Protected Shallow Foundations for Unheated Structures*, National Association of Home Builders, Upper Marlboro, MD.

NAIR, KESHAVAN — *see* Luscher, Ulrich

NEGUSSEY, DAWIT (1997), *Properties and Applications of Geofoam*, Society of the Plastics Industry.

NELSON, JOHN D., AND MILLER, DEBORAH J. (1992), *Expansive Soils: Problems and Practice in Foundation and Pavement Engineering*, John Wiley, New York.

NEWELL, CHARLES J. — *see* Bedient, Philip B.

NEWMARK, NATHAN M. (1935), *Simplified Computation of Vertical Pressures in Elastic Foundations*, Engineering Experiment Station Circular No. 24, University of Illinois, Urbana.

NEWMARK, N. (1965), "Effects of Earthquakes on Dams and Embankments," *Géotechnique*, Vol. 15, No. 2, pp. 139–160.

NIV (1984), *The Holy Bible, New International Version*, International Bible Society, Colorado Springs, CO.

NIXON, IVAN K. (1982), "Standard Penetration Test State-of-the-Art Report," *Second European Symposium on Penetration Testing* (ESOPT II), Amsterdam, Vol. 1, pp. 3–24, A. Verruijt et al., Eds.

NOORANY, IRAJ, SWEET, JOEL A., AND SMITH, IAN M. (1992), "Deformation of Fill Slopes Caused by Wetting," *Stability and Performance of Slopes and Embankments II*, R. B. Seed and R. W. Boulanger, Eds., pp. 1244–1257, ASCE.

NOORANY, IRAJ, AND STANLEY, JEFFREY (1994), "Settlement of Compacted Fills Caused by Wetting," *Vertical and Horizontal Deformations of Foundations and Embankments*, pp. 1516–1530, A. T. Yeung and G. Y. Felio, Eds., ASCE.

NOORANY, IRAJ — *see* Forrest, Carol

NORRIS, ROBERT D. ET AL. (1994), *Handbook of Bioremediation*, CRC Press.

NRC (1994), *Alternatives for Ground Water Cleanup*, National Research Council, National Academy Press, Washington, DC.

NUNNALLY, S. W. (1993), *Construction Methods and Management*, 3rd ed., Prentice Hall, Upper Saddle River, NJ.

O'ROURKE, T. D., AND JONES, C. J. F. P. (1990), "Overview of Earth Retention Systems: 1970–1990," *Design and Performance of Earth Retaining Structures*, Geotechnical Special Publication No. 25, pp. 22–51, P. C. Lambe and L. A. Hansen, Eds., ASCE.

O'ROURKE, T. D. — *see* Kulhawy, F. H.

OLSHANSKY, ROBERT B. (1993), "Selling Seismic Building Codes in the Central United States," *Proceedings, 1993 National Earthquake Conference*, Vol. I, pp. 649–658.

OTA (1984), *Protecting the Nation's Groundwater from Contamination*, Office of Technology Assessment.

OTTO, W. C. — *see* Wallace, G. B.

PANARESE, WILLIAM C. — *see* Kosmatka, Steven H.

PARCHER, J. V. — *see* Means, R. E.

PARK, JAE K. – *see* Kim, Jae Y.

PCA (1991), "Durability of Concrete in Sulfate-Rich Soils," *Concrete Technology Today*, Vol. 12, No. 3, pp. 6–8, Portland Cement Association, Skokie, IL.

PECK, RALPH B. (1969), "Advantages and Limitations of the Observational Method in Applied Soil Mechanics," *Géotechnique*, Vol. 19, pp. 171–187 (reprinted in Dunnicliff and Deere (1984), *Judgement in Geotechnical Engineering*, pp. 122–127).

PECK, RALPH Q. — *see* Terzaghi, Karl

PETERSON, JOHN Q. — *see* Troxell, Harold C.

PETTERSON, KNUT E. (1955), "The Early History of Circular Sliding Surfaces," *Géotechnique*, Vol. 5, pp. 275–296.

PHUKAN, ARVIND (1985), *Frozen Ground Engineering*, Prentice Hall, Upper Saddle River, NJ.

PLAFKER, G., AND ERICKSEN, G. E. (1978), "Nevados Huascarán Avalanches, Peru," *Rockslides and Avalanches, 1: Natural Phenomena*, B. Voight, Ed., pp. 277–314, Elsevier, Amsterdam.

POULOS, H. G., AND DAVIS, E. H. (1974), *Elastic Solutions for Soil and Rock Mechanics*, John Wiley, New York.

POWER, M. S. — *see* Seed, R. B.

POWERS, J. PATRICK (1992), *Construction Dewatering: New Methods and Applications*, 2nd ed., John Wiley, New York.

POWRIE, W. — *see* Clayton, C. R. I.

PRICE, V. E. — *see* Morgenstern, N. R.

PRIEST, STEPHEN D. (1993), *Discontinuity Analysis for Rock Engineering*, Chapman and Hall, London.

PRIOR, DAVID B. — *see* Brunsden, Denys

PROCTOR, R. R. (1933), "Fundamental Principles of Soil Compaction" pp. 245–248; "Description of Field and Laboratory Procedures," pp. 286–289; "Field and Laboratory Verification of Soil Suitability," pp. 348–351; and "New Principles Applied to Actual Dam-Building," pp. 372–376; *Engineering News Record*, Vol. 111, No. 9.

PUFAHL, D. E. — *see* Fredlund, D. G.

RAD, NADER SHAFII — *see* Clough, G. Wayne

RAHARDJO, H. — *see* Fredlund, D. G.

RAHN, PERRY H. (1996), *Engineering Geology: An Environmental Approach*, 2nd ed., Prentice Hall, Upper Saddle River, NJ.

RANKINE, W. J. M. (1857), "On the Stability of Loose Earth," *Philosophical Transactions of the Royal Society*, Vol. 147, London.

REDDI, LAKSHMI N., AND BONALA, MOHAN V. S. (1997), "Analytical Solution for Fine Particle Accumulation in Soil Filters," *Journal of Geotechnical and Geoenvironmental Engineering*, Vol. 123, No. 12, pp. 1143–1152.

REDDY, KRISHNA R., AND ADAMS, JEFFREY A. (1996), "In-Situ Air Sparging: A New Approach for Groundwater Remediation," *Geotechnical News*, Vol. 14, No. 4, pp. 27–32.

REIMER, M. F. — *see* Seed, R. B.

RICHTER, C. F. (1935), "An Instrumental Earthquake Magnitude Scale," *Seismological Society of America Bulletin* 251, pp. 1–32.

RICHTER, CHARLES F. (1958), *Elementary Seismology*, W. H. Freeman, San Francisco.

RIFAI, HANADI S. — *see* Bedient, Philip B.

RIZVI, RAIS — *see* Hicks, Randall

ROBERTS, DON V., AND DARRAGH, ROBERT D. (1962), "Areal Fill Settlements and Building Foundation Behavior at the San Francisco Airport," *Field Testing of Soils*, Special Publication 322, pp. 211–230, ASTM.

ROBERTSON, P. K., AND CAMPANELLA, R. G. (1983), "Interpretation of Cone Penetration Tests: Parts 1 and 2," *Canadian Geotechnical Journal*, Vol. 20, pp. 718–745.

ROBERTSON, P. K., AND CAMPANELLA, R. G. (1989), *Guidelines for Geotechnical Design Using the Cone*

Penetrometer Test and CPT With Pore Pressure Measurement, 4th ed., Hogentogler & Co., Columbia, MD.

RODINE, J. R. — *see* Johnson, A. M.

ROGERS, J. DAVID (1992a), "Mechanisms of Seismically-Induced Slope Movements," Unpublished manuscript presented at Transportation Research Board meeting.

ROGERS, J. DAVID (1992b), "Recent Developments in Landslide Mitigation Techniques," Chapter 10 in *Landslides/Landslide Mitigation*, Reviews in Engineering Geology, Vol. IX, Geological Society of America.

ROGERS, J. DAVID (1992c), "Long Term Behavior of Urban Fill Embankments," *Stability and Performance of Slopes and Embankments II*, R. B. Seed and R. W. Boulanger, Eds., pp. 1258–1273, ASCE.

ROGERS, J. DAVID (1995), "A Man, A Dam, and A Disaster: Mulholland and the St. Francis Dam," *The St. Francis Dam Disaster Revisited*, Doyce B. Nunnis, Jr., Ed., pp. 1–109, Historical Society of Southern California, Los Angeles.

ROLLINS, KYLE M., AND KIM, JI–HYOUNG (1994), "U.S. Experience with Dynamic Compaction of Collapsible Soils," *In-Situ Deep Soil Improvement*, Special Geotechnical Publication 45, pp. 26–43, Kyle M. Rollins, Ed., ASCE.

ROTH, MARY JOEL S. — *see* Kulhawy, Fred H.

ROVERS, FRANK A. — *see* McBean, Edward A.

RUBRIGHT, R., AND WELSH, J. (1993), "Compaction Grouting," Chapter 6 in *Ground Improvement*, M. P. Moseley, Ed., Chapman and Hall, London.

SALLBERG, J. R. — *see* Johnson, A. W.

SALOMONE, L. A. — *see* Kovacs, W. D.

SANDFORD, THOMAS – *see* Whetton, Nathan

SANTAMARINA, J. CARLOS (1997), "Cohesive Soil: A Dangerous Oxymoron," *Electronic Journal of Geotechnical Engineering*, published at web site geotech.civen.okstate.edu/magazine/oxymoron/dangeoxi.htm.

SARMA, S. K. (1973), "Stability Analysis of Embankments and Slopes," *Géotechnique*, Vol. 23, No. 3, pp. 423–433.

SCHAEFER, VERNON R. (1997), *Ground Improvement, Ground Reinforcement, Ground Treatment Developments 1987–1997*, Geotechnical Special Publication No. 69, ASCE.

SCHEIL, THOMAS J. (1979), "Long Term Monitoring of a Building Over the Deep Hackensack Meadowlands Varved Clays," presented at the 1979 Converse Ward Davis Dixon, Inc. Technical Seminar, Pasadena, CA.

SCHIFFMAN, R. L. — *see* De Boer, R.

SCHMERTMANN, JOHN H. (1955), "The Undisturbed Consolidation Behavior of Clay," *Transactions of the American Society of Civil Engineers*, Vol. 120, pp. 1201–1233, ASCE.

SCHMERTMANN, JOHN H. (1978), *Guidelines for Cone Penetration Test: Performance and Design*, Report FHWA–TS–78–209, Federal Highway Administration, Washington, DC.

SCHMERTMANN, J. H. (1986a), "Suggested Method for Performing the Flat Dilatometer Test," *Geotechnical Testing Journal*, Vol. 9, No. 2, pp. 93–101.

SCHMERTMANN, JOHN H. (1986b), "Dilatometer to Compute Foundation Settlement," *Use of In-Situ Tests in Geotechnical Engineering*, pp. 303–319, Samuel P. Clemence, Ed., ASCE.

SCHMERTMANN, JOHN H. (1988a), "Dilatometers Settle In," *Civil Engineering*, Vol. 58, No. 3, pp. 68–70, March 1988, ASCE.

SCHMERTMANN, JOHN H. (1988b), *Guidelines for Using the CPT, CPTU and Marchetti DMT for Geotechnical Design*, Vol. I–IV, Federal Highway Administration, Washington, DC.

SCHNEIDER, J. F. — *see* Toro, G. R.

SCHUSTER, ROBERT L. (1996), "Socioeconomic Significance of Landslides," Chapter 2 in *Landslides: Investigation and Mitigation*, A. Keith Turner and Robert L. Schuster, Eds., Transportation Research Board.

SCHUSTER, ROBERT L. — *see* Turner, A. Keith

SCHUYLER, JAMES DIX (1905), *Reservoirs for Irrigation, Water-Power, and Domestic Water-Supply*, John Wiley, New York.

SCS (1986), *Guide for Determining the Gradation of Sand and Gravel Filters*, Soil Mechanics Note No. 1, 210–VI, U.S. Dept. of Agriculture, Soil Conservation Service, Lincoln, NE.

SCULLIN, C. MICHAEL (1983), *Excavation and Grading Code Administration, Inspection, and Enforcement*, Prentice Hall, Upper Saddle River, NJ.

SEED, H. BOLTON, AND WILSON, STANLEY D (1964), *The Turnagain Heights Landslide in Anchorage, Alaska*, Department of Civil Engineering, University of California, Berkeley.

SEED, H. BOLTON, AND IDRISS, I. M. (1966), *An Analysis of Soil Liquefaction in the Niigata Earthquake*, Department of Civil Engineering, University of California, Berkeley.

SEED, H. BOLTON, AND IDRISS, I. M. (1967), "Analysis of Soil Liquefaction: Niigata Earthquake," *Journal of the Soil Mechanics and Foundations Division*, Vol. 93, No. SM3, pp. 83–108, ASCE.

SEED, H. BOLTON (1970), "Soil Problems and Soil Behavior," Chapter 10 in *Earthquake Engineering*, Robert L. Wiegel, Ed., Prentice Hall, Upper Saddle River, NJ.

SEED, H. B., AND IDRISS, I. M. (1971) "Simplified Procedure for Evaluating Soil Liquefaction Potential," *Journal of the Soil Mechanics and Foundations Division*, Vol. 107, No. SM9, pp. 1249–1274, ASCE.

SEED, H. B., AND BOOKER, JOHN R. (1976), *Stabilization of Potentially Liquefiable Sand Deposits Using Gravel Drain Systems*, Report No. 76–10, Earthquake Engineering Research Center, University of California at Berkeley.

SEED, H. BOLTON, AND IDRISS, I. M. (1982), "Ground Motions and Soil Liquefaction During Earthquakes," Earthquake Engineering Research Institute.

SEED, H. BOLTON, TOKIMATSU, K., HARDER, L. F., AND CHUNG, RILEY M. (1985), "Influence of SPT Procedures in Soil Liquefaction Resistance Evaluations," ASCE *Journal of Geotechnical Engineering*, Vol. 111, No. 12, pp. 1425–1445.

SEED, H. BOLTON — *see* Mitchell, James K.; Harder, Leslie F.; Tokimatsu, K.

SEED, R. B., DICKENSON, S. E., REIMER, M. F., BRAY, J. D., SITAR, N., MITCHELL, J. K., IDRISS, I. M., KAYEN, R. E., KROPP, A., HARDER, L. F., AND POWER, M. S. (1990), *Preliminary Report on the Principal Geotechnical Aspects of the October 17, 1989 Loma Prieta Earthquake*, Report No. UCB/EERC–90/05, Earthquake Engineering Research Center, University of California at Berkeley.

SEED, RAYMOND B., AND BOULANGER, ROSS W. (1992), *Stability and Performance of Slopes and Embankments– II*, Geotechnical Special Publication No. 31, ASCE.

SEED, RAYMOND B. — *see* Mitchell, James K.

SHARMA, SUNIL, AND MOUDUD, ABDUL (1992), "Interactive Slope Analysis Using Spencer's Method," *Stability and Performance of Slopes and Embankments–II*, Geotechnical Special Publication No. 31, pp. 506–520, R. B. Seed and R. W. Boulanger, Eds., ASCE.

SHARMA, SUNIL — *see* Abramson, Lee W.

SHEN, C. K. — *see* Greenfield, Steven J.

SHERARD, JAMES L., WOODWARD, RICHARD J, GIZIENSKI, STANLEY F., AND CLEVENGER, WILLIAM A. (1963), *Earth and Earth-Rock Dams*, John Wiley, New York.

SHERARD, J. L., DUNNIGAN, L. P., AND TALBOT, J. R. (1984a), Basic Properties of Sand and Gravel Filters," *Journal of Geotechnical Engineering*, Vol. 110, No. 6, pp. 684–700, ASCE.

SHERARD, J. L., DUNNIGAN, L.P., AND TALBOT, J. R. (1984b), Filters for Silts and Clays," *Journal of Geotechnical Engineering*, Vol. 110, No. 6, pp. 701–718, ASCE.

SHERARD, J. L., AND DUNNIGAN, L.P. (1985), "Filters and Leakage Control in Embankment Dams," *Seepage and Leakage Control in Embankment Dams*, pp. 1–29, ASCE.

SHERARD, J. L., AND DUNNIGAN, L.P. (1989), "Critical Filters for Impervious Soils," *Journal of Geotechnical Engineering*, Vol. 115, No. 7, pp. 927–947, ASCE.

SHIELDS, D. H. — *see* Baguelin, F.

SHUIRMAN, GERARD, AND SLOSSON, JAMES E. (1992), *Forensic Engineering: Environmental Case Histories for Civil Engineers and Geologists*, Academic Press, San Diego.

SICHART, W., AND KYRIELEIS, W. (1930), *Grundwasser Absekungen bei Fundierungsarbeiten* (in German), as quoted in Powers (1992).

SINGH, SUKHMANDER (1994), "Liquefaction Characteristics of Silts," *Ground Failures Under Seismic Conditions*, Geotechnical Special Publication No. 44, pp. 105–116, S. Prakash and P. Dakoulas, Eds., ASCE.

SINTON, L. W. (1980), "Two Antibiotic-Resistant Strains of Escherichia Coli For Tracing the Movement of Sewage in Ground Water," *Journal of Hydrology* (New Zealand), Vol. 19, pp. 119–129.

SITAR, NICHOLAS — *see* Clough, G. Wayne; Seed, R. B.

SKEMPTON, A. W. (1949), "Alexandre Collin, A Note on His Pioneer Work in Soil Mechanics," *Géotechnique*, Vol. 1, No. 4, pp. 216–221.

SKEMPTON, A. W. (1954), "The Pore-Pressure Coefficients *A* and *B*," *Géotechnique*, Vol. IV, pp. 143–147.

SKEMPTON, A. W. (1960), "Significance of Terzaghi's Concept of Effective Stress," *From Theory to Practice in Soil Mechanics*, pp. 42–53.

SKEMPTON, A. W. (1979), "Landmarks in Early Soil Mechanics," *Seventh European Conference on Soil Mechanics and Foundation Engineering*, Vol. 5, pp. 1–26.

SKEMPTON, A. W. (1986), "Standard Penetration Test Procedures and the Effects in Sands of Overburden Pressure, Relative Density, Particle Size, Aging, and Overconsolidation," *Géotechnique*, Vol. 36, No. 3, pp. 425–447.

SKINNER, BRIAN J. — *see* Flint, Richard F.

SLOSSON, JAMES E. — *see* Shuirman, Gerard

SMITH, RONALD E., Ed. (1987), *Foundations and Excavations in Decomposed Rock of the Piedmont Province*, Geotechnical Special Publication No. 9, ASCE.

SNETHEN, D. R. (1984), "Evaluation of Expedient Methods for Identification and Classification of Potentially Expansive Soils," *Proceedings of the Fifth International Conference on Expansive Soils*, pp. 22–26, National Conference Publication 84/3, The Institution of Engineers, Australia.

SOIL SURVEY STAFF (1975), *Soil Taxonomy*, Agriculture Handbook No, 436, Soil Conservation Service, US Department of Agriculture, Washington, DC.

SOIL CONSERVATION SERVICE (1980), *Soil Survey of San Bernardino County, Southwestern Part, California*, US Department of Agriculture, Washington, DC.

SOOYSMITH, W. (1892), "The Building Problem in Chicago From an Engineering Standpoint," *The Technograph*, No. 6, pp. 9–19, University of Illinois.

SOUTHWEST BUILDER AND CONTRACTOR (1936), "Tamping Feet of a Flock of Sheep Gave Idea for Sheepsfoot Roller," Aug 7, 1936, p. 13.

SOWERS, GEORGE F. (1979), *Introductory Soil Mechanics and Foundations: Geotechnical Engineering*, 4th ed., Macmillan, New York.

SOWERS, GEORGE F. (1992), "Natural Landslides," *Stability and Performance of Slopes and Embankments – II*, Geotechnical Special Publication No. 31, Vol. I, pp. 804–833, R. B. Seed and R. W. Boulanger, Eds., ASCE.

SPANN, STEVE W. (1986), "Available Compaction Equipment," *Earthmoving and Heavy Equipment*, Garold D. Oberlender, Ed., pp. 10–13, ASCE.

SPENCER, C. B. — *see* Thornley, J. H.

SPENCER, E. (1967), "A Method of Analysis of the Stability of Embankments Assuming Parallel Interslice Forces," *Géotechnique*, Vol. 17, No. 1, pp. 11–26.

SPENCER, E. (1973), "The Thrust Line Criterion in Embankment Stability Anaysis," *Géotechnique*, Vol. 23, No. 1, pp. 85–100.

STAMATOPOULOS, ARIS C., AND KOTZIAS, PANAGHIOTIS C. (1985), *Soil Improvement by Preloading*, John Wiley, New York.

STANLEY, JEFFREY — *see* Noorany, Iraj

STARK, TIMOTHY D., AND DUNCAN, J. MICHAEL (1991), "Mechanisms of Strength Loss in Stiff Clays," *Journal of Geotechnical Engineering*, Vol. 117, No. 1, pp. 139–154, ASCE.

STARK, TIMOTHY D. — *see* Duncan, J. Michael

STATENS JÄRNVÄGARS GEOTEKNISKA KOMMISSION: SLUTBETÄNKANDE (1922), (The State Railways Geotechnical Commission: Final Report), Stockholm (In Swedish).

STEARNS, R. G., AND WILSON, C. W. (1972), *Relationships of Earthquakes and Geology in West Tennessee and Adjacent Areas*, Tennessee Valley Authority, Knoxville, TN.

STEINBRUGGE, KARL V. (1970), "Earthquake Damage and Structural Performance in the United States," Chapter 9 in *Earthquake Engineering*, Robert L. Wiegel, Ed., Prentice Hall, Upper Saddle River, NJ.

STEVENS, MICHAEL G — *see* Dise, Karl

STEWART, JONATHAN P., BRAY, JONATHAN D., MCMAHON, DAVID J., AND KROPP, ALAN L. (1995), "Seismic Performance of Hillside Fills," *Landslides Under Static and Dynamic Conditions — Analysis, Monitoring, and Mitigation*, Geotechnical Special Publication No. 52, pp. 76–95, David K. Keefer and Carlton L. Ho, Eds., ASCE.

SUROS, OSCAR — *see* York, Donald

TALBOT, J. R. — *see* Sherard, J. L.

TANG, WILSON H. — *see* Wu, Tien H.

TAYLOR, DONALD W. (1948), *Fundamentals of Soil Mechanics*, John Wiley, New York.

TELFORD, THOMAS (1830), "Inland Navigation," in *Edinburgh Encyclopedia*, Vol 15, pp. 209–315.

TERZAGHI, KARL (1920), "New Facts About Surface-Friction," *Physical Review*, Vol. 16, No. 1, pp. 54–61 (Reprinted in *From Theory to Practice in Soil Mechanics*, John Wiley, New York, 1960).

TERZAGHI, KARL (1921), "Die physikalischen Grundlagen der technischgeologischen Gutachtens," *Österreichischer Ingenieur und Architekten-Verein Zeitschrift*, Vol. 73, No. 36/37, pp. 237–241 (in German).

TERZAGHI, KARL (1923a), "Die Beziehungen zwischen Elastizität und Innendruck," *Akademic der Wissenschaften in Wien. Sitzungsberichte.* Mathematisch-naturwissenschaftliche Klasse., Part IIa, Vol. 132, No. 3/4, pp. 105–124 (in German).

TERZAGHI, KARL (1923b), "Die Berechnung der Durchlässigkeitsziffer des Tones aus dem Verlauf der hydrodynamischen Spannungserscheinungen" (A Method of Calculating the Coefficient of Permeability of Clay from the Variation of Hydrodynamic Stress with Time), *Akademic der Wissenschaften in Wien. Sitzungsberichte.* Mathematisch-naturwissenshaftliche Klasse, Part IIa, Vol. 132, No. 3/4, pp. 125–138 (in German) (reprinted in *From Theory to Practice in Soil Mechanics*, John Wiley, New York, 1960, pp. 133–146); Translated into English by Clayton et al. (1995).

TERZAGHI, KARL (1924), "Die Theorie der hydrodynamischen Spannungerscheinungen und ihr erdbautechnisches Anwendungsgebiet," *Proceedings International Congress on Applied Mechanics*, Delft, pp. 288–294 (in German).

TERZAGHI, KARL (1925a), *Erdbaumechanik auf bodenphysikalischer Grundlage*, Deuticke, Vienna (in German) (Forward translated into English in *From Theory to Practice in Soil Mechanics*, John Wiley, New York, 1960, pp. 58–61). *Also see* Clayton et al., (1995).

TERZAGHI, KARL (1925b), "Principles of Soil Mechanics," *Engineering News Record*, Vol. 95, No. 19–23 and 25–27, pp. 742–746, 796–800, 832–836, 874–878, 912–915, 987–990, 1026–1029, 1064–1068.

TERZAGHI, KARL (1925c), "Modern Conceptions Concerning Foundation Engineering," *Journal of the Boston Society of Civil Engineers*, Vol. 12, No. 10, pp. 397–439.

TERZAGHI, KARL (1929), "The Mechanics of Shear Failures on Clay Slopes and the Creep of Retaining Walls," *Public Roads*, Vol. 10, No. 10, pp. 177–192.

TERZAGHI, KARL (1934a), "Die Ursachen der Schiefstellung des Turmes von Pisa," *Der Bauingenieur*, Vol. 15, No. 1/2, pp. 1–4 (Reprinted in *From Theory to Practice in Soil Mechanics*, L. Bjerrum et al.., Eds, pp. 198–201, John Wiley, New York, 1960) (in German).

TERZAGHI, KARL (1934b), "Large Retaining Wall Tests," a series of articles in *Engineering News-Record*, Vol. 112; 2/1/34, 2/22/34, 3/8/34, 3/29/34, 4/19/34, and 5/17/34.

TERZAGHI, KARL (1936), "Discussion on Instruction in Soil Mechanics," *Proceedings, International Conference on Soil Mechanics and Foundation Engineering*, Vol. III, pp. 261–263.

TERZAGHI, KARL (1939), "Soil Mechanics — A New Chapter in Engineering Science," *Journal of the Institution of Civil Engineers*, Vol. 12, pp. 106–141.

TERZAGHI, KARL (1943) *Theoretical Soil Mechanics*, John Wiley, New York.

TERZAGHI, KARL, AND PECK, RALPH B. (1967), *Soil Mechanics in Engineering Practice*, 2nd ed., John Wiley, New York.

TERZAGHI, KARL, PECK, RALPH B., AND MESRI, GHOLAMREZA (1996), *Soil Mechanics in Engineering Practice*, 3rd ed., John Wiley, New York .

THORNLEY, J. H., SPENCER, C. B., AND ALBIN, PEDRO (1955), "Mexico's Palace of Fine Arts Settles 10 ft," *Civil Engineering*, pp. 356–360, 576, 616, and 707, ASCE.

THORP, JAMES (1936), *Geography of the Soils of China*, National Geological Survey of China, Nanking.

THORSON, BRUCE M., AND BRAUN, J. S. (1975), "Frost Heaves— A Major Dilemma for Ice Arenas," *Civil Engineering*, Vol. 45, No. 3, pp. 62–64, ASCE.

TOKIMATSU, K., AND SEED, H. B. (1987), "Evaluation of Settlements in Sands Due to Earthquake Shaking," *Journal of Geotechnical Engineering*, Vol. 113, No. 8, pp. 861–878, ASCE.

TOKIMATSU, K. — *see* Seed, H. Bolton

TORO, G. R., ABRAHAMSON, N. A., AND SCHNEIDER, J. F. (1995), "Engineering Model of Strong Ground Motions from Earthquakes in the Central and Eastern United States," *Earthquake Spectra* (*also see* Kramer, 1996).

TOTH, P. S. (1993), "In-Situ Soil Mixing," Chapter 9 in *Ground Improvement*, M.P. Moseley, Ed., Chapman and Hall, London.

TRAUTMANN, C. H. — *see* Kulhawy, F. H.

TROXELL, HAROLD C., AND PETERSON, JOHN Q. (1937), *Flood in La Cañada Valley, California*, US Geological Survey Water Supply Paper 796C.

TRUSHEL, ANTHONY D. — *see* Mercer, James W.

TSCHEBOTARIOFF, GREGORY P. (1951), *Soil Mechanics, Foundations, and Earth Structures*, McGraw Hill, New York.

TSENG, DAR–JEN — *see* Mitchell, James K.

TURNER, A. KEITH, AND SCHUSTER, ROBERT L., EDITORS (1996), *Landslides — Investigation and Mitigation*, National Research Council, Transportation Research Board, Washington, DC.

U.S. ARMY (1996), World Wide Web site www.pmrma-www.army.mil.

U.S. ARMY, NAVY, AND AIR FORCE (1971), *Dewatering and Groundwater Control for Deep Excavations*, Army TM 5–818–5, Navy NAVFAC P–418, Air Force AFM 88–5.

U.S. NAVY (1982), *Soil Mechanics*, NAVFAC Design Manual 7.1, Naval Facilities Engineering Command, Arlington, VA.

VAID, Y.P. (1994), "Liquefaction of Silty Soils," *Ground Failures Under Seismic Conditions*, Geotechnical Special Publication No. 44, pp. 1–16, S. Prakash and P. Dakoulas, Eds., ASCE.

VARNES, D. J. (1958), "Landslide Types and Processes," *Landslides and Engineering Practice*, Special Report 29, pp. 20–47, E. B. Eckel, Ed., Highway Research Board, National Research Council.

VARNES, DAVID J. (1978), "Slope Movement Types and Processes," Chapter 2 in *Landslides: Analysis and Control*, Special Report 176, Robert L. Schuster and Raymond J. Krizek, Eds., Transportation Research Board, Highway Research Council.

VARNES, DAVID J. — *see* Cruden, David M.

VESIĆ, ALEKSANDAR S. (1973), "Analysis of Ultimate Loads of Shallow Foundations," *Journal of the Soil Mechanics and Foundations Division*, Vol. 99, No. SM1, pp. 45–73, ASCE.

VOIGHT, BARRY (1990), "The 1985 Nevado del Ruiz Volcano Catastrophe: Anatomy and Retrospection," *Journal of Volcanology and Geothermal Research*, Vol 42, pp. 151–188.

VON THUN, J. LAWRENCE — *see* Dise, Karl

WALLACE, G. B., AND OTTO, W. C. (1964), "Differential Settlement at Selfridge Air Force Base," *Journal of the Soil Mechanics and Foundations Division*, Vol. 90, No. SM5, pp. 197–220, ASCE.

WEAVER, JAMES – *see* Whetton, Nathan

WELSH, J.P. (1986), "Construction Considerations for Ground Modification Projects," *Proceedings International Conference on Deep Foundations*, Beijing, Deep Foundations Institute, Englewood Cliffs, NJ.

WELSH, J. — *see* Rubright, R.

WEST, TERRY R. (1995), *Geology Applied to Engineering*, Prentice Hall, Upper Saddle River, NJ.

WESTERGAARD, H. M. (1938), "A Problem of Elasticity Suggested by a Problem of Soil Mechanics: Soft Material Reinforced by Numerous Strong Horizontal Sheets," *Contributions to the Mechanics of Solids*, MacMillan, New York.

WHETTON, NATHAN, WEAVER, JAMES, HUMPHREY, DANA, AND SANDFORD, THOMAS (1997), "Rubber Meets the Road in Maine," *Civil Engineering*, Vol. 67, No. 9, pp. 60–63, ASCE.

WHITMAN, ROBERT V. — *see* Lambe, T. William; Liao, S. S. C.

WIEGEL, ROBERT L., Ed. (1970), *Earthquake Engineering*, Prentice Hall, Upper Saddle River, NJ.

WIEGEL, ROBERT L. (1970), "Tsunamis," Chapter 11 in *Earthquake Engineering*, Robert L. Wiegel, Ed., Prentice Hall, Upper Saddle River, NJ.

WILLIAMS, T. (1994), *Magnitude Scaling Factors for Analysis of Liquefaction Hazard*, PhD Dissertation, Department of Civil Engineering, Brigham Young Univ., Provo, UT.

WILSON, C. W. — *see* Stearns, R. G.

WILSON, R. C., AND KEEFER, D. K. (1985), "Predicting Areal Limits of Earthquake-Induced Landsliding," *Evaluating Earthquake Hazards in the Los Angeles Region*, USGS Professional Paper 1360, pp. 317–345, J. I. Ziony, Ed.

WILSON, STANLEY D. — *see* Seed, H. Bolton

WOLFF, THOMAS F. (1996), "Probabilistic Slope Stability in Theory and Practice," *Uncertainty in the Geologic Environment*, Vol. 1, pp. 419–433.

WONG, KAI S. — *see* Duncan, James M.

WOOD, STUART (1977), *Heavy Construction: Equipment and Methods*, Prentice Hall, Upper Saddle River, NJ.

WOOD, W. A. — *see* Kulhawy, F. H.

WOODWARD, RICHARD J. — *see* Sherard, James L.

WRIGHT, STEPHEN G. (1985), "Limit Equilibrium Slope Analysis Procedures," *Design of Non-Impounding Waste Dumps*, pp. 63–77, American Institute of Mining Engineers.

WRIGHT, STEPHEN G. — *see* Duncan, James M.

WU, TIEN H., TANG, WILSON H., AND EINSTEIN, HERBERT H. (1996), "Landslide Hazard and Risk Assessment," Chapter 6 in *Landslides: Investigation and Mitigation*, Special Report 247, Transportation Research Board, National Research Council, A. K. Turner and R. L. Schuster, Eds.

WU, GUOXI — *see* Finn, W. D. Liam

YANG, DAVID S. (1994), "The Applications of Soil Mix Walls in the United States," *Geotechnical News*, Dec. 1994, pp. 44–47, BiTech Publishers, Vancouver, BC.

YATES, MARYLYNN V., AND YATES, S. R. (1989), "Septic Tank Setback Distances: A Way to Minimize Virus Contamination of Drinking Water," *Ground Water*, Vol. 27, No. 2, pp. 202–208.

YATES, S. R. — *see* Yates, Marylynn V.

YODEL, F. Y. — *see* Kovacs, W. D.

YORK, DONALD L., AND SUROS, OSCAR (1989), "Performance of a Building Foundation Designed to Accommodate Large Settlements," *Foundation Engineering: Current Principles and Practices*, Vol. 2, pp. 1406–1419, F. H. Kulhawy, Ed., ASCE.

YOUD, T. L. (1973), "Factors Controlling Maximum and Minimum Densities of Sands," *Evaluation of Relative Density and Its Role in Geotechnical Projects Involving Cohesionless Soils*, STP 523, pp. 98–112, ASTM.

YOUD, T. L. (1980), "Ground Failure Displacement and Earthquake Damage to Buildings," *Proceedings, Second ASCE Conference on Civil Engineering and Nuclear Power*, pp. 7–6–1 to 7–6–26.

YOUD, T. L. — *see* Bartlett, S. F.

ZACCHEO, P. F. — *see* Lowe, J.

ZEAVAERT, LEONARDO (1983), *Foundation Engineering for Difficult Subsoil Conditions*, 2nd ed., Van Nostrand Reinhold.

ZHOU, SHENGPING — *see* Evans, Mark D.

Name Index

Subject Index